Environmental Contamination

Health Risks and Ecological Restoration

Environmental Contamination

Health Risks and Ecological Restoration

Edited by **Ming H. Wong**

CRC Press
Taylor & Francis Group
Boca Raton London New York

CRC Press is an imprint of the
Taylor & Francis Group, an **informa** business

CRC Press
Taylor & Francis Group
6000 Broken Sound Parkway NW, Suite 300
Boca Raton, FL 33487-2742

First issued in paperback 2019

© 2013 by Taylor & Francis Group, LLC
CRC Press is an imprint of Taylor & Francis Group, an Informa business

No claim to original U.S. Government works

ISBN-13: 978-1-4398-9238-1 (hbk)
ISBN-13: 978-0-367-38103-5 (pbk)

Library of Congress Cataloging-in-Publication Data

Environmental contamination : health risks and ecological restoration / editor, Ming H. Wong.
    p. cm.
   From the Croucher Advanced Study Institute workshop 2010, entitled Remediation of contaminated land bioavailability and health risk, held at the Hong Kong Baptist University, between December 9 and 13, 2010.
   Includes bibliographical references and index.
   ISBN 978-1-4398-9238-1 (hardback)
   1. Pollution--Congresses. 2. Restoration ecology--Congresses. I. Wong, Ming H. II. Croucher Foundation.

TD172.5.E5585 2012
363.73--dc23   2012017901

**Visit the Taylor & Francis Web site at
http://www.taylorandfrancis.com**

**and the CRC Press Web site at
http://www.crcpress.com**

Contents

Part III   Ecological Restoration of Contaminated
Sites: Bioremediation

Part IV Ecological Restoration of Contaminated Sites: Phytoremediation

Preface

The chapters in this book are a collection of the lectures and discussion held during the Croucher Advanced Study Institute (ASI) workshop 2010, entitled *Remediation of Contaminated Land—Bioavailability and Health Risk*, held at the Hong Kong Baptist University, between December 9 and 13, 2010.

Croucher ASI is a new funding initiative of the Croucher Foundation catering to the interests of established scientists, with the main objective of regularly bringing to Hong Kong the leading international experts in specific fields to conduct refresher programs for a limited number of established scientists on highly focused scientific topics. The workshop series also provided an excellent opportunity for young scientists from Hong Kong, mainland China, and South-East Asian countries to learn from the leading scientists from the region and overseas.

As a result, we hope to disseminate to the readers the knowledge exchanged on the latest remediation technologies for pollutant cleanup of contaminated lands, the issues of bioavailability of chemicals, and their associated human risks.

This book covers the timely scientific research of 22 distinguished, leading experts, specializing on different aspects relevant to the topic.

I would like to take this opportunity to express my sincere thanks to Dr. Ji Dong Gu of Hong Kong University for co-organizing this workshop, the authors for their contribution, and the Croucher Foundation for their continued support, and last but not least, to Sue Fung for her expert editorial assistance.

Ming Hung Wong

Editor

Professor Ming Hung Wong, BSc (CUHK), MSc, PhD, DSc (Durham), MBA, DSc (Strathclyde), is dean of School of Environmental and Resource Sciences at Zhejiang Agriculture and Forestry University, and chair professor of biology and honorary director of the Croucher Institute for Environmental Sciences at Hong Kong Baptist University. His major research areas include ecotoxicological assessment and remediation of sites contaminated with toxic metals and persistent organic pollutants, and bioconversion of waste. He has been awarded a DSc each from the University of Durham and the University of Strathclyde based on papers published during the period 1977–1990 and 1991–2002, respectively.

Professor Wong served as the regional coordinator of Central and North-East Asia for the project entitled "Regionally Based Assessment of Persistent Toxic Substances" and recently joined a panel of three to review "Emerging Chemicals Management Issues in Developing Countries and Countries with Economies in Transition," which was sponsored by the United Nations Environment Programme (UNEP) and the Global Environment Facility (GEF). Currently, he is editor-in-chief of *Environmental Geochemistry and Health* (Springer) and editorial board member of five other international scientific journals.

Professor Wong has published over 450 peer-reviewed papers and 24 book chapters and has served as the editor of 22 books/special issues of scientific journals. He is currently the most cited Chinese scientist in the world in the field of environment/ecology, according to the Web of Science.

Contributors

Alan J.M. Baker
School of Botany
The University of Melbourne
Melbourne, Victoria, Australia
and
Centre for Contaminant Geoscience
Environmental Earth Sciences
    International Pty Ltd
North Sydney, New South Wales,
    Australia

Hao Ban
Analytic Cytology Laboratory
and
Key Immunopathology Laboratory
    of Guangdong Province
Shantou University Medical College
Shantou, Guangdong, People's
    Republic of China

Nicholas T. Basta
School of Environment and Natural
    Resources
The Ohio State University
Columbus, Ohio

Amanda Black
Department of Soil and Physical
    Sciences
Lincoln University
Christchurch, New Zealand

Zhihong Cao
Institute of Soil Science
Chinese Academy of Sciences
and
Joint Laboratory on Soil and
    Environment co-organized by
    Institute and Hong Kong Baptist
    University
Nanjing, Jiangsu, People's Republic
    of China

Cameron S. Crook
Cameron S Crook & Associates
Preston, United Kingdom

Richard P. Dick
School of Environment and Natural
    Resources
The Ohio State University
Columbus, Ohio

Nicholas M. Dickinson
Department of Ecology
Lincoln University
Christchurch, New Zealand

Augustine Doronila
School of Chemistry
The University of Melbourne
Parkville, Victoria, Australia

J. Bryan Ellis
Urban Pollution Research Centre
Middlesex University
London, United Kingdom

Kathryn A. Gogolin
Department of Biology
Colorado State University
Fort Collins, Colorado

Ji-Dong Gu
Laboratory of Environmental
    Microbiology and Toxicology
School of Biological Sciences
The University of Hong Kong
Hong Kong, People's Republic of
    China

Guanyu Guan
National Research Center for
   Environmental Toxicology
   (EnTox)
The University of Queensland
Brisbane, Queensland, Australia

and

Cooperative Research Centre for
   Contamination Assessment and
   Remediation of the Environment
Adelaide, South Australia, Australia

Yongyong Guo
Analytic Cytology Laboratory
and
Key Immunopathology Laboratory
   of Guangdong Province
Shantou University Medical College
Shantou, Guangdong, People's
   Republic of China

William Hartley
Department of Computing, Science
   and Engineering
University of Salford
Greater Manchester, United
   Kingdom

Jinrong Huang
Analytic Cytology Laboratory
and
Key Immunopathology Laboratory
   of Guangdong Province
Shantou University Medical College
Shantou, Guangdong, People's
   Republic of China

Yeqing Huang
Croucher Institute for
   Environmental Sciences
Hong Kong Baptist University
Hong Kong, People's Republic of
   China

Xia Huo
Analytic Cytology Laboratory
and
Key Immunopathology Laboratory
   of Guangdong Province
Shantou University Medical College
Shantou, Guangdong, People's
   Republic of China

Liping Jiao
Key Laboratory of Global Change
   and Marine-Atmospheric
   Chemistry
Third Institute of Oceanography,
   State Oceanic Administration
Xiamen, Fujian Province, People's
   Republic of China

Lin Ke
College of Environmental Science
   and Engineering
South China University of
   Technology
Guangzhou, Guangdong, People's
   Republic of China

and

Department of Biology and
   Chemistry
City University of Hong Kong
Hong Kong, People's Republic of
   China

Peeranart Kiddee
Centre for Environmental Risk
   Assessment and Remediation
University of South Australia
Mawson Lakes, South Australia,
   Australia

and

Cooperative Research Centre for
   Contamination Assessment and
   Remediation of the Environment
Adelaide, South Australia, Australia

Ki-Rak Kim
School of Environmental Science
and Engineering
Gwangju Institute of Science and
Technology
Gwangju, Korea

and

School of Engineering and Science
Stevens Institute of Technology
Hoboken, New Jersey

Kyoung-Woong Kim
School of Environmental Science
and Engineering
Gwangju Institute of Science and
Technology
Gwangju, Korea

Lillian Y.Y. Ko
Hong Kong Child Development
Centre
Kowloon, Hong Kong, People's
Republic of China

Myoung-Soo Ko
School of Environmental Science
and Engineering
Gwangju Institute of Science and
Technology
Gwangju, Korea

Paul K.L. Lam
Croucher Institute for
Environmental Sciences
Hong Kong Baptist University
Hong Kong, People's Republic of
China

Anna O.W. Leung
Croucher Institute for
Environmental Sciences
Hong Kong Baptist University
Hong Kong, People's Republic of
China

Clement K.M. Leung
IVF Centre Limited
Hong Kong Sanatorium and
Hospital
Hong Kong, People's Republic of
China

Weitang Liao
Analytic Cytology Laboratory
and
Key Immunopathology Laboratory
of Guangdong Province
Shantou University Medical College
Shantou, Guangdong, People's
Republic of China

Junxiao Liu
Analytic Cytology Laboratory
and
Key Immunopathology Laboratory
of Guangdong Province
Shantou University Medical College
Shantou, Guangdong, People's
Republic of China

Wei Liu
Analytic Cytology Laboratory
and
Key Immunopathology Laboratory
of Guangdong Province
Shantou University Medical College
Shantou, Guangdong, People's
Republic of China

Lian Lundy
Urban Pollution Research Centre
Middlesex University
London, United Kingdom

Ming Man
Croucher Institute for
    Environmental Sciences
Hong Kong Baptist University
Hong Kong, People's Republic of
    China

Mallavarapu Megharaj
Centre for Environmental Risk
    Assessment and Remediation
University of South Australia
Mawson Lakes, South Australia,
    Australia

and

Cooperative Research Centre for
    Contamination Assessment and
    Remediation of the Environment
Adelaide, South Australia, Australia

Ravi Naidu
Centre for Environmental Risk
    Assessment and Remediation
University of South Australia
Mawson Lakes, South Australia,
    Australia

and

Cooperative Research Centre for
    Contamination Assessment and
    Remediation of the Environment
Adelaide, South Australia, Australia

Jack C. Ng
The University of Queensland
National Research Center for
    Environmental Toxicology
    (EnTox)
Brisbane, Queensland, Australia

and

Cooperative Research Centre for
    Contamination Assessment and
    Remediation of the Environment
Adelaide, South Australia, Australia

Cheng Peng
The University of Queensland
National Research Center for
    Environmental Toxicology
    (EnTox)
Brisbane, Queensland, Australia

and

Cooperative Research Centre for
    Contamination Assessment and
    Remediation of the Environment
Adelaide, South Australia, Australia

Kongkea Phan
Soil Environment Laboratory
School of Environmental Science
    and Engineering
Gwangju Institute of Science and
    Technology
Gwangju, Republic of Korea

and

Resource Development
    International-Cambodia
Kandal, Cambodia

Marinus Pilon
Department of Biology
Colorado State University
Fort Collins, Colorado

Elizabeth A.H. Pilon-Smits
Department of Biology
Colorado State University
Fort Collins, Colorado

Yan Yan Qin
Croucher Institute for
    Environmental Sciences
Hong Kong Baptist University
Hong Kong, People's Republic of
    China

D. Michael Revitt
Urban Pollution Research Centre
Middlesex University
London, United Kingdom

R. Brian E. Shutes
Urban Pollution Research Centre
Middlesex University
London, United Kingdom

Suthipong Sthiannopkao
Department of Environmental
    Engineering
College of Engineering
Dong-A University
Busan, Republic of Korea

Nora F.Y. Tam
Department of Biology and
    Chemistry
City University of Hong Kong
Hong Kong, People's Republic of
    China

Palanisami Thavamani
Centre for Environmental Risk
    Assessment and Remediation
University of South Australia
Mawson Lakes, South Australia,
    Australia

and

Cooperative Research Centre for
    Contamination Assessment and
    Remediation of the Environment
Adelaide, South Australia, Australia

Kadiyala Venkateswarlu
Centre for Environmental Risk
    Assessment and Remediation
University of South Australia
Mawson Lakes, South Australia,
    Australia

and

Cooperative Research Centre for
    Contamination Assessment and
    Remediation of the Environment
Adelaide, South Australia,
    Australia

and

Department of Microbiology
Vikrama Simhapuri University
Nellore, India

Jian Ping Wang
The University of Queensland
National Research Center for
    Environmental Toxicology
    (EnTox)
Brisbane, Queensland, Australia

and

Cooperative Research Centre for
    Contamination Assessment and
    Remediation of the Environment
Adelaide, South Australia,
    Australia

Yuanping Wang
Analytic Cytology Laboratory
and
Key Immunopathology Laboratory
    of Guangdong Province
Shantou University Medical
    College
Shantou, Guangdong, People's
    Republic of China

Yuping Wang
Laboratory of Environmental
    Microbiology and Toxicology
School of Biological Sciences
The University of Hong Kong
Hong Kong, People's Republic of
    China

Ron T. Watkins
Department of Applied Geology
Curtin University
Perth, Western Australia, Australia

Chris K.C. Wong
Croucher Institute for
    Environmental Sciences
and
Department of Biology
Hong Kong Baptist University
Hong Kong, People's Republic of
    China

Ming Hung Wong
School of Environmental and
    Resource Sciences
Zhejiang Agriculture and Forestry
    University
Linan, People's Republic of China

and

Croucher Institute for
    Environmental Sciences
Hong Kong Baptist University
Hong Kong, People's Republic of
    China

Kusheng Wu
Analytic Cytology Laboratory
and
Key Immunopathology Laboratory
    of Guangdong Province
Shantou University Medical College
Shantou, Guangdong, People's
    Republic of China

Shengchun Wu
Croucher Institute for
    Environmental Sciences
Hong Kong Baptist University
Hong Kong, People's Republic of
    China

Qin Wu
School of Environment and Natural
    Resources
The Ohio State University
Columbus, Ohio

Qiongna Xiao
Analytic Cytology Laboratory
and
Key Immunopathology Laboratory
    of Guangdong Province
Shantou University Medical College
Shantou, Guangdong, People's
    Republic of China

Xijin Xu
Analytic Cytology Laboratory
and
Key Immunopathology Laboratory
    of Guangdong Province
Shantou University Medical College
Shantou, Guangdong, People's
    Republic of China

Hui Yang
Analytic Cytology Laboratory
and
Key Immunopathology Laboratory
    of Guangdong Province
Shantou University Medical College
Shantou, Guangdong, People's
    Republic of China

Tingting You
School of Life Sciences
Sun Yat-Sen University
Guangzhou, Guangdong Province,
    People's Republic of China

Gene J.S. Zheng
Croucher Institute for
    Environmental Sciences
Hong Kong Baptist University
Hong Kong, People's Republic of
    China

Guina Zheng
Analytic Cytology Laboratory
and
Key Immunopathology Laboratory
    of Guangdong Province
Shantou University Medical College
Shantou, Guangdong, People's
    Republic of China

1

Environmental Contamination: Health Risks and Ecological Restoration

Ming Hung Wong

CONTENTS

Introduction

The major objective of this chapter is to provide basic background on the topic, that is, "environmental contamination, health risks, bioavailability, and ecological remediation." Basic background on environmental pollution of public concern by listing some known cases of major environmental disasters around the world since the 1940s. This is followed by some examples of environmental degradation in our region, the Pearl River Delta (PRD), which has undergone rapid socioeconomic development during the past 30 years, leading to excessive levels of contaminants in crops and fish produced, and consequently, suspected to be associated with some of the health problems observed in the region, such as high mercury (Hg) and lead (Pb) levels in autistic children. We need to fully understand the sources, fates, and effects of these environmental contaminants, in order to avoid their entry into our food production systems and minimize the chance of these contaminants transforming to more toxic forms, for example, from inorganic Hg to organic form.

There are different types of environmental contaminants, including heavy metals, persistent organic pollutants (POPs), and emerging chemicals of concern. The latter is a moving target, especially because a large amount of chemicals have not been fully tested for their toxicity before placing them on the market, and eventually adverse environmental and health effects are revealed. One of the very good examples is flame retardants, which has become a global contaminant after DDT and PCB, due to their endocrine-disruptive nature. In the PRD, recycling of electronic waste (e-waste) imported from the United States, Japan, and European Communities by taking advantage of less stringent environmental regulations and cheaper labor cost aggravated the regional environmental and health problems. Ecological restoration based on bioremediation and phytoremediation of these contaminated sites would be necessary to clean up these sites, especially if they will be allocated for other usages, after the industries have been moved to other regions.

Twenty-two chapters in total are included in this book, addressing different topics related to environmental contamination, health risks, and ecological restoration. They are listed under four major parts: Part I: Health Impacts and Risk Assessment (5 chapters), Part II: Emerging Chemicals and Electronic Waste (5 chapters), Part III: Ecological Restoration of Contaminated Sites: Bioremediation (5 chapters), and Part IV: Ecological Restoration of Contaminated Sites: Phytoremediation (6 chapters).

Environmental Pollution of Public Concern: Some Examples

Population growth, urbanization, and industrialization have contributed to the emission of different pollutants into different ecological compartments, that is, air, water, and soil. Environmental pollution drew public concern in industrialized countries in early years, for example, the great smog in London in 1952 that killed at least 4000 people (Bell et al. 2004); the Superfund established by USEPA in 1978 due to long-term contamination of Love Canal (notably contamination by dioxins) since 1947 (USEPA 1996); the radioactive contamination associated with disposal of nuclear waste throughout the 1950s and 1960s in Lake Karachay, the accidents that occurred in the nuclear power plant at Chelyabinsk, USSR, in 1957, and also the nuclear accident at the Chernobyl nuclear plant in Ukraine, 1986 (Williams 2002); the overuse and abuse of organochlorine pesticides in the 1950s and 1960s, leading to the significant loss of biodiversity, notably carnivorous birds (Carson 1962); mercury poisoning of residents and their offspring in Minamata Bay (a fishing village) in Japan, due to consumption of seafood containing very high concentrations of methyl mercury during the mid- to late 1950s (Walker 2010); and PCB dumping (from discharges of paper mills) in Hudson Bay during 1954–1971 led to the ban of consumption of fish caught in the area in 1980 (NYSDEC 2011).

With the recent socioeconomic development, both in speed and scale, in some developing countries, such as China, with less stringent environmental regulations, it is evident that different industrial activities have given rise to tremendous environmental degradation (affecting air, water, and land) and severe health problems. China contributed 22% of the global emissions of greenhouse gases (Nersesian 2010). Environmental pollution has rendered cancer the major cause of death in 30 cities and 78 counties in China, according to the Ministry of Health (China Daily 2007). Furthermore, lead poisoning is prevalent in China, with one-third of children suffering from elevated blood lead levels (New York Times 2011). This is related to different industrial emissions containing lead, for example, manufacturing of lead-acid batteries for cars and electric bikes, and mining and smelting of lead.

Deterioration of Environmental Quality due to Rapid Socioeconomic Development in the Pearl River Delta

Increasing amounts of toxic chemicals are discharged into our environment via different anthropogenic activities, which impose adverse impacts on biota and human beings. This is especially true in regions such as the PRD in South China, one of the regions in China that has undergone rapid industrialization and urbanization without due regard for environmental protection, in the past two decades. As a consequence, serious environmental problems have arisen including, (1) Air: power plants, cement plants, factories, and vehicles are the major sources of air pollutants. The high demand for electric power due to export activities has given rise to the increased use of coal, which aggravated the acid rain problem (due to sulfur), in addition to the emission of other pollutants such as Hg. (2) Solid waste/land: Huge amounts of wastes have been generated, without efficient waste management, resulting in severe land contamination and dereliction, especially those caused by mining activities. (3) Water: Only 16% of sewage is treated in Guangdong Province, compared with other provinces, for example, 62% in Shanghai and 42% in Jiangsu (Ng 2003).

Moreover, the PRD has become the power house of modern appliances, producing electrical and electronic products, textiles, furniture, shoes, and so on, emitting a large amount of POPs and emerging chemicals of concern (such as flame retardants, bisphenol A [BPA]). There is also an urgent need to manage the illegal import of e-waste from other countries as well as the ever-rising amount of e-waste generated domestically. The harmful substances emitted during uncontrolled recycling of e-waste (notably open burning) have imposed serious adverse effects on the environment and the health of workers and residents of e-waste recycling sites, which is evidenced from the rising morbidity of major diseases, such as respiratory and cardiovascular diseases and malignant tumors (Wong et al. 2007).

Soil Pollution Affecting Crops Grown in the Region

Because of the favorable weather, fertile soil, and abundant water resources, PRD has served as "homeland for crops and fish" in the past. However, the environmental quality of our region has deteriorated due to recent rapid socioeconomic changes indicated earlier. As a consequence, domestic and industrial discharges containing nutrients, heavy metals, pesticides, PAHs, and oil are prevalent in freshwater and coastal marine water in the region. These led to eutrophication and formation of algal bloom and red tides, depletion of oxygen and contamination of water, as well as soil erosion, depletion of soil fertility, and contamination of soil (Wong and Wong 2004).

Heavy metals are released from different industries, for example, electroplating factories and leather tanneries, but mining and smelting activities are the major sources of these toxic metals detected in farm soils. In addition, overuse of agrochemicals and phosphate fertilizers in farm soils that contain different heavy metals, notably Cd, is also a major concern. Leachate generated from waste landfills also contained a wide range of toxic chemicals, heavy metals, as well as POPs.

Both illegal use of technical DDTs and the use of dicofol containing DDTs (as impurities) are also a major fresh input of DDTs in soils (and crops) in the PRD (Li et al. 2006). This is mainly due to the fact that China had produced and used DDT in the past, resulting in high background levels in soils. In addition, DDT has been exempted to be used for vector control in China (although agricultural application has been nonexistent since 1983). It is also expected that the illegal use of technical DDTs and the use of dicofol, an approved pesticide that contains DDT as impurities, are sources of fresh input into different ecological compartments (Wong et al. 2005).

Soils in the PRD are grossly contaminated by heavy metals, such as Cu, Zn, Ni, and especially Pb, Cd, Hg, with concentrations higher than the background level of Guangdong Province, with about 30% of soil examined having more Hg than the country's Grade II standard (Cai et al. 2008). Elevation of both Cd and Pb in crops and farm soils in PRD was observed in the region, as a result of the environmental release of heavy metals and the use of agrochemicals (Wong et al. 2006).

Pesticide residues (including residuals of DDTs, HCHs, and dieldrin) are frequently detected in grains, grapes, vegetables, fruits, tea, and medicinal herbs throughout China. A survey showed that 41 out of 81 vegetable samples collected from vegetable and fruit markets contained pesticide residuals, with the residuals in leek and cabbage exceeding the national standard of China by 80% and 60%, respectively (Zhang et al. 2011). In addition, it has been observed that both soils and vegetables produced in PRD, using wastewater for irrigation, were contaminated with phthalic acid esters (PAEs), which are widely used in the plastic production industry in the region (Zhang et al. 2009).

Water Pollution Affecting Aquaculture Products Produced in the Region

It has been shown that major species of freshwater fish and marine fish available in local markets (Cheung et al. 2007: DDTs and PAHs, Cheung et al. 2008: PBDEs) were grossly contaminated. Attempts were subsequently made to trace the sources of these pollutants, which were detected in fish cultivated in the region. Results indicated that the organic matter consisting of fecal materials from the cultivated fish and unconsumed fish feeds, deposited at the seabed underneath fish cages, and the trash fish used as fish feeds were potential sources of these pollutants entering into the cultivated fish (Leung et al. 2010: DDTs, Wang et al. 2010a,b: PAHs, Liang et al. 2011 and Shao et al. 2011: Hg, Wang et al. 2011: PCBS). This is especially true for Hg, and possibly for most of these POPs, where bioaccumulation and biomagnification are commonly observed along aquatic food chains with very high concentrations of these toxic chemicals present in organisms of higher trophic levels, including carnivorous fish such as grouper cultured in the region given a high market demand (Cheng et al. 2011).

Environmental Pollutants, Food Intake, and Human Health

It is commonly known that the major food sources for Hong Kong residents are imported from the PRD and other parts of China, which are often contaminated with undesirable chemicals. A health risk assessment conducted based on recent published data has revealed that the daily intake of DDTs may pose health threat to local residents in the coastal regions of China, as the EDI of DDTs (33.39 ng/kg, body weight per day) by residents of South China through consumption of seafood in South China well exceeded the acceptable limit (20 ng/kg, body weight per day) imposed by FAO/WHO (Yatawara et al. 2009).

Qin et al. (2011) examined some major food items available in Hong Kong and revealed that food materials such as goose liver and chicken skin, which contained more fat, had the highest levels of POPs, as most of the POPs are lipophilic and tend to store in adipose tissues. An early investigation indicated that higher concentrations of DDTs and PCBs were detected in human milk samples collected from Guangzhou and Hong Kong, and significant correlations were found to exist between the age of donors and the frequency of their fish consumption (Wong et al. 2002).

With assistance from the Red Cross (Hong Kong), blood samples were collected from the general public, and it was noted that the concentrations of fluorene (a congener of PAHs), pp-DDE, pp-DDT, DDT, PCB126, As, and Hg were significantly correlated with the frequency of seafood consumption (Qin et al. 2010). There seems to be a lack of information related to how environmental contaminants cause disease incidence, as this may not be a straightforward exposure–outcome paradigm. Nevertheless, it has been shown that the concentrations of certain POPs (DDTs, PAHs) and heavy metal/metalloid (Hg, Pb, As) in adipose tissues of patients with uterine leiomyomas were closely linked with seafood diet, BMI, and age (Qin et al. 2010).

Sources, Fates, and Effects of Toxic Chemicals

When monitoring a toxic chemical, it is essential to identify the sources, trace its fates in different ecological compartments, and assess its effects on biota and human beings. An excellent example to illustrate the sources, fates, and effects of toxic chemicals would be the mercury poisoning case mentioned earlier. The Chisso factory at Minamata Bay produced a large amount of acetaldehyde, using mercury sulfate as a catalyst, which was discharged into Minamata Bay from the factory between 1932 and 1968 (Walker 2010). Methyl mercury, an organic mercury compound, is formed through methylation of the metallic mercury in the sediment by bacteria. Methyl mercury can be efficiently bioaccumulated (stored in lipids) in aquatic animals (i.e., shrimps, fish) and biomagnified in food chains, leading to extremely high concentrations in carnivorous fish. Adverse effects were first observed in cats and birds around the region and later on noticeable symptoms were observed in hundreds of people who had high mercury levels, greater than 50 ppm, in their hair (with an extreme case of 920 ppm) (Walker 2010). It is, therefore, extremely important to inhibit the entry of environmental contaminants into our food production systems (e.g., crop growth and fish culture) and also to avoid the chance for these contaminants transforming into more toxic forms.

Different Types of Environmental Pollutants

Heavy Metals

Heavy metals are naturally occurring in the environment. However, human activities have greatly altered their biochemical and geochemical cycles, and their balance in the nature. They are mainly emitted through industrial activities, such as fuel combustion (diffusion point sources, e.g., from traffic), mineral extraction, and metal smelting (industrial point sources). This is especially true as China is full of mineral resources. China is the largest producer of coal and also possesses ample reserves of tungsten, tin, stibium, zinc, titanium, tantalum, thallium, lead, nickel, mercury, molybdenum, niobium, and aluminum (State Statistical Bureau 1998). Mining of these reserves has given rise to the large-scale destruction of land surface leading to soil erosion and emissions of heavy metals into the immediate environment, enhancing their entry into food production systems, for example, crop and fish cultivation.

Environmental contamination and exposure (through oral intake, inhalation, and dermal contact) to heavy metals has become a serious public health concern worldwide. Some of the heavy metalloid/metals, for example, arsenic (As), cadmium (Cd), mercury (Hg), and lead (Pb) are highly toxic, and they have become topical issues since the mid-1970s. In fact, the Priority List of the "Top 20 Hazardous Substances" included As (first place), Pb (second place), Hg (third place), and Cd (seventh place), a list released by the Agency for Toxic Substances and Disease Registry (ATSDR). Some of these heavy metal elements such as Cu and Ni are needed in trace amounts in biota, including humans, but would become toxic under high concentrations. Most others are toxic or carcinogenic, imposing adverse effects on different systems: As, Hg, and Pb on central nervous system; Cd, Hg, and Pb on kidneys or liver; and Cd, Cu, Cr, and Ni on skin, bones, and teeth (ATSDR 2001).

Persistent Organic Pollutants

Stockholm Convention POPs

The original list of the 12 POPs listed in The Stockholm Convention on POPs included 7 chlorinated hydrocarbon pesticides (aldrin, chlordane, DDT, dieldrin, endrin, heptachlor, hexachlorobenzene, mirex, toxaphene), 2 industrial chemicals (PCB and hexachlorobenzene), and 2 unintended byproducts (dioxins and furans), and 10 chemicals have been included in the control list recently, which included some flame retardants (such as polybrominated diphenyl ethers, PBDEs) and pesticides (such as lindane) (Secretariat of the Stockholm Convention 2009). The 172 countries that signed the legally binding treaty have to prepare their national plan to ensure reduction or elimination of these chemicals in the environment.

There is sufficient information related to the sources, fates, and effects of these chemicals, notably pesticides such as DDT (Wong et al. 2005), and dioxins/furans (Zhang et al. 2011). Basically, these POPs share four common characteristics. (1) They are highly persistent and resist biological, chemical, and photolytic degradation. (2) Being fat soluble (lipid seeking), they can be efficiently stored in fatty tissues of living organisms, and some of them can be biomagnified in food chains, resulting in extremely high concentrations in top carnivores such as tuna, sword fish, and shark. (3) They are highly toxic, causing adverse health effects by disrupting normal developmental functions (serving as endocrine disruptors), and lowering immunity, leading to lower reproductive success, especially in some more sensitive living organisms (and hence lowering biological diversity). (4) Due to their semivolatile nature, they can be transported long distance through air and water, resulting in high concentrations of body loadings of some of these chemicals (DDT and PCBs) in animals (polar bear) and residents (Eskimos) (Whylie et al. 2003).

Flame Retardants

PBDEs are added into daily products (e.g., furniture, carpets, clothes, electrical, and electronic products) to slow down burning process and, thus, to save human life. Due to the fact that PBDEs are not chemically bound to the products, they can be leached into the environment easily. Human beings are exposed to low levels of PBDEs through oral consumption of contaminated food and inhalation; even from indoor household items. This has become a more recent public health concern, as PBDEs have the same persistent and bioaccumulative and endocrine-disruptive properties as other POPs (Harley et al. 2010).

Polycyclic Aromatic Hydrocarbons

Although not included in the list, polycyclic aromatic hydrocarbons (PAHs) are one of the most widespread air pollutants, mainly derived from fuel and biomass (such as crop residues) combustion. They are also found in cooked food (especially meat cooked under high temperature, and smoked fish). Some of these compounds, in particular benzo(a)pyrene, are carcinogenic, teratogenic, and mutagenic, hence causing adverse health effects (Luch 2005).

Emerging Chemicals of Concern

The term "emerging chemicals" is indeed a moving target. A very good example is polybrominated diphenyl ethers (PBDEs), which have been recently included in the Stockholm Convention's list of chemicals given the increasing evidence presented in recent research works that shows their adverse effects on the environment and human health. Like PBDEs, there are other chemicals that have not been tested fully concerning their long-term ecotoxicological, biological, and human health effects before launching the products containing these chemicals into markets. Listed in the following are some of the common emerging chemicals that have received increasing public attention.

Bisphenol A

Bisphenol A is used as an intermediate in the production of polycarbonate plastics and epoxy resins and is widely used in all daily products such as digital media (such as CDs, DVDs), electronic equipment, automobiles, construction materials, plastic bottles (including baby bottles), children's toy, food storage containers, and so on. Epoxy resins are also applied as international coating for cans that protects food and drink from direct contact with metals. There is a genuine concern about the endocrine effects of BPA on humans, based on animal data. Given the fact that South China has become a power house for manufacturing electronic equipment and other daily appliances containing BPA, their possible release of toxic materials into the environment should not be overlooked (Huang et al. 2011).

Phthalates

Phthalates are added to plastics to strengthen their flexibility, transparency, and durability, in a number of products, such as coatings of pharmaceutical pills and nutritional supplements, and stabilizers, dispersants, lubricants, and adhesives in various industrial usages. The concern for human uptake is mainly derived from its use in food packaging and processing (such as gloves used in food handling). It is known that phthalates interfere with androgen production and thereby affect male reproduction (Lomenick et al. 2009). It is extremely alarming to find out that phthalates have been added to drinks and syrups (used in drugs) as cloudy agents illegally (May 2011) instead of edible oil for 30–40 years in Taiwan, which may link to the finding that Taiwan is one of the regions in the world with a very low birthrate (8.9 births/1000, 2010 estimation) (Central Intelligence Agency 2011).

Pharmaceuticals and Personal Care Products

This group of chemicals includes more than 3000 different chemicals used as medicines (painkillers, antibiotics, contraceptives, lipid regulators, tranquilizers, impotence drugs) and personal care products (dental care and hair care products, soaps, sunscreen lotions, etc.) (Ternes et al. 2004), with antibiotics being the largest single group of drugs traded in developing countries (WHO 2001). In general, municipal sewage is the major source of most pharmaceuticals and personal care products (PPCPs), except antibiotics, which are mainly derived from farming and aquaculture.

There is a lack of information on the harmful effects of PPCPs, except some standard acute toxicity data reported by some pharmaceuticals. However, there is sufficient evidence showing the adverse health effects arising from overuse and abuse of some antibiotics, especially those that are illegally used (e.g., tetracycline) for both humans and animals (in farming livestock and fish) in some regions, including South China (Wei et al. 2010). It is also known that high concentrations of certain antibiotics would alter microbial community structure, which would then affect components of higher food web, imposing genotoxicity (Richardson et al. 2005).

Mixed Pollutants: Toxic Chemicals Emitted via Open Burning of E-Waste

E-waste contains valuable products, in particular precious metals such as gold, silver, and platinum, with a relatively high market value once isolated from the waste. However, it also contains a wide range of chemicals such as lead, chromium, flame retardants, and so on, which are extremely hazardous. Some of the primitive techniques used in recycling of e-waste will enhance the release of toxic chemicals, and especially in more toxic forms (with high bioavailability), to living organisms inhabiting the e-waste recycling site (Wong et al. 2007).

Among different primitive techniques used in recycling e-waste, open burning of materials derived from e-waste that are not of economic value (plastics, printed circuit boards, etc.) and cable wires to extract copper is most destructive to the environment and human health, as some substances contained in e-waste will give rise to incomplete combustion. Printed circuit boards are also baked on open fire in semienclosed workshops by workers without wearing any health protection. The toxic fumes such as dioxins produced from burning (especially materials that contain PVC) under relatively low temperature would be a major health concern of these workers. In addition, acid tripping of printed circuit boards to collect precious metals also produce waste materials and effluents with high acidity and heavy metals (with high bioavailability under extremely acidic conditions) that are disposed along river banks (Leung et al. 2007).

Extremely high concentrations of brominated flame retardants (PBDEs), dioxins/furans (PCDD/Fs), PAHs, and Pb were detected in different ecological compartments, that is, air, soil/sediment, water, and biota. Health risk assessments based on bioassay toxicity tests estimated the daily intake of contaminants through food consumption survey and food basket analysis, the actual body loadings (human tissues such as milk, placenta, and hair) of workers engaged in e-waste recycling and local residents, and the epidemiological data provided by the local hospital (from Taizhou, one of the two major e-waste recycling sites in China), all of which indicated the detrimental effects of uncontrolled e-waste recycling (Wong et al. 2007).

In addition to the huge expenditure in treating and curing the workers and local residents who have suffered from acute and chronic toxicity due to their immediate environment, remediation of the contaminated sites with mixed pollutants will be extremely difficult.

The Need to Remediate Contaminated Sites

It is recognized that mineral exploitation including (1) metals, (2) industrial minerals (such as lime or soda ash), (3) construction materials, and (4) energy minerals (i.e., coal, natural gas, oil, etc.) has been a major cause of land damage in China, imposing adverse effects on the local ecosystems and inhabitants (including human beings), and through the release of heavy metals and organic pollutants (notably oil), underground and nearby farmland and fishponds will also be affected (Wong and Bradshaw 2002a). Other industrial activities such as air emissions from power plants and incinerators, uncontrolled discharges of different types of industrial effluents, and the use of contaminated irrigation water are potential sources of soil pollution.

Remediation of sites contaminated with different environmental pollutants, including heavy metal/metalloids (such as Cu, Pb, and Zn) and POPs (such as PAHs and DDT), is essential to avoid spreading of pollutants that may have adverse impacts on plants, animals, and human beings and restore the ecosystems by establishing soil and associated plant, animal, and microbial communities (Wong and Bradshaw 2002b).

Bioremediation of Contaminated Sites

Bioremediation refers to the use of microbes (i.e., bacteria and fungi) for degrading POPs (such as PAHs and DDT), and it has become an effective and low-cost technology for cleaning up sites contaminated by these organic contaminants. This can be achieved by isolating and subsequently enhancing the growth of indigenous microbes through provision of nutrients, carbon sources, or electron donors, either *in situ* (e.g., bioventing, biosparging, biostimulation, and some composting methods) or *ex situ* (e.g., bioreactors, biofilters, land farming, and some composting methods). The former is more attractive because it requires less equipment and is therefore more cost-effective, and causes less disturbance to the immediate environment (Diaz 2008).

Phytoremediation of Contaminated Sites

Phytoremediation is an emerging green technology that involves the use of green plants to decontaminate different pollutants from soils/waters. For soils/waters contaminated by heavy metals, (1) heavy metal tolerance plants (which avoid taking up toxic metals into their tissues) are used to stabilize (phytostabilization) the contaminated sites containing high levels of heavy metals such as Pb and Zn; (2) heavy metal hyperaccumulating plants (which are able to absorb and translocate toxic metals from roots into their upper-ground tissues) are used to remove (phytoextraction) a substantial amount of toxic metals from the sites, upon harvest; and (3) wetland plants (such as *Typha* sp. and *Phragmites* sp.) are used to remove toxic metals from wastewaters by precipitating toxic metals in the rhizosphere soils (Wong 2003).

In terms of treating sites contaminated by organic contaminants (such as DDT and PAHs), methods used in phytoremediation include (1) phytodegradation, which relies on the internal and external metabolic processes of plants to break down organic contaminants; (2) rhizodegradation is the use of microbes associated with rhizosphere soils to degrade soil contaminants, and the process could be enhanced by certain plants that activate microbial activity; and (3) phytovolatilization refers to the uptake of organic contaminants that are soluble in water and their subsequent release into the atmosphere through transpiration by plants (Alkorta and Garbisu 2001).

Issues Addressed in This Book

Health Impacts and Risk Assessment

In Part I, "Health Impacts and Risk Assessment," there are five chapters that further illustrate different timely topics in depth. Chapter 2 is aimed at reviewing the current environmental problems in the PRD region, which may link with health impacts encountered in Hong Kong, and calling for health care reform in the region by adopting more effective management, based on dietary modification, nutrition supplementation, and detoxification of contaminants accumulated in our bodies. The investigation presented in Chapter 3 is based on the data concerning autistic children collected from Mainland and Hong Kong, demonstrating the close association between environmental heavy metal poisoning and the development of autism. Unfortunately, there seems to be a continuous rise on the number of cases disclosed in South China. The risk and potential effects of endocrine-disruptive contaminants to the reproductive and developmental health are reviewed in Chapter 4. Some of these POPs (such as DDTs, PBDEs, and PFOA/PFOS) share similar structure and behavior with our hormones and, therefore, can adversely affect our normal physiology and development. Chapter 5 assesses health risks from arsenic (As) intake by residents in Cambodia, by analyzing As concentrations in the underground water, which has been used as drinking water by local residents, and in the hair samples of local residents with the prevalence of arsenicosis. Chapter 6 is an excellent example to indicate the need for a more pragmatic approach for health risk assessment, as part of an overall effective management strategy of potential toxic mine waste storage facility such as bauxite storage facility.

Emerging Chemicals and E-Waste

There are five chapters in Part II. In Chapter 7, an attempt has been made to review the removal efficiencies of some emerging chemicals in publicly owned sewage treatment works. However, these sewage treatment works were originally designed to treat organic materials (by aerobic and anaerobic decomposition) contained in domestic sewage, and are not adequate for treating emerging chemicals of concern, such as antibiotics and PPCPS. There is a danger of these pollutants entering into food chains via the discharge of sewage outfalls. Chapter 8 outlines the sources, trends, and adverse health effects of flame retardants used in China. The stringent fire regulations in North America and European Communities resulted in the use of a large amount of flame retardants in these regions. It is, therefore, timely to review the situation in China, which has witnessed a rapid industrial development in the past 10 years. Chapter 9 demonstrates the adverse health problems of children residing in Guiyu (Guangdong Province), the mega e-waste recycling site in China, based on their manganese levels in blood and renal tubular dysfunction. Chapter 10 describes the health risks associated with uncontrolled recycling of e-waste according to the studies conducted in Guiyu as well as another important e-waste recycling location (recycling a large number of obsolete transformers), Taizhou (Zhejiang Province), and reviewed the effectiveness of existing regulations in controlling the problem, that is, transboundary movement of e-waste and toxic substances contained in electronic and electrical equipment. Chapter 11 provides an overview on the problems faced in most countries and possible decision-making support tools for managing e-waste.

Ecological Restoration of Contaminated Sites

Part III, "Ecological Restoration of Contaminated Sites: Bioremediation," contains five chapters. Chapter 12 illustrates the potential of As biomethylation as a cost-effective and feasible method for bioremediation of sites contaminated by As, which is prevalent in several Asian countries/regions. Future research on field applications along this line should be encouraged as most studies on As biomethylation are limited to laboratory settings. Chapter 13 examines the mechanisms of homeostasis of copper (Cu), an important trace element, which is essential for living organisms, but can be toxic under high concentrations. A detailed account is given to copper homeostasis mechanisms, which includes responses to deficiency as well as toxic excess. Similarly, Chapter 14 focuses on another important element, selenium (Se), which is an essential nutrient for living organisms, but is toxic at excessive levels. It deals with the current understanding of the genetic and biochemical mechanism that controls Se tolerance and accumulation in plants. Chapter 15 provides a comprehensive review on the occurrence and fate of phthalate esters in different ecological compartments, with emphasis on its microbial transformation, which includes aerobic as well as anaerobic degradation. Chapter 16 describes the toxicity and biological degradation of mixed contamination: PAHs and toxic metals (such as Cd, Cu, Ni, and Pb) that are commonly encountered at manufactured gas plants (MGP) sites. It highlights the importance of isolating and characterizing suitable microbes that are capable of detoxifying both organic contaminants and toxic metals and their survival and efficiency under field conditions.

There are six chapters in Part IV, "Ecological Restoration of Contaminated Sites: Phytoremediation." Chapter 17 also deals with the same problem observed in mangrove sites that are contaminated by both PAHs and heavy metals. The major aim of this investigation is to screen most suitable mangrove plant species based on their structural modifications in roots and physiological and biochemical changes in different plant tissues for restoring these sites. Chapter 18 describes the origin of paddy fields in urbanized areas of Yangtze River Delta and their potential function in phytoremediation of contaminants, which is important to ensure sustainable development by provision of food security and environmental safety in the region. Chapter 19 further focuses on the use of constructed wetlands for flood prevention and pollution control, by introducing the concept of Integrated Urban Water Management (IUWM). Elevated levels of As are commonly found in mine tailings derived from extraction of gold. Chapter 20 highlights the successful application of phytostabilization in the Victorian Goldfields of southeastern Australia, by means of plants that are As tolerant with low As accumulation (four species of *Eucalyptus* and some native grass species), for firewood production, extraction of cineole (*Eucalyptus* oil), and a drought-resistant and low-maintenance pasture. Chapter 21 also provides a successful case for stabilizing As-contaminated sites in Korea, using stabilizing agents, including cement, pozzolans, lime, iron(hydr)oxide, and fly ash, in order to reduce the mobility and toxicity of As. Chapter 22 deals with the restoration of brownfield sites in northwest England, by focusing on the importance of reducing bioavailability of As, Cd, Cu, Ni, and Zn by applying green waste compost and phosphate, in terms of phytoremediation and/or natural attenuation for risk assessment.

Conclusion

Being one of the world's mega river deltas, the PRD has been the most important homeland for supplying fish, meat, and vegetables to Hong Kong due to its abundant water resource and fertile soil. Unfortunately, due to its rapid urbanization, population growth and industrial development, the area is grossly contaminated by overuse and abuse of pesticides, fertilizers, and antibiotics, threatening food security and safety in the region.

In addition, the area has become an important sector for manufacturing electronic and electrical equipment and other daily appliances, which may release a large amount of emerging chemicals (such as flame retardants) into the environment. Uncontrolled recycling of e-waste, both imported from other countries and generated domestically, also presented additional problems, which called for international cooperation. In time, each country has to deal with its problem with environmental pollution due to increase of environmental awareness, improved living standard, and tightened environmental regulations, especially pressing are those governing import of e-waste.

There seems to be an urgent need to safeguard environmental quality and human health, by remediating the contaminated sites, using cost-effective and green technology, especially after some of the industries are moved outside the PRD; a recent government policy was imposed addressing the problem.

Acknowledgments

The author thanks the Research Grants Council of Hong Kong and Croucher Foundation for the continued support for the past 20 years and also his colleagues and research students who have provided the information described in this chapter.

References

Alkorta I, Garbisu C. 2001. Phytoremediation of organic contaminants in soils. *Bioresource Technology* 79: 273–276.

ATSDR, Agency for Toxic Substances and Disease Registry. 2001. The priority list of hazardous substances that will be the subject of toxicological profiles. http://www.atsdr.cdc.gov/spl/

Bell ML, Davis DL, Fletcher T. 2004. A retrospective assessment of mortality from the London smog episode of 1959: The role of influenza and pollution. *Environmental Health Perspectives* 112: 6–8.

Cai LM, Ma J, Zhou YZ et al. 2008. Multivariate geostatistics and GIS-based approach to study the spatial distribution and sources of heavy metals in agricultural soil in the Pearl River Delta, China. *Environmental Science* 29: 3496–3502.

Carson R. 1962. *Silent Spring*. Boston: Houghton Mifflin Harcourt.

Central Intelligence Agency. 2011. *The World Fact Book Taiwan*.

Cheng Z, Liang P, Shao DD et al. 2011. Mercury biomagnification in the aquaculture pond ecosystem in the Pearl River Delta. *Archives of Environmental Contamination and Toxicology* 61: 491–499.

Cheung KC, Leung HM, Kong KY, Wong MH. 2007. Residual levels of DDTs and PAHs in freshwater and marine fish from Hong Kong markets and their health risk assessment. *Chemosphere* 66: 460–468.

Cheung KC, Zheng JS, Leung HM, Wong MH. 2008. Exposure to polybrominated diphenyl ethers associated with consumption of marine and freshwater fish in Hong Kong. *Chemosphere* 70: 1707–1720.

China Daily. 2007. Pollution makes cancer the top killer. *Xie Chuanjiao (China Daily)*. May 21, 2007. http://www.chinadaily.com.cn/china/2007-05/21/content_876476.htm

Diaz E (Ed.). 2008. *Microbial Biodegradation: Genomics and Molecular Biology* (1st edn.). Caister Academic Press, New York.

Harley K, Marks A, Chevrier J, Bradman A, Sjödin A, Eskenazi B. 2010. PBDE concentrations in women's serum and fecundability. *Environmental Health Perspectives* 118: 699–704.

Huang YQ, Wong CKC, Zheng JS et al. 2011. Bisphenol A in China: Sources, environmental levels, and potential human health impacts. *Environmental International* 42: 91–99.

Leung SY, Kwok CK, Nie XP, Cheung KC, Wong MH. 2010. Risk assessment of residual DDTs in freshwater and marine fish cultivated around the Pearl River Delta, China. *Archives of Environmental Contamination and Toxicology* 58: 415–430.

Leung AOW, Luksemburg WJ, Wong AS, Wong MH. 2007. Spatial distribution of polybrominated diphenyl ethers and polychlorinated dibenzo-p-dioxins and dibenzofurans in soil and combusted residue at Guiyu, an electronic waste recycling site in southeast China. *Environmental Science and Technology* 41: 2730–2737.

Li J, Zhang G, Oi SH, Li XD, Peng XZ. 2006. Concentrations, enantiomeric compositions, and sources of HCH, DDT and chlordane in soils from the Pearl River Delta, South China. *Science of the Total Environment* 372: 215–224.

Liang P, Shao DD, Wu SC, Shi JB, Wu FY, Lo SCL, Wang WX, Wong MH. 2011. The influence of mariculture on mercury distribution in sediments and fish around Hong Kong and adjacent mainland China waters. *Chemosphere* 82: 1038–1043.

Lomenick JP, Calafat AM, Melguizo Castro MS et al. 2009. Phthalates exposure and precocious puberty in females. *Journal of Pediatrics* 156: 221–225.

Luch A. 2005. *The Carcinogenic Effects of Polycyclic Aromatic Hydrocarbons*. Imperial College Press, London, U.K.

Nersesian RL. 2010. *Energy for the 21st Century: A Comprehensive Guide to Conventional and Alternative Sources* (2nd edn.). M.E. Sharpe, Inc., New York.

New York Times. 2011. Lead poisoning in china: The hidden scourge. June 15, 2011. https://www.nytimes.com/2011/06/15/world/asia/15lead.html

Ng SL. 2003. Guangdong's environment and China's accession to the WTO In: *Guangdong, Preparing for the WTO Challenge*. Ed. Cheng JYS. The Chinese University Press, Hong Kong, pp. 307–326.

Qin YY, Leung CKM, Leung AOW, Wu SC, Zheng JS, Wong MH. 2010. Persistent organic pollutants and heavy metals in adipose tissues of patients with uterine leiomyomas and the association of these pollutants with seafood diet, BMI, and age. *Environmental Science and Pollution Research* 17: 229–240.

Qin YY, Leung CKM, Leung AOW, Zheng JS, Wong MH. 2011. Persistent organic pollutants in food items collected in Hong Kong. *Chemosphere* 82: 1329–1336.

Richardson BJ, Lam PKS, Martin M. 2005. Emerging chemicals of concern: Pharmaceuticals and personal care products (PPCPs) in Asia, with particular reference to Southern China. *Marine Pollution Bulletin* 50: 913–920.

Secretariat of the Stockholm Convention. 2009. Measures to reduce or eliminate POPs, Geneva, Switzerland. http://chm.pops.int/Portals/0/docs/publications/sc_factsheet_001.pdf

Shao DD, Liang P, Kang Y, Wang HS, Cheng Z, Wu SC, Shi JB, Lo SCL, Wang WX, Wong MH. 2011. Mercury species of sediment and fish in freshwater fish ponds around the Pearl River Delta, PR China: Human health risk assessment. *Chemosphere* 83: 443–448.

State Statistical Bureau. 1998. *China Energy Statistical Yearbook, 1991–1996*. Statistical Publishing House, Beijing, China.

Ternes TA, Joss A, Siegrist H. 2004. Scrutinizing pharmaceuticals and personal-careproducts in wastewater treatment. *Environmental Science & Technology* 38: 392A–399A.

USEPA. 1996. Soil Screening Guidance: User's Guide. EPA Document Number: EPA540/R-96/018, United States Environmental Protection Agency, Washington, DC.

USEPA. 2011. Hudson River PCBs. http://www.epa.gov/hudson/ (accessed September 2011).

Walker B. 2010. *Toxic Archipelago: A History of Industrial Disease in Japan*. University of Washington Press, Washington, DC.

Wang HS, Du J, Leung HM, Leung AOW, Liang P, Giesy JP, Wong CKC, Wong MH. 2011. Distribution and source apportionments of polychlorinated biphenyls (PCBs) in mariculture sediments from the Pearl River Delta, South China. *Marine Pollution Bulletin* 61: 101–114.

Wang HS, Liang P, Kang Y, Shao DD, Zheng GJ, Wu SC, Wong CKC, Wong MH. 2010a. Enrichment of polycyclic aromatic hydrocarbons (PAHs) in mariculture sediments of Hong Kong. *Environmental Pollution* 158: 3298–3308.

Wang HS, Man YB, Wu FY, Zhao YG, Wong CKC, Wong MH. 2010b. Oral bioaccessibility of polycyclic aromatic hydrocarbons (PAHs) through fish consumption, based on an *in vitro* digestion model. *Journal of Agricultural and Food Chemistry* 58: 11517–11524.

Wong SC, Li XD, Zhang G, Qi SH, Min YS. 2002. Heavy metals in agricultural soils of the Pearl River Delta, South China. *Environmental Pollution* 119: 33–44.

Wei X, Ching LY, Cheng H, Wong MH, Wong CKC. 2010. The detection of dioxin- and estrogen-like pollutants in marine and freshwater fishes cultivated in Pearl River Delta, China. *Environmental Pollution* 158: 1–8.

WHO. 2001. Antibiotics resistance: Synthesis of recommendation by expert policy group. World Health Organization, Geneva, Switzerland. WHO/CDS/CSR/DRS/2001.10.

Whylie P, Albaiges J, Barra R, Bouwman H, Dyke D, Wania F, Wong MH. 2003. Regionally based assessment of persistent toxic substances. Global Report. UNEP/GEF.

Williams D. 2002. Cancer after nuclear fallout: Lessons from the Chernobyl accident. *Nature Reviews* 2: 543–549.

Wong MH. 2003. Ecological restoration of mine degraded soils, with emphasis on metal contaminated soils. *Chemosphere* 50: 775–780.

Wong MH, Bradshaw AD. 2002a. China: Progress in the reclamation of degraded land. In: *Handbook of Ecological Restoration, Vol 2, Restoration in Practice*. Eds. Perrow MR, Davy AJ. Cambridge University Press, Cambridge, U.K.

Wong MH, Bradshaw AD. 2002b. *The Restoration and Management of Derelict Land: Modern Approaches*. World Science, London, U.K.

Wong MH, Leung A, Choi M, Cheung KC. 2005. Usage of DDT in China, with emphasis on the use of human milk as an indicator of environmental contamination. *Chemosphere* 60: 740–772.

Wong CKC, Leung KM, Poon BHT, Lan CY, Wong MH. 2002. Organochlorine hydrocarbons in human breast milk collected in Hong Kong and Guangzhou. *Archives of Environmental Contamination and Toxicology* 43: 364–372.

Wong AWM, Wong MH. 2004. Recent socio-economic changes in relation to environmental quality of the Pearl River Delta. *Regional Environmental Change* 4: 28–38.

Wong MH, Wu SC, Deng WJ, Yu XZ, Luo Q, Leung AOW, Wong CSC, Lukesmburg WJ, Wong AS. 2007. Export of toxic chemicals—A review of the case of uncontrolled electronic-waste recycling. *Environmental Pollution* 149: 131–140.

Yatawara M, Devi NL, Qi SH. 2009. Persistent organic pollutants (POPs) in sea food of China—A review. *American Journal of Science* 5: 164–174.

Zhang WJ, Jiang FB, Pi JF. 2011. Global pesticide consumption and pollution: With China as a focus. *Proceedings of the International Academy of Ecology and Environmental Sciences* 1: 125–144.

Zhang MS, Li MY, Wang JY. 2009. Occurrence of phthalic acid esters (PAEs) in vegetable fields of Gongguan City. *Guangdong Academy of Agricultural Sciences* June 2009: 172–175.

Part I

Health Impacts and Risk Assessment

Part I

Fundamental Bioethics

2

Health Impacts of Toxic Chemicals in the Pearl River Delta: The Need for Healthcare Reform?

Paul K.L. Lam and Ming Hung Wong

CONTENTS

Introduction: Background on Toxic Chemicals in the Pearl River Delta Region

On behalf of the Hong Kong government, a consulting company conducted a study of toxic chemicals pollution in Hong Kong (CH2M-IDC 2003). The major aim of the study was to identify and characterize toxic chemicals of potential concern to Hong Kong and their ecological and incremental human health risk assessments. Among the preliminary Priority Toxic Substance List of 135 chemicals, two organic compounds (DDD and DDE) and one metalloid (inorganic arsenic) were identified as chemicals of potential concern, based on the human health risk assessment. However, most of the Stockholm Convention POPs and emerging chemicals of concern have not been covered in the study.

The Pearl River Delta has undergone a rapid socioeconomic development over the past 30 years since the economic reform, with environmental degradation resulting as a consequence (Enright et al. 2005). During the progression of different stages of industrial development, different toxic chemicals have been emitted through human activities. These could be grouped under (1) heavy metals, (2) Stockholm Convention–classified persistent organic pollutants (POPs), and (3) chemicals of emerging concern.

Heavy Metals

It is commonly known that arsenic (As, a metalloid), cadmium (Cd), mercury (Hg), and lead (Pb) are widespread in the environment. A substantial amount of information concerning the sources, fates, and effects in Hong Kong and the Pearl River Delta region has been available since the 1970s.

Based on the bioaccessible heavy metal results from our recent investigation (Man et al. 2010), it was revealed that the concentrations of As and other heavy metals found in the soil of former agricultural lands are highly detrimental to the human health (especially to children) due to electronic waste storing and recycling, car dismantling and repairing activities that causes drastic increases in concentrations. In addition, high concentrations of As, Cd, Hg, and Pb in water and sediment giving rise to elevated levels of their presence in fish would also impose potential health risk.

According to a study on the levels of Hg in six fish species cultured in the same pond, there is a clear indication that Hg is bioaccumulated and biomagnified in the aquatic food chain, with the highest concentration (56.7 ng/g dry wt.) detected in the black bass, a top carnivore (Zhou and Wong 2000; Cheng et al. 2010).

An early study indicated that individuals consuming four or more meals of fish per week had a hair Hg of 4.07 mg/kg dry weight while those consuming fish less frequently had significantly lower levels (2.56 mg/kg), and men with high hair levels were twice as likely to be subfertile, when compared with men with low levels (Dickman et al. 1998, Dickman and Leung 1998). Secondary school students in Hong Kong have been advised not to consume excessive amount of predatory fish (such as tuna), which may contain higher Hg, as well as shellfish, which contains higher As and Cd concentrations (HKFEHD 2002).

In fact, our previous study testing heavy metals in common freshwater and marine fish available in Hong Kong markets indicated that a few fish species had average concentrations greater than the international standards for Cd and Pb established by the European Union and the China National Standard Management Department. Total Hg concentrations in 10 of 20 fish species available in the market were generally greater than those of the World Health Organization's recommended limit of 0.2 mg/kg for at-risk groups, such as children and pregnant women (Cheung et al. 2007).

Due to the rapid industrial development in our region, Hg emission from coal combustion, metal smelting, and relevant industries seems to be a public concern. There is a severe lack of information concerning the concentrations and speciation of Hg in different ecological compartments of our region (Zhang and Wong 2007). Our recent study demonstrated that there is a potential danger that the organic matter accumulated in freshwater fish ponds (around the Pearl River Delta region) and beneath mariculture rafts (along the coastal areas of South China) would provide an ideal opportunity for sulfate bacteria to transform inorganic mercury to methyl mercury and then enter cultured fish (Liang et al. 2010, Shao et al. 2011).

Stockholm Convention POPs

The original 12 POPs included 8 organochlorine pesticides (OCPs): aldrin, chlordane, DDT, dieldrin, endrin, heptachlor, mirex, and toxaphene; 2 industrial chemicals: polychlorinated biphenyls (PCBs) and hexachlorobenzene (which is also a pesticide); and 2 combustion by-products: dioxins and furans (PCDDs and PCDFs). Additional POPs, including flame retardants (polybrominated diphenyl ethers [PBDEs]), have been added to the list recently. Each country that signed the legally binding treaty is required to develop a National Implementation Plan (NIP) to fulfill the obligations under the Stockholm Convention (Article 7), either to eliminate or reduce the production and usage of the 12 POPs in the near future. Each country also needs a strategy to fulfill monitoring obligation. Regular monitoring of these chemicals in different ecological compartments (i.e., air, soil/sediment, water), biota, and human tissues is required.

It is observed that levels of PCBs and dioxins/furans in all environmental media were relatively low in China, when compared with other industrialized countries (Fiedler et al. 2002, Xing et al. 2005). Although OCPs, including eight of the original nine pesticides listed in the Stockholm Convention on POPs, and PAHs, detected in Hong Kong air samples were considered to be at a low level and that they would not pose any health hazards (Choi et al. 2008, 2009). A number of studies indicated the gross contamination of DDTs in the estuarine system in South China; for example, Iwata et al. (1993) tested the concentrations of DDT in the suspended particulate matter in surface seawater around the world and noted the highest concentration of DDT in the South China Sea, which is very near to Hong Kong and Macao.

Our early studies indicated that the inland river systems and fish ponds in the Pearl River Delta were grossly polluted by PCBs, HCHs, and DDTs, which resulted in higher concentrations of these chemicals in fish collected from inland rivers as well as fish ponds (Liang et al. 1999, Zhou et al. 1999a, Zhou and Wong 2000). The uptake of these chemicals seemed to depend on feeding modes of fish, with carnivores having higher concentrations (Zhou et al. 1999b), an indication of bioaccumulation and biomagnification through the aquatic food chain. Our recent survey showed that trash fish used in both freshwater and marine fish farms contained significantly higher levels of DDTs ($86.5–641\,\mu g/kg$ lipid wt.) than in commercial pellets, resulting in higher loadings of DDTs in fish cultivated in our region (Leung et al. 2010).

Our recent study investigating the body burdens of POPs in residents of Hong Kong showed that there are significant correlations: Σ DDTs and p,p'-DDE in human milk with consumption of freshwater and marine fish, and maternal age, and Σ PAHs in human milk with maternal age, respectively. The estimated daily intakes of DDTs by infants indicated that 7 out of 29 of the human milk samples exceeded $20\,ng/g/day$, the tolerable daily intake (TDI) proposed by the Health Canada Guideline in terms of DDTs levels (Tsang et al. 2010). Qin et al. (2010) also demonstrated that DDTs, HCHs, PCBs, PAHs, PBDEs, As, Cd, Pb, and Hg detected in adipose tissues may have some correlation with uterine leiomyomas (a kind of tumor), and their accumulation in the body is positively correlated with regular consumption of seafood and other factors such as body mass index and age.

Based on source analyses, it has been demonstrated that the PAHs in indoor dust were derived from pyrogenic (combustion) origins (Kang et al. 2010). Wang et al. (2010a,b) further showed that surface sediments of freshwater fishponds and mariculture rafts around the PRD were grossly contaminated by PAHs derived from combustion sources. Even in Mai Po Marshes, a more remote area designated for biological conservation, it was found that the sediments contained high concentrations of PAHs and various chlorinated hydrocarbon pesticides that would impose toxicity on the environment, based on the results of toxicity tests using different trophic organisms (Microtoxs solid-phase test, Daphnia mortality test, algal growth inhibition test, and rye grass seed germination test) and biomarker studies (tilapia hepatic metallothionein; glutathione [GSH], and EROD activity using H4IIE rat hepatoma cell) (Kwok et al. 2010).

By analyzing 10 common types of marine fish and 10 common types of freshwater fish available in local markets of Hong Kong, it was found that PBDEs contained in these fish were rather high, and because of the higher fish consumption rate by Hong Kong residents, there is higher daily PBDE intake ($222–1198\,ng/day$ for marine fish and $403–2170\,ng/day$ in freshwater fish) than those reported from the United States ($8.94–15.7\,ng/day$) and Europe ($14–23.1\,ng/day$) (Cheung et al. 2008). Due to the rapid industrial development in the Pearl River Delta, PBDEs can be commonly detected in the environment, and a rather high concentration of $21.4\,ng/day$ dry wt. total PBDEs was detected in the sediment from wastewater discharge from the vehicle-repairing workshop in Lo Uk Tusen, a former farm land in the New Territories of Hong Kong (Luo et al. 2007).

Emerging Chemicals of Concern

A review on "Emerging chemicals issues in developing countries and countries with economy in transition" commissioned by the United Nation Environment Programme (UNEP) and Global Environment Facility (GEF) has been completed (Wong et al. 2010), which identifies a number of emerging chemicals of concern. The major aim of this project was to support GEF's immediate goal in its chemicals program, that is, to promote effective management of chemicals throughout their lifecycle, in ways that lead to the minimization of significant adverse effects on human health and the global environment. The drafting group, with the assistance of the Scientific and Technical Advisory Panel (STAP) of UNEP, identified a preliminary list of emerging chemicals issues based on numerous policy and guidance documents, combined knowledge, and active screening of recent literature.

In addition to the newly added PBDEs into the Stockholm Convention on POPs, the recent view has identified some emerging chemicals of worldwide concern, such as Bisphenol A, PFOA/PFOS, pharmaceuticals, and personal care products, such as antibiotics. Like POPs, these chemicals are highly persistent, are concentrated in the food chain, are accumulated in body lipids, and impose human health hazards. They may lead not only to losses but also to the appearance of new genes and ecotypes, resulting in changes of structural and functional biodiversity. Thus, they influence not only monospecies but also populations.

Bisphenol A is a high-production-volume chemical, as there is an increasing demand for polycarbonates and epoxy resins. Suspected of being hazardous to humans, concerns about the use of BPA in consumer products were regularly reported in the news media since 2008 after several governments questioned its safety. Some retailers have removed products containing BPA from their shelves. The U.S. Food and Drug Administration raised further concerns regarding exposure of fetuses, infants, and young children to BPA (USFDA 2010).

Perfluorinated compounds (PFCs) are man-made fluorinated hydrocarbons, which have been manufactured since late 1940s (Paul et al. 2009). Their products possess distinctive surface-active properties, which allow them to be widely used in coating formulations, fire-fighting foams, and lubricants. Their exceptional chemical stability and inertness also render them suitable for extensive use in surface-protecting agents (e.g., textile, paper products, and carpet). These substances are now widespread in different ecological compartments, biota, and human tissues (Fromme et al. 2009, Paul et al. 2009). Our recent study analyzing 20 species of common marine and freshwater fish available in Hong Kong markets showed that mandarin fish (a freshwater carnivore, commonly cultivated in South China, under dense culture) contained the highest concentrations of both bisphenol A and PFCs (Wei et al. 2010, Zhao et al. 2010).

It has been indicated that the agricultural use of animal manure or fishpond sediment containing considerable amounts of antibiotics such as tetracycline, which is commonly used in both humans and in farming of shrimps and livestocks in our region, may give rise to ecological risks (Wei et al. 2009). In the past, Chinese shrimp and prawns were banned in Europe and the United States due to the high residual antibiotics detected (Chao 2003). However, the most detrimental effect is the provocation of resistance in pathogens through long-term exposure to the residues of these antibiotics because of genetic variation, and the transfer to plasmids containing an antibiotic-resistant gene, either directly or indirectly from nonpathogenic to pathogenic microorganisms (Wegener 1999).

Health Impacts of Toxic Chemicals

Chronic Diseases

With the dramatic rise of the incidence of chronic diseases globally (Yach et al. 2004), diseases such as allergies, neuropsychiatric diseases, metabolic disorders, gynecological and obstetric disorders, cardiovascular disorders, and cancers have become more difficult to manage. Modern medical practice often emphasizes on a single procedure or a pharmaceutical agent to cure a disease by conducting randomized placebo-controlled trials. Despite technological advances, many diseases have become intractable. The patterns of diseases have changed dramatically with many patients having multiple medical problems. Functional abnormalities like insomnia, anxiety, and fatigue have become more prevalent. In the quest for a rational explanation and more efficacious management, a novel approach engendered from the perspectives of "functional medicine" has been tried. This approach has been shown to be effective and is based on scientific advances in the field of studies on nutrition and environmental pollution (Galland 2006).

Over a period of 8 years, an observational research in a private pediatric clinic in Hong Kong has been very active in unraveling the myths of intractable illnesses in infants and children. Many cases of severe allergies, autism, apparent hereditary neurodevelopmental disorders, and compromised growth could be reversed through nutritional therapy. This health management idea has been extended to treat both fetus and adult with health problems. A new model of diseases can be explained by the concept of genetic polymorphism, nutrient insufficiencies, toxic pollutants overload, and hormonal imbalance. Diagnostic approach involves a detailed history of dietary habit, lifestyle, drug history, and occupational exposure to potential pollutants. Investigations would include a hair mineral analysis, a provocative urine challenge test for toxic metals, and a test of morning urine for porphyrins (Fowler 2001, Nataf et al. 2006).

Environmental Pollution and Health Impacts

The environment of Hong Kong is fast becoming polluted, and with a sizeable level of fish consumption and frequent use of dental amalgams and Chinese Herbal medicine, literally all people have toxic overload. The U.S. Environmental Protection Agency has monitored human exposure to toxic environmental chemicals since 1972 when they began the National Human Adipose Tissue Survey (Committee on National Monitoring of Human Tissues 1991). This study evaluated the levels of various toxins in the fat tissues from cadavers and elective surgeries. More than 90% of these samples had five or more toxic organic pollutants. A survey of the newborn infants' cord blood mercury in Hong Kong has shown that more than 70% of the infants had mercury levels above the acceptable levels (Fok et al. 2007).

The local health authority declared fish consumption can be associated with raised blood mercury in the general population, but this would bear no relation to ill-health. The local health authority has educated the public to apply "Chlorine Bleach" widely in the home environment after the SARS and influenza epidemic, thus aggravating the problem of pollutant overload. Pregnant women in Hong Kong stand a high chance of having toxic overload. The fetus *in utero* would share a portion of the mother's toxic burden (Scandborgh-Englund et al. 2001). After delivery, the level of toxic load present in the infants would further increase when they are fed with breast milk (Grandjean et al. 2003). The protean manifestation of the children growing up with toxic overload are poor sleep, allergies, anorexia, stunted growth, attention deficit, hyperactivity, and mood disorders. Women with toxic overload can be asymptomatic or may have chronic fatigue, migraine, menstrual disorders, recurrent infection, insomnia, autoimmune diseases, severe allergies, and gynecological diseases.

The pathophysiological mechanism of mercury overload is the blockade of ATP production and mediation of excessive oxidative stress (Goyer 2000, Clarkson 2002). The immune function would be severely affected, and the affected patients would have "gut dysbiosis": Pathogenic viruses, fungi, and bacteria would flourish inside the gastrointestinal tract. The concomitant overload of POPs acts as xenoestrogen. Through excessive stimulation of estrogen receptors, these women are prone to develop endometriosis, breast cancer, gynecological cancers, menorrhagia, and premature labor. For example, the perchlorate, a particular organic pollutant, can block thyroid receptors causing obesity, fatigue, constipation, and menorrhagia (Porterfield 1994).

A spectrum of illnesses have been observed in infants born with toxic overload. They may have a latent period of normal health in the first few weeks of life. The symptoms are hyperirritability with excessive inconsolable crying, slow feeding, intractable constipation, fidgety sleep, stunted growth, early atopic eczema, respiratory allergies, and repeated infection. The teeth would show dark staining. Teeth grinding or bruxism occurs in sleep. The more severely affected would have delay in neurocognitive function and gross motor function. The affected children tend to have speech delay, clumsiness, poor eye contact, impulsivity, and hyperactivity. If infants with these symptoms were not treated, some of them would develop autism, asperger syndrome, or dyslexia. Treatments of these children include identification of the sources of toxic input. These usually were consumption of too much fish and excessive cleaning of the home with Chlorine Bleach. Elimination of the toxic overload involved the use of supplementary antioxidants, probiotic, and essential fatty acid. Chelation therapy would be required for the more severely affected children (Masters et al. 2008). In contrast with conventional medical practice, judicious use of chelation therapy has been found to be safe and effective in restoring the health of these toxin-laden children with no adverse side effects.

Nutritional Therapy

In the course of this novel therapy, a phenomenal catch up growth was frequently observed in the treated children during the recovery period. This implies the toxic overload has compromised normal growth and has never been recognized in conventional medicine. Once it was relieved, the normal growth trajectory would be resumed. From the perspective of maximal clinical benefit, the younger the patient started on detoxification therapy, the better the outcome. This can be explained by the higher potential of the inherent stem cell in the younger patients to enable compensatory repair of the damaged organs. According to our experience, a child presenting with obvious signs of cerebral palsy and an infant presenting with signs of hemiplegia due to intrauterine stroke have both responded favorably to the detoxification regime. In both cases, a dramatic recovery from the neurological handicap was observed, specifically because the intervention was started at an early age. The inherent plasticity of the infants' brain should be given the chance of compensating for any damage inflicted by the passage of toxin from asymptomatic yet toxin-overloaded mothers to infants during pregnancy.

Adults with multiple medical problems have been tested for toxic heavy metal overload. Literally all of these patients have been shown to have mercury and other toxic heavy metal overload. After they have been started on intensive chelation and nutritional therapy, these medical problems would be significantly ameliorated or relieved. Examples include hypertension, diabetes mellitus, neurosis, anxiety-depression, peripheral arterial insufficiency, irritable bowel syndrome, reflux esophagitis, chronic pain, fatigue, metabolic syndrome, chronic active hepatitis, autoimmune diseases, dementia, chronic bronchitis, retinal degeneration, glaucoma, chronic asthma, and other chronic allergies. Detoxification and antioxidant therapy has been utilized to support the immune dysfunction of cancer survivors, with varying degrees of success.

Nutritional management can likely have a role in reducing the incidence of women's diseases. From a functional medicine approach, micronutrient deficiencies and toxic overload should be identified and rectified. A modified dietary plan and lifestyle should be advised. Nutritional therapy includes recommendation on a diet that is rich in antioxidants, contains good-quality protein, fiber, and sufficient in water. For treating diseases specific to women as mentioned previously, nutritional supplements such as Indole-3-Carbinol and Diindolylmethane (DIM) are recommended to be made a part of these women's diet, as these supplements can help in hastening the metabolism of excessive xenoestrogens (Auborn et al. 2003).

Oral or intravenous magnesium chloride or sulfate can be orally or intravenously administered. Vitamin C, N Acetylcysteine, or Glutathione can be given intravenously to combat the excessive oxidative stress. Selenium and zinc are essential for toxin excretion and are usually required to be supplemented. Plant-derived bioidentical progesterone can be used to counteract the excessive estrogen stimulation of these patients (Holtorf 2009). This is available as topical cream for absorption through the skin, sparing the first pass effect to the liver.

To facilitate elimination of toxic pollutants, constipation should not be allowed. If supplemented dietary fiber is not adequate to prevent constipation, graded doses of Armour thyroid may be required. Many health problems in women can be resolved through this novel regime. These problems are menstrual disorders, endometriosis, polycystic ovaries, infertility, premature labor, breast nodules, fibroids, and gynecological cancers. The use of interventional nutritional therapy and the concept of toxin overload in relation to diseases have long been regarded as "Alternative Medicine," which is unfamiliar among conventional medical practitioners.

Conflict of Interest and Political Reasons for Downplaying Mercury Toxic Effects

Implantation of dental amalgams, consumption of seafood, and Chinese herbal remedy are common practice of the people of Hong Kong. Yet the adverse health effects of getting overloaded with toxic heavy metals, especially mercury, has never received much attention by the medical profession. It has been shown that the passing of fish-derived mercury from mother to the fetus *in utero* and its long-term adverse effects on the heart and the brain are of greater significance (Steuerwald et al. 1999). Conventional medicine never recognizes the adverse health effect of chronic low-dose toxicity by mercury.

For fear of public outcry and litigation, the three main groups with vested financial interest have to be protected at the sacrifice of denying the cure for intractable illnesses. These three groups are dentistry, coal burning electric power plant, and the vaccines industry. Thimerosal, a form of Ethylmercury used in influenza vaccine, is a product associated with huge financial interest worldwide (Pfab et al. 1996).

The impact of worsening environmental pollution on the physical and neurocognitive health of the future generations of China would never be curtailed if such negative or unethical entities operating with interests are not identified and controlled or eliminated. It appears that the medical practitioner community is powerless when it comes to fighting the rapidly rising levels of autism, severe allergies, autoimmune diseases, cancers, psychiatric, and metabolic diseases in children. With the fast-growing industrialization of China and the increasing affluence of its citizens, there exists a clear possibility that toxin-mediated ill-health would wreak havoc on child health, starting *in utero*. Our present healthcare and medical system have not paid sufficient attention to the adverse effect of environmental degradation on human health.

Conclusion

There seems to be rather substantial evidence linking metal (especially Hg) overload and its long-term low-dose effects related to human health, especially in developing fetus and young children. It is expected that other toxic chemicals that include various POPs (such as chlorinated hydrocarbon pesticides and flame retardants) and emerging chemicals of concern (such as bisphenol A and PFOA/PFOS) would also exert their harmful effects on human health. As there are powerful entities with vested financial interests operating in the pharmaceutical and medical healthcare industries, any efforts in disseminating knowledge of environmental degradation among general public and designing an appropriate nutritional approach to curtail the diseases arising out of such environmental degradation would face insurmountable political barriers. Unfortunately, it appears that only with further dramatic rise in disease burden globally with worsening environmental degradation would this reform of the healthcare system be felt imperative.

Acknowledgments

Financial support from the Special Equipment Grant of the Research Grants Council of Hong Kong (SEG_HKBU09) and the Mini-AOE Grant from Hong Kong Baptist University (RC/AOE/08–09/01) is gratefully acknowledged.

References

Auborn KJ, Fan SJ, Rosen EM, Goodwin L, Chandraskaren A, Williams DE, Chen DZ, Carter TH. 2003. Indole-3-carbinol is a negative regulator of estrogen. *Journal of Nutrition* 133: 2470s–2475s.

CH2M-IDC. 2003. A study of toxic substances pollution in Hong Kong. Agreement No. CE22/99. HKEPD.

Chao J. 2003. Shrimp and free trade: China's production booming but antibiotics cause concern. *The Atlanta Journal-Constitution* February, F6.

Cheng Z, Liang P, Shao DD, Wu SC, Nie XP, Chen KC, Li KB, Wong MH. 2010. Mercury biomagnification in the aquaculture pond ecosystem in the Pearl River Delta. *Archives of Environmental Contamination and Toxicology* 61: 491–499.

Cheung KC, Leung HM, Wong MH. 2007. Metal concentrations of common freshwater and marine fish from the Pearl River Delta, South China. *Archives of Environmental Contamination and Toxicology* 54: 705–715.

Cheung KC, Zheng JS, Leung HM, Wong MH. 2008. Exposure to polybrominated diphenyl ethers associated with consumption of marine and freshwater fish in Hong Kong. *Chemosphere* 70: 1707–1720.

Choi MPK, Ho SK, So BK, Cai Z, Lau AK, Wong MH. 2008. PCDD/F and dioxin-like PCB in Hong Kong air in relation to their regional transport in the Pearl River Delta region. *Chemosphere* 71: 211–218.

Choi MPK, Kang YH, Peng XL, Ng KW, Wong MH. 2009. Stockholm convention organochlorine pesticides and polycyclic aromatic hydrocarbons in Hong Kong air. *Chemosphere* 77: 714–719.

Clarkson TW. 2002. The three modern faces of mercury. *Environmental Health Perspectives* 110 (Suppl-1): 11–23.

Committee on National Monitoring of Human Tissues. 1991. *Monitoring Human Tissues for Toxic Substances*. National Research Council, The National Academies Press, Washington, DC.

Dickman MD, Leung CKM. 1998. Mercury and organochlorine exposure from fish consumption in Hong Kong. *Chemosphere* 37: 991–1015.

Dickman MD, Leung CKM, Leong MKH. 1998. Hong Kong male subfertility links to mercury in human hair and fish. *Science of the Total Environment* 214: 165–174.

Enright MJ, Scott EE, Chang KM. 2005. *Regional Powerhouse: The Greater Pearl River Delta and the Rise of China*. John Wiley & Sons (Asia) Pte Ltd, Singapore.

Fiedler H, Cheung KC, Wong MH. 2002. PCDD/PCDF, chlorinated pesticides and PAHs in Chinese teas. *Chemosphere* 46: 1429–1433.

Fok TF, Lam HS, Ng PC, Yip ASK, Sin NC, Chan HIS, Gu GJS, So HK, Wong EMC, Lam CWK. 2007. Fetal methylmercury exposure as measured by cord blood mercury concentrations in mother–infant cohort in Hong Kong. *Environmental International* 33: 84–92.

Fowler BA. 2001. Porphyrinurias induced by mercury and other metals. *Toxicological Sciences* 61: 197–198.

Fromme H, Tittlemier S, Volkel W, Wilhelm M, Twardella D. 2009. Perfluorinated compounds—Exposure assessment for the general population in western countries. *International Journal of Hygiene and Environmental Health* 212: 239–270.

Galland L. 2006. Patient-centered care: Antecedents, triggers, and mediators. *Alternative Therapies in Health and Medicine* 12: 62–70.

Goyer RA. 2000. *Toxicological Effects of Methylmercury*. National Research Council, National Academy Press, Washington, DC.

Grandjean P, Budtz-Jorgensen E, Steuerwald U, Heinziw B, Needham LL, Jorgensen PJ, Weihe P. 2003. Attenuated growth of breast-fed children exposed to increased concentration of methylmercury and polychlorinated biphenyls. *The FASEB Journal* 17: 699–701.

HKFEHD. 2002. Food and Environmental Hygiene Department. Hong Kong Government.

Holtorf K. 2009. The bioidentical hormone debate: Are bioidentical hormone safer or more efficacious than commonly used synthetic versions in hormone replacement therapy? *Postgraduate Medicine* 121: 1949–1971.

Iwata H, Tanabe S, Sakai N, Tatsukawa R. 1993. Distribution of persistent organochlorines in the oceanic air and surface seawater and the role of ocean on their global transport and fate. *Environmental Science Technology* 27: 1080–1092.

Kang Y, Cheung KC, Wong MH. 2010. Polycyclic aromatic hydrocarbons (PAHs) in different indoor dusts and their potential cytotoxicity based on two human cell lines. *Environmental International* 36: 542–547.

Kwok CK, Yang MS, Mak NK, Wong CKC, Liang Y, Leung SY, Young L, Wong MH. 2010. Ecotoxicological study on sediments of Mai Po marshes, Hong Kong using organisms and biomarkers. *Ecotoxicology Environmental Safety* 73: 541–549.

Leung SY, Kwok CK, Nie XP, Cheung KC, Wong MH. 2010. Risk assessment of residual DDTs in freshwater and marine fish cultivated around the Pearl River Delta, China. *Archives of Environmental Contamination and Toxicology* 58: 415–430.

Liang P, Shao DD, Wu SC, Shi JB, Wu FY, Lo SCL, Wang WX, Wong MH. 2010. The influence of mariculture on mercury distribution in sediments and fish around Hong Kong and adjacent mainland China waters. *Chemosphere* 82: 1038–1043.

Liang Y, Wong MH, Shutes RBE, Revitt DM. 1999. Ecological risk assessment of polychlorinated biphenyl contamination in the Mai Po marshes nature reserve, Hong Kong. *Water Research* 33: 1337–1346.

Luo Q, Wong MH, Cai ZW. 2007. Determination of polybrominated diphenyl ethers in freshwater fishes from a river polluted by e-wastes. *Talanta* 72: 1644–1649.

Man YB, Sun XL, Zhao YG, Lopez BN, Chung SS, Wu SC, Cheung KC, Wong MH. 2010. Health risk assessment of abandoned agricultural soils based on heavy metal contents in Hong Kong, the world's most populated city. *Environmental International* 36: 570–576.

Masters SB, Trevor AJ, Katzung BG. 2008. *Katzung and Trevor's Pharmacology: Examination and Board Review*. McGraw Hill Medical, New York.

Nataf R, Skorupka C, Amet L, Lam A, Springbett A, Lathe R. 2006. Porphyrinuria in childhood autistic disorder: Implications for environmental toxicity. *Toxicology and Applied Pharmacology* 214: 99–108.

Paul AG, Jones KC, Sweetman AJ. 2009. A first global production, emission, and environmental inventory for perfluorooctane sulfonate. *Environmental Science and Technology* 43: 386–392.

Pfab R, Muckter H, Roider G, Zilker T. 1996. Clinical course of severe poisoning with thiromersal. *Journal of Toxicology—Clinical Toxicology* 34: 453–460.

Porterfield SP. 1994. Vulnerability of the developing brain to thyroid abnormalities: Environmental insults to the thyroid system. *Environmental Health Perspectives* 102 (Suppl-2): 125–130.

Qin YY, Leung CKM, Leung AOW, Wu SC, Zheng JS, Wong MH. 2010. Persistent organic pollutants and heavy metals in adipose tissues of patients with uterine leiomyomas and the association of these pollutants with seafood diet, BMI, and age. *Environmental Science and Pollution Research* 17: 229–240.

Scandborgh-Englund G, Elinder CG, Johanson G, Lind B, Skare I, Ekstrand J. 2001. Mercury exposure in utero and during infancy. *Journal of Toxicology Environmental Health* 63: 317–320.

Shao DD, Liang P, Kang Y, Wang HS, Cheng Z, Wu SC, Shi JB, Lo SCL, Wang WX, Wong MH. 2011. Mercury species of sediment and fish in freshwater fish ponds around the Pearl River Delta, PR China: Human health risk assessment. *Chemosphere* 83: 443–448.

Steuerwald U, Weihe P, Jorgensen PJ, Bjerve K, Brock J, Neinzow B, Budtz-Jorgensen E, Grandjean P. 1999. Maternal seafood diet, methylmercury exposure, and neonatal neurologic function. *Journal of Pediatrics* 136: 599–605.

Tsang HL, Wu SC, Leung CKM, Tao S, Wong MH. 2010. Body burden of POPs of Hong Kong residents, based on human milk, maternal and cord serum. *Environmental International* 37: 142–151.

U.S. Food and Drug Administration (USFDA). 2010. Update on bisphenol A for use in food contact applications. Available online at: http://www.fda.gov/NewsEvents/PublicHealthFocus/ucm197739.htm

Wang HS, Cheng Z, Liang P, Shao DD, Kang Y, Wu SC, Wong CKC, Wong MH. 2010a. Characterization of PAHs in surface sediments of aquaculture farms around the Pearl River Delta. *Ecotoxicology Environmental Safety* 73: 900–906.

Wang HS, Liang P, Kang Y, Shao DD, Zheng GJ, Wu SC, Wong CKC, Wong MH. 2010b. Enrichment of polycyclic aromatic hydrocarbons (PAHs) in mariculture sediments of Hong Kong. *Environment Pollution* 158: 3298–3308.

Wegener HC. 1999. The consequences for food safety of the use of fluoroquinolones in food animals. *New England Journal of Medicine* 340: 1581–1582.

Wei X, Ching LY, Cheng SH, Wong MH, Wong CKC. 2010. The detection of dioxin- and estrogen-like pollutants in marine and freshwater fishes cultivated in Pearl River Delta, China. *Environment Pollution* 158: 2302–2309.

Wei X, Wu SC, Nie XP, Yediler A, Wong MH. 2009. The effects of residual tetracycline on soil enzymatic activities and plant growth. *Journal of Environmental Science Health B* 44: 461–471.

Wong MH, Bouwman H, Barra R, Wahlström B, Neretin L. 2010. Preliminary identification of emerging chemicals issues in developing countries and countries with economy in transition. Abstract, *SETC Meeting*, Guangzhou, PR China, June 2010.

Xing Y, Lu Y, Dawson RW, Shi Y, Zhang H, Wang T, Liu W, Ren H. 2005. A spatial temporal assessment of pollution from PCBs in China. *Chemosphere* 60: 731–739.

Yach D, Hawkes C, Gould CL, Hofman KJ. 2004. The global burden of chronic diseases. *JAMA* 291: 2616–2622.

Zhang L, Wong MH. 2007. Environmental mercury contamination in China: Sources and impacts. *Environmental International* 33: 108–121.

Zhao YG, Wan HT, Law AYS, Wei X, Huang YQ, Giesy JP, Wong MH, Wong CKC. 2011. Risk assessment for human consumption of perfluorinated compound-contaminated freshwater and marine fish from Hong Kong and Xiamen. *Chemosphere* 85: 277–283.

Zhou HY, Cheung RYH, Wong MH. 1999a. Residues of organochlorines in sediments and tilapia collected from inland water systems of Hong Kong. *Archives of Environmental Contamination and Toxicology* 36: 424–431.

Zhou HY, Cheung RYH, Wong MH. 1999b. Bioaccumulation of organochlorines in freshwater fish with different feeding modes cultured in treated wastewater. *Water Research* 33: 2747–2756.

Zhou HY, Wong MH. 2000. Accumulation of sediment-sorbed PCBs in tilapia. *Water Research* 34: 2905–2914.

3

Heavy Metal Overloads and Autism in Children from Mainland China and Hong Kong: A Preliminary Study

Lillian Y.Y. Ko, Yan Yan Qin, and Ming Hung Wong

CONTENTS

Introduction

Health Effects of Heavy Metals

Heavy metals refer to elements with a high (>5.0) relative density. Heavy metal poisoning is the term used to describe acute or chronic poisoning, by substances such as mercury, lead, cadmium, and arsenic (a metalloid, with some metallic properties). These may enter into human beings by ingestion and inhalation or by absorption through the skin or mucous membranes and stored in the soft tissues of the body. Once absorbed, the heavy metals would compete with other ions and bind to proteins, subsequently impair enzymatic activity, and damage many organs (Stine and Brown 2006).

Mercury occurs naturally in the environment. It can also be released into the air through industrial pollution, such as coal-fired power plants. Mercury falls from the air and accumulates in aquatic systems. The sulfate-reducing bacteria in water or sediment can transform inorganic mercury into methyl mercury. Fish living in contaminated water absorbs methyl mercury present in it. Methyl mercury is further absorbed through the food chains and food webs, including via consumption of carnivorous fish types such as sword fish, tuna, and shark that carry extremely high concentrations of this highly toxic form of mercury (USFDA 2006). Mercury is cytotoxic, as it can bind to – SH (sulfhydryl) groups, which exist in almost every enzymatic process in the body, and disturb every metabolic process. In addition, it can be absorbed directly into the brain, by crossing the blood–brain barrier. Mercury can also bind to hemoglobin in the red blood cell, reduce its oxygen-carrying capacity, and damage blood vessels by reducing blood supply to the tissues. Through placenta transfer, mercury can be stored in the fetus and infant up to eight times higher than that of the mother. It can also be passed from the mother to the infant through breastfeeding at later stage (Steuerwald et al. 2000).

Lead occurs naturally in the earth's crust, and human activities such as burning fossil fuels, mining, and manufacturing have spread it throughout our homes, workplaces, and the environment. Deteriorated lead-based paint in older homes, house dust, and water pipe and gasoline containing lead are the most common sources of lead poisoning throughout the world. Lead is toxic to many of the tissues and enzymes in the human body, and children are particularly more vulnerable to lead as it can accumulate in their nervous system, which in turn results in lower intelligence and poor school performance. Lead poisoning may go undetected because, frequently, there are no obvious signs or symptoms (National Library of Medicine 2004).

Cadmium is also a natural element in the earth's crust, and it is commonly found as a mineral combined with other elements such as oxygen (cadmium oxide), chlorine (cadmium chloride), or sulfur (cadmium sulfate, cadmium sulfide). All soils and rocks, including coal and mineral fertilizers, contain cadmium. It is produced during extraction of other metals such as zinc, lead, and copper. It has a wide range of industrial usages, including batteries, pigments, metal coatings, and plastics. Cadmium enters into the atmosphere from mining and other industrial activities such as burning coal and from household wastes. All biota, including plants, animals, and human beings, take up cadmium from the environment. Cadmium can stay in the human body for a very long time and can build up from many years of exposure to low levels (ATSDR 1999).

Arsenic is widely distributed in the earth's crust and may enter the air, water, and land from wind-blown dust and reach the aquatic systems from runoff and leaching. It combines with oxygen, chlorine, and sulfur to form inorganic arsenic compounds in the environment. In animals and plants, it combines with carbon and hydrogen to form organic arsenic compounds. Fish and shellfish can accumulate arsenic. Fortunately, most of this arsenic is in an organic form called arsenobetaine, which is much less harmful. Inorganic arsenic compounds are mainly used to preserve wood; for example, copper-chromated arsenic (CCA) has been used in "pressure-treated" lumber. Although CCA is no longer allowed for residential uses in the United States, it is still used in industrial applications. Organic arsenic compounds are used as pesticides, mainly on cotton plants. There is some evidence suggesting that long-term exposure to arsenic in children may result in lower IQ scores. Arsenic can cross the placenta and has been found in fetal tissues. Arsenic is also found at low levels in breast milk (ATSDR 2005).

Oxidative Stress

Oxidative stress (OS) is a general term used to describe the steady-state level of oxidative damage to a cell, tissue, or organ caused by the reactive oxygen species (ROS). This damage can affect a specific molecule or the entire organism. ROS, such as free radicals and peroxides, represent a class of molecules that are derived from the metabolism of oxygen and exist inherently in all aerobic organisms (Granot and Kohen 2004). Most ROS are derived from the endogenous sources as by-products of normal and essential metabolic reactions, such as energy generation from mitochondria or detoxification reactions involving liver cytochrome P-450 enzyme system. Exogenous sources include environmental pollutants such as emission from automobiles and industries, exposure to cigarette smoke, consumption of excess alcohol, asbestos, exposure to ionizing radiation, and bacterial, fungal, or viral infections (Stohs and Bagchi 1995).

Autism

Autism is a complex developmental disability that typically appears during the first 3 years of life, resulting from a neurological disorder. This subsequently affects not only the normal functioning of the brain but also the

development in social interaction and communication skills, including difficulties in verbal and nonverbal communication and leisure or play activities
(Bernard 2001, Autism Society of America 2003).

Children's exposure to heavy metals in immunizations has been a long-
term concern of autism advocates. All the heavy metals mentioned earlier
are highly toxic to brain cells and other body systems affected by autism.
Several classic features of autism, such as speech loss and loss of social and
communication skills, are signature traits of heavy metal poisoning (Bernard
2001). Minamata disease, a form of acute mercury poisoning in the inhabitants of a fishing village located at Minamata Bay (Japan), was a result of the
exposure of fetus to mercury-contaminated fish consumed by the mother.
Children with Minamata disease had symptoms indistinguishable from
mental retardation or cerebral palsy (Kondo 2000). In addition, acrodynia, a
disease caused by mercury in infant teething powder, was reported in the
early 1900s. Children with acrodynia suffered peeling and reddened skin
on their hands and feet and heightened sensitivity to light (Warkany 1966,
Autism Society of America 2003).

Heavy Metals in Hair, Urine, and Blood

A hair sample can provide an accurate and powerful means of evaluating
the effects of cumulative, long-term exposure to toxins. As the hair follicle
grows, toxic elements in the blood are absorbed into the growing hair protein. Due to the fact that scalp hair grows at an average of 1–2 cm in a month,
hair elements analysis can provide a temporal record of element metabolism
that has occurred during the previous 1–10 months (Bencko 1995, DTHM
2004, McDowell et al. 2004).

Analyses of urine samples carried out to detect the presence of heavy
metal content are believed to reveal heavy metal exposure over the previous
2–3 months. It is believed that heavy metal excretion after a dose of chelating agent tends to reflect the body burden better than basal excretion. In the
present study meso-2, 3-dimercapto-succinicacid (DMSA, metal chelating
agent) was used to remove heavy metals (Baselt and Cravey 1989).

In general, whole blood element analysis is used as a diagnostic method
that could determine imbalance, insufficiency, or excess of certain elements
associated with essential functions. Whole blood element analyses are useful for testing elevated or excessive levels of potentially toxic elements such
as cadmium, lead, mercury, nickel, and uranium but are not reliable as an
indicator of total body burden in long-term exposures (Klaassen 1996).

Aims and Objectives of Present Research

Since toxic heavy metals can enter into different ecological compartments
of our environment and persist for a long period, they could be transferred
into food chains, biomagnified in species of higher trophic levels and finally

reach human beings. This study was aimed at assessing toxic heavy metal loads in different tissues (blood, hair, and urine) of children with autism, in relation to their health and performance in activities requiring physical and mental skills. More specifically, the objectives of the present project were as follows:

1. To investigate heavy metal loading in different tissues of children with autism.
2. To compare heavy metal poisoning of children from Hong Kong and Mainland China.
3. To test and analyze the changes due to nutrient and food allergy in the body of children with autism.
4. To analyze the gender and district (residential locations) difference, in terms of heavy metal loading in children with autism.

Materials and Methods

Collection of Samples

The experimental plan and procedures were approved by the Ethical Committee of Hong Kong Baptist University. The studied subjects were children with autism who came to Hong Kong Child Development Center in recent years from different provinces of Mainland China and Hong Kong. With the consent of their parents, samples of their blood (22 from Mainland China, and 11 from Hong Kong), hair (100 and 50), and urine (86 and 15) were collected 6 h after administrating meso-2, 3-dimercapto-succinicacid (DMSA) to the children, in order to test the concentration level of heavy metals present in the samples (arsenic, cadmium, lead, and mercury).

Analyses of Heavy Metals

The samples were sent to Doctor's Data Lab (the United States) where, using the inductively coupled plasma mass spectroscopy (ICP-MS), they were tested for detecting the presence of four heavy metals.

Blood IgG Test

Blood IgG is an immunoglobulin associated with delayed allergic (hypersensitivity) reactions. The Blood IgG Test provided by The Great Plains Laboratory, Inc., Lenexa, Kansas, was used to test food allergy of these children with autism.

Results and Discussion

Hair Test

According to Table 3.1, the concentrations of 4 of the 12 heavy metals tested all exceeded their respective reference ranges, especially mercury in Hong Kong samples (exceeded 7.47 times), lead in Mainland China samples (5.26 times), and arsenic in Mainland China samples (4.87 times). Mainland China samples contained significantly higher ($p < 0.05$) lead and arsenic but lower mercury than Hong Kong samples.

Urine Test

Table 3.2 shows that 6h after the children were administered DMSA, a substantial amount of the four heavy metals were detected in the urine. All metals exceeded their respective reference ranges, especially lead (9.77 times and 3.58 times in Mainland China and Hong Kong samples, respectively) and mercury (2.03 times and 2.65 times in Mainland China and Hong Kong samples, respectively). A significantly higher ($p < 0.05$) concentration of lead was observed in Mainland China samples than in Hong Kong samples, whereas no significant differences were noted in concentrations of the other three metals between Mainland China and Hong Kong samples.

Blood Test

Table 3.3 shows the concentrations of mercury, cadmium, and lead which all exceeded their respective ranges, especially lead (0.61 and 0.3 times in

TABLE 3.1

Concentrations of Heavy Metals in the Hair Samples of the Children with Autism

Heavy Metal	District	Concentrations (μg/g)	Reference Range (μg/g)	Excess Multiple
Mercury	Mainland China	0.917±1.018*		2.29
	Hong Kong	2.99±3.18	<0.4	7.47
Lead	Mainland China	5.26±7.96*		5.26
	Hong Kong	3.27±3.15	<1	3.27
Cadmium	Mainland China	0.238±0.768		1.59
	Hong Kong	0.215±0.531	<0.15	1.43
Arsenic	Mainland China	0.389±0.443*		4.87
	Hong Kong	0.177±0.133	<0.08	2.21

* Indicates significant difference $p < 0.05$, according to student's t-test.
Reference Range is collected according to Doctor's Data, Inc. (DDI) from America.

TABLE 3.2

Concentrations of Heavy Metals in the Urine Samples of Children with Autism

Heavy Metal	District	Concentrations (µg/g)	Reference Range (µg/g)	Excess Multiple
Mercury	Mainland China	10.2 ± 7.44		2.03
	Hong Kong	13.2 ± 12.7	<5	2.65
Lead	Mainland China	48.9 ± 31.2*		9.77
	Hong Kong	17.9 ± 11.01	<5	3.58
Cadmium	Mainland China	0.543 ± 0.297		0.27
	Hong Kong	0.515 ± 0.389	<2	0.26
Arsenic	Mainland China	138 ± 153		1.06
	Hong Kong	127 ± 104	<130	0.98

* Indicates significant difference $p < 0.05$, according to student's t-test.
Reference Range is collected according to Doctor's Data, Inc. (DDI) from America.

TABLE 3.3

Concentrations of Mercury, Lead, and Cadmium in Blood Samples Children with Autism

Heavy Metal	District	Concentrations (µg/g)	Reference Range (µg/g)	Excess Multiple
Mercury	Mainland China	0.00336 ± 0.00272		0.26
	Hong Kong	0.00417 ± 0.00283	<0.013	0.32
Lead	Mainland China	0.0336 ± 0.0106*		0.61
	Hong Kong	0.0165 ± 0.00926	<0.055	0.30
Cadmium	Mainland China	0.00336 ± 0.00272		0.07
	Hong Kong	0.000533 ± 0.000132	<0.014	0.04

* Indicates significant difference $p < 0.05$, according to student's t-test.
Reference Range is collected according to Doctor's Data, Inc. (DDI) from America.

Mainland China and Hong Kong samples, respectively) and mercury (0.26 and 0.32 times in Mainland China and Hong Kong samples, respectively). A significantly higher ($p < 0.05$) concentration of lead was noted in Mainland China samples than in Hong Kong samples, whereas no significant differences were found in mercury and arsenic concentrations.

District Difference

It seems apparent that the children with autism from Hong Kong contained higher concentrations of mercury and cadmium, whereas those from Mainland China had higher lead and arsenic. An attempt was also made to divide the samples into two groups: coastal districts (Guangdong, Fujian, Zhejiang, Tianjin, Shanghai, Jiangsu, Shandong) and inland districts

(Beijing, Hubei, Hunan, Shaanxi, Sichuan, Jiangxi) for comparing heavy metal loadings in children with autism. A general trend was observed showing that samples (hair, blood, and urine) from coastal districts tended to contain higher mercury and cadmium whereas those from inland districts higher lead and arsenic (although no statistical differences were observed due to small size of some samples).

Gender Difference

Among the 111 children with autism from Mainland China, 101 (90.99%) were male and 10 (9.01%) were female, and among the 71 children with autism from Hong Kong, 64 (90.14%) were male and 10 (14.08%) were female. In general, male children with autism made up 90% of all the children with autism from both Mainland China and Hong Kong.

Food Allergy

The patients' blood IgG (from The Great Plains Laboratory, Inc., Lenexa, Kansas) was tested to analyze the food allergy of these children with autism. It was observed that these children were sensitive to grains (e.g., wheat, rye, oats, barley, gluten), dairy products (e.g., milk, cheese, whey, yogurt), and other food items, such as egg white, egg yolk, sugar cane, asparagus, carrot, and citrus fruit.

Discussion

Body Loadings of Heavy Metals

The children with autism, both from Mainland China and Hong Kong, accumulated extremely high concentrations of heavy metals in their bodies. In Mainland China, heavy metal pollution caused by both metal mining and smelting and other industrial activities is very serious. In addition, the use of gasoline and paints containing lead and pesticides containing arsenic explains why higher levels of these two elements were detected in children from Mainland China. A survey conducted by WHO in 2004 indicated lead poisoning of children in many parts of China. Among 17,141 children examined, it was revealed that 1,791 children had blood lead content that exceeded 100 μg/L, with lead poisoning rate reaching 10.45% (Sina News Centre 2005). On the other hand, the high concentrations of mercury and cadmium observed in children with autism from Hong Kong and, to certain extent, from other coastal districts, may reflect the intake of these two heavy metals through consumption of fish and shellfish as these two fish types can efficiently absorb mercury and cadmium from the environment (Cheng et al. 2011). In addition, it has been noted that the hair mercury level of the subfertile men in Hong Kong (with problems related to sperm volume and motility) is linked with their high consumption rate of fish (Dickman et al. 1998).

Food Allergies

It is well known that people with autism are more susceptible to allergies and food sensitivities than the average person. This is likely linked to their impaired immune system (Edelson 2000). Food sensitivities may be responsible for numerous physical and behavioral problems as observed in most of the children with autism who took part in this study. These included a wide range of symptoms such as headaches, stomachaches, feeling of nausea, bed-wetting, excessive whining and crying, sleeping problems, hyperactivity, aggression, temper tantrums, fatigue, depression, intestinal problems, ear infections, and even seizures. In order to reduce allergy symptoms, vitamins, in particular vitamin C, are used (Edelson 2000).

Heavy Metals and Oxidative Stress

OS is caused by oxygen radicals produced by the body in manageable amounts as byproducts of normal body metabolism. However, their prevalence can be exacerbated by exposure to environmental chemicals, such as heavy metals. Oxygen radicals damage cells (by reacting with proteins, DNA, carbohydrates, and fats), and interfere with signals sent between cells in the body, which can lead to autoimmunity (Klein and Ackerman 2003). Glutathione is one of the body's most important mechanisms of heavy metal detoxification and excretion, as it can bind with heavy metals such as mercury, lead, cadmium, and nickel (Stohs and Bagchi 1995) and can be more easily filtered out of the body.

Due to limited access to the bulk of antioxidants produced by the body, the brain and nervous system are particularly vulnerable to OS, with neurons being the first cells to be affected (Shulman et al. 2004). More importantly, children are more vulnerable than adults to OS due to their naturally low glutathione levels (Ono 2001, Erden-Inal et al. 2002). As has been revealed that deficits in glutathione can cause degeneration of the jejunum and colon, glutathione is vital to control proper function of the intestines (Martensson 1990). In the case of disabled gut, undigested proteins pass through the gut and cause oxidative damage to the brain and nervous system. In addition, intestinal disorder can reduce the intake of nutrients such as zinc, selenium, and cobalt, which are important to eliminate heavy metals in the bodies, and thereby aggravate the poisoning effect of heavy metals (White 2003).

Gender Difference

It has been suggested that autism represents an extreme form of the way in which men's brains differ from those of women. This so-called male brain theory is linked with the "empathizing/systemizing (E-S)" theory, which states that men are better at systemizing than women and that women are better at empathizing than men due to physical differences between male and female brains. However, the concept of differing types of intelligence between men and women is controversial and remains speculative (Frequency of Autism 2004)

By measuring testosterone levels in the amniotic fluid of mothers while pregnant, it was observed that the babies with higher fetal testosterone levels had a smaller vocabulary and less eye contact when they were a year old. The original 58 children were examined again, at age 4, and it was further noted that the children with higher testosterone in the womb were less developed socially and the interests of boys were more restricted than girls (Knickmeyer et al. 2005). However, there is still no concrete evidence indicating testosterone levels would affect brain development or autism.

The present results showed that men contributed to 90% of all autism cases and the children were diagnosed with learning disabilities and attention-deficit disorder. This seems associated with weaker antioxidant capacity in young males, with greater vulnerability in their brain and nervous systems. This may render them more vulnerable to heavy metals and autism. On the contrary, women and girls possess lower levels of inactive antioxidant chemicals, and estrogen is a powerful antioxidant that could counteract free radical–mediated damage in aging (Rush and Sandiford 2003).

Conclusion

The present study demonstrated the connection between heavy metal poisoning and autism. Children are more susceptible to a vast number of common pollutants, for example, arsenic in drinking water, lead in paint and dust, and mercury in fish. Based on the limited information available, there seems to be an indication that in Hong Kong children with autism (and other coastal districts) are associated with high levels of mercury and cadmium, possibly through oral intake of fish and shellfish, and inland children with autism are associated with high levels of arsenic and lead, due to environmental pollution. The present results also show that men are far more susceptible to heavy metal poisoning than women. Children with autism should be treated under a Heavy Metal Detoxification Program, providing them with nutrients, antioxidants, and chelating agents.

Acknowledgments

The authors would like to thank all the children and their parents participating in this survey. Financial support from the Special Equipment Grant of the Hong Kong Research Grants Council, and Mini-AoE (Area of Excellence) Fund and General Research Fund of Hong Kong Baptist University is gratefully acknowledged.

References

ATSDR, Agency for Toxic Substances and Disease Registry. 1999. Cadmium. http://www.atsdr.cdc.gov/tfacts5.html

ATSDR, Agency for Toxic Substances and Disease Registry. 2005. Arsenic. http://www.atsdr.cdc.gov/tfacts2.html

Autism Society of America. 2003. Autism cost to economy in billions. [Online available] http://www.autism-society.org/site/

Baselt RC, Cravey RH. 1989. *Disposition of Toxic Drugs and Chemicals in Man*, 3rd edn. Year Book Medical Publishers, New York.

Bencko V. 1995. Use of human hair as bio-marker in the assessment of exposure to pollutants in occupational and environmental settings. *Toxicology* 101: 29–39.

Bernard S. 2001. Autism: A novel form of mercury poisoning. *Medical Hypotheses* 56: 462–471.

Cheng Z, Liang P, Shao DD, Wu SC, Nie XP, Chen KC, Li KB, Wong MH. 2011. Mercury biomagnifications in the aquaculture pond ecosystem in the Pearl River Delta. *Archives of Environmental Contamination and Toxicology* 61: 491–499.

Dickman MD, Leung CK, Leong MK. 1998. Hong Kong male sub-fertility links to mercury in human hair and fish. *Science of the Total Environment* 214: 165–174.

DTHM, Detoxification of Toxic Heavy Metals. 2004. Autism and detoxification. http:// www.extremehealthusa.com/autism-detoxification.html

Edelson SM. 2000. Center for the Study of Autism, Allergies and Food Sensitivities. http://www.autism.org/allergy.html

Erden-Inal M, Sunal E, Kanbak G. 2002. Age-related changes in the glutathione redox system. *Cell Biochemistry and Function* 1: 61–66.

Frequency of Autism. 2004. http://en.wikipedia.org/wiki/Autism_epidemic

Granot E, Kohen R. 2004. Oxidative stress in childhood health and disease states. *Clinical Nutrition* 23: 3–11.

Klaassen CD. 1996. *Casarett and Doull's Toxicology: The Basic Science of Poisons.* McGraw-Hill, New York.

Klein JA, Ackerman SL. 2003. Oxidative stress, cell cycle, and neurodegeneration. In Redox signaling in biology and disease. *Journal of Clinical Investigation* 111: 785–793.

Knickmeyer R, Baron-Cohen S, Raggatt P, Taylor K. 2005. Foetal testosterone, social relationships, and restricted interests in children. *Journal of Child Psychology and Psychiatry* 46: 123–130.

Kondo K. 2000. Congenital Minamata disease: Warning from Japan's experience. *Journal of Child Neurology* 15: 458–464.

Martensson J. 1990. Glutathione is required for intestinal function. *Proceedings of National Academy of Sciences* 87: 1715–1719.

McDowell MA, Dillon CF, Osterloh J et al. 2004. Hair mercury levels in U.S. children and women of childbearing age: Reference range data from NHANES 1999–2000. *Environmental Health Perspectives* 112: 1165–1167.

National Library of Medicine. 2004. Lead. http:// www.nlm.nih.gov/medlineplus/leadpoisoning

Ono H. 2001. Plasma total glutathione concentrations in healthy pediatric and adult subjects. *Clinical Chimica Acta* 312: 227–229.

Rush JW, Sandiford SD. 2003. Plasma glutathione peroxidase in healthy young adults: Influence of gender and physical activity. *Clinical Biochemistry* 36: 341–351.

Shulman RG, Rothman DL, Behar KL, Hyder F. 2004. Energetic basis of brain activity: Implications for neuroimaging. *Trends Neuroscience* 27: 489–495.

Sina News Centre. 2005. The problem about lead poisoning of Chinese children. http://www.news.sina.com.cn

Steuerwald U, Weihe P, Jorgensen PJ et al. 2000. Maternal seafood diet, methyl mercury exposure, and neonatal neurologic function. *The Journal of Pediatrics* 136: 599–605.

Stine KE, Brown TM. 2006. *Principles of Toxicology*, 2nd edn. Taylor & Francis, London, U.K.

Stohs SJ, Bagchi D. 1995. Oxidative mechanisms in the toxicity of metal ions. *Free Radical Biology and Medicine* 18: 321–336.

USFDA. 2006. U.S. Food and Drug Administration. Mercury levels in commercial fish and shellfish. http:// www.cfsan.fda.gov/~frf/sea-mehg.html

Warkany J. 1966. Acrodynia—Postmortem of a disease. *American Journal of Diseases of Children* 112: 147–156.

White JF. 2003. Intestinal pathophysiology in autism. *Society for Experimental Biology and Medicine* 228: 639–649.

4

Endocrine-Disrupting Contaminants and Their Effects on Reproductive and Developmental Health

Chris K.C. Wong

CONTENTS

Introduction

Industrial revolution unleashed a vast variety of new chemical compounds into the environment. Over the past 60 years, more than 80,000 synthetic chemical compounds have been made, and recently more than 3,000 new chemicals are produced each year (Landrigan et al. 2002). Since the adoption of the Stockholm Convention on Persistent Organic Pollutants (POPs) in the year 2001, the environmental and health impacts of environmental contaminants have drawn more attention from scientists, policy makers, industries, NGOs, and the general public. Without doubt these chemicals are ubiquitous and are widely dispersed in air, water, soil, and daily necessity. For the identification and quantification of all these contaminants, substantial labor and financial cost are required. However, by the year 2000, less than 7% of the synthetic chemical compounds were tested for their chronic and developmental toxicity (Goldman and Koduru 2000).

Presumably, dietary food intake is the major route to increase the body burden of environmental contaminants. Hence, a positive correlation between age and body burdens is expected. In 1979, a mass poisoning of more than 2000 people occurred in central Taiwan due to consumption of rice-bran oil contaminated with polychlorinated biphenyls (PCBs) and their heat-degraded by-product (Guo et al. 1997). In North Vietnam, a study in 1994 showed that the serum level of DDT in urban dwellers (32.3 ng/mL) was threefold higher than in the rural residents (11.7 ng/mL), suggesting the major route of DDT exposure was through the ingestion of contaminated foods, such as fatty meats and poultry (Schecter et al. 1997). In addition, it has been reported that consumption of fatty fish is the main exposure route for POPs (Grimvall et al. 1997, Hu et al. 2009, Järnberg et al. 1997, Svensson et al. 1995). In 1997, the dioxin episode in Belgium has led to Europe's worst food contamination crisis since mad cow disease. The original source of the contamination that entered the food chain in Belgium now appears to be wastes of POPs (i.e., PCB oils), which have possibly been illegally disposed of and found their way into food oils.

It is known that these chemicals affect growth and development, immune function, neurological function, reproduction, and induce mutations and cancers (Langer et al. 2009, Li et al. 2006, Safe and Zacharewski 1997, White et al. 1994). The levels of chemical contaminants detected in human breast milk are of particular concern because of the potential health risk to the nursing baby (LaKind et al. 2008). During the lactation period, PCB transfer via breast milk was shown to lower the body pollutant burden of the mother but cause a simultaneous increase in the body pollutant burden of the infant (Duarte-Davidson and Jones 1994). More importantly, the detection of the pollutants in human milk and placenta samples revealed the risk of pollutant transfer from mothers to infants (Chan et al. 2007, Doucet et al. 2009, Wong et al. 2005). Developmental, neurological, musculoskeletal, and behavioral abnormalities have been documented in the children born after their mother's consumption of PCB/PCDF–contaminated rice oil (Chen and Hsu 1994, Guo et al. 1994a,b, Honda et al. 2009, Kanagawa et al. 2008, Lai et al. 1994, Tsukimori et al. 2008).

Environmental Pollution and Reproductive Health

It is generally believed that the exposure to environmental pollutants is one of the culprits behind reproductive problems worldwide. This relationship has long been established on the basis of the body pollutant burdens' association with reproductive dysfunction as found in wildlife, animal, and human studies. Adverse biological effects on male reproductive function were first reported in wild animals, whereas an accidental exposure to estrogenic pollutants caused feminization or changes in reproductive behavior in the animals (Vos et al. 2000). For example, in the 1980s, the adult male alligators in Apopka Lake that were exposed to agricultural wastes produced low testosterone levels and presented with micropenis and disorganized testes (Guillette et al. 1994, 1995, Guillette and Guillette 1996). Other reports also highlighted the pathophysiological consequences of chemical exposures that affect reproductive functions in mammals, bird, amphibian, and fish (Aravindakshan et al. 2004, Barnhoorn et al. 2004, De Guise et al. 1994, Fry 1995, Hayes et al. 2003, Jobling et al. 2002, Mansfield and Land 2002, Oskam et al. 2003).

In human epidemiologic study, a significant reduction in the ratio of "male birth to total number of birth" was recorded in the Aamjiwnaang First Nation community (areas close to industrial areas) in Canada (Mackenzie et al. 2005). During the episode of PCBs and dioxin food contamination in Belgium in the year 1999, a study of reproductive status in local male population indicated that the rise of serum dioxin levels were associated with a decrease of serum testosterone levels and a reduction of semen volume (Dhooge et al. 2006). Occupational study also revealed that the exposure to phthalate (substances added to plastics to increase their flexibility) caused significant reduction of serum testosterone level (Pan et al. 2006). In addition, epidemiological studies reported an increased risk of genital malformations and cryptorchidism in children of workers who were exposed to pesticides (Garcia-Rodriguez et al. 1996, Weidner et al. 1998). Collectively, the reports have confirmed the association between reproductive disorders and the accumulation of environmental pollutants.

Endocrine Disruptors and Their General Mechanistic Actions

Most of the pollutants are known as endocrine disruptors (EDs) (Crews and McLachlan 2006). EDs can affect the hormonal system via (but not limited to) estrogenic, androgenic, antiandrogenic, and antithyroid mechanisms and/or modulation of steroidogenesis/steroid metabolism (Phillips et al. 2008, Phillips and Foster 2008b). With the benefit of hindsight, the endocrine-disrupting effects of environmental contaminants have been shown to impose acute and/or long-term effect on animal development. As the most toxic man-made pollutant, dioxins are known to impose biological effects via the aryl hydrocarbon receptor (AhR), which belongs to a member of the basic helix-loop-helix/Per-Arnt-Sim (bHLH/PAS) family of transcription factors. AhR exhibits its transcriptional activity primarily via ligand-dependent nuclear translocation (Mimura and Fujii-Kuriyama 2003). The role of AhR in mediating dioxins-elicited developmental toxicity has been demonstrated using zebrafish model (Carney et al. 2006, Prasch et al. 2003). Other AhR-mediated regulatory functions include the modulation of other transcriptional factors, including Rb/E2F, NFκB, and the estrogen (ERα and ERβ) and androgen receptors (Beischlag and Perdew 2005, Matthews et al.

2005, Ohtake et al. 2003, 2007, Puga et al. 2000, Vogel et al. 2007, Wormke et al. 2003). Due to the unavailability of the AhR x-ray crystal structure, quantitative structure–activity relationship (QSAR) models have been used for the binding prediction for virtual screening. Comparative molecular field analysis (CoMFA), VolSurf, and Hologram QSAR (HQSAR) models have been constructed using a training set of 84 AhR ligands (Aravindakshan et al. 2004, Lo et al. 2006, Tuppurainen and Ruuskanen 2000). The results showed that CoMFA, VolSurf, HQSAR, and the hybrid models give good correlation. Since the techniques analyzed show a good prediction quality against an external test set, particularly the HQSAR and the hybrid model, these models were concluded to be able to predict AhR binding in virtual screening.

Compared to dioxin/AhR-mediated actions, contaminants that cause estrogenic and/or antiestrogenic activities were shown to have striking effect on animals, demonstrating the power of estrogen in environmental signaling (McLachlan 2001). It is particularly true when we look at various domains in an evolutionary perspective; the DNA-binding domain and the ligand-binding domain of estrogen receptor-α (ERα) are conserved across metazoans (McLachlan 1993, Thornton et al. 2003). Global environmental contaminants, POPs (i.e., dichlorodiphenyl-trichloroethane [DDT], hydroxylated PCBs, bis-phenol A, p-nonylphenol, and dioxins), and heavy metals (i.e., cadmium and mercury) can exhibit either or both estrogenic and androgenic activities (Johnson et al. 2003, Martin et al. 2003, McLachlan 2001). Some newly identified emerging contaminants, like perfluorinated compounds (PFOA) and flame retardants, were also reported to possess estrogenic activities (Maras et al. 2006, Meerts et al. 2001). More importantly, nongenomic rapid xenoestrogen actions were demonstrated when the xenoestrogens bind to a common membrane binding site that has the pharmacological profile of the γ-adrenergic receptor (Nadal et al. 2000).

Endocrine Disruptors and Their Effects on Animal Fertility

Mammalian spermatogenesis and folliculogenesis are characterized by a complicated cascade of processes that occur due to the influence of the hypothalamus–pituitary–gonadal axis as well as the *de novo* auto/paracrine circuit. The fundamental role of the hormones involved is to enable a coordinated regulation of the processes that allow for the development of highly differentiated spermatozoa and the selection of the fitness oocyte for ovulation. Any interruption of the hormonally mediated regulation, the constituents of the microenvironments in seminiferous tubules, and developing follicles may result in a transient/long-term modification of the hormonal feedback circuitry, disturbance of gametogenesis, and possibly the epigenetic modification of the gametes/germ cells.

Using *in utero* and lactational exposure studies in rodent models, toxicities of TCDD (μg/kg) at the early stage of animal development were reported. In female progenies, TCDD disrupted regular estrous cycles and inhibited the onset of ovulation (Li et al. 1995, Salisbury and Marcinkiewicz 2002). In male offsprings, a reduction in sperm count per cauda epididymis and an increase in the number of abnormal sperm production at adulthood were observed (Bjerke and Peterson 1994, Faqi et al. 1998, Mably et al. 1992). Most of the male progenies were characterized by reduced size of sex accessory glands (Bjerke and Peterson 1994, Gray et al. 1995, Mably et al. 1992, Theobald and Peterson 1997). In addition to these, there are other studies that scrutinized the biological consequences in postnatal animals that received intraperitoneal injection of TCDD (μg/kg body weight). Those studies demonstrated that TCDD altered the process of testicular steroidogenesis and caused a reduction of Leydig cell volume and number (Johnson et al. 1992, 1994, Wilker et al. 1995). Detrimental effects on Sertoli and germ cells of rat testes, such as reduction of intercellular contact of neighboring cells, disruption of germ cell development, decreased of spermatogenesis, depletion of antioxidant enzymes, and increase in the levels of lipid peroxidation were observed (Chahoud et al. 1992, Latchoumycandane et al. 2002a,b, Mably et al. 1992, Peterson et al. 1993, Rune et al. 1991). It is generally believed that the adverse effects exerted by TCDD on the male reproductive functions are manifold and pleiotropic. At the molecular levels, EDs interrupt intracellular signaling pathways, leading to the misinterpretation of cellular signals and, therefore, the modification of cell functions. We have demonstrated effects of dioxin in modulating the expressions of rat Sertoli cell secretory products and protein markers for cell–cell interaction (Lai et al. 2005a). In addition, the synthesis and secretion of progesterone and testosterone were considerably suppressed in dioxin-treated Leydig cells (Lai et al. 2005b).

The presence of dioxins in human follicular fluid was reported (Tsutsumi et al. 1998). Adverse effects of EDs in spermatogenesis and folliculogenesis were demonstrated in many studies (Baldridge et al. 2003, Guillette and Moore 2006, Phillips and Tanphaichitr 2008, Sheweita et al. 2005, Uzumcu and Zachow 2007). Effect of EDs to meiotic nondisjunction in human oocytes was identified (Czeizel et al. 1993, Hunt et al. 2003). More significantly, the expression of AhR, estrogen receptor-α (ERα), and estrogen receptor-β (ERβ) have been detected in mouse preimplantation embryos, indicating that EDs present in reproductive fluids can modulate the genomic actions via intracellular AhR and ERs, affecting embryonic development (Hiroi et al. 1999, Peters and Wiley 1995). In this regard, the adverse biological effects may be reflected in the processes of fertilization, blastocyst development, and implantation. Probably, these effects can possibly be manifested transgenerationally (Hunt et al. 2003, Nomura 2008, Phillips and Tanphaichitr 2008, Taylor 2008, Weselak et al. 2008). In organism levels, the disturbance may lead to infertility, birth defects, precocious puberty and reproductive cancers in the next generations, and hence animal perpetuation (Henley and

Korach 2006, Phillips and Foster 2008a,b). According to the 2001 WHO report, at least 80 million worldwide were estimated to be affected by infertility. Among those, more than 10% of infertility cases cannot be explained medically, which is probably attributed to the chronic exposure to low level of environmental pollutants, as early as *in utero*. Therefore, research is needed to elucidate the role and the mechanistic actions of pollutants in wildlife and humans, for the protection of our future generations from reproductive problems.

Endocrine Disruptors and Their Effects on Embryo Development and Epigenomic Modification

There is increasing epidemiological evidence to suggest that environmental exposure to EDs early in the fetal development has a role in susceptibility to reproductive dysfunction and disease in later life (Jirtle and Skinner 2007). Interestingly, numerous protection strategies have been identified in animal embryos to provide robustness for buffering the developmental processes in response to changing stress conditions (Hamdoun and Epel 2007). Although cellular mechanisms in embryos can buffer the stress encountered, exposure to highly bioactive chemicals/pollutants may overwhelm this intrinsic robustness. We have learned the lesson from human exposure to the synthetic estrogen, diethylstilbestrol, and thalidomide; these can induce changes in the development of reproductive tract in male and female offspring, lower sperm counts, increase incidence of vaginal cancer, and cause severe developmental malformations (Finnell et al. 2002, Rubin 2007). In animal study that used rodents reported the link between prenatal exposure to several phthalates and a shortening of the anogenital index (Swan et al. 2005). These examples illustrate that the placenta cannot provide complete protection to the developmental fetus. The leakage of environmental toxins via placenta barriers to developmental fetus has been reported (Couture et al. 1990, Safe 1990).

Using *in utero* exposure studies in rodent models as well, transient exposure of gestating female rats to a fungicide (i.e., vinclozolin) or a pesticide (methoxychlor, a substitute for DDT), between E8 and E15, decreased spermatogenic capacity and increased infertility in male offspring. These effects were found to be manifested via an alternation of DNA methylation patterns in germ line, leading to changes in embryonic testis transcriptome in subsequent generations (Anway et al. 2005, Anway and Skinner 2008, Jirtle and Skinner 2007). Other reports demonstrated that the exposure of preimplantation embryo to dioxin caused an increase in the cellular methyltransferase activity (Wu et al. 2004), establishing a possible link between EDs and epigenetic modification. More recently, Jirtle's group demonstrated that rodents' maternal oral exposure to bisphenol A shifted the coat color distribution in mouse offspring by decreasing CpG methylation in the *Agouti* gene (Dolinoy et al. 2007). Although such observations are evident of the adverse effects caused by EDs, the mechanistic role of EDs in the modification of epigenome remains largely not known (Crews and McLachlan 2006).

Since DNA methylation and chromatin patterning is programmed during early development, the most susceptible window falls in the methylation state of the genome, which in general leads to reprogramming (Warnecke et al. 1998). The time windows include (1) the development of preimplantation embryo, and (2) the development of germ cells closer to indifferent gonads. Only certain genomic regions (i.e., imprinted genes) are protected from demethylation and are maintained during embryogenesis (Bartolomei and Tilghman 1997, Olek and Walter 1997, Sanford et al. 1987, Tremblay et al. 1997, Warnecke et al. 1998). The remethylation of the germ cells appears to be dependent on the germ cells' association with gonadal somatic cells (i.e., precursor Sertoli cells and tubular myoid cells). Although steroid hormones are not produced in gonads at this early stage of development, steroid hormone receptors, like androgen receptor, ERα, and/or ERβ, are expressed in Leydig cells, Sertoli cells, and germ cells (Delbes et al. 2006, Greco et al. 1992, Jefferson et al. 2000, Jirtle and Skinner 2007, O'Donnell et al. 2001, Saunders et al. 1998, 2001, van Pelt et al. 1999). Therefore, any exogenous factor acting on these cells at the time of reprogramming of the genome methylation state inevitably alters the epigenetic modification of the germ line. Consequently, transgenerational transmission of an altered phenotype or genetic trait appears.

Conclusion and Future Recommendation

Environmental pollutant exposure poses a potential risk for human health. Epidemiological and laboratory studies have demonstrated adverse effects of such exposure on animal and human fertility as well as in embryonic development. A recent study has indicated that pollutant exposure can affect the reproductive health of future generations (Anway et al. 2005). It will be beneficial to establish the analytical and computational techniques, together with the *in vitro* and animal models, to provide an analysis platform that can reveal the "structure–activity" relationship of common and emerging hazardous chemical pollutants and to establish risk assessment systems to understand the mechanistic actions of endocrine-disrupting contaminants that affect human reproductive and developmental potentials. It has been proposed that dioxins and pollutants with estrogenic- or antiestrogenic-like activities can have the most striking effect on reproductive health. The newly emerged ED contaminants (i.e., bisphenol A, flame retardants, and perfluorooctanoic acid) are ubiquitous and possess dioxin-like and/or estrogenic activities. Although a considerable number of studies have reported their environmental contamination profiles as well as investigated their general toxicities, the information on the "structure–activity" relationship of most of the identified hazardous chemical pollutants is not yet known systematically. One way to gain a deeper understanding of the structural aspects of ED, in order to characterize its mechanisms, is to use an approach commonly known as quantitative structure–activity relationship (QSAR). As is well-known, QSAR is considerably dependent on the exact training/test sets and the particular bioassay being monitored, and we attempted to build a 2D-/3D-model based on a selected training/test set of chemicals of regional importance and with particular bioactivities.

Among different food items, it was found that it is fish that has highest accumulated levels of pollutants in the food chain and in the environment. Thus, dietary intake of fish is the major source of endocrine-disrupting contaminants (Brustad et al. 2008, Dovydaitis 2008, Genuis 2008). It is anticipated that the body burden of EDs can interfere with the gonadotropin–ovary–placenta axis of female animals, inducing adverse changes via modulation of processes such as gametogenesis, fertilization, and embryo implantation and development. It is important to investigate the effects of ED exposure particularly on animal fertility and embryo development. These approaches provide invaluable information to evaluate the most susceptible stage of the embryos at which normal embryonic development is significantly affected by EDs. Hopefully this can provide insights into possible transgeneration effects of *in utero* ED exposure.

References

Anway MD, Cupp AS, Uzumcu M, Skinner MK. 2005. Epigenetic transgenerational actions of endocrine disruptors and male fertility. *Science* 308: 1466–1469.

Anway MD, Skinner MK. 2008. Epigenetic programming of the germ line: Effects of endocrine disruptors on the development of transgenerational disease. *Reproductive BioMedicine Online* 16: 23–25.

Aravindakshan J, Paquet V, Gregory M, Dufresne J, Fournier M, Marcogliese DJ, Cyr DG. 2004. Consequences of xenoestrogen exposure on male reproductive function in spottail shiners (*Notropis hudsonius*). *Toxicology Sciences* 78: 156–165.

Baldridge MG, Stahl RL, Gerstenberger SL, Tripoli V, Hutz RJ. 2003. Modulation of ovarian follicle maturation in Long-Evans rats exposed to polychlorinated biphenyls (PCBs) in-utero and lactationally. *Reproductive Toxicology* 17: 567–573.

Barnhoorn IE, Bornman MS, Pieterse GM, van Vuren JH. 2004. Histological evidence of intersex in feral sharptooth catfish (*Clarias gariepinus*) from an estrogen-polluted water source in Gauteng, South Africa. *Environmental Toxicology* 19: 603–608.

Bartolomei MS, Tilghman SM. 1997. Genomic imprinting in mammals. *Annual Review of Genetics* 31: 493–525.

Beischlag TV, Perdew GH. 2005. ER alpha-AHR-ARNT protein–protein interactions mediate estradiol-dependent transrepression of dioxin-inducible gene transcription. *Journal of Biological Chemistry* 280: 21607–21611.

Bjerke DL, Peterson RE. 1994. Reproductive toxicity of 2,3,7,8-tetrachlorodibenzo-p-dioxin in male rats: Different effects of in utero versus lactational exposure. *Toxicology and Applied Pharmacology* 127: 241–249.

Brustad M, Sandanger TM, Nieboer E, Lund E. 2008. 10th anniversary review: When healthy food becomes polluted-implications for public health and dietary advice. *Journal of Environmental Monitoring* 10: 422–427.

Carney SA, Prasch AL, Heideman W, Peterson RE. 2006. Understanding dioxin developmental toxicity using the zebrafish model. *Birth Defects Research Part A Clinical and Molecular Teratology* 76: 7–18.

Chahoud I, Hartmann J, Rune GM, Neubert D. 1992. Reproductive toxicity and toxicokinetics of 2,3,7,8-tetrachlorodibenzo-p-dioxin. 3. Effects of single doses on the testis of male rats. *Archives of Toxicology* 66: 567–572.

Chan JK, Xing GH, Xu Y et al. 2007. Body loadings and health risk assessment of polychlorinated dibenzo-p-dioxins and dibenzofurans at an intensive electronic waste recycling site in China. *Environmental Science and Technology* 41: 7668–7674.

Chen YJ, Hsu CC. 1994. Effects of prenatal exposure to PCBs on the neurological function of children: A neuropsychological and neurophysiological study. *Developmental Medicine and Child Neurology* 36: 312–320.

Couture LA, Abbott BD, Birnbaum LS. 1990. A critical review of the developmental toxicity and teratogenicity of 2,3,7,8-tetrachlorodibenzo-p-dioxin: Recent advances toward understanding the mechanism. *Teratology* 42: 619–627.

Crews D, McLachlan JA. 2006. Epigenetics, evolution, endocrine disruption, health, and disease. *Endocrinology* 147: S4–S10.

Czeizel AE, Elek C, Gundy S et al. 1993. Environmental trichlorfon and cluster of congenital abnormalities. *Lancet* 341: 539–542.

De Guise S, Lagace A, Beland P. 1994. True hermaphroditism in a St. Lawrence beluga whale (*Delphinapterus leucas*). *Journal of Wildlife Diseases* 30: 287–290.

Delbes G, Levacher C, Habert R. 2006. Estrogen effects on fetal and neonatal testicular development. *Reproduction* 132: 527–538.

Dhooge W, van Larebeke N, Koppen G et al. 2006. Serum dioxin-like activity is associated with reproductive parameters in young men from the general Flemish population. *Environmental Health Perspectives* 114: 1670–1676.

Dolinoy DC, Huan D, Jirtle RL. 2007. Maternal nutrient supplementation counteracts bisphenol A-induced DNA hypomethylation in early development. *Proceedings of the National Academy of Science USA* 104: 13056–13061.

Doucet J, Tague B, Arnold DL, Cooke GM, Hayward S, Goodyer CG. 2009. Persistent organic pollutant residues in human fetal liver and placenta from Greater Montreal, Quebec: A longitudinal study from 1998 through 2006. *Environmental Health Perspectives* 117: 605–610.

Dovydaitis T. 2008. Fish consumption during pregnancy: An overview of the risks and benefits. *The Journal of Midwifery Women's Health* 53: 325–330.

Duarte-Davidson R, Jones KC. 1994. Polychlorinated biphenyls (PCBs) in the UK population: Estimated intake, exposure and body burden. *Science of the Total Environment* 151: 131–152.

Faqi AS, Dalsenter PR, Merker HJ, Chahoud I. 1998. Reproductive toxicity and tissue concentrations of low doses of 2,3,7,8-tetrachlorodibenzo-p-dioxin in male off-spring rats exposed throughout pregnancy and lactation. *Toxicology and Applied Pharmacology* 150: 383–392.

Finnell RH, Waes JG, Eudy JD, Rosenquist TH. 2002. Molecular basis of environmen-tally induced birth defects. *Annual Review of Pharmacology and Toxicology* 42: 181–208.

Fry DM. 1995. Reproductive effects in birds exposed to pesticides and industrial chemicals. *Environmental Health Perspectives* 103 Suppl 7: 165–171.

Garcia-Rodriguez, J, Garcia-Martin M, Nogueras-Ocana M, de Dios Luna-del-Castillo, Espigares GM, Olea N, Lardelli-Claret P. 1996. Exposure to pesticides and crypt-orchidism: Geographical evidence of a possible association. *Environmental Health Perspectives* 104: 1090–1095.

Genuis SJ. 2008. To sea or not to sea: Benefits and risks of gestational fish consump-tion. *Reproductive Toxicology* 26: 81–85.

Goldman LR, Koduru S. 2000. Chemicals in the environment and developmental tox-icity to children: A public health and policy perspective. *Environmental Health Perspectives* 108 Suppl 3: 443–448.

Gray LE Jr, Kelce WR, Monosson E, Ostby JS, Birnbaum LS. 1995. Exposure to TCDD during development permanently alters reproductive function in male Long Evans rats and hamsters: Reduced ejaculated and epididymal sperm numbers and sex accessory gland weights in offspring with normal androgenic status. *Toxicology Applied Pharmacology* 131: 108–118.

Greco TL, Furlow JD, Duello TM, Gorski J. 1992. Immunodetection of estrogen receptors in fetal and neonatal male mouse reproductive tracts. *Endocrinology* 130: 421–429.

Grimvall E, Rylander L, Nilsson-Ehle P, Nilsson U, Stromberg U, Hagmar L, Ostman C. 1997. Monitoring of polychlorinated biphenyls in human blood plasma: Methodological developments and influence of age, lactation, and fish consumption. *Archives of Environmental Contamination and Toxicology* 32: 329–336.

Guillette LJ Jr, Gross TS, Gross DA, Rooney AA, Percival HF. 1995. Gonadal steroido-genesis in vitro from juvenile alligators obtained from contaminated or control lakes. *Environmental Health Perspectives* 103 Suppl 4: 31–36.

Guillette LJ Jr, Gross TS, Masson GR, Matter JM, Percival HF, Woodward AR. 1994. Developmental abnormalities of the gonad and abnormal sex hormone concen-trations in juvenile alligators from contaminated and control lakes in Florida. *Environmental Health Perspectives* 102: 680–688.

Guillette LJ Jr, Guillette EA. 1996. Environmental contaminants and reproductive abnormalities in wildlife: Implications for public health? *Toxicology and Industrial Health* 12: 537–550.

Guillette LJ Jr, Moore BC. 2006. Environmental contaminants, fertility, and multi-oocytic follicles: A lesson from wildlife? *Seminars in Reproductive Medicine* 24: 134–141.

Guo YL, Chen YC, Yu ML, Hsu CC. 1994a. Early development of Yu-Cheng children born seven to twelve years after the Taiwan PCB outbreak. *Chemosphere* 29: 2395–2404.

Guo YL, Lin CJ, Yao WJ, Ryan JJ, Hsu CC. 1994b. Musculoskeletal changes in chil-dren prenatally exposed to polychlorinated biphenyls and related compounds (Yu-Cheng children). *Journal of Toxicology and Environmental Health* 41: 83–93.

Guo YL, Ryan JJ, Lau BP, Yu ML, Hsu CC. 1997. Blood serum levels of PCBs and PCDFs in Yucheng women 14 years after exposure to a toxic rice oil. *Archives of Environmental Contamination and Toxicology* 33: 104–108.

Hamdoun A, Epel D. 2007. Embryo stability and vulnerability in an always changing world. *Proceedings of the National Academy of Science USA* 104: 1745–1750.

Hayes T, Haston K, Tsui M, Hoang A, Haeffele C, Vonk A. 2003. Atrazine-induced hermaphroditism at 0.1 ppb in American leopard frogs (*Rana pipiens*): Laboratory and field evidence. *Environmental Health Perspectives* 111: 568–575.

Henley DV, Korach KS. 2006. Endocrine-disrupting chemicals use distinct mechanisms of action to modulate endocrine system function. *Endocrinology* 147: S25–S32.

Hiroi H, Momoeda M, Inoue S et al. 1999. Stage-specific expression of estrogen receptor subtypes and estrogen responsive finger protein in preimplantational mouse embryos. *Endocrine Journal* 46: 153–158.

Honda T, Wada M, Nakashima K. 2009. PCBs and PCDD/DFs in waste oil illegally dumped and neglected for more than 20 years. *Journal of Environmental Science and Health A* 44: 654–660.

Hu G, Sun C, Li J, Zhao Y, Wang H, Li Y. 2009. POPs accumulated in fish and benthos bodies taken from Yangtze River in Jiangsu area. *Ecotoxicology* 18: 647–651.

Hunt PA, Koehler KE, Susiarjo M et al. 2003. Bisphenol a exposure causes meiotic aneuploidy in the female mouse. *Current Biology* 13: 546–553.

Järnberg U, Asplund L, de Wit C, Egebäck A, Wideqvist U, Jakobsson E. 1997. Distribution of polychlorinated naphthalene congeners in environmental and source-related samples. *Archives of Environmental Contamination and Toxicology* 32: 232–245.

Jefferson WN, Couse JF, Banks EP, Korach KS, Newbold RR. 2000. Expression of estrogen receptor beta is developmentally regulated in reproductive tissues of male and female mice. *Biology of Reproduction* 62: 310–317.

Jirtle RL, Skinner MK. 2007. Environmental epigenomics and disease susceptibility. *Nature Reviews Genetics* 8: 253–262.

Jobling S, Coey S, Whitmore JG et al. 2002. Wild intersex roach (*Rutilus rutilus*) have reduced fertility. *Biology of Reproduction* 67: 515–524.

Johnson L, Dickerson R, Safe SH, Nyberg CL, Lewis RP, Welsh TH Jr. 1992. Reduced Leydig cell volume and function in adult rats exposed to 2,3,7,8-tetrachlorodibenzo-p-dioxin without a significant effect on spermatogenesis. *Toxicology* 76: 103–118.

Johnson MD, Kenney N, Stoica A et al. 2003. Cadmium mimics the in vivo effects of estrogen in the uterus and mammary gland. *Nature Medicine* 9: 1081–1084.

Johnson L, Wilker CE, Safe SH, Scott B, Dean DD, White PH. 1994. 2,3,7,8-Tetrachlorodibenzo-p-dioxin reduces the number, size, and organelle content of Leydig cells in adult rat testes. *Toxicology* 89: 49–65.

Kanagawa Y, Matsumoto S, Koike S et al. 2008. Association of clinical findings in Yusho patients with serum concentrations of polychlorinated biphenyls, polychlorinated quarterphenyls and 2,3,4,7,8-pentachlorodibenzofuran more than 30 years after the poisoning event. *Environmental Health* 7: 47.

Lai TJ, Guo YL, Yu ML, Ko HC, Hsu CC. 1994. Cognitive development in Yucheng children. *Chemosphere* 29: 2405–2411.

Lai KP, Wong MH, Wong CK. 2005a. Effects of TCDD in modulating the expression of Sertoli cell secretory products and markers for cell–cell interaction. *Toxicology* 206: 111–123.

Lai KP, Wong MH, Wong CK. 2005b. Inhibition of CYP450scc expression in dioxin-exposed rat Leydig cells. *Journal of Endocrinology* 185: 519–527.

LaKind JS, Berlin CM, Mattison DR. 2008. The heart of the matter on breastmilk and environmental chemicals: Essential points for healthcare providers and new parents. *Breastfeeding Medicine* 3: 251–259.

Landrigan PJ, Schechter CB, Lipton JM, Fahs MC, Schwartz J. 2002. Environmental pollutants and disease in American children: Estimates of morbidity, mortality, and costs for lead poisoning, asthma, cancer, and developmental disabilities. *Environmental Health Perspectives* 110: 721–728.

Langer P, Kocan A, Tajtakova M et al. 2009. Multiple adverse thyroid and metabolic health signs in the population from the area heavily polluted by organochlorine cocktail (PCB, DDE, HCB, dioxin). *Thyroid Research* 2: 3.

Latchoumycandane C, Chitra C, Mathur P. 2002a. Induction of oxidative stress in rat epididymal sperm after exposure to 2,3,7,8-tetrachlorodibenzo-p-dioxin. *Archives of Toxicology* 76: 113–118.

Latchoumycandane C, Chitra KC, Mathur PP. 2002b. The effect of 2,3,7,8-tetrachloro-dibenzo-p-dioxin on the antioxidant system in mitochondrial and microsomal fractions of rat testis. *Toxicology* 171: 127–135.

Li X, Johnson DC, Rozman KK. 1995. Reproductive effects of 2,3,7,8-tetrachlorod-ibenzo-p-dioxin (TCDD) in female rats: Ovulation, hormonal regulation, and possible mechanism(s). *Toxicology and Applied Pharmacology* 133: 321–327.

Li QQ, Loganath A, Chong YS, Tan J, Obbard JP. 2006. Persistent organic pollutants and adverse health effects in humans. *Journal of Toxicology and Environmental Health A* 69: 1987–2005.

Lo PE, Koehler K, Chana A, Benfenati E. 2006. Virtual screening for aryl hydrocarbon receptor binding prediction. *Journal of Medicinal Chemistry* 49: 5702–5709.

Mably TA, Bjerke DL, Moore RW, Gendron-Fitzpatrick A, Peterson RE. 1992. In utero and lactational exposure of male rats to 2,3,7,8-tetrachlorodibenzo-p-dioxin. 3. Effects on spermatogenesis and reproductive capability. *Toxicology and Applied Pharmacology* 114: 118–126.

Mackenzie CA, Lockridge A, Keith M. 2005. Declining sex ratio in a first nation community. *Environmental Health Perspectives* 113: 1295–1298.

Mansfield KG, Land ED. 2002. Cryptorchidism in Florida panthers: Prevalence, features, and influence of genetic restoration. *Journal of Wildlife Diseases* 38: 693–698.

Maras M, Vanparys C, Muylle F et al. 2006. Estrogen-like properties of fluorotelomer alcohols as revealed by mcf-7 breast cancer cell proliferation. *Environmental Health Perspectives* 114: 100–105.

Martin MB, Reiter R, Pham T et al. 2003. Estrogen-like activity of metals in MCF-7 breast cancer cells. *Endocrinology* 144: 2425–2436.

Matthews J, Wihlen B, Thomsen J, Gustafsson JA. 2005. Aryl hydrocarbon receptor-mediated transcription: Ligand-dependent recruitment of estrogen receptor alpha to 2,3,7,8-tetrachlorodibenzo-p-dioxin-responsive promoters. *Molecular and Cellular Biology* 25: 5317–5328.

McLachlan JA. 1993. Functional toxicology: A new approach to detect biologically active xenobiotics. *Environmental Health Perspectives* 101: 386–387.

McLachlan JA. 2001. Environmental signaling: What embryos and evolution teach us about endocrine disrupting chemicals. *Endocrine Reviews* 22: 319–341.

Meerts IA, Letcher RJ, Hoving S. 2001. In vitro estrogenicity of polybrominated diphenyl ethers, hydroxylated PDBEs, and polybrominated bisphenol A compounds. *Environmental Health Perspectives* 109: 399–407.

Mimura J, Fujii-Kuriyama Y. 2003. Functional role of AhR in the expression of toxic effects by TCDD. *Biochimica et Biophysica Acta* 1619: 263–268.

Nadal A, Ropero AB, Laribi O, Maillet M, Fuentes E, Soria B. 2000. Nongenomic actions of estrogens and xenoestrogens by binding at a plasma membrane receptor unrelated to estrogen receptor alpha and estrogen receptor beta. *Proceedings of the National Academy of Science USA* 97: 11603–11608.

Nomura T. 2008. Transgenerational effects from exposure to environmental toxic substances. *Mutation Research* 659: 185–193.

O'Donnell L, Robertson KM, Jones ME, Simpson ER. 2001. Estrogen and spermatogenesis. *Endocrine Reviews* 22: 289–318.

Ohtake F, Baba A, Takada I. 2007. Dioxin receptor is a ligand-dependent E3 ubiquitin ligase. *Nature* 446: 562–566.

Ohtake F, Takeyama K, Matsumoto T. 2003. Modulation of oestrogen receptor signalling by association with the activated dioxin receptor. *Nature* 423: 545–550.

Olek A, Walter J. 1997. The pre-implantation ontogeny of the H19 methylation imprint. *Nature Genetics* 17: 275–276.

Oskam IC, Ropstad E, Dahl E. 2003. Organochlorines affect the major androgenic hormone, testosterone, in male polar bears (*Ursus maritimus*) at Svalbard. *Journal of Toxicology and Environmental Health A* 66: 2119–2139.

Pan G, Hanaoka T, Yoshimura M et al. 2006. Decreased serum free testosterone in workers exposed to high levels of di-n-butyl phthalate (DBP) and di-2-ethylhexyl phthalate (DEHP): A cross-sectional study in China. *Environmental Health Perspectives* 114: 1643–1648.

Peters JM, Wiley LM. 1995. Evidence that murine preimplantation embryos express aryl hydrocarbon receptor. *Toxicology and Applied Pharmacology* 134: 214–221.

Peterson RE, Theobald HM, Kimmel GL. 1993. Developmental and reproductive toxicity of dioxins and related compounds: Cross-species comparisons. *Critical Reviews in Toxicology* 23: 283–335.

Phillips KP, Foster WG. 2008a. Endocrine toxicants with emphasis on human health risks. *Journal of Toxicology and Environmental Health B* 11: 149–151.

Phillips KP, Foster WG. 2008b. Key developments in endocrine disrupter research and human health. *Journal of Toxicology and Environmental Health B* 11: 322–344.

Phillips KP, Foster WG, Leiss W et al. 2008. Assessing and managing risks arising from exposure to endocrine-active chemicals. *Journal of Toxicology and Environmental Health B* 11: 351–372.

Phillips KP, Tanphaichitr N. 2008. Human exposure to endocrine disrupters and semen quality. *Journal of Toxicology and Environmental Health B* 11: 188–220.

Prasch AL, Teraoka H, Carney SA et al. 2003. Aryl hydrocarbon receptor 2 mediates 2,3,7,8-tetrachlorodibenzo-p-dioxin developmental toxicity in zebrafish. *Toxicological Sciences* 76: 138–150.

Puga A, Barnes SJ, Dalton TP, Chang C, Knudse ES, Maier MA. 2000. Aromatic hydrocarbon receptor interaction with the retinoblastoma protein potentiates repression of E2F-dependent transcription and cell cycle arrest. *The Journal of Biological Chemistry* 275: 2943–2950.

Rubin MM. 2007. Antenatal exposure to DES: Lessons learned...future concerns. *Obstetrical and Gynecological Survey* 62: 548–555.

Rune GM, de Souza P, Krowke R, Merker HJ, Neubert D. 1991. Morphological and histochemical pattern of response in rat testes after administration of 2,3,7,8-tetrachlorodibenzo-p-dioxin (TCDD). *Histology and Histopathology* 6: 459–467.

Safe S. 1990. Polychlorinated biphenyls (PCBs), dibenzo-p-dioxins (PCDDs), dibenzo-furans (PCDFs), and related compounds: Environmental and mechanistic considerations which support the development of toxic equivalency factors (TEFs). *Critical Reviews in Toxicology* 21: 51–88.

Safe SH, Zacharewski T. 1997. Organochlorine exposure and risk for breast cancer. *Progress in Clinical and Biological Research* 396: 133–145.

Salisbury TB, Marcinkiewicz JL. 2002. In utero and lactational exposure to 2,3,7,8-tetra-chlorodibenzo-p-dioxin and 2,3,4,7,8-pentachlorodibenzofuran reduces growth and disrupts reproductive parameters in female rats. *Biology of Reproduction* 66: 1621–1626.

Sanford JP, Clark HJ, Chapman VM, Rossant J. 1987. Differences in DNA methylation during oogenesis and spermatogenesis and their persistence during early embryogenesis in the mouse. *Genes and Development* 1: 1039–1046.

Saunders PT, Fisher JS, Sharpe RM, Millar MR. 1998. Expression of oestrogen receptor beta (ER beta) occurs in multiple cell types, including some germ cells, in the rat testis. *Journal of Endocrinology* 156: R13–R17.

Saunders PT, Sharpe RM, Williams K et al. 2001. Differential expression of oestrogen receptor alpha and beta proteins in the testes and male reproductive system of human and non-human primates. *Molecular Human Reproduction* 7: 227–236.

Schecter A, Toniolo P, Dai LC, Thuy LT, Wolff MS. 1997. Blood levels of DDT and breast cancer risk among women living in the north of Vietnam. *Archives of Environmental Contamination and Toxicology* 33: 453–456.

Sheweita SA, Tilmisany AM, Al Sawaf H. 2005. Mechanisms of male infertility: Role of antioxidants. *Current Drug Metabolism* 6: 495–501.

Svensson BG, Nilsson A, Jonsson E, Schutz A, Akesson B, Hagmar L. 1995. Fish consumption and exposure to persistent organochlorine compounds, mercury, selenium and methylamines among Swedish fishermen. *Scandinavian Journal of Work, Environment and Health* 21: 96–105.

Swan SH, Main KM, Liu F et al. 2005. Decrease in anogenital distance among male infants with prenatal phthalate exposure. *Environmental Health Perspectives* 113: 1056–1061.

Taylor HS. 2008. Endocrine disruptors affect developmental programming of HOX gene expression. *Fertility and Sterility* 89: e57–e58.

Theobald HM, Peterson RE. 1997. In utero and lactational exposure to 2,3,7,8-tetra-chlorodibenzo-rho-dioxin: Effects on development of the male and female reproductive system of the mouse. *Toxicology and Applied Pharmacology* 145: 124–135.

Thornton JW, Need E, Crews D. 2003. Resurrecting the ancestral steroid receptor: Ancient origin of estrogen signaling. *Science* 301: 1714–1717.

Tremblay KD, Duran KL, Bartolomei MS. 1997. A 5' 2-kilobase-pair region of the imprinted mouse H19 gene exhibits exclusive paternal methylation throughout development. *Molecular and Cellular Biology* 17: 4322–4329.

Tsukimori K, Tokunaga S, Shibata S et al. 2008. Long-term effects of polychlorinated biphenyls and dioxins on pregnancy outcomes in women affected by the Yusho incident. *Environmental Health Perspectives* 116: 626–630.

Tsutsumi O, Uechi H, Sone H et al. 1998. Presence of dioxins in human follicular fluid: Their possible stage-specific action on the development of preimplantation mouse embryos. *Biochemical and Biophysical Research Communications* 250: 498–501.

Tuppurainen K, Ruuskanen J. 2000. Electronic eigenvalue (EEVA): A new QSAR/QSPR descriptor for electronic substituent effects based on molecular orbital energies. A QSAR approach to the Ah receptor binding affinity of polychlorinated biphenyls (PCBs), dibenzo-p-dioxins (PCDDs) and dibenzofurans (PCDFs). *Chemosphere* 41: 843–848.

Uzumcu M, Zachow R. 2007. Developmental exposure to environmental endocrine disruptors: Consequences within the ovary and on female reproductive function. *Reproductive Toxicology* 23: 337–352.

van Pelt AM, de Rooij DG, van der BB, van der Saag PT, Gustafsson JA, Kuiper GG. 1999. Ontogeny of estrogen receptor-beta expression in rat testis. *Endocrinology* 140: 478–483.

Vogel CF, Sciullo E, Li W, Wong P, Lazennec G, Matsumura F. 2007. RelB, a new partner of aryl hydrocarbon receptor-mediated transcription. *Molecular Endocrinology* 21: 2941–2955.

Vos JG, Dybing E, Greim HA et al. 2000. Health effects of endocrine-disrupting chemicals on wildlife, with special reference to the European situation. *Critical Reviews in Toxicology* 30: 71–133.

Warnecke PM, Mann JR, Frommer M, Clark SJ. 1998. Bisulfite sequencing in preimplantation embryos: DNA methylation profile of the upstream region of the mouse imprinted H19 gene. *Genomics* 51: 182–190.

Weidner IS, Moller H, Jensen TK, Skakkebaek NE. 1998. Cryptorchidism and hypospadias in sons of gardeners and farmers. *Environmental Health Perspectives* 106: 793–796.

Weselak M, Arbuckle TE, Walker MC, Krewski D. 2008. The influence of the environment and other exogenous agents on spontaneous abortion risk. *Journal of Toxicology and Environmental Health B* 11: 221–241.

White R, Jobling S, Hoare SA, Sumpter JP, Parker MG. 1994. Environmentally persistent alkylphenolic compounds are estrogenic. *Endocrinology* 135: 175–182.

Wilker CE, Welsh TH Jr, Safe SH, Narasimhan TR, Johnson L. 1995. Human chorionic gonadotropin protects Leydig cell function against 2,3,7,8-tetrachlorodibenzo-p-dioxin in adult rats: Role of Leydig cell cytoplasmic volume. *Toxicology* 95: 93–102.

Wong MH, Leung AOW, Chan JK, Choi MP. 2005. A review on the usage of POP pesticides in China, with emphasis on DDT loadings in human milk. *Chemosphere* 60: 740–752.

Wormke M, Stoner M, Saville B, Walker K, Abdelrahim M, Burghardt R, Safe S. 2003. The aryl hydrocarbon receptor mediates degradation of estrogen receptor alpha through activation of proteasomes. *Molecular and Cellular Biology* 23: 1843–1855.

Wu Q, Ohsako S, Ishimura R, Suzuki JS, Tohyama C. 2004. Exposure of mouse preimplantation embryos to 2,3,7,8-tetrachlorodibenzo-p-dioxin (TCDD) alters the methylation status of imprinted genes H19 and Igf2. *Biology of Reproduction* 70: 1790–1797.

5

Assessing Health Risks from Arsenic Intake by Residents in Cambodia

Suthipong Sthiannopkao, Kongkea Phan,
Kyoung-Woong Kim, and Ming Hung Wong

CONTENTS

Introduction

The widespread switch from microbiologically unsafe surface water to microbiologically safe groundwater has led to the unanticipated poisoning of large numbers of people in the developing world who have consumed various toxic trace elements. In particular, elevated concentrations of arsenic in groundwater have been reported in Taiwan (Tseng 1977, Tseng et al. 1968), West Bengal (India), and Bangladesh (Das et al. 1994, Mandal et al. 1998, Nickson et al. 1998), resulting in a major public health issue. In Bangladesh and West Bengal (India), it is estimated that approximately 40 million people are suffering from drinking naturally occurring arsenic-rich shallow groundwater (Gault et al. 2008). Despite these concerns, groundwater is still a major source of drinking water in the developing world, especially in Southeast Asia. Recently, unsafe levels of arsenic have also been revealed in

Vietnam (Berg et al. 2001, 2007, Buschmann et al. 2007, Nguyen et al. 2009) and Cambodia (JICA 1999, Polya et al. 2003). Individuals can be exposed to arsenic through several pathways, but the most critical one is daily diet and drinking water ingestion. Toxicological studies show that both trivalent and pentavalent soluble arsenic compounds are rapidly absorbed from the gastrointestinal tract and can be further metabolized. Reduction of As (V) to As (III) followed by oxidative methylation of As (III) takes place to form mono-, di-, or trimethylated products (Hughes 2002). Oral pathway exposure of organic arsenic compounds is less toxic since organic arsenicals are less extensively metabolized and more rapidly eliminated in urine than inorganic arsenicals (ATSDR 2007, WHO 2004). There is no evidence that arsenic is essential in human bodies. In contrast, chronic oral consumption of arsenic is considered to cause an adverse impact on human beings, known as "arsenicosis" or "arsenic poisoning disease." Arsenicosis can cause skin lesions, pigmentation of the skin, and the development of hard patches of skin on the palm of the hands and soles of the feet. Arsenic poisoning finally leads to skin, bladder, kidney, and lung cancers, as well as diseases of the blood vessels of the legs and feet. Diabetes, high blood pressure, and reproductive disorders may also be the side effects of chronic arsenic exposure (ATSDR 2007, Tseng 1977, WHO 2004). In Cambodia, unsafe levels of arsenic in shallow groundwater were first reported by JICA (1999) in its first unpublished draft report named "The study on groundwater development in Southern Cambodia" to the Cambodia Ministry of Rural Development. Consequently, numerous studies have been conducted and documented. Polya et al. (2003, 2005), Stanger et al. (2005), Berg et al. (2007), Buschmann et al. (2007), Quicksall et al. (2008), Sthiannopkao et al. (2008), and Luu et al. (2009) have described the distribution of arsenic in shallow Cambodian groundwater. The chemical, biological, and physical processes that control the heterogeneous arsenic distribution in groundwater have also been widely studied (Benner et al. 2008, Berg et al. 2007, Buschmann et al. 2007, Kocar et al. 2008, Lear et al. 2007, Polizzotto et al. 2008, Polya et al. 2003, 2005, Robinson et al. 2009, Rowland et al. 2008). Arsenic treatment systems, modified from traditional sand filters, have also been developed to enhance arsenic removal from groundwater following seasonal and spatial variations in groundwater composition (Chiew et al. 2009). In addition, studies of baseline concentrations of As in human hairs, nails, and urine have been used to assess potential biomarkers of As exposure (Berg et al. 2007, Gault et al. 2008, Kubota et al. 2006, Sampson et al. 2008). The development of visual arsenicosis symptoms have been generally assumed to follow 8–10 years of consumption of water with unsafe level of arsenic; however, new cases discovered in Cambodia have followed exposure times as short as 3 years, due to extremely elevated arsenic levels (3500 µg/L), socioeconomic status, and malnutrition (Sampson et al. 2008). In Kandal alone, by using groundwater quality and population data, Sampson et al. (2008) have estimated that 100,000 people are at high risk of chronic arsenic exposure. The objectives of the present study were (1) to determine the distribution of toxic trace elements in groundwater of the Mekong River basin of Cambodia, (2) to assess carcinogenic and noncarcinogenic risks among the population exposed to arsenic through groundwater drinking pathway, and (3) to determine the arsenic content in scalp hair of the people living in the Mekong River basin.

Materials and Methods

Study Area

Sampling was carried out within three purposely selected areas with different arsenic exposure scenarios in the Mekong River basin of Cambodia. Kampong Kong commune (Preak Russey and Lvea Toung villages) in Kandal province was selected as an extremely contaminated area. Khsarch Andaet commune (Preak Samrong I and II villages) in Kratie province was selected as a moderately contaminated area, and Ampil commune (Andoung Chros and Veal Sbov villages) in Kampong Cham province was selected as an uncontaminated area. Kratie and Kampong Cham provinces are located along the Mekong River, upstream of Phnom Penh, whereas Kandal province is located between the Mekong and the Bassac Rivers, downstream of Phnom Penh (Figure 5.1).

Field Sampling

Groundwater samples were collected from the study areas of Kandal (n = 46) and Kampong Cham (n = 18) in February 2009 and Kratie province (n = 12) in August 2009. Concurrently, in the first batch of sampling, some scalp hair specimens were also sampled from the study areas of Kandal and Kampong Cham provinces, and the remainders were collected in the second batch. Sampling was conducted based on the accessibilities to tube wells, the willingness of respondents to provide hair samples, and respondents' claims of tube-well use for a certain period of time. Each groundwater sample was collected from a tube well after 5–10 min of flushing to remove any standing water from the tube. Groundwater was filled in two separate polyethylene bottles for different purposes of analyses. Raw samples (no pretreatment) were analyzed for total arsenic. Filtered water samples (0.45 μm pore sized membrane filter) were analyzed for the trace elements. Simultaneously, onsite measurements for additional parameters were conducted by using a *HORIBA pH/Cond meter D-54* for pH and Eh. During field sampling, all of the collected water samples were kept in an ice box and were then transferred to a refrigerator where they were stored at 4°C until delivery to GIST, South Korea, for analysis. Hair samples were randomly collected from several

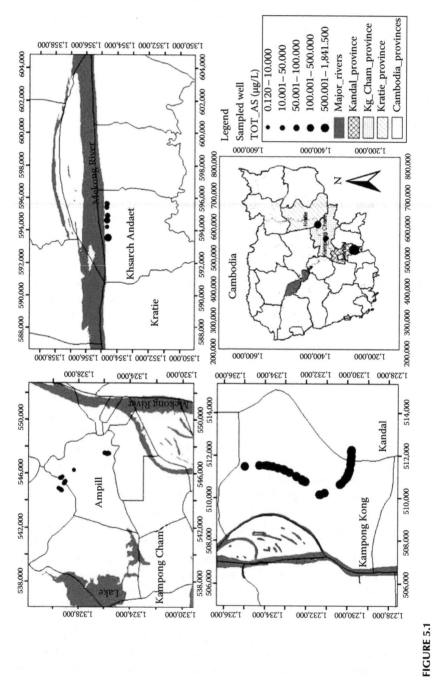

FIGURE 5.1
Map of study areas. (From Sthiannopkao, S. et al., *Appl. Geochem.*, 23, 1086, 2008.)

members of each household where people claimed to routinely use a tube well. Hair was cut from the nape of the head, as near as possible to the scalp, using stainless steel scissors. The collected hair samples were kept in labeled ziplock bags and stored in darkness until analyses.

Sample Analyses

Groundwater samples from the Kandal province study area were diluted to analyze the concentrations of total As, Mn, and Ba. Dilution (1:25) was made for the final concentration of aliquots to meet the standard calibration curve as recommended by ICP-MS analytical technique. Dilution was done with 2% HNO_3 (prepared by 18.2 MΩ *MilliQ* deionized water with 70% HNO_3). Similarly, some samples from the Kratie province study area were treated in the same manner, whereas all of groundwater samples from the Kampong Cham study area were analyzed without any treatment, aside from centrifuging for total arsenic analysis. Hair specimens were cut into small pieces (~0.3 cm) and washed with the recommended method (Ryabukhin 1978). Washed hair specimens were dried at 60°C overnight prior to digestion. Acid digestion was performed using a slightly modified version of the method described by Gault et al. (2008). Approximately 100 ± 5 mg of two replicated subsamples of each dry-washed hair sample were weighed into acid-cleaned polyethylene tubes. One milliliter of concentrated HNO_3 (70% HNO_3) was added to each sample, and tubes were capped and left at room temperature. After 48 h, the digestate was diluted with 9 mL of deionized water and centrifuged at 4500 rpm for 10 min, after which the supernatant was transferred into a fresh acid-cleaned polyethylene tube. A human hair standard reference material (GBW07601) was treated in the same manner as the samples to check the precision and accuracy of digestion method. Calibration standard solutions (0.1, 1, 5, 10, 20, 50, and 100 µg/L) were prepared from a stock solution (multielement 2A) with the above 2% HNO_3. Concentrations of total arsenic, manganese, and barium were analyzed by inductively coupled plasma mass spectrometry (ICP-MS, Agilent 7500ce). Iron concentrations were determined by flame atomic absorption spectrometry (Flame-AAS, Perkin Elmer 5100).

Statistical Analysis

An SPSS for Windows (Version 13.0) was used to perform all statistical analyses on the data. Since the experimental data sets were not normally distributed, nonparametric tests were performed. Kruskal–Wallis test was used to investigate the regional differences of the arsenic and trace element concentrations in groundwater (As_w), the average daily dose (ADD) of arsenic, and the arsenic contents in scalp hair (As_h). Concurrently, Mann–Whitney's U test was performed to verify the gender differences of ADD and As_h. In addition, Spearman's rho correlation was conducted to determine the intercorrelations between As_w, As_h, ADD, hazard quotient (HQ), cancer risk (R), body weight (BW), IR, exposure duration (ED), gender, and age groups. Significance was considered in circumstances where $p < 0.05$.

Results and Discussion

Chemistry of Cambodia Groundwater

The results from chemical measurements of Cambodia groundwater are presented in Table 5.1. In the Kampong Cham province study area, groundwater was slightly acidic with a pH range of 6.21–6.96. High Eh values (164.5–319 mV) indicated that the groundwater was under oxidizing conditions. In contrast, groundwater from Kandal and Kratie study areas was quite similar; it had circumneutral pH and was under reducing conditions (low Eh values). Reducing condition might favor the dissolution of toxic trace elements from sediments to pore water. A significant regional difference was observed in As, Mn, Fe, and Ba concentrations in groundwater among the three study areas (Kruskal–Wallis test, $p < 0.0001$). In the Kandal province study area, analytical results indicate that groundwater

TABLE 5.1

Mean, Median, Standard Deviation, and Min and Max of Groundwater Analyzed in "Kandal," "Kratie," and "Kampong Cham" Provinces

Study Area	Statistics	pH	Eh (mV)	Ba (µg/L)	Fe (µg/L)	Mn (µg/L)	Tot As (µg/L)
Kandal (n = 46)	Mean	7.17	−151.67	1028.26	5,901.9	584.23	846.14
	Median	7.13	−154.75	872.38	5,564.2	405.75	822.63
	SD	0.26	20.27	477.74	3,017.6	516.49	298.11
	Min	6.58	−189.00	446.25	1,367.4	88.18	247.08
	Max	7.85	−55.00	2652.50	17,134.5	3045.00	1841.50
Kratie (n = 12)	Mean	6.48	−116.70	102.38	899.2	588.71	22.22
	Median	6.36	−101.35	64.45	74.8	274.45	1.30
	SD	0.73	37.34	102.96	1,895.7	750.98	43.89
	Min	5.53	−169.85	0.25	0.7	0.33	0.12
	Max	7.37	−57.95	359.90	5,114.0	2139.00	140.60
Kampong Cham (n = 18)	Mean	6.75	221.44	18.56	16.5	4.31	1.28
	Median	6.84	200.50	18.06	15.8	0.81	1.22
	SD	0.23	46.27	10.08	13.1	8.16	0.58
	Min	6.21	164.50	4.04	2.2	0.22	0.12
	Max	6.96	319.00	47.40	59.8	26.26	2.37

Source: Sthiannopkao, S. et al., *Appl. Geochem.*, 23, 1086, 2008.
Tot As, Total As; SD, Standard deviation; Max, Maximum; Min, Minimum.

was extremely polluted. This might be due to the low-relief topography with organic-rich Holocene alluvial sediment deposits (Buschmann et al. 2007). ^{14}C dating and regional history showed that the age of aquifer, and associated sedimentary organic carbon, was greater than 6000 years, and the average annual clay layer deposit was found to be 3.3 mm (Polizzotto et al. 2008). Arsenic concentrations in groundwater ranged from 247.08 to 1841.50 μg/L (n = 46, average 846.14 μg/L), with all of the observed wells exceeding the Cambodian drinking water standard of 50 μg/L; Mn concentrations were 584.23 ± 516.49 μg/L (mean ± σ) with 52% > 400 μg/L; Fe concentrations were 5901.93 ± 3017.64 μg/L (mean ± σ), with 100% exceeding the 300 μg/L regulation; and Ba concentrations were 1028.26 ± 477.75 μg/L (mean ± σ), with 74% > 700 μg/L. Out of the 12 observed wells in the Kratie province study area, a quarter were found with elevated arsenic levels and another quarter had Mn > 400 μg/L, whereas groundwater samples from the Kampong Cham study area were relatively clean, with arsenic concentrations less than WHO's guideline of 10 μg/L and no toxic trace elements found with elevated levels.

Common manganese minerals are secondary deposits of oxides MnO_2 (pyrolusite) and Mn_3O_4 (hausmanite) and carbonates $MnCO_3$ (rhodochrosite) (Greenwood and Earnshaw 1997). However, the mechanisms involved in manganese release to the aqueous phase remain unclear. Recently, it has been reported that, under anaerobic conditions, some microorganisms can utilize Mn (IV) as an electron acceptor to oxidize elemental sulfur to sulfate (Prescott et al. 2002). This microbial activity might link the sulfur cycle to the Mn cycle. This finding supports the notion that MnO_2 is reduced; Mn (II) is consequently released to pore water under the reducing conditions, which could be harmful to human beings, as in the cases of the Kandal and Kratie province study areas. Moreover, human and other animals can acquire Mn, an essential element for metabolism pathways, through many food sources (WHO 2004). However, an excess or deficiency of Mn can cause adverse effects. Neurological disorders resulting from drinking very high levels of Mn have been reported in epidemiological studies (WHO 2004). Barium is a trace element present in igneous and sedimentary rocks. The most common barium mineral is $BaSO_4$ (Barite) (Greenwood and Earnshaw 1997). To date, barium has not been proven to be carcinogenic or mutagenic, although drinking barium-contaminated water might lead to hypertension (WHO 2004).

Arsenic Risk Assessment

A health risk assessment model derived from the USEPA (Integrated Risk Information System—IRIS: arsenic, inorganic, CASRN 7440-38-2, 1998) was applied to compute the noncarcinogenic and carcinogenic effects on individuals who consume groundwater as their drinking water source:

$$\text{ADD} = \frac{\text{As}_w \times \text{IR} \times \text{EF} \times \text{ED}}{\text{AT} \times \text{BW}} \qquad (5.1)$$

where
    ADD is the average daily dose from ingestion (mg/kg/day)
    As_w is the arsenic concentration in water (mg/L)
    IR is the water ingestion rate (L/day)
    EF is the exposure frequency (days/year)
    ED is the exposure duration (year)
    AT is the average time/life expectancy (days)
    BW is the body weight (kg)

$$\text{HQ} = \frac{\text{ADD}}{\text{RfD}} \qquad (5.2)$$

where
    HQ is the hazard quotient (risk is considered occurring if HQ > 1.00)
    RfD is the oral reference dose (RfD $= 3 \times 10^{-4}$ mg/kg/day)

$$R = 1 - \exp(-\text{SF} \times \text{ADD}) \qquad (5.3)$$

where SF is a slope factor, equal to 1.5 mg/kg/day.

The survey results, which were used in risk computation, are shown in Table 5.2, and calculated results of risk assessments are presented in Table 5.3. Although there was a significantly regional difference in arsenic uptake of residents in each of the study areas (Kruskal–Wallis test, $p < 0.0001$), no significant difference in gender (Mann–Whitney's U test, $p = 0.06 > 0.05$) and age groups (Kruskal–Wallis test, $p = 0.24 > 0.05$) were observed, suggesting that individual ADD of arsenic was not affected by the different gender and age groups of Cambodia residents; this is likely due to the regional differences of arsenic levels in groundwater. A positive significant correlation between average arsenic daily dose (ADD) and arsenic levels in groundwater (As_w) (r_s (568) $= 0.92$, $p < 0.01$) and a positive significant correlation of ADD with ED (r_s (568) $= 0.43$, $p < 0.0001$) were obtained, revealing that ADD was correlated with As_w and ED. In addition, negatively significant correlation between ADD and body weight (BW) (r_s (568) $= -0.12$, $p < 0.01$) was also observed. However, no significant association between ADD and IR was found. This is more likely due to high IR in low arsenic contaminated areas (Table 5.2). Recently, Nguyen et al. (2009) reported that Vietnam residents who consumed untreated groundwater (As > 100 µg/L) ingested arsenic at the rate of 1.1×10^{-3} – 4.3×10^{-3} mg As/kg/day. Saipan and Ruangwises (2009) used a duplicate diet study to show that Ronphibun residents of Thailand ingested an average 2.1×10^{-3} mg As/kg/day. The present study clearly indicates that residents in the Kandal province study area of Cambodia ingest

TABLE 5.2

Summary of Specific Study Area, Sex and Age Groups of Body Weight, Age, Ingestion Rate, and Exposure Duration

Sex	Age Groups		Kandal				Kratie				Kampong Cham			
			BW	Age	IR	ED	BW	Age	IR	ED	BW	Age	IR	ED
Female	1–9	N	12				3				15			
		Mean	15.0	6.4	0.9	3.8	18.7	7.7	1.0	5.7	15.5	5.7	0.9	4.0
		Median	15.0	6.5	1.0	3.0	19.0	8.0	1.0	6.0	15.0	6.0	1.0	5.0
		SD	3.7	1.8	0.2	1.7	4.5	1.5	0.0	1.5	2.8	2.0	0.2	1.6
		Min	7.0	3.0	0.5	1.0	14.0	6.0	1.0	4.0	12.0	2.0	0.5	1.0
		Max	20.0	9.0	1.0	7.0	23.0	9.0	1.0	7.0	20.0	9.0	1.0	6.0
	10–19	N	38				12				18			
		Mean	32.6	13.3	1.3	7.1	41.1	15.0	1.3	10.8	45.2	16.2	1.6	4.5
		Median	31.5	13.0	1.0	8.0	43.5	16.0	1.0	11.0	48.0	17.0	1.5	5.0
		SD	10.6	2.9	0.4	3.6	11.3	2.8	0.3	1.8	10.5	3.1	0.4	2.8
		Min	19.0	10.0	1.0	1.0	23.0	10.0	1.0	8.0	21.0	10.0	1.0	1.0
		Max	59.0	19.0	2.5	14.0	57.0	18.0	2.0	13.0	58.0	19.0	3.0	10.0
	20–29	N	35				7				24			
		Mean	48.6	24.7	1.6	10.0	48.7	24.9	1.5	11.3	48.8	24.6	1.9	6.0
		Median	48.0	25.0	1.5	11.0	49.0	27.0	1.5	12.0	48.0	25.0	2.0	5.0
		SD	5.2	3.1	0.5	4.4	5.6	4.3	0.3	1.4	7.2	2.6	0.5	3.0
		Min	39.5	20.0	1.0	1.0	40.0	20.0	1.0	9.0	40.0	20.0	1.0	1.0
		Max	59.0	29.0	3.0	19.0	55.0	29.0	2.0	13.0	73.0	29.0	3.0	12.0
	30–39	N	25				3				16			
		Mean	47.8	35.1	1.8	6.9	54.7	32.3	1.7	10.0	50.6	33.4	1.8	5.1
		Median	48.0	36.0	2.0	6.0	52.0	30.0	1.5	10.0	48.0	33.5	2.0	4.5

(continued)

TABLE 5.2 (continued)

Summary of Specific Study Area, Sex and Age Groups of Body Weight, Age, Ingestion Rate, and Exposure Duration

Sex	Age Groups		Kandal				Kratie				Kampong Cham			
			BW	Age	IR	ED	BW	Age	IR	ED	BW	Age	IR	ED
		SD	6.4	3.5	0.5	3.8	6.4	4.0	0.3	2.0	10.0	2.9	0.5	3.7
		Min	38.0	30.0	1.0	2.0	50.0	30.0	1.5	8.0	37.0	30.0	1.0	1.0
		Max	61.0	39.0	3.0	15.0	62.0	37.0	2.0	12.0	68.0	38.0	2.5	12.0
	40–49	N		20				10				12		
		Mean	53.2	44.1	1.8	8.7	53.7	43.4	1.3	10.6	54.3	43.3	2.0	4.5
		Median	52.8	43.5	1.5	9.0	55.0	43.0	1.3	12.0	56.0	42.5	2.0	4.5
		SD	11.8	2.9	0.6	5.2	6.4	1.8	0.3	2.3	8.8	3.1	0.5	2.6
		Min	35.0	40.0	1.0	1.0	44.0	41.0	1.0	5.0	38.0	40.0	1.5	1.0
		Max	89.0	48.0	3.5	19.0	65.0	47.0	2.0	12.0	63.0	49.0	3.0	9.0
	50–59	N		28				9				14		
		Mean	47.5	54.6	1.7	8.6	48.2	54.2	1.6	10.7	57.5	54.1	2.1	6.6
		Median	46.0	55.0	1.5	8.0	47.0	55.0	1.5	11.0	55.5	54.0	2.0	6.5
		SD	7.6	3.6	0.5	4.2	6.1	2.2	0.4	2.4	9.8	3.1	0.5	3.8
		Min	34.0	50.0	1.0	1.0	41.0	50.0	1.0	5.0	43.0	50.0	1.5	1.0
		Max	64.0	59.0	3.0	16.0	58.0	57.0	2.0	13.0	73.0	58.0	3.0	12.0
	60–69	N		4				7						
		Mean	43.4	63.8	1.4	6.5	43.6	64.9	1.4	11.0	57.7	64.3	2.4	5.0
		Median	44.0	64.0	1.5	6.5	40.0	65.0	1.5	12.0	56.0	65.0	2.0	5.0

Sex	Age	Stat												
	70+	SD	6.8	3.8	0.3	4.7	8.0	3.0	0.2	1.6	10.2	3.7	0.9	3.8
		Min	34.5	60.0	1.0	2.0	35.0	60.0	1.0	9.0	47.0	60.0	1.5	1.0
		Max	51.0	67.0	1.5	11.0	56.0	69.0	1.5	13.0	78.0	69.0	4.0	12.0
		N	7				5				8			
Male	1–9	Mean	44.9	75.4	1.4	7.1	45.0	75.2	1.4	11.6	51.6	75.4	1.9	2.9
		Median	44.5	73.0	1.0	7.0	44.0	74.0	1.5	12.0	49.5	75.5	2.0	1.5
		SD	5.8	5.0	0.5	4.5	3.7	4.5	0.4	0.9	11.4	5.3	0.6	2.4
		Min	39.0	71.0	1.0	1.0	41.0	71.0	1.0	10.0	35.0	70.0	1.0	1.0
		Max	55.0	84.0	2.0	14.0	51.0	80.0	2.0	12.0	67.0	85.0	3.0	7.0
		N	21				5				17			
	10–19	Mean	17.2	7.1	1.1	4.9	17.8	6.6	1.0	4.6	16.9	6.2	1.1	3.6
		Median	17.0	7.0	1.0	5.0	18.0	7.0	1.0	5.0	17.0	6.0	1.0	3.0
		SD	2.8	1.3	0.3	1.4	1.1	0.5	0.0	0.5	3.3	1.6	0.2	2.2
		Min	12.0	5.0	0.5	2.0	16.0	6.0	1.0	4.0	11.0	3.0	1.0	1.0
		Max	23.0	9.0	2.0	7.0	19.0	7.0	1.0	5.0	24.0	9.0	1.5	8.0
		N	33				10				19			
	20–29	Mean	35.2	14.4	1.5	7.9	42.5	14.3	1.6	9.8	33.9	13.6	1.7	5.4
		Median	35.0	15.0	1.5	8.0	42.5	14.5	1.5	10.0	33.0	14.0	1.5	5.0
		SD	9.1	2.6	0.5	3.7	12.8	2.9	0.4	2.4	9.2	2.3	0.5	2.7
		Min	22.0	10.0	1.0	2.0	25.0	10.0	1.0	5.0	20.0	10.0	1.0	1.0
		Max	52.0	18.0	2.5	15.0	59.0	19.0	2.0	13.0	52.0	18.0	3.0	10.0
		N	17				N/A				7			
		Mean	53.9	23.5	1.9	9.4	N/A	N/A	N/A	N/A	54.0	22.1	2.1	6.0
		Median	54.0	22.0	2.0	12.0	N/A	N/A	N/A	N/A	53.0	22.0	2.0	7.0
		SD	6.8	2.9	0.4	4.6	N/A	N/A	N/A	N/A	4.6	2.1	0.2	3.1
		Min	39.0	20.0	1.5	2.0	N/A	N/A	N/A	N/A	50.0	20.0	2.0	2.0
		Max	67.0	29.0	3.0	16.0	N/A	N/A	N/A	N/A	62.0	26.0	2.5	10.0

(continued)

TABLE 5.2 (continued)

Summary of Specific Study Area, Sex and Age Groups of Body Weight, Age, Ingestion Rate, and Exposure Duration

Sex	Age Groups		Kandal				Kratie				Kampong Cham			
			BW	Age	IR	ED	BW	Age	IR	ED	BW	Age	IR	ED
	30–39	N	15				2				9			
		Mean	54.2	33.1	2.0	8.1	65.0	33.5	1.8	12.0	58.4	32.9	2.6	4.7
		Median	54.0	32.0	2.0	6.0	65.0	33.5	1.8	12.0	55.0	32.0	3.0	5.0
		SD	4.9	3.5	0.2	4.2	12.7	4.9	0.4	0.0	8.6	2.8	0.8	2.3
		Min	45.0	30.0	1.5	2.0	56.0	30.0	1.5	12.0	50.0	30.0	1.5	1.0
		Max	63.0	39.0	2.5	15.0	74.0	37.0	2.0	12.0	78.0	37.0	4.0	7.0
	40–49	N	13				6				8			
		Mean	55.0	44.9	2.2	8.3	60.2	45.7	2.1	9.7	56.4	44.4	2.5	4.9
		Median	55.0	46.0	2.0	8.0	60.5	46.5	2.0	10.0	55.0	44.0	2.5	4.5
		SD	10.3	3.4	0.6	3.5	6.0	2.8	0.4	2.7	10.5	2.8	0.7	3.0
		Min	35.0	40.0	1.0	2.0	52.0	41.0	1.5	5.0	45.0	41.0	1.5	1.0
		Max	79.0	49.0	3.0	14.0	68.0	49.0	2.5	13.0	72.0	49.0	3.5	10.0
	50–59	N	14				3				3			
		Mean	54.4	53.1	2.1	9.2	58.7	56.0	1.7	12.0	66.0	52.7	3.2	6.0
		Median	53.5	52.0	2.0	8.0	54.0	56.0	1.5	12.0	60.0	53.0	3.0	5.0
		SD	6.5	2.5	0.5	5.6	13.6	1.0	0.3	1.0	12.2	1.5	0.8	2.6
		Min	45.5	50.0	1.0	2.0	48.0	55.0	1.5	11.0	58.0	51.0	2.5	4.0
		Max	70.0	57.0	3.0	19.0	74.0	57.0	2.0	13.0	80.0	54.0	4.0	9.0

		9				5					4		
60–69	N												
	Mean	56.1	64.0	2.3	8.6	58.8	63.2	1.7	9.4	61.3	64.8	2.6	4.5
	Median	55.0	63.0	2.0	8.0	63.0	63.0	2.0	10.0	62.0	64.5	2.5	5.0
	SD	7.9	2.8	0.8	4.7	8.2	2.6	0.4	2.9	10.4	2.5	0.8	2.5
	Min	45.0	60.0	1.5	2.0	50.0	60.0	1.0	5.0	48.0	62.0	2.0	1.0
	Max	68.5	68.0	4.0	15.0	67.0	67.0	2.0	13.0	73.0	68.0	3.5	7.0
		6				2					3		
70+	N												
	Mean	51.8	73.8	1.6	8.2	67.5	82.0	1.8	7.5	52.7	72.3	1.8	4.7
	Median	51.3	74.5	1.5	7.5	67.5	82.0	1.8	7.5	46.0	71.0	2.0	1.0
	SD	4.6	2.1	0.6	3.3	17.7	1.4	0.4	3.5	16.1	3.2	0.3	6.4
	Min	47.0	70.0	1.0	5.0	55.0	81.0	1.5	5.0	41.0	70.0	1.5	1.0
	Max	59.5	76.0	2.5	13.0	80.0	83.0	2.0	10.0	71.0	76.0	2.0	12.0

BW, Body weight (kg); Age (year); IR, Ingestion rate (L/day); ED, Exposure duration (Year); N/A, Not applicable.

TABLE 5.3

Summary of Specific Study Area, Sex, and Age Groups of Average Daily Dose of Arsenic, Hazard Quotient, and Cancer Risk Probability

Sex	Age Groups	Statistics	Kandal			Kratie			Kampong Cham		
			ADD	HQ	R	ADD	HQ	R	ADD	HQ	R
Female	1–9	N	12			3			15		
		Mean	2.3E–03	7.6E+00	3.4E–03	1.8E–05	5.9E–02	2.7E–05	5.4E–06	1.8E–02	8.1E–06
		Median	2.3E–03	7.8E+00	3.5E–03	2.3E–05	7.8E–02	3.5E–05	4.6E–06	1.5E–02	6.8E–06
		SD	1.4E–03	4.7E+00	2.1E–03	1.2E–05	3.9E–02	1.8E–05	3.8E–06	1.3E–02	5.7E–06
		Min	2.0E–04	6.7E–01	3.0E–04	4.2E–06	1.4E–02	6.2E–06	5.8E–07	1.9E–03	8.7E–07
		Max	5.8E–03	1.9E+01	8.7E–03	2.6E–05	8.6E–02	3.9E–05	1.2E–05	4.1E–02	1.8E–05
	10–19	N	38			12			18		
		Mean	3.2E–03	1.1E+01	4.8E–03	5.2E–05	1.7E–01	7.8E–05	4.1E–06	1.4E–02	6.2E–06
		Median	2.2E–03	7.5E+00	3.4E–03	6.1E–06	2.0E–02	9.2E–06	4.6E–06	1.5E–02	6.9E–06
		SD	2.4E–03	8.1E+00	3.6E–03	6.6E–05	2.2E–01	9.9E–05	2.9E–06	9.8E–03	4.4E–06
		Min	1.9E–04	6.3E–01	2.8E–04	8.8E–07	2.9E–03	1.3E–06	6.2E–07	2.1E–03	9.2E–07
		Max	8.4E–03	2.8E+01	1.2E–02	1.6E–04	5.3E–01	2.4E–04	1.1E–05	3.5E–02	1.6E–05
	20–29	N	35			7			24		
		Mean	3.8E–03	1.3E+01	5.6E–03	1.0E–04	3.4E–01	1.5E–04	6.2E–06	2.1E–02	9.3E–06
		Median	3.5E–03	1.2E+01	5.3E–03	2.8E–05	9.2E–02	4.2E–05	6.5E–06	2.2E–02	9.7E–06
		SD	2.3E–03	7.8E+00	3.5E–03	1.8E–04	5.9E–01	2.6E–04	3.9E–06	1.3E–02	5.8E–06
		Min	3.4E–04	1.1E+00	5.1E–04	1.7E–06	5.6E–03	2.5E–06	4.5E–07	1.5E–03	6.8E–07
		Max	9.1E–03	3.0E+01	1.4E–02	4.8E–04	1.6E+00	7.2E–04	1.4E–05	4.6E–02	2.1E–05
	30–39	N	25			3			16		
		Mean	3.3E–03	1.1E+01	4.9E–03	1.4E–04	4.8E–01	2.2E–04	3.5E–06	1.2E–02	5.2E–06
		Median	2.8E–03	9.4E+00	4.2E–03	1.7E–04	5.7E–01	2.6E–04	1.9E–06	6.5E–03	2.9E–06

	C1	C2	C3	C4	C5	C6	C7	C8	C9
SD	2.7E-03	9.1E+00	4.1E-03	1.3E-04	4.4E-01	2.0E-04	3.9E-06	1.3E-02	5.8E-06
Min	4.0E-04	1.3E+00	5.9E-04	1.6E-06	5.4E-03	2.4E-06	5.3E-07	1.8E-03	7.9E-07
Max	1.1E-02	3.6E+01	1.6E-02	2.6E-04	8.7E-01	3.9E-04	1.5E-05	5.0E-02	2.3E-05
40-49 N		20			10			12	
Mean	3.7E-03	1.2E+01	5.5E-03	1.7E-04	5.5E-01	2.5E-04	5.1E-06	1.7E-02	7.6E-06
Median	3.0E-03	1.0E+01	4.5E-03	7.8E-05	2.6E-01	1.2E-04	4.7E-06	1.6E-02	7.0E-06
SD	2.8E-03	9.3E+00	4.2E-03	2.1E-04	7.1E-01	3.2E-04	4.2E-06	1.4E-02	6.2E-06
Min	2.1E-04	7.1E-01	3.2E-04	8.3E-07	2.8E-03	1.3E-06	3.1E-07	1.0E-03	4.7E-07
Max	8.6E-03	2.9E+01	1.3E-02	6.3E-04	2.1E+00	9.4E-04	1.1E-05	3.8E-02	1.7E-05
50-59 N		28			9			14	
Mean	3.3E-03	1.1E+01	5.0E-03	3.2E-05	1.1E-01	4.7E-05	5.8E-06	1.9E-02	8.7E-06
Median	2.4E-03	8.1E+00	3.7E-03	2.7E-06	9.2E-03	4.1E-06	6.0E-06	2.0E-02	9.0E-06
SD	2.7E-03	9.0E+00	4.0E-03	7.9E-05	2.6E-01	1.2E-04	4.1E-06	1.4E-02	6.2E-06
Min	2.8E-04	9.3E-01	4.2E-04	1.6E-06	5.3E-03	2.4E-06	6.0E-07	2.0E-03	9.1E-07
Max	1.1E-02	3.5E+01	1.6E-02	2.4E-04	8.1E-01	3.6E-04	1.5E-05	5.0E-02	2.3E-05
60-69 N		4			7			7	
Mean	3.2E-03	1.1E+01	4.8E-03	1.9E-04	6.2E-01	2.8E-04	4.8E-06	1.6E-02	7.1E-06
Median	2.3E-03	7.6E+00	3.4E-03	3.7E-05	1.2E-01	5.5E-05	2.2E-06	7.5E-03	3.4E-06
SD	2.9E-03	9.7E+00	4.3E-03	2.6E-04	8.6E-01	3.9E-04	4.9E-06	1.6E-02	7.4E-06
Min	9.7E-04	3.2E+00	1.5E-03	3.9E-07	1.3E-03	5.8E-07	5.9E-07	2.0E-03	8.9E-07
Max	7.3E-03	2.4E+01	1.1E-02	5.7E-04	1.9E+00	8.5E-04	1.3E-05	4.3E-02	1.9E-05
70+ N		7			5			8	
Mean	3.4E-03	1.1E+01	5.1E-03	1.2E-04	4.2E-01	1.9E-04	3.0E-06	1.0E-02	4.6E-06
Median	2.2E-03	7.5E+00	3.3E-03	3.3E-05	1.1E-01	4.9E-05	1.0E-06	3.5E-03	1.6E-06
SD	3.1E-03	1.0E+01	4.6E-03	1.6E-04	5.2E-01	2.4E-04	3.3E-06	1.1E-02	5.0E-06
Min	3.5E-04	1.2E+00	5.2E-04	2.5E-06	8.2E-03	3.7E-06	7.3E-07	2.4E-03	1.1E-06
Max	9.3E-03	3.1E+01	1.4E-02	3.7E-04	1.2E+00	5.5E-04	9.6E-06	3.2E-02	1.4E-05

(continued)

TABLE 5.3 (continued)

Summary of Specific Study Area, Sex, and Age Groups of Average Daily Dose of Arsenic, Hazard Quotient, and Cancer Risk Probability

Sex	Age Groups	Statistics	Kandal ADD	Kandal HQ	Kandal R	Kratie ADD	Kratie HQ	Kratie R	Kampong Cham ADD	Kampong Cham HQ	Kampong Cham R
Male	1–9	N		21			5			17	
		Mean	3.6E–03	1.2E+01	5.4E–03	4.7E–05	1.6E–01	7.1E–05	5.0E–06	1.7E–02	7.5E–06
		Median	3.3E–03	1.1E+01	5.0E–03	2.5E–05	8.3E–02	3.7E–05	2.7E–06	9.0E–03	4.1E–06
		SD	2.2E–03	7.2E+00	3.2E–03	4.3E–05	1.4E–01	6.4E–05	6.2E–06	2.1E–02	9.3E–06
		Min	7.7E–04	2.6E+00	1.2E–03	1.3E–06	4.4E–03	2.0E–06	6.8E–07	2.3E–03	1.0E–06
		Max	7.0E–03	2.3E+01	1.0E–02	9.8E–05	3.3E–01	1.5E–04	2.2E–05	7.4E–02	3.3E–05
	10–19	N		33			10			19	
		Mean	3.8E–03	1.3E+01	5.7E–03	1.3E–04	4.3E–01	1.9E–04	7.0E–06	2.3E–02	1.1E–05
		Median	3.1E–03	1.0E+01	4.6E–03	6.0E–06	2.0E–02	8.9E–06	7.4E–06	2.5E–02	1.1E–05
		SD	2.6E–03	8.7E+00	3.9E–03	2.2E–04	7.2E–01	3.2E–04	5.3E–06	1.8E–02	7.9E–06
		Min	6.6E–04	2.2E+00	1.0E–03	1.9E–06	6.2E–03	2.8E–06	9.0E–07	3.0E–03	1.4E–06
		Max	1.0E–02	3.5E+01	1.6E–02	5.7E–04	1.9E+00	8.5E–04	2.0E–05	6.6E–02	3.0E–05
	20–29	N		17			N/A			7	
		Mean	4.0E–03	1.3E+01	6.0E–03	N/A	N/A	N/A	5.6E–06	1.9E–02	8.5E–06
		Median	3.5E–03	1.2E+01	5.3E–03	N/A	N/A	N/A	4.3E–06	1.4E–02	6.5E–06
		SD	2.5E–03	8.4E+00	3.7E–03	N/A	N/A	N/A	4.0E–06	1.3E–02	6.0E–06
		Min	8.0E–04	2.7E+00	1.2E–03	N/A	N/A	N/A	1.3E–06	4.4E–03	2.0E–06
		Max	8.9E–03	3.0E+01	1.3E–02	N/A	N/A	N/A	1.1E–05	3.5E–02	1.6E–05
	30–39	N		15			2			9	
		Mean	3.7E–03	1.2E+01	5.5E–03	7.8E–05	2.6E–01	1.2E–04	7.8E–06	2.6E–02	1.2E–05
		Median	1.9E–03	6.4E+00	2.9E–03	7.8E–05	2.6E–01	1.2E–04	7.2E–06	2.4E–02	1.1E–05
		SD	3.0E–03	1.0E+01	4.5E–03	7.0E–05	2.3E–01	1.0E–04	6.0E–06	2.0E–02	9.0E–06

Age group	Statistic									
	Min	5.9E−04	2.0E+00	8.8E−04	2.9E−05	9.6E−02	4.3E−05	6.5E−07	2.2E−03	9.8E−07
	Max	9.4E−03	3.1E+01	1.4E−02	1.3E−04	4.2E−01	1.9E−04	2.0E−05	6.6E−02	3.0E−05
40–49	N	13				6			8	
	Mean	3.6E−03	1.2E+01	5.4E−03	1.6E−04	5.4E−01	2.4E−04	5.2E−06	1.7E−02	7.8E−06
	Median	3.9E−03	1.3E+01	5.8E−03	6.0E−06	2.0E−02	9.0E−06	4.0E−06	1.3E−02	6.0E−06
	SD	2.0E−03	6.7E+00	3.0E−03	2.6E−04	8.6E−01	3.9E−04	4.0E−06	1.3E−02	6.0E−06
	Min	8.0E−04	2.7E+00	1.2E−03	1.8E−06	6.1E−03	2.8E−06	1.1E−06	3.8E−03	1.7E−06
	Max	7.1E−03	2.4E+01	1.1E−02	6.1E−04	2.0E+00	9.1E−04	1.2E−05	3.9E−02	1.7E−05
50–59	N	14				3			3	
	Mean	3.7E−03	1.2E+01	5.5E−03	5.1E−05	1.7E−01	7.7E−05	7.3E−06	2.4E−02	1.1E−05
	Median	2.7E−03	9.1E+00	4.1E−03	4.0E−06	1.3E−02	5.9E−06	8.0E−06	2.7E−02	1.2E−05
	SD	2.9E−03	9.6E+00	4.3E−03	8.3E−05	2.8E−01	1.3E−04	3.2E−06	1.1E−02	4.8E−06
	Min	4.7E−04	1.6E+00	7.0E−04	2.0E−06	6.8E−03	3.1E−06	3.8E−06	1.3E−02	5.7E−06
	Max	9.7E−03	3.2E+01	1.4E−02	1.5E−04	4.9E−01	2.2E−04	1.0E−05	3.3E−02	1.5E−05
60–69	N	9				5			4	
	Mean	3.4E−03	1.1E+01	5.0E−03	6.6E−05	2.2E−01	9.9E−05	6.7E−06	2.2E−02	1.0E−05
	Median	3.0E−03	1.0E+01	4.5E−03	3.5E−06	1.2E−02	5.2E−06	7.8E−06	2.6E−02	1.2E−05
	SD	1.7E−03	5.7E+00	2.5E−03	1.4E−04	4.7E−01	2.1E−04	3.9E−06	1.3E−02	5.9E−06
	Min	1.2E−03	4.2E+00	1.9E−03	4.8E−07	1.6E−03	7.2E−07	1.0E−06	3.5E−03	1.6E−06
	Max	5.9E−03	2.0E+01	8.8E−03	3.2E−04	1.1E+00	4.8E−04	1.0E−05	3.3E−02	1.5E−05
70+	N	6				2			3	
	Mean	2.8E−03	9.3E+00	4.2E−03	3.4E−06	1.1E−02	5.1E−06	8.2E−07	2.7E−03	1.2E−06
	Median	2.9E−03	9.6E+00	4.3E−03	3.4E−06	1.1E−02	5.1E−06	7.9E−07	2.6E−03	1.2E−06
	SD	6.1E−04	2.0E+00	9.1E−04	4.8E−07	1.6E−03	7.1E−07	2.3E−07	7.5E−04	3.4E−07
	Min	1.8E−03	5.9E+00	2.6E−03	3.1E−06	1.0E−02	4.6E−06	6.1E−07	2.0E−03	9.2E−07
	Max	3.4E−03	1.1E+01	5.1E−03	3.8E−06	1.3E−02	5.6E−06	1.1E−06	3.5E−03	1.6E−06

ADD, Average daily dose (mg/kg/day); HQ, Hazard quotient; R, Cancer risk probability; SD, Standard deviation; Min, Minimum; Max, Maximum; N/A, Not applicable.

higher amount of arsenic than those in Vietnam and Thailand. The higher IRs of arsenic might result in more significantly adverse health effects. Computational results of risk revealed that 98.65% of respondents of Kandal posed the potential noncancer effect. In addition, cancer risk index was found to be 13.8% > 1 in 100 and 92.59% > 1 in 1000, which exceeded the highest safe standard for cancer risk, 1 in 10,000. The calculation also indicated that, in Kratie, 13.48% of respondents were affected by noncancer and 33.71% were threatened by cancer in comparison to the highest safe standard. It was also found that 93.48% of respondents of Kampong Cham appeared to be at risk for carcinogenic effects in comparison to the safe standard for cancer, that is, 1 in 1,000,000 (Figure 5.2). The present results corresponded with the rapid manifestation of arsenicosis symptoms discovered by Sampson et al. (2008). Nguyen et al. (2009) found that approximately 42% of Vietnam residents in the contaminated areas could be toxically affected by arsenic and that carcinogenic effect had the highest risk index, 5 in 1000. It is apparent that the residents in the study area of Kandal province, Cambodia, are suffering a much higher risk of noncancer and cancer effects. The highest risk indices were found 2 in 100, 9 in 10,000, and 3 in 100,000 in Kandal, Kratie, and Kampong Cham provinces, respectively. Saipan and Ruangwises (2009) through duplicate diet study reported that residents of Ronphibun, Thailand, might suffer significant health impact because the risk indices for cancer and noncancer were found to be 1.26×10^{-3} and 6.98, which exceeds the safe standard for cancer, 1 in 10,000, and typical toxic risk at 1.00, respectively. However, cases discovered in the study area of Kandal province, Cambodia, were much more risky because the residents in the study area of Kandal province were exposed to carcinogenic effect on an average of 5 in 1000, with toxic risk index at 11.67.

Arsenic Content in Scalp Hair

The results of ICP-MS analyses of acid-digested hair samples revealed that, in the Kandal province study area (n = 270), arsenic content in scalp hair (As_h) ranged from 0.27 to 57.21 µg/g, with mean and median of 6.40 and 4.03 µg/g, respectively. Approximately, 78.1% of this group had As_h greater than typical arsenic concentration in scalp hair (1.00 µg/g), indicating significant arsenic toxicity. Concurrently, in the Kratie province study area (n = 84), As_h ranged from 0.05 to 1.42 µg/g with mean of 0.29 µg/g and median of 0.24 µg/g, whereas As_h in the Kampong Cham province study area (n = 172) ranged from 0.01 to 1.01 µg/g with mean of 0.12 µg/g and median of 0.09 µg/g (Table 5.4). The upper end of the ranges for the last two groups were higher than the typical arsenic concentration, 1.00 µg/g; approximately 1.2% and 0.6% of residents in the Kratie and Kampong Cham province study areas, respectively, indicated arsenic toxicity. However, arsenic content in the scalp hair samples collected from the Kandal province residents can be found in other studies as well. Gault et al. (2008) reported that arsenic concentration in the

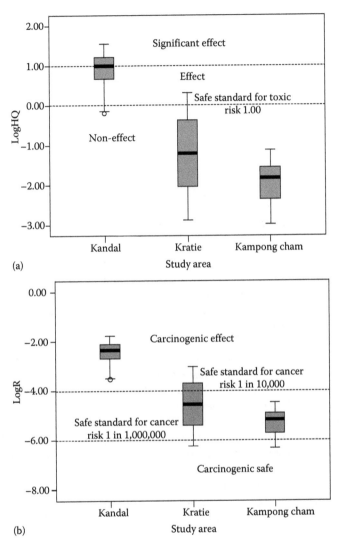

FIGURE 5.2
Risk incidence in the three study area provinces in the Mekong River basin of Cambodia through groundwater drinking pathway: (a) toxic risk and (b) carcinogenic risk. (From Sthiannopkao, S. et al., *Appl. Geochem.*, 23, 1086, 2008.)

scalp hair samples of Kandal province residents ranged from 0.10 to 7.95 µg/g (n = 40, median = 0.54 µg/g, mean = 1.41 µg/g); Sampson et al. (2008) reported that arsenic concentration in hair samples in Preak Russey village in Kandal province, where residents affected with arsenicosis patients were found, ranged from 2.1 to 13.94 µg/g (n = 36), with geometric mean of 5.64 µg/g; and Mazumder et al. (2009) reported that the arsenic level in scalp hair of the same Preak Russey residents, Kandal province, ranged from 0.92 to

TABLE 5.4

Mean, Median, Standard Deviation, and Min and Max of Arsenic Content in Scalp Hair in Each of the Study Areas (μg/g)

	Kandal	Kratie	Kampong Cham
N	270	84	173
Mean	6.40	0.29	0.12
Median	4.03	0.24	0.09
SD	8.01	0.21	0.10
Min	0.27	0.05	0.01
Max	57.21	1.42	1.01

Source: Sthiannopkao, S. et al., *Appl. Geochem.*, 23, 1086, 2008.
SD, Standard deviation; Min, Minimum; Max, Maximum.

25.60 μg/g (n = 93). The present study reveals elevated arsenic content not only in scalp hair samples of Preak Russey residents but also in the samples collected from their neighborhood village, Lvea Toung village, in Kampong Kong commune, Koh Thom district, Kandal province, which indicates that the residents in the Kandal province study area are at high risk of arsenico- sis. Similarly, Kubota et al. (2006) reported that the mean value of arsenic content in Kratie residents' scalp hair samples was 1.77 ± 2.94 μg/g (mean ± σ); in this case, 42.6% of hair samples exceeded the level of possible indication of arsenic toxicity (1.00 μg/g), whereas in the present study only 1.2% of the samples indicated an elevated level of arsenic.

The variations of arsenic content in scalp hair (As_h) are more likely due to differences in arsenic levels in groundwater (As_w) and ADD, which would lead to significantly different arsenic accumulation levels. Indeed, ground- water arsenic levels in the Kandal province study area in the present study ranged from 247.08 to 1841.5 μg/L (Table 5.1), a finding that is in agreement with some other studies that reported arsenic levels in comparable range; Gault et al. (2008) reported that the arsenic concentration in the groundwater of Kandal province was in the range of 0.21–943 μg/L, and Mazumder et al.'s (2009) findings show that to be in the range of 148–286 μg/L. Similarly, in the present study, groundwater arsenic levels in the Kratie province study area ranged from 0.12 to 140.60 μg/L (Table 5.1), whereas a previous study reported that the presence of groundwater arsenic was in the range of <1–886 μg/L (Kubota et al. 2006).

Positive significant correlations between arsenic content in scalp hair (As_h) and arsenic level in groundwater (As_w) (r_s (525) = 0.75, $p < 0.0001$) (Figure 5.3, Table 5.5) and ADD (r_s (525) = 0.74, $p < 0.0001$) (Figure 5.4, Table 5.5) were observed. No significant differences in gender (Mann–Whitney's U test, $p = 0.45 > 0.05$) and age groups (Kruskal–Wallis test, $p = 0.92 > 0.05$) were found, but significant regional differences (Kruskal–Wallis test, $p < 0.0001$) in the arsenic content of scalp hair were observed, undoubtedly suggesting

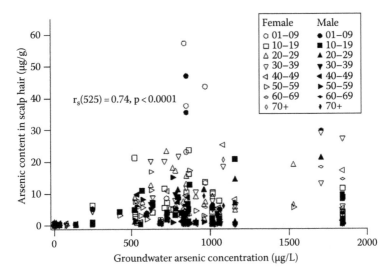

FIGURE 5.3
Variation of arsenic content in scalp hair (As$_h$) with arsenic concentration in groundwater (As$_w$). (From Sthiannopkao, S. et al., *Appl. Geochem.*, 23, 1086, 2008.)

that the arsenic accumulation in residents' scalp hair, which happens mainly through a groundwater drinking pathway, appeared uninfluenced by individual gender and age. A positive significant correlation between arsenic content in scalp hair (As$_h$) and ED (r$_s$ (525)=0.15, p<0.01) and a positive, nonsignificant correlation with gender (r$_s$ (525)=0.03, p=0.45>0.05) and age group (r$_s$ (525)=0.01, p=0.84>0.05) (Table 5.5) demonstrated that arsenic accumulation in scalp hair might be induced by ED rather than individual gender and age. However, negative significant correlations of As$_h$ with body weight (r$_s$ (525)=−0.10, p=0.027<0.05) and IR (r$_s$ (525)=−0.10, p=0.026<0.05) were obtained. The negative association between As$_h$ and IR might be due to high ingestion in low-arsenic-contaminated areas. Our survey questionnaire results showed that the residents in the Kampong Cham province study area drink groundwater from 0.5 to 4.0 L/day (Table 5.2). In West Bengal (India), Samanta et al. (2004) reported that the arsenic content in the scalp hair of study participants ranged from 0.17 to 14.4 μg/g (n=44, mean=3.43 μg/g, and median=2.29 μg/g), whereas Mandal et al. (2003) reported arsenic content in scalp hair samples to be in a range of 0.70–16.2 μg/g (n=47, mean=4.50 μg/g). In Bangladesh, Karim (2000) presented that arsenic content found in scalp hair samples ranged from 1.1 to 19.84 μg/g. Agusa et al. (2006) reported that the arsenic content in scalp hair of Vietnam residents ranged from 0.09 to 2.77 μg/g. The ranges of arsenic content in scalp hair samples analyzed in Bangladesh and West Bengal (India) are comparable to that of our present study in Kandal province, whereas the case of Vietnam is comparable to our present study in Kratie and Kampong Cham provinces.

TABLE 5.5

Intercorrelation, Mean, and Standard Deviation for As_h, As_w, ADD, HQ, R, BW, IR, ED, Specific Sex, and Specific Age Groups

Variables	As_w	ADD	HQ	R	BW	IR	ED	Sex	Age Groups	Mean	SD
As_h	0.75**	0.74**	0.74**	0.74**	−0.10*	−0.10*	0.15*	0.03	0.01	3.36	6.52
As_w	—	0.92**	0.92**	0.92**	−0.06	−0.07	0.19**	0.01	0.01	507.98	564.76
ADD	—	—	1.00**	1.00**	−0.12**	−0.01	0.43**	0.08	−0.03	0.0019	0.0025
HQ	—	—	—	1.00**	−0.12**	−0.01	0.43**	0.08	−0.03	6.20	8.35
R	—	—	—	—	0.12**	−0.01	0.43**	0.08	−0.03	0.0028	0.0037
BW	—	—	—	—	—	0.60**	0.06	0.02	0.62**	44.03	15.21
IR	—	—	—	—	—	—	−0.05	0.14**	0.46**	1.66	0.62
ED	—	—	—	—	—	—	—	−0.01	0.12**	7.43	4.08
Sex	—	—	—	—	—	—	—	—	−0.12**	0.41	0.49
Age groups	—	—	—	—	—	—	—	—	—	2.78	2.06

Source: Sthiannopkao, S. et al., *Appl. Geochem.*, 23, 1086, 2008.

As_h, hair arsenic; ADD, average daily dose of arsenic; HQ, hazard quotient; R, cancer risk; BW, body weight; IR, ingestion rate; ED, exposure duration; SD, standard deviation.

* $p < 0.05$.

** $p < 0.01$.

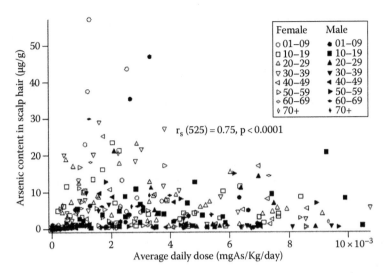

FIGURE 5.4
Variation of arsenic content in scalp hair (As$_h$) with ADD (average daily dose) of arsenic. (From Sthiannopkao, S. et al., *Appl. Geochem.*, 23, 1086, 2008.)

Mazumder et al. (2009) found that approximately 72% of Preak Russey residents showed evidences of arsenical skin lesion (pigmentation and/or keratosis) and the largest number of cases belonged to those in the age group of 31–45 years. Concurrently, 37% of children aged less than 16 years had skin lesions resulting from arsenicosis. The youngest child having the evidence of Keratosis and pigmentation was 8 years old, although features such as redness and mild thickening of the palms were found only in children aged between 4 and 5 years (Mazumder et al. 2009). This finding corresponds to our present study results that arsenic contents in scalp hair were found highest in children in the age range of 1–9 followed by adults in the age group of 30–39 years.

Although there was a statistically significant positive correlation between As$_h$ and ADD, the scatter plot of Figure 5.4 indicates that the arsenic accumulation rate varies among individuals. Such variations—for example, "low ADD, but high arsenic content in scalp hair" or "high ADD, but low arsenic content in scalp hair"—might be due to family socioeconomic status, malnutrition level, individual health status, hygiene, and/or habits in drinking water maintenance and storage before consumption. Sampson et al. (2008) reported that higher family socioeconomic status and better nutrition could reduce physical susceptibility and delay the manifestation of arsenicosis symptoms. Moreover, healthier residents with sufficient macro- and micronutrients might more efficiently remove arsenic from their bodies (Sampson et al. 2008). However, some residents might ingest an excessive amount of arsenic, not only through groundwater drinking pathway but also through their daily diet. In fact, some residents in the study areas have used arsenic-rich groundwater to irrigate their farms as well as their rice fields. Therefore, the actual ingestion of arsenic content of this group would be much higher than that of the groundwater drinking pathway alone. Concurrently, in practice, some residents did not consume instantly pumped groundwater. Pumped groundwater may be stored for a period of time in traditional water storage containers, such as open rainwater jars, which may lead to natural oxidation and precipitation processes that lower arsenic levels; moreover, those who have more jars might store rainwater for use throughout the dry season. In short, a number of factors may play a role in arsenic accumulation in the bodies of residents in the Mekong River basin of Cambodia.

Conclusions

Analytical results demonstrated that groundwater in Kandal was more significantly enriched with As, Mn, Fe, and Ba than that in Kratie and Kampong Cham provinces. Consequently, the computation of risks through drinking water pathway indicated that the residents of Kandal might be exposed to more elevated toxic and carcinogenic risks than those of Kratie and Kampong Cham. Positive significant correlation between arsenic content in scalp hair and arsenic level in groundwater and individual ADD of arsenic undoubtedly suggested that arsenic accumulation in Cambodia residents' bodies was mainly through groundwater drinking pathway. Rapid development of arsenicosis symptoms found during a field sampling was closely correlated with other risk factors such as extremely high arsenic level in groundwater (As_w), ADD, and ED. Arsenic accumulation rate might vary among individuals and/or households owing to family socioeconomic status and numerous other factors, including malnutrition, individual health status, hygiene, and good habits in drinking water maintenance and storage, which could lower the arsenic levels through natural oxidation and precipitation processes. Deleterious As concentration found in groundwater sources in the Mekong river basin of Cambodia might lead to thousands of cases of arsenicosis in the near future if remedial actions are not taken.

Acknowledgments

This project was financially supported by a joint research project of UNU&GIST, UNU-IIGH, and Hong Kong Baptist University.

References

Agusa T, Kunito T, Fujihara J, Kubota R, Minh TB, Kim Trang PT et al. 2006. Contamination by arsenic and other trace elements in tube-well water and its risk assessment to humans in Hanoi, Vietnam. *Environmental Pollution* 139: 95–106.

ATSDR. 2007. *Toxicological Profile for Arsenic.* US Department of Health and Human Services, Public Health Service, Agency for Toxic Substances and Disease Registry (ATSDR), Atlanta, GA.

Benner SG, Polizzotto ML, Kocar BD et al. 2008. Groundwater flow in an arsenic-contaminated aquifer, Mekong Delta, Cambodia. *Applied Geochemistry* 23: 3072–3087.

Berg M, Stengel C, Pham TK, Pham HV, Sampson ML, Leng M et al. 2007. Magnitude of arsenic pollution in the Mekong and Red River Deltas—Cambodia and Vietnam. *The Science of the Total Environment* 372: 413–425.

Berg M, Tran HC, Nguyen TC, Pham HV, Schertenleib R, Giger W. 2001. Arsenic contamination of groundwater and drinking water in Vietnam: A human health threat. *Environmental Science and Technology* 35: 2621–2626.

Buschmann J, Berg M, Stengel C, Sampson ML. 2007. Arsenic and manganese contamination of drinking water resources in Cambodia: Coincidence of risk areas with low relief topography. *Environmental Science and Technology* 41: 2146–2152.

Chiew H, Sampson ML, Huch S, Ken S, Bostick BC. 2009. Effect of groundwater iron and phosphate on the efficacy of arsenic removal by iron-amended BioSand filters. *Environmental Science and Technology* 43: 6295–6300.

Das D, Chatterjee A, Samanta G, Mandal B, Chowdhury TR, Samanta G, Chowdhury PP et al. 1994. Arsenic contamination in groundwater in six districts of West Bengal, India: The biggest arsenic calamity in the world. *Analyst* 119: 168N–170N.

Gault AG, Rowland HA, Charnock JM, Wogelius RA, Gomez-Morilla I, Vong S et al. 2008. Arsenic in hair and nails of individuals exposed to arsenic-rich groundwaters in Kandal province, Cambodia. *The Science of the Total Environment* 393: 168–176.

Greenwood NN, Earnshaw A. 1997. *Chemistry of the Elements* (2nd edn.). Butterworth-Heinemann, Boston, MA.

Hughes MF. 2002. Arsenic toxicity and potential mechanisms of action. *Toxicology Letters* 133: 1–16.

Japanese International Cooperation Agency (JICA). 1999. The Study of Groundwater Development in Southern Cambodia. Unpublished Draft Report by Kokusai Kogyo Co Ltd to Ministry of Rural Development, Phnom Penh, Cambodia.

Karim M. 2000. Arsenic in groundwater and health problems in Bangladesh. *Water Research* 34: 304–310.

Kocar BD, Polizzotto ML, Benner SG, Ying SC, Ung M, Ouch K, Samreth S et al. 2008. Integrated biogeochemical and hydrologic processes driving arsenic release from shallow sediments to groundwaters of the Mekong delta. *Applied Geochemistry* 23: 3059–3071.

Kubota R, Kunito T, Agusa T, Fujihara J, Monirith I, Iwata H et al. 2006. Urinary 8-hydroxy-2'-deoxyguanosine in inhabitants chronically exposed to arsenic in groundwater in Cambodia. *Journal of Environmental Monitoring* 8: 293–299.

Lear G, Song B, Gault AG, Polya DA, Lloyd JR. 2007. Molecular analysis of arsenate-reducing bacteria within Cambodian sediments following amendment with acetate. *Applied and Environmental Microbiology* 73: 1041–1048.

Luu TT, Sthiannopkao S, Kim KW. 2009. Arsenic and other trace elements contamination in groundwater and a risk assessment study for the residents in the Kandal Province of Cambodia. *Environment International* 35: 455–460.

Mandal BK, Chowdhury TR, Samanta G, Mukherjee DP, Chanda CR, Saha KC et al. 1998. Impact of safe water for drinking and cooking on five arsenic-affected families for 2 years in West Bengal, India. *The Science of the Total Environment* 218: 185–201.

Mandal BK, Ogra Y, Suzuki KT. 2003. Speciation of arsenic in human nail and hair from arsenic-affected area by HPLC-inductively coupled argon plasma mass spectrometry. *Toxicology and Applied Pharmacology* 189: 73–83.

Mazumder DN, Majumdar KK, Santra SC, Kol H, Vicheth C. 2009. Occurrence of arsenicosis in a rural village of Cambodia. *Journal of Environmental Science and Health Part A—Toxic/Hazardous Substances and Environmental Engineering* 44: 480–487.

Nguyen VA, Bang S, Viet PH, Kim KW. 2009. Contamination of groundwater and risk assessment for arsenic exposure in Ha Nam province, Vietnam. *Environment International* 35: 466–472.

Nickson R, McArthur J, Burgess W, Ahmed KM, Ravenscroft P, Rahman M. 1998. Arsenic poisoning of Bangladesh groundwater. *Nature* 395: 338.

Polizzotto ML, Kocar BD, Benner SG, Sampson M, Fendorf S. 2008. Near-surface wetland sediments as a source of arsenic release to ground water in Asia. *Nature* 454(7203): 505–508.

Polya DA, Gault AG, Bourne NJ, Lythgoe PR, Cooke DA. 2003. Coupled HPLC-ICP-MS analysis indicates highly hazardous concentrations of dissolved arsenic species in Cambodian groundwaters. In: Holland, J.G., Tanners, S.D. (Eds.), *Plasma Source Mass Spectrometry: Applications and Emerging Technologies*, vol. 288. The Royal Society of Chemistry Special Publication, Cambridge, U.K., pp. 127–140.

Polya DA, Gault AG, Diebe N, Feldman P, Rosenboom JW, Gilligan E, Fredericks D et al. 2005. Arsenic hazard in shallow Cambodian groundwaters. *Mineralogical Magazine* 69(5):807–823.

Prescott LM, Harley JP, Klein DA. 2002. *Microbiology* (5th edn.). McGraw-Hill, New York.

Quicksall AN, Bostick BC, Sampson ML. 2008. Linking organic matter deposition and iron mineral transformations to groundwater arsenic levels in the Mekong Delta Cambodia. *Applied Geochemistry* 23: 3088–3098.

Robinson DA, Lebron I, Kocar B et al. 2009. Time-lapse geophysical imaging of soil moisture dynamics in tropical deltaic soils: An aid to interpreting hydrological and geochemical processes. *Water Resources Research* 45: 1–12.

Rowland HAL, Gault AG, Lythgoe PR, Polya DA. 2008. Geochemistry of aquifer sediment and arsenic-rich groundwater from Kandal Province, Cambodia. *Applied Geochemistry* 23: 3029–3046.

Ryabukhin YS. 1978. Activation analysis of hair as an indicator of contamination of man by environmental trace element pollutants. IAEA report, IAEA/RL/50, Vienna, Austria.

Saipan P, Ruangwises S. 2009. Health risk assessment of inorganic arsenic intake of Ronphibun residents via duplicate diet study. *Journal of The Medical Association of Thailand* 92: 849–855.

Samanta G, Sharma R, Roychowdhury T, Chakraborti D. 2004. Arsenic and other elements in hair, nails, and skin-scales of arsenic victims in West Bengal, India. *The Science of the Total Environment* 326: 33–47.

Sampson ML, Bostick B, Chiew H, Hangan JM, Shantz A. 2008. Arsenicosis in Cambodia: Case studies and policy response. *Applied Geochemistry* 23: 2976–2985.

Stanger G, Truong TV, Ngoc KS, Luyen TV, Thanh TT. 2005. Arsenic in groundwaters of the Lower Mekong. *Environmental Geochemistry and Health* 27: 341–357.

Sthiannopkao S, Kim K-W, Seing S, Choup S. 2008. Arsenic and manganese in tube well waters of Prey Veng and Kandal Provinces, Cambodia. *Applied Geochemistry* 23: 1086–1093.

Tseng WP. 1977. Effects and dose–response relationships of skin cancer and blackfoot disease with arsenic. *Environmental Health Perspectives* 19: 109–119.

Tseng WP, Chu HM, How SW, Fong JM, Lin CS, Yeh S. 1968. Prevalence of skin cancer in an endemic area of chronic arsenicism in Taiwan. *Journal of the National Cancer Institute* 40: 453–463.

United State Environmental Protection Agency (USEPA). 1998. Integrated Risk Information System (IRIS): arsenic, inorganic, CASRN 7440-38-2. http://www.epa.gov/iris/subst/0278.htm

World Health Organization (WHO). 2004. *Guideline for Drinking Water Quality*, vol. 1. World Health Organization, Geneva, Switzerland.

6

Pragmatic Approach for Health Risk Assessment of Bauxite Tailing Waste Materials

Jack C. Ng, Jian Ping Wang, Cheng Peng, and Guanyu Guan

CONTENTS

Introduction

Alumina mining and refining is a major industry in some countries. It contributes significant benefit to the general world economy. With globally increasing demand for mineral resources, a concomitant and ongoing increase in mining activities has been taking place globally in recent years. This spurt in growth is not without problems; mining operations inevitably lead to dust fall-out and chemical exposure, which in turn has given rise to community complaints and concerns all over the world.

Bauxite is screened and washed after mining, in order to remove organics and under- or over-sized ore. This ore extraction process generates tailings predominantly comprised of over- and under-sized bauxite, which may constitute up to 30% of the mined ore tonnage. Tailings dams are, therefore, required to be expanded as a routine part of the mining process. Like most mining operations, including dam expansion, there will be occasions when elevated levels of dust occur. A key driver for this study was the complaints raised by local residents living in closer proximity to such a bauxite storage facility on account of dust fall-out and chemical exposure. In addition to the harmful effects of dust particulates on human health, some metals and metalloids can also cause adverse health effects if the exposure is excessive, particularly in the long term. This study is aimed to identify and prioritize potential environmental hazard that may exist at a bauxite tailings storage facility, as part of an overall effective management strategy.

Criteria for the rehabilitation of mine sites in Australia have become more stringent, given the increasing awareness of potential detrimental environmental effects caused by exposure to metals and metalloids that form part of mine waste. Relevant to this report under the risk assessment frame (enHealth 2004), we will evaluate all the pathways that could lead to exposure to bauxite/soil waste.

Where there are no specific health investigation levels (HILs) for mined land, NEPM guideline values exist for soil. EPHC (the Environmental Health Council, Australia) has set the National Environmental Protection Measure (NEPM) for health-based investigation levels (HIL) in soil for residential settings (NEPC 1999); the investigation involves detecting metal content in soil, including detection of metals such as arsenic (As) (100 mg/kg), beryllium (Be) (20 mg/kg), cadmium (Cd) (20 mg/kg), chromium (Cr^{III}) (12%), chromium (Cr^{VI}) (100 mg/kg), cobalt (Co) (100 mg/kg), copper (Cu) (1000 mg/kg), lead (Pb) (300 mg/kg), manganese (Mn) (1500 mg/kg), mercury (Hg) (15 mg/kg inorganic mercury), nickel (Ni) (600 mg/kg), and Zinc (Zn) (7000 mg/kg), and reported urban background level for vanadium (V) (20–500 mg/kg) (NEPC 1999). The NEPM guideline recommends that further investigation be conducted if total elemental concentration in soil exceeds the HIL. In the absence of site-specific bioavailability data, it is usually assumed to be 100% for risk assessment purposes. The advantage of using site-specific bioavailability data is that they can be used for a more refined risk assessment (Ng et al. 2010).

For risk assessment, if the bioavailability of a contaminant is unknown, then the standard precautionary practice encourages the use of 100% bioavailability. This is recognized as a very conservative approach. A realistic risk-based approach is to measure bioavailability (BA) or, at the very least, the bioaccessibility (BAC) of the metal of interest. The concept of BA and BAC is discussed in the following.

The bioavailability of heavy metals and metalloids from contaminated land has been assessed using mammals, including rodents (Ng and Moore 1996, Ng et al. 1998), meadow voles (Pascoe et al. 1994), dogs (Groen et al. 1994), pigs (Juhasz et al. 2007), cattle (Bruce et al. 2003), monkeys (Freeman et al. 1995), humans (Ziegler et al. 1978), and more typically guinea pigs or rabbits (Freeman et al. 1993). As part of an environmental health risk assessment of metals (including lead and arsenic) in a residential area in Canberra (Ng et al. 1998), arsenic contamination at dip sites, and copper chrome arsenate–contaminated site in NSW (Ng and Moore 1996), a rat model was successfully utilized to estimate the potential uptake (bioavailability). This can then be applied to provide a realistic risk assessment in calculating the potential exposure route via soil ingestion. For example, the health investigation level (HIL) for arsenic in soil is set at 100 mg/kg in Australia (NEPC 1999). If the arsenic in soil is only 10% bioavailable for absorption, then the real value for justifying further health investigation could be 1000 mg As/kg soil and not 100 mg As/kg. The conventional approach assuming that metals are 100% bioavailable for absorption is very conservative and often gives an overestimation of the real risk. Animal models have widely been used for determining the bioavailability of metal contamination at super-fund sites in the United States, by environmental scientists and engineers. Although animal models are regarded as the ultimate tool for the assessment of contaminant uptake, *in vitro* physiologically based extraction test (PBET) for bioaccessibility measurement has been gaining momentum and acceptance as a surrogate test for predicting bioavailability. The application of several such *in vitro* bioaccessibility assays and their validation against an *in vivo* assay has been recently reviewed (Ng et al. 2010).

In this study we utilized bioaccessibility data as a pragmatic approach for a more structured and holistic assessment of health risk posed by bauxite tailings dust.

Methods

We collected 13 bauxite tailings samples from 3 adjoining bauxite tailings storage dams identified as BT1, BT2, and BT3 and veranda fall-out dust from 2 houses combined into 1 composite dust sample in the residential area downwind from the bauxite tailings storage facility (see sample description in Table 6.1). The dust samples collected from the two houses were combined in equal weight to yield a composite house dust sample. The residential area studied is just a few hundred meters downwind from the closest tailings dam (Figure 6.1).

Particle-size distribution profiles of the bulk bauxite tailings were obtained using a Malvern Mastersizer (Mastersizer 2000, Malvern Instruments Ltd., Malvern, United Kingdom). The scanning range was found to be 0.02–2000 μm, based on a default RI (refractive index) of 1.52, which is closer to the value of 1.53 attributed to gibbsite mineral.

TABLE 6.1

EnTox Identification Codes for Bauxite Tailings
and Respective PM_{10} ID

Location of Dam	EnTox ID—Bulk Tailings	EnTox—PM_{10}
BT1-North NE	EnTox 11	PM_{10}-11
BT1-North NW	EnTox 7	PM_{10}-7
BT1-North SE	EnTox 8	PM_{10}-8
BT1-North SW	EnTox 6	PM_{10}-6
BT2-South Dune	EnTox 3	PM_{10}-3
BT2-South NE	EnTox 9	PM_{10}-9
BT2-South NW	EnTox 2	PM_{10}-2
BT2-South SE	EnTox 5	PM_{10}-5
BT2-South SW	EnTox 4	PM_{10}-4
BT3-West NE	EnTox 12	PM_{10}-12
BT3-West NW	EnTox 13	PM_{10}-13
BT3-West SE	EnTox 1	PM_{10}-1
BT3-West SW	EnTox 10	PM_{10}-10
House Dust	EnTox 14	PM_{10}-14

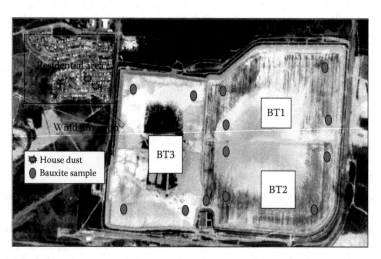

FIGURE 6.1
A schematic diagram of the 3 adjoining bauxite tailings dams and a nearby residential area showing 13 bauxite sampling locations and 2 house dust samples to form a composite sample.

Separation of $\leq PM_{10}$ particles (henceforth PM_{10}) was carried out using a Cyclone separator (SKC-225-69-25, SKC, Somerset, PA) fitted with membrane disc filters (GLA-5000, Pall Corporation, East Hills, New York).

Mineralogy of the bulk tailings and house dust was examined using the XRD technique. Each of the $\leq PM_{10}$ samples (about 100 mg) was lightly pressed into shallow stainless steel sample holders for x-ray diffraction analysis. XRD patterns were recorded with a PANalytical X'Pert Pro Multi-purpose Diffractometer using Co Kα radiation, variable divergence slit, postdiffraction graphite monochromator, and fast X'Celerator Si strip detector. The diffraction patterns were recorded in steps of 0.05° 2θ with a 0.5 s counting time per step and logged to data files for analysis.

Total elemental concentrations (strong acid-solubilized metals/metalloids) in the bulk tailings and house dust were measured by ICP-MS (7500CS, Agilent Technologies, Santa Clara, CA) from multiple diluted acid extracts following a reverse Aqua Regia digestion process in accordance with the Australian Standard AS4479.2 (Standards Australia 1997). This method is consistent with NEPM requirements as well (NEPC 1999). Briefly, about 1 g of bauxite and house dust was accurately weighed into a Pyrex beaker, covered in reverse Aqua Regia acid (10 mL) and allowed to stand at room temperate with a watch glass before a gentle reflux for 2 h at 120°C on a temperature-controlled heating block. The digest was allowed to cool before dilution to 100 mL. Further multiple dilutions were made to this digest (to make digested solution to final dilutions of 500 times and 500,000 times) before analysis, in order to accommodate high and low concentrations of various analytes in the bauxite sample. The final ICP-MS analyzing solution contained 2% nitric acid and various internal standards. All solutions were analyzed for Al, Ti, V, Cr, Mn, Fe, Co, Ni, Cu, Zn, As, Se, Sr, Ag, Cd, Pb, and U. The risk assessment focuses exclusively on elements that have high concentrations and can adversely affect human health.

Certified reference materials, including TM28.3 (water, National Water Research Institute, Environment Canada, Canada), BCSS-1, and PCSS-1 (sediment, NRC-CNRC, Canada), were analyzed in the same manner with an optimum sample weight of 0.25 to 0.5 g. These QC standards allow for the evaluation of instrument performance and assay precision. They also allow for the selection of best dilution factor for each specific certified elemental concentration. This provides a guide for selection of dilution applicable to individual elements in the sample. The certified reference material results are shown in Table 6.2.

Bulk tailings and house dust as collected and their respective PM_{10} fractions were measured for bioaccessibility (BAC), using a physiologically based extract test (PBET) method (Ruby et al. 1996). Physiologically based extract test methods have been widely adopted for the estimation of bioavailability internationally and in Australia, for risk assessment of metal-contaminated sites (Bruce et al. 2007, Diacomanolis et al. 2007, Juhasz et al. 2007, 2009a,b). BAC has been validated against *in vivo* models and accepted by USEPA (2007). It has also been recently recommended for use by an NEPM review (Ng et al. 2010).

Briefly, each sample was extracted using a synthetic gastrointestinal fluid in a solid:liquid ratio of 1:100 at pH values of 1.5, 2.5, 4.0, and 7.0 over a number of hours, simulating the GI tract pH conditions representing fasting, semi-fed, and fully fed stomach (Phase I or stomach phase) and the small intestine, respectively, over a period of 4 h. Each sample was assessed across all pH values and 3 different time intervals to derive a set for 12 extracts.

TABLE 6.2

Quality Control Results of Certified Reference Materials TM28.3, BCSS-1, and PCSS-1

ID	Concentration	Cr	Mn	Ni	Cu	Zn	As	Cd	Hg	Pb
TM28.3	Certified (µg/L)	4.83	6.9	9.8	6.15	—	—	—	—	—
	Found (µg/L)									
	Run 1 (n=3)	4.96±0.3	6.97±0.3	9.72±0.14	6.44±0.03					
	Run 2 (n=3)	4.93±0.04	6.77±0.08	9.54±0.15	6.14±0.1					
BCSS-1	Certified (mg/kg)	123±14	—	—	18.5±2.7	119±12	11.1±1.4	0.25±0.04	4.5	22.7±3.4
	Found (mg/kg)	121±6	—	—	15±0.4	102±1.3	9.4±4.6	0.49±0.01	3.3±0.15	31.9±2.6
PCSS-1	Certified (mg/kg)	113±8	—	—	452±16	824±22	211±11	2.38±0.2	77	404±20
	Found (mg/kg)	139±23	—	—	440±4	886±34	182±8	2.12±0.07	91±0.7	404±44

TM28.3 (n=3 for each run) is a certified water sample; BCSS-1 (n=3) is sediment-soil; PCSS-1 (n=3) is sediment-soil; "—" indicates no certified value for this element. The standard deviations are those obtained in this study from the number of CRMs (as indicated by the n value) assayed.

Metals/metalloids were measured in the 0.45 μm filtrate of the extracts. The amount found in each extract was then compared to the actual amount (calculated from the initial total elemental concentration) put in the GI digestion tube to derive the BAC value (%) of each metal/metalloid targeted. The overall BAC is the average of 12 measurements representing the various physiological states of a human being during his or her active period of a day.

Results and Discussion

Particle Distribution Profile

Particle size distribution profiles of the samples are shown in Figure 6.2. Nanoparticles were not observed in any of the samples. Most of the particles were in the range of <250 μm. Particle size of <250 μm is the fraction of soil/dust likely to stick to human hands, resulting in exposure via hand-to-mouth activities (Duggan et al. 1985).

Accumulative PM_{10} ranged from 14% to 64% of the composition of dust particles up to PM250 in three bauxite tailings dams. PM_{10} ranged from 14%

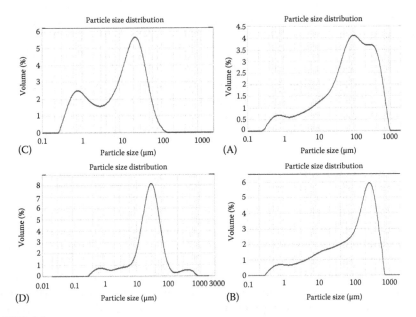

FIGURE 6.2
Representative particle distribution profiles measured in samples collected from the bauxite tailings dams RT1, RT2, RT3, and the composite house dust. (A)=RT1 (north dam, southwest location), EnTox 6; (B)=RT2 (south dam, northwest location), EnTox 2; (C)=RT3 (west dam, northwest location), EnTox 13; and (D)=composite house dust, EnTox14, in close proximity to RT3.

TABLE 6.3

Cumulative % for a Specific Range of Bauxite
Tailing Particle Size

EnTox ID	$PM_{2.5}$	PM_{10}	PM_{250}
EnTox 11	8.585074	17.55623	91.42739
EnTox 7	6.473802	14.36745	64.72323
EnTox 8	11.0446	23.42975	84.91839
EnTox 6	7.874993	16.24238	76.6757
EnTox 3	9.205302	19.92355	87.02744
EnTox 9	6.817268	15.11208	70.76206
EnTox 2	8.202545	18.01139	72.06393
EnTox 5	11.18186	22.96217	91.87329
EnTox 4	7.115728	15.22794	69.98528
EnTox 12	25.70403	42.02388	100
EnTox 13	26.29055	45.37628	100
EnTox 1	39.08025	64.09891	100
EnTox 10	23.58167	38.98474	93.75553
EnTox 14	7.599203	16.583025	96.700611

to 20% in the BT1 and BT3 dams, whereas BT3 had 39%–64% of PM_{10}. $PM_{2.5}$ was about 50% of the PM_{10} concentration (Table 6.3). It is noted that tailings collected from the west dam (BT3) have a higher volume in finer particle sizes of both $PM_{2.5}$ and PM_{10}. $PM_{2.5}$ and PM_{10} particles are considered to be relevant for inhalation exposure, whereas PM250 particles are ingestion relevant via hand-to-mouth activities (Duggan et al. 1985). It is interesting to note that the size distribution profile of the house dust is similar to those obtained for the tailings in the west dam (Figure 6.2) though PM_{10} and $PM_{2.5}$ are lower in percentages. This is expected, as lighter particles will remain airborne and likely travel longer distances before deposition takes place.

Mineralogical Composition and Elemental Concentrations of Bauxite Tailings

Quantitative analysis was performed on the XRD data, using the commercial package SIROQUANT from Sietronics Pty Ltd (Belconnen, ACT, Australia). The results are normalized to 100% and, hence, do not include estimates of unidentified or amorphous materials. The XRD results are summarized in Table 6.4. XRD shows that the major minerals found in decreasing concentrations include gibbsite, kaolinite, and boehmite. The south dam (BT2) has higher quartz content averaging 6.3%, with the highest being 8.4%. The west dam (BT3) tailings have significantly less quartz (crystalline silica), with an average of 0.7%, slightly less gibbsite but more boehmite. Although it is difficult to confirm with only one house dust sample, its mineral composition is similar to that of the tailings from the west Dam. This is also supported

TABLE 6.4

Mineral Composition Normalized to 100% PM_{10} Present in Bauxite Tailings and House Dust

EnTox ID	Quartz SiO_2	Hematite α-Fe_2O_3	Goethite α-FeOOH	Gibbsite α-$Al_2O_3 \cdot 3H_2O$	Boehmite α-$Al_2O_3 \cdot H_2O$	Anatase TiO_2	Rutile TiO_2	Kaolinite $Al_2O_3 \cdot 2SiO_2 \cdot 2H_2O$
PM_{10}-11	1.2	1.5	1.1	43.6	6.9	1.5	0.8	43.4
PM_{10}-7	6.5	1.8	0.9	48.8	13.6	1.5	0.6	26.4
PM_{10}-8	6.4	1.4	0.6	46.3	9.1	1.4	0.6	34.3
PM_{10}-6	3.4	1.4	0.6	49.4	7.8	1.7	0.6	35
PM_{10}-3	7.5	1.4	1	43.5	9.5	1.2	0.5	35.3
PM_{10}-9	8.4	1.7	0.3	53	5.5	1.6	0.7	28.9
PM_{10}-2	4.5	1.6	1	42	13.2	1.5	0.6	35.6
PM_{10}-5	4.3	1.1	0.6	47.6	7.8	1.5	0.6	36.5
PM_{10}-4	6.6	2	0.9	43.7	6.3	1.4	0.6	38.7
PM_{10}-12	0.4	5.5	1.3	30.3	42	2.2	0.7	17.7
PM_{10}-13	0.8	2.2	1.3	33	40.7	2.1	0.7	19.3
PM_{10}-1	1.2	6	1.5	33	42.4	1.9	0.6	13.5
PM_{10}-10	0.5	6.7	1.5	28.6	42.4	2	0.7	17.7
BT1-average	4.4	1.5	0.8	47.0	9.4	1.5	0.7	34.8
BT2-average	6.3	1.6	0.8	46.0	8.5	1.4	0.6	35.0
BT3-average	0.7	5.1	1.4	31.2	41.9	2.1	0.7	17.1
PM_{10}-14	1.9	2.3	1.8	32.2	29.9	1.8	0.6	29.5

by similarity of particle distribution profiles between the house dust and BT3 tailings.

Metal concentrations in bulk tailings and PM_{10} fractions are shown in Table 6.5. The metal concentrations of the house dust sample are similar to those found in the tailings for some key elements. PM_{10} fractions had higher Al levels than the bulk tailings, but there were exceptions. It is, therefore, difficult to conclude which storage dam the house dust derived from, based on the concentrations of metals alone. The metal concentrations of house dust could have been mixed with other sources of contamination, for example, garden soil and road dust.

Bioaccessibility

All bauxite samples (bulk tailings and PM_{10} tailings) and the house dust were tested for solubilized fraction (bioaccessibility) under various pH values in a synthetic gastric fluid system, using a PBET method (Ruby et al. 1996). PBETs were analyzed in duplicates across pH values of 1.3, 2.5, and 4.0 over various time intervals for up to 3h when each pH was adjusted to 7.0 to simulate the intestinal phase. Average percentages of bioaccessibility (based on the 24 individual readings that form part of the duplicate set of results) of various targeted elements are shown in Table 6.6. Only the PBET results for harmful metal/metalloids and/or elements with significant concentrations in the original samples are shown here and considered for risk assessment (see the following section).

Risk Assessment

Aluminum

There is no National Environmental Protection Measure (NEPM) health investigation level (HIL) for aluminum (Al) (NEPC 1999). However, because of its high concentration in bauxite tailings and in PM_{10} dust, this element is taken into consideration for the risk assessment activity.

A pragmatic approach is to treat the house dust as the same as bauxite tailings in terms of risk assessment. In addition, there are several reasons to believe that west dam tailings (BT3) are a major contributor to the house dust based on characterization of its particle size profile, mineralogy, and metal concentrations. Aluminum is the third most abundant element in the earth's crust, exceeded only by oxygen and silicon. U.S. Geological Survey reported soil aluminum ranged from 700 to 100,000 mg/kg with an average of 72,000 mg/kg (Shacklette and Bolerngen 1984).

Aluminum (Al) is poorly absorbed (0.1%–0.3%) via the gastrointestinal tract, with soluble citrate complexes reaching as high as 1% bioavailability. The bioavailability of Al is about 3% in water of 100 μg/L. The addition of silica can reduce Al bioavailability by sevenfolds (IPCS 1997).

TABLE 6.5

Total Elemental Concentrations of Bauxite Tailing Samples and Their Respective PM$_{10}$ Particles

EnTox ID	Al (mg/kg)	Ti (mg/kg)	V (mg/kg)	Cr (mg/kg)	Mn (mg/kg)	Fe (mg/kg)	Co (mg/kg)	Ni (mg/kg)	Cu (mg/kg)	Zn (mg/kg)	As (mg/kg)	Se (mg/kg)	Sr (mg/kg)	Ag (mg/kg)	Cd (mg/kg)	Pb (mg/kg)	U (mg/kg)
EnTox 11	70,500	126.6	3898	32.6	70.6	11,435	0.5	0	1.6	4.4	0	0	20.2	0.5	0	2.9	0.6
EnTox 7	109,650	626	1141	114.9	102.8	18,185	1.0	0	2.7	6.5	0	0	23.9	1.0	0.	4.0	1.3
EnTox 8	120,700	648	1445.5	44.5	158.1	14,705	1.1	0.5	6.3	101.9	0.5	0	20.7	0.7	1.0	110.2	1.0
EnTox 6	120,600	785.5	830.5	52.1	122.8	14,310	1.4	1.3	2.4	2.8	0	0	32.5	0.7	0	5.7	1.0
EnTox 3	51,800	481.85	359.3	31.3	49	9,190	0.5	0	1.7	4.2	0	0	46.3	0.5	0	2.9	0.6
EnTox 9	60,950	977.5	1368.5	37.3	37.9	9,560	0.3	0	196.9	17.4	0.3	0	20.8	0.5	0.2	12.4	0.6
EnTox 2	98,350	137	117	45.0	161.2	15,980	1.7	1.0	3.2	17.6	0	0	195.8	0.8	0.1	13.0	0.9
EnTox 5	68,150	819	472	38.7	96.1	10,295	0.7	0	1.6	3.6	0.2	0	59.1	0.5	0	3.0	0.7
EnTox 4	54,800	751.5	398.5	45.0	42.5	10,885	0.34	0	1.8	4.9	0	0	40.9	0.5	0.	2.5	0.6
EnTox 12	100,000	515	3786.5	61	215.2	39,895	2.3	1.4	11.2	19.9	0	0	15.5	1.4	0	7.2	0.8
EnTox 13	105,800	229.4	120.1	44.8	183.4	21,460	1.5	0.9	5.2	8.2	0.	0	11.4	1.3	0.1	5.6	1.0
EnTox 1	117,050	830.5	224.5	85.9	168.4	45,025	2.0	0.8	12.6	29.3	0.1	0	50.9	1.5	0	9.1	0.8
EnTox 10	144,800	1174	1672	78.1	180.5	43,935	2.4	1.4	11.7	24.8	0.1	0	22.3	1.4	0	7.5	0.8
EnTox 14	133,835	808	289.6	105.1	206.0	35,301	3.6	0	22.1	0	1.1	0.3	13.7	0	0.1	0	1.2
PM$_{10}$'11	101,500	137	144.9	43.7	113.2	20,380	0.8	0	3.9	15.9	0.0	0	1.5	1.4	0.1	9.2	1.0
PM$_{10}$'7	150,400	171.6	141.7	89.5	160	26,200	2.2	0.8	5.5	23.3	0.4	0	1.7	1.8	0.2	15.6	1.8
PM$_{10}$'8	184,400	167.4	148.3	59.2	191.2	18,750	1.8	0.4	3.6	12.8	0.0	0	1.4	1.2	0.1	10.9	1.3
PM$_{10}$'6	156,000	120.2	155.6	65.1	209.8	21,730	3.1	4.8	4.2	17.3	0.2	0	1.9	2.0	0.1	17.7	1.2

(continued)

TABLE 6.5 (continued)

Total Elemental Concentrations of Bauxite Tailing Samples and Their Respective PM$_{10}$ Particles

EnTox ID	Al (mg/kg)	Ti (mg/kg)	V (mg/kg)	Cr (mg/kg)	Mn (mg/kg)	Fe (mg/kg)	Co (mg/kg)	Ni (mg/kg)	Cu (mg/kg)	Zn (mg/kg)	As (mg/kg)	Se (mg/kg)	Sr (mg/kg)	Ag (mg/kg)	Cd (mg/kg)	Pb (mg/kg)	U (mg/kg)
PM$_{10}$-3	141,200	133.7	144.5	62.1	116.5	21,550	1.3	0	4.4	24.7	0.5	0	1.6	1.7	0.1	16.3	1.3
PM$_{10}$-9	154,900	191.6	150.7	94.3	79	23,210	1.0	0.2	6.8	22.2	0.4	0	1.4	1.6	0.1	10.5	1.4
PM$_{10}$-2	122,400	142.5	145.1	62.1	220.9	19,700	2.5	0.9	7.9	42.0	0.6	0	1.5	1.6	0.3	29.6	1.0
PM$_{10}$-5	177,500	161.9	153.2	62.9	176.5	21,480	1.7	2.5	5.0	30.2	0.7	0	1.6	1.6	0.4	20.8	1.5
PM$_{10}$-4	190,000	216.8	198	87.2	139.5	35,890	1.8	3.7	8.6	32.1	0.8	0	2.1	2.0	0.2	20.9	1.9
PM$_{10}$-12	115,600	453.9	177.8	68.1	248.5	40,680	3.2	13.3	13.4	50.8	18.9	0.3467	1.4	2.4	1.1	31	0.9
PM$_{10}$-13	104,100	210.2	134.4	46.1	213.8	25,160	2.2	0	6.8	20.2	0	0	1.4	1.8	0.1	10.7	1.1
PM$_{10}$-1	126,900	931.9	254.7	92.7	205	52,790	2.6	4.0	15.4	56.7	0.1	0	1.5	2.4	0.1	18.8	0.9
PM$_{10}$-10	100,500	571	213.2	68.7	189.3	49,950	2.8	0	14.5	43.7	0.01	0	1.4	2.1	0	12.	0.8
PM$_{10}$-14	140,070	263	157.7	95.3	200.4	29,420	5.4	24.1	21.8	0	5.2	0.3	0	0.5	0	20.4	N.D.

TABLE 6.6

Average Percentages of Bioaccessibility
Measured by a Physiologically Based Extraction
Test of Various Targeted Elements from the Bulk
Bauxite Tailings and PM_{10} Fraction of the Tailings

EnTox ID	Al	V	EnTox ID	Al	V
EnTox11	0.04	0.01	PM_{10} 11	0.39	1.96
EnTox 7	0.04	0.13	PM_{10} 7	0.37	3.61
EnTox 8	0.04	0.09	PM_{10} 8	0.30	2.32
EnTox 6	0.01	0.16	PM_{10} 6	0.16	2.90
EnTox 3	0.04	0.23	PM_{10} 3	0.13	4.87
EnTox 9	0.06	0.04	PM_{10} 9	0.53	2.29
EnTox 2	0.03	0.58	PM_{10} 2	0.07	4.10
EnTox 5	0.02	0.32	PM_{10} 5	0.13	2.96
EnTox 4	0.04	0.16	PM_{10} 4	0.28	3.51
EnTox 12	0.03	0.11	PM_{10} 12	0.21	1.35
EnTox 13	0.02	2.01	PM_{10} 13	0.19	2.36
EnTox 1	0.01	1.04	PM_{10} 1	0.01	2.53
EnTox1 0	0.01	0.08	PM_{10} 10	0.15	1.15
EnTox 14	0.83	2.23	PM_{10} 14	1.07	3.64

Systemically absorbed Al is rapidly excreted in the urine by those with normal kidney function. Aluminum toxicosis is expressed largely as a secondary phosphorus deficiency, presumably because it binds phosphorus, forming an unabsorbable Al-phosphorus complex in the intestine. No studies were located regarding death following acute- or intermediate-duration inhalation exposure to various forms of Al in humans. Chronic exposure to very high concentration of fine powder metallic Al had led to fatalities in workers in the past. One chronic exposure case was related to total dust Al of 615–685 mg/m³, of which 51 mg Al/m³ was found to be in the respirable form. Chemical speciation revealed that the respirable dust contained 81% metallic Al and 17% of oxides and hydroxide of aluminum (IPCS 1997).

Pulmonary fibrosis is the most commonly reported respiratory disorder observed in workers exposed to fine aluminum dust (pyropowder), alumina (aluminum hydroxide), or bauxite. However, conflicting reports exist on the fibrogenic potential of aluminum. It is very likely that there was simultaneous exposure to silica; thus, it is possible to conclude that it is the ingestion of silica that could have caused this disorder and not that of dust or other toxic impurities derived from aluminum (ATSDR 1999).

Aluminum is a neurotoxicant. However, healthy people have a naturally high barrier to aluminum penetrating the nervous system except under specific serious illnesses, for example, in patients presenting with renal (kidney) failure, who had probably used dialysis fluid with high aluminum content.

Acute poisoning due to aluminum is very rare, even though it is widespread in the environment, including food, drinking water, and many antacid medications. Aluminum compound has been widely used as a flocculent in water treatment for many decades. Restrictive pulmonary disease (pulmonary fibrosis) was reported to exist in many workers in the aluminum industry, particularly in the 1940s when the industry hygiene standards were below par compared to the current scenario (IPCS 1997).

The PTWI for Al is 420 mg for a 60 kg adult (490 mg for 70 kg adult) (IPCS 1997). This equates to a TDI of 1 mg Al/kg b.w. Males consume 2.4 mg Al/day while females consume 1.9 mg Al/day as reported by the Australian market basket survey (NFA 1993).

Dermal absorption of Al is insignificant (IPCS 1997); therefore, our risk assessment for Al focused on the oral ingestion and inhalation routes.

The highest concentration of bulk bauxite tailings is 144,800 mg/kg (slightly higher than that of house dust). For an adult who ingests 25 mg of tailing dust a day with a bioaccessibility of about 1%, this represents a daily intake of 36.2 µg of absorbed aluminum.

PM_{10} fine dust has the highest concentration, at 190,000 mg/kg (pg/µg). Assume an NEPM dust value of 50 µg/m^3 (the actual medium value was 48.5 µg/m^3 while undertaking tasks such as handling heavy-duty earthmoving equipment; data are not presented in this chapter). An adult breathes about 23 m^3/day of air (NEPC 1999) and if we assume the bioaccessibility of Al in air-borne dust to be 1% (the actual maximum value was 0.53%, according to PBET), then the quantity of Al absorbed in dust inhaled would be 2.2 µg/day. The total absorbed intake of Al from all the three exposure routes added up to 0.549 µg Al/kg b.w./day, as shown by the following calculations:

$$Al_{air} = 50 \ \mu g/m^3 \times 190,000 \ pg \ Al/\mu g = 9,500,000 \ pg \ Al/m^3 = 9.5 \ \mu g \ Al/m^3$$

$$Al(\text{inhaled from air}) = 9.5 \ \mu g/m^3 \times 20 \ m^3/day = 218.5 \ \mu g/day$$

$$\text{Bioaccessible Al (absorbed intake)} = 218.5 \ \mu g \times 1\% = 2.2 \ \mu g/day$$

Total daily intake of Al from the three exposure routes:

$$TI = 0 + 36.2 + 2.2 = 38.4 \ \mu g/day \text{ or } 0.549 \ \mu g$$
$$/kg \ b.w./day \text{ for an adult of } 70 \ kg \ b.w.$$

$$\text{Hazardous index (HI)} = \frac{TI}{TDI} = \frac{0.549}{(1 \times 1000)}$$
$$= 5.49 \times 10^{-4} \text{ or a safety factor of 1821 times}$$

When considering other sources of Al intakes, food will account for an intake of 2.4 mg Al/day (i.e., 0.034 mg/kg b.w./day, assuming 100% bio-availability); 2 L water consumption per day with a 3% bioavailable Al will account for a daily intake of Al at 6 μg/day. Clearly, the major source of Al intake is food and not the dust. If one adds all sources to the aforementioned exposure routes to dust, the total intake per day (2.444 mg) is still much lower than the TDI of 70 mg/day (or 1 mg Al/kg b.w./day for an adult) with a HI of 0.035.

Similarly, for a young child of 14 kg b.w. the equivalent TDI of 14 mg of Al from all exposure routes is allowed. Children breathe 12 m^3/day and ingest 100 mg soil/day via hand-to-mouth activity (NEPC 1999). The absorbed daily intakes of Al from dust ingestion and dust inhalation were calculated to be 144.8 and 1.14 μg, respectively, with a HI of 0.01 (a safety factor of 96). Assuming a total Al dietary intake for a child is 50% of that of an adult (1.2 mg Food Al, 0.003 mg water Al), the HI would be 0.096. It can be concluded that the risk of Al due to its presence in bauxite dust is low.

Vanadium

Mild respiratory distresses, such as cough, wheezing, chest pain, runny nose, or sore throat, have been associated with acute V exposure. In one case study, 0.06 mg V/m^3 as vanadium pentoxide induced coughing and mucus formation. Based on the 0.06 mg V/m^3 and a safety (uncertainty) factor applies, the MRL (minimum risk level) is 2 μg/m^3 (ATSDR 1992). Factory workers exposed chronically to V dusts complained of nausea and vomiting. Exposure to vanadium via inhalation could be considered to be more significant than acute oral exposure. For this reason, we will focus on this route of exposure, with special reference to PM$_{10}$ vanadium concentrations as found during this study.

PM$_{10}$ vanadium ranged from 134.4 to 254.7 mg/kg in the bauxite tailing fine dust, with the highest concentration found in the west dam (SE and NE) samples. Assuming PM$_{10}$ is in the upper end of NEPM value of 50 μg/m^3 and the dust has the highest concentration of 255 mg V/kg, the HI for vanadium exposure would be 0.015 when compared with the MRL as shown as follows:

$$255 \text{ mg/kg} = 255 \text{ μg/g or } 255 \text{ ng/mg or } 255 \text{ pg/μg}$$

$$V_{air} = 50 \text{ μg/m}^3 \times 255 \text{ pg/μg} = 12{,}750 \text{ pg/m}^3 \text{ or } 12.75 \text{ ng/m}^3 \text{ or } 0.013 \text{ μg/m}^3$$

$$\textbf{HI} = 0.013 \text{ μg/m}^3 / 2 \text{ μg/m}^3 = \textbf{0.015 (a safety factor of 154)}$$

We conclude that there is a safety factor of 154 times when compared to the MRL even assuming that vanadium in the dust is more soluble (e.g., in the form of vanadium pentoxide). Although the form of vanadium in the bauxite dust has not been determined, it is reasonable to believe that at least some of it will be present in relatively insoluble forms, such as sulfide and other mineral-bound species. This is reflected in a relatively low bioaccessibility of less than 5%.

Other Metal/Metalloids

Concentrations of other metals/metalloids tested (see Table 6.5) are relatively low and below the recommended NEPM HIL values, which warrants further detailed risk assessment (NEPC 1999). It is worth noting that certain individuals could be allergic to nickel metal (ATSDR 1997). However, the risk of this occurring is considered low because of the relatively low concentration of nickel found in the samples.

Particulate Matter with Special Reference to Crystalline Silica (Quartz)

Particulate matter levels in relation to their effects on human health are commonly expressed on the basis of mass concentration of inhalable particles, which is defined as containing particles with an aerodynamic diameter equal to or less than 10 μm. This is because only these small particles can penetrate into the airways and lungs.

In Australia, the NEPM air-quality standards for particulate concentrations are similar to those recommended by WHO. The current 1 day average for particles up to PM_{10} is $50 \mu g/m^3$, with a maximum allowable exceedance of 5 days a year, and for particles up to $PM_{2.5}$, it is $25 \mu g/m^3$ for 1 day (NEPC 1999). In the United States, the PM_{10} standard is set at $150 \mu g/m^3$ at 24 h average and $50 \mu g/m^3$ at 1 year average. The respective values for $PM_{2.5}$ are 15 and $65 \mu g/m^3$ and these values will become effective from 2012 (USEPA 2004). As one can see, setting health standards is complicated. However, it is understood that standards have to be conservative but not overly so.

Inhalation of particulate matter can exacerbate existing pulmonary and cardiac vascular diseases, oxidative stress, and inflammation (CARB 2002, CEPA 1998, IAEA 2008, Pope and Dockery 2006, WHO 2007). For example, Ghio et al. (2000) measured exposure to concentrated ambient PM2.5 levels in the range of $23-311 \mu g/m^3$. Exposures for a period of 2 h were found to induce transient, mild pulmonary inflammatory reactions in healthy human volunteers exposed to the highest concentrations (average of $200 \mu g/m^3$). The PM_{10} at the residential area was found to peak at $149 \mu g/m^3$ once with an average of $48.5 \mu g/m^3$ (detailed data not shown) during a survey period of 3 months when earthmoving activities were at the highest scale. This suggests that exposure to dust *per se* would present a relatively low health risk to healthy individuals. However, discomfort might have been experienced by asthmatics and patients with compromised pulmonary or cardiac functions, as shown by anecdotal evidence gathered from the local residents who were subjected to high-dust activities.

$PM_{2.5}$ concentrations were not measured directly in our study. However, the particle size distribution analysis showed that less than 10% of the bulk tailings were $PM_{2.5}$, and of the PM_{10} fraction, there was about half of the fine dust that was $PM_{2.5}$. We extrapolated that the average $PM_{2.5}$ was half of $48.5 \mu g/m^3$ to yield a value of $24.25 \mu g/m^3$. This value is within the NEPM $PM_{2.5}$ standard of $25 \mu g/m^3$ and, hence, poses no risk.

It is generally considered that respirable particles have an aerodynamic diameter of $<3–4 \mu m$, whereas most particles larger than $5 \mu m$ may be deposited in the tracheobronchial airways and, thus, may not reach the alveolar (IARC 1997). The clearance rates for respirable particles deposited in the respiratory bronchioles and proximal alveoli (deep lung) are slower and more likely to injure the lung. The clearance kinetics of quartz particles in humans remains inconclusive to date (IARC 1997).

Exposure to excessive concentrations of respirable ambient air particles can adversely affect human health. The specific particle size fractions of the chemical or biological components are the most dominant sources that might be responsible for adverse health effects and are not fully studied for their potential to cause health hazards. For bauxite tailings tested thus far, it appears that the metal and metalloid levels present limited health risk as discussed earlier. We then turned our focus on air particulates, with special reference to crystalline silica (e.g., quartz) component present in bauxite waste materials.

There are many studies both epidemiologically and in animal experimentations linking silica (Quartz: CAS No. 14808-60-7) exposure to cancer and noncancer health effects (WHO 2000). Silicosis, lung cancer, and pulmonary tuberculosis are associated with occupational exposure to quartz dust at relatively large doses over a long time. However, to date, there is no known adverse health effect associated with nonoccupational exposure to quartz dust (WHO 2000). Some of the inherent problems with many epidemiological studies include lack of accurate exposure data and confounding factors, which may make it difficult to interpret accurately the observations made during such studies. Quartz was found to induce pulmonary tumors in rats but not in hamsters or in mice. This demonstrates species differences in their response to exposure to chemicals.

The risk estimates for silicosis prevalence for a working lifetime of exposure to respirable quartz dust concentration of about 50 or $100 \mu g/m^3$ in the occupational environment vary widely (i.e., 2%–90%). Using methods described by USEPA (1996), the BMD (bench mark dose) analysis predicted that the silicosis risk for a continuous 70 year lifetime exposure to $8 \mu g/m^3$ is less than 3% of healthy individuals not compromised by other conditions such as respiratory diseases and pollution in ambient environment. The highest average quartz content in bauxite tailings was found to be about 6% of the set value of $48.5 \mu g/m^3$, leading to a calculated concentration of less than $3 \mu g/m^3$

of quartz. The dust sample collected from the BT3 has a much lower quartz concentration. Therefore, it can be concluded that chronic quartz health risk in healthy individuals, including silicosis at ambient air quality as calculated ($<3\,\mu g/m^3$ of quartz), is considered low.

Conclusions

Gibbsite, kaolinite, and boehmite were the major minerals found in the bauxite tailings, and Al, Fe, and V were major metal loadings containing trace amounts of arsenic, cadmium, lead, and other elements. The quartz concentrations were relatively low, with the highest found in BT2. BT3 had higher $PM_{2.5}$ and PM_{10} concentrations, compared with BT1 and BT2. Mineral characteristics and particle size distribution profile of the house dust resembled that of the tailings in BT3. This led to the conclusion that the fall-out dust that impacted the residential area was primarily a result of increased earthmoving activities at BT3. The bioaccessibility for the major contaminants in bauxite tailings is relatively low. The risk of both acute and chronic toxicity was considered to be low at this study site. However, some individuals with asthmatic and pulmonary (lung) conditions may experience discomfort during periods of high dust in the ambient air. It is highly recommended that precautionary measures should be taken in order to minimize impact on the susceptible individuals during elevated dust event.

References

ATSDR. 1992. Toxicological profile for vanadium. US Department of Health and Human Services, Washington, DC.

ATSDR. 1997. *Toxicological Profile for Nickel.* US Department of Health and Human Services, Washington, DC.

ATSDR. 1999. *Toxicological Profile for Aluminum.* US Department of Health and Human Services, Washington, DC.

Bruce SL, Noller BN, Grigg AH, Mullen BF, Mulligan DR, Ritchie PJ, Currey N, Ng JC. 2003. A field study conducted at Kidston Gold Mine, to evaluate the impact of arsenic and zinc from mine tailing to grazing cattle. *Toxicology Letters* 137: 23–34.

Bruce SL, Noller BN, Matanitobua V, Ng JC. 2007. In-vitro physiologically-based extraction test (PBET) and bioaccessibility of arsenic and lead from various mine waste materials. *Journal of Toxicology and Environmental Health, Part A* 70: 1700–1711.

CARB. 2002. Staff report: Public hearing to consider amendments to the ambient air quality standards for particulate matter and sulfates, Prepared by the Staff of the California Air Resources Board and the Office of Environmental Health Hazard Assessment, May 2002.

CEPA. 1998. National ambient air quality objectives for particulate matter. Part 1: Science assessment document, A report by the Canadian Environmental Protection Agency (CEPA) Federal-Provincial Advisory Committee (FPAC) on air quality objectives and guidelines.

Diacomanolis V, Ng JC, Noller BN. 2007. Development of mine site close-out criteria for arsenic and lead using a health risk approach. In: *Mine Closure 2007. Proceedings of the Second International Seminar on Mine Closure*, Eds. A. Fourie, M. Tibbett, J. Wiertz, October 16–19, 2007, Santiago, Chile, pp. 191–198.

Duggan MJ, Inskip MJ, Rundle SA, Moorcroft JS. 1985. Lead in playground dust and on the hands of school children. *The Science of the Total Environment* 44: 65–79.

enHealth. 2004. *Environmental Health Risk Assessment: Guidelines for Assessing Human Health Risks from Environmental Hazards*. Department of Health and Aged Care and NH&MRC Canberra, ACT, Australia, pp. 1–227.

Freeman GB, Johnson JD, Killinger JM, Liao SC, Davis AO, Ruby MV, Chaney RL, Lovre SC, Bergstrom PD. 1993. Bioavailability of arsenic in soil impacted by smelter activities following oral administration in rabbits. *Fundamental and Applied Toxicology* 21: 83–88.

Freeman GB, Schoof RA, Ruby MV, Davis AO, Dill JA, Liao SC, Lapin CA, Bergstrom PD. 1995. Bioavailability of arsenic in soil and house dust impacted by smelter activities following oral administration in cynomolgus monkeys. *Fundamental and Applied Toxicology* 28: 215–222.

Ghio AJ, Kim C, Devlin RB. 2000. Concentrated ambient air particles induce mild pulmonary inflammation in healthy human volunteers. *American Journal of Respiratory and Critical Care Medicine* 162 (3 Pt 1): 981–988.

Groen K, Vaessen HAMG, Klies JJG, Deboer JLM, Vanooik T, Timmerman A, Vlug RF. 1994. Bioavailability of inorganic arsenic from bog containing soil in the dog. *Environmental Health Perspectives* 102: 182–184.

IAEA. 2008. Assessment of levels and "health-effects" of airborne particulate matter in mining, metal refining and metal working industries using nuclear and related analytical techniques, International Atomic Energy Agency, Report IAEA-TECDOC-1576, January 2008.

IARC. 1997. *Monographs on the Evaluation of Carcinogenic Risks to Humans, Vol. 68: Silica, Some Silicates, Coal Dust and Pararamid Fibrils*. International Agency for Research on Cancer, Lyon, France, pp. 1–242.

IPCS. 1997. Environmental health criteria 194—Aluminium. International Programme on Chemical Safety. WHO, Geneva, Switzerland, 282 pp.

Juhasz AL, Smith E, Weber J, Naidu R, Rees M, Rofe A, Kuchel T, Sansom L. 2009a. Assessment of four commonly employed in vitro arsenic bioaccessibility assays for predicting in vivo arsenic bioavailability in contaminated soils. *Environmental Science and Technology* 43: 9487–9494.

Juhasz AL, Smith E, Weber J, Rees M, Rofe A, Kuchel T, Sansom L, Naidu R. 2007. Comparison of in vivo and in vitro methodologies for the assessment of arsenic bioavailability in contaminated soils. *Chemosphere* 69: 961–966.

Juhasz AL, Weber J, Smith E, Naidu R, Marschner B, Rees M, Rofe A, Kuchel T, Sansom L. 2009b. Evaluation of SBRC-gastric and SBRC-intestinal methods for the prediction of in vivo relative lead bioavailability in contaminated soils. *Environmental Science and Technology* 43: 4503–4509.

NEPC. 1999. Draft national environmental protection measure for the assessment of site contamination. National Environmental Protection Council (NEPC), Adelaide, SA, Australia.

NFA. 1993. *The 1992 Australian Market Basket Survey—A Total Diet Survey of Pesticides and Contaminants*. Australian National Food Authority, Canberra, ACT, Australia, 96p.

Ng JC, Juhasz A, Smith E, Naidu R. 2010. Contaminant bioavailability and bioaccessibility. Part I: Scientific and technical review, Technical Report No. 14. CRC for Contamination Assessment and Remediation of the Environment Pty Ltd., Adelaide, SA, Australia.

Ng JC, Kratzmann SM, Qi L, Crawley H, Chiswell B, Moore MR. 1998. Speciation and bioavailability: Risk assessment of arsenic contaminated sites in a residential suburb in Canberra. *The Analyst* 123: 889–892.

Ng JC, Moore MR. 1996. Bioavailability of arsenic in soils from contaminated sites using a 96 hour rat blood model. In: *The Health Risk Assessment and Management of Contaminated Sites*. Contaminated Sites Monograph Series No. 5, Eds. A. Langley, B. Markey, H. Hill. Commonwealth Department of Human Services and Health and the Environmental Protection Agency, Washington, DC, pp. 355–363.

Pascoe GA, Blanchet RJ, Linder G. 1994. Bioavailability of metals and arsenic to small mammals at a mining waste-contaminated wetland. *Archives of Environmental Contamination and Toxicology* 27(1): 44–45.

Pope III CA, Dockery DWC. 2006. Health effects of fine particulate air pollution: Lines that connect. *Journal of the Air and Waste Management Association* 56: 709–742.

Ruby MV, Davis A, Schoof R, Eberle S, Sellstone CM. 1996. Estimation of lead and arsenic bioavailability using a physiologically based extraction test. *Environmental Science and Technology* 30: 422–430.

Shacklette HT, Bolerngen JG. 1984. Elemental concentrations in soils and other surficial materials of the conterminous United States. US Department of Interior, Geological Survey, Washington, DC, 105p (Geological Survey Professional Paper No. 1270).

Standards Australia. 1997. AS4479.2—Analysis of soils, Part 2: Extraction of heavy metals and metalloids from soil by aqua regia—Hotplate digestion method.

USEPA. 1996. Ambient levels and noncancer health effects of inhaled crystalline and amorphous silica: Health issue assessment. US Environmental Protection Agency, Washington, DC. Office of Research and Development (Publication NO. EPA600/R-95/115; National Technical information Service Publication No. PB97-1881220).

USEPA. 2004. Air quality criteria for particulate matter. Vol 1. EPA/600/P-99/002aF, October 2004. National Center for Environmental Assessment-RTP Office, Office of Research and Development, U.S. Environmental Protection Agency, Research Triangle Park, NC.

USEPA. 2007. Estimation of relative bioavailability of lead in soil and soil-like materials using in vivo and in vitro methods. USEPA OSWER 9285.7-77, Office of Solid Waste and Emergency Response U.S. Environmental Protection Agency, Washington, DC.

WHO. 2000. Crystalline silica, quartz. Concise International Chemical Assessment Document 24. WHO, Geneva, Switzerland. pp. 1–50.

WHO. 2007. Health relevance of particulate matter from various sources. Report on a World Health Organisation Workshop, Bonn, Germany, March 26–27, 2007.

Ziegler EE, Edwards BB, Jensen RL, Mahaffy KR, Fomon S. 1978. Absorption and retention of lead by infants. *Pediatric Research* 12: 29–34.

Part II

Emerging Chemicals and Electronic Waste

Part II

7

Polybrominated Diphenyl Ethers in China: Sources, Trends, and Their Adverse Impacts on Human Health

Gene J.S. Zheng, Liping Jiao, Anna O.W. Leung,
Yeqing Huang, Tingting You, and Ming Hung Wong

CONTENTS

Introduction

There are four main types of flame retardants: chlorinated flame retardants (such as chlorinated paraffin wax), phosphoric flame retardants (such as phosphoric esters), inorganic flame retardants (such as $Mg(OH)_2$, $Al(OH)_2$), and brominated flame retardants (such as PBBs, HBCD, and polybrominated diphenyl ethers [PBDEs]) that are produced on a global scale. The brominated flame retardants have the highest efficiency for retarding flames, at a lower cost. They are especially suitable for organic polymer products and are widely used in electrical and electronic products (Liu et al. 2005a).

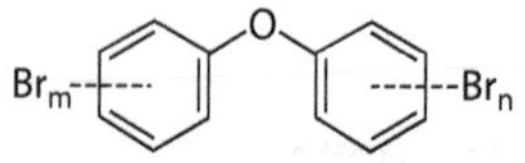

FIGURE 7.1
Molecular structure of PBDEs $m = 1–5$, $n = 0–5$.

In the case of fire, PBDEs release bromine atoms, which are strong reducing agents for capturing burning ions such as –OH and =O at high temperatures. This causes a release of high-density unburning gases that separate air away from the burning center, thereby retarding the fire.

The molecular formula of PBDEs is $C_{12}H_{(0–9)}Br_{(1–10)}$, and the structure is shown in Figure 7.1. It is composed of two rings (phenyl rings) linked by an oxygen bridge (ether linkage). "Polybromine" means many "bromine" atoms linked to carbons on the phenyl rings. There are up to 10 locations where the bromine atom can attach to the carbons on the rings; for example, if a PBDE has 10 bromines, it is called a deca-BDE; if it has 5 bromines, it is called penta-BDE. As a result, there are 209 possible congeners of PBDEs, all of which, like polychlorinated biphenyls (PCBs), are numbered by the International Union of Pure and Applied Chemistry (IUPAC).

PBDE products have been widely used for global commercial requirements and are produced with three degrees of bromination: penta-, octa-, and deca-BDE technical mixture (La Guardia et al. 2006). Penta-BDEs are added to foam mainly for manufacturing furniture, car cabins, and chairs. Octa-BDEs are mainly used in textiles and plastic products such as the outer covering for TVs, computers, and DVDs. Deca-BDEs consist of 80% of the total PBDE production and are primarily used in electronic products (La Guardia et al. 2006). The amount of brominated flame retardants produced was around 239,000 t per annum during the twentieth century, with nearly 30% consisting of PBDEs. The estimated world market demand was 67,125 t in 1999 (Liu et al. 2005b).

PBDEs possess physiochemical properties similar to that of PCBs, which are chlorinated congeners with teratogenic, mutagenic, and carcinogenic properties. The fate and behavior of PBDEs and PCBs are comparable. PBDEs have been associated with endocrine disruption, reproductive-developmental toxicity, including neurotoxicity and even carcinogenicity. The toxicological end points of concern for environmental levels of PBDEs are likely to be thyroid hormone disruption, neurodevelopmental deficits, and cancer (Manchester et al. 2001, McDonald 2002).

Prior to 2004, there was a paucity of published articles (both in Chinese and English) about PBDE contamination in mainland China. There were only a few articles written in English, concerning Hong Kong and Taiwan. Since 2007, the number of articles on PBDEs (in both languages) related to mainland China has steadily increased. This review summarizes the sources, levels, trends, and adverse impacts of PBDEs on human health in mainland China, Hong Kong, Taiwan, and Macau. Comparisons are made between data generated in China with those from other parts of the world.

Main Sources of PBDEs in the Environment of China

As flame retardants, PBDEs are additives that are physically mixed with polymers in most products and, therefore, may separate and leach out from the surface of their product applications, entering into air, and eventually depositing onto water, sediments, and soils (Deng et al. 2007, Liu et al. 2005a). Therefore, most household items, especially furniture, TVs, VCD players, washing machines, refrigerators, computers, and children's toys, which contain flame retardants, are common sources of PBDEs (Pamade et al. 2002). It is envisaged that company offices, municipal administration offices, and school classrooms, which often contain a high density of electronic equipment, emit a substantial amount of PBDEs into the indoor air.

The concentrations of PBDEs in the air and dust of homes and offices in Taiwan and Guangzhou, South China, were analyzed. Table 7.1 shows that the concentrations of PBDEs in office dust were 1.2–1.6 times higher than that of house dust in both Taiwan and Guangzhou. In addition, PBDEs in indoor air were three times higher than that detected in outdoor air (Huang et al. 2009b, Wang et al. 2011). No doubt, all electrical and electronic equipment are the most common sources of PBDEs. Although China has historically been an agricultural country, the rapid industrialization and modernization in the past two decades have resulted in rapid production and widespread usage of electronic equipment and, hence, the high PBDEs detected in indoor air and dust.

The Pearl River Delta (approximately 41,000 km^2) has experienced rapid development in the past decades. The gross industrial output in 2001 was U.S.$145 billion, accounting for 13% of the total gross industrial output for the whole of China (excluding Hong Kong). With a size bigger than the Pearl River Delta, the Yangtze River Delta has also undergone rapid socioeconomic development during the past two decades. In addition, the Chinese government–designated areas such as Shantou and Xiamen as Economic Special Developing Areas, as well as the newly developed region of "Bei Bu Wan Developing Areas," which specifically focuses on trading with Southern Asian countries.

TABLE 7.1

Concentrations of PBDEs in Dust and Air of Family Houses and Offices in Hsinchu (Taiwan) and Guangzhou (South China) (Dust ng/g dry wt., Air pg/m^3)

Site	Office Dust (ng/g)	House Dust (ng/g)	Office (pg/m^3)	Family (pg/m^3)	Indoor (pg/m)	Outdoor (pg/m^3)	References
Hsinchu	3758	2326	160	95	141	49	Wang et al. (2011)
Guangzhou	3179	2662	/	/	/	/	Huang et al. (2009b)

In China, 10,000 metric tons of brominated flame retardants is used in electrical and electronic products annually (Guan et al. 2007). For example, one third of the gross industrial output in the Pearl River Delta is associated with electronic and telecommunication equipment (Liu et al. 2005b), leading to large usage of PBDEs (1300 t) in the region (Decision Etudes & Conseil 2004, Guan et al. 2007). In addition, the Chinese government had recently encouraged the transfer of industrial products to the countryside, in order to satisfy domestic needs, which subsequently resulted in widespread contamination of PBDEs in more remote areas of China.

The second source of PBDEs emitted into the environment is from uncontrolled recycling of electrical and electronic waste (e-waste), which has become a global problem in the twenty-first century, with about 20–25 million ton of e-waste generated annually in the world in recent years (Robinson 2009). In developed countries, it has become one of the fastest growing waste streams comprising more than 5% of municipal solid waste. From 1997 to 2004, there were between 315 and 680 million sets of computers that became obsolete in the world, amounting to around 2–4 million ton of e-waste and accounting for approximately 150,000–900,000 t of PBDEs separated and emitted into the surrounding environment of e-waste recycling sites (Liu et al. 2007, 2008).

Moreover, e-waste is being transported in massive quantities around the world with about 80% exported to Asia, and 90% of the e-waste is sent illegally to China for recycling (UNEP DEWA/GRID-Europe 2005). The recycling of e-waste in Guiyu, a small town in the southeastern part of Guangdong Province, China (Figure 7.1), has been undertaken by migrant workers using primitive and environmentally unacceptable techniques to separate recoverable electronic components, mainly from computers, since the late 1980s and early 1990s (Leung et al. 2007). This entailed the burning of cables and wires for copper recovery, heating of printed circuit boards for recovery of solder and computer chips, and acid stripping of printed circuit boards for removal of remaining precious metals such as gold and platinum. During the crude recycling stage, persistent toxic substances, including PBDEs, are emitted into the environment when e-wastes are heated and burnt. The total PBDE concentrations found in Guiyu are the highest in terms of combusted residue of plastic chips and cables collected from a riverbank close to village houses in Guiyu, China (63,300 and 48,600 ng/g dry wt. for total PBDEs and BDE-209, respectively). BDE-209 was the most dominant congener (35%–82%) among the sites indicating the prevalence of commercial Deca-BDE; however, congeners from commercial Penta- and Octa-BDE mixtures were also present (Leung et al. 2007).

In addition to Guiyu, there are more than 12 e-waste recycling locations, distributed over 7 provinces in China, including Qingyuan, Longtang (northwestern part of Guangzhou), Zhao Guang, Dali in Guangdong Province,

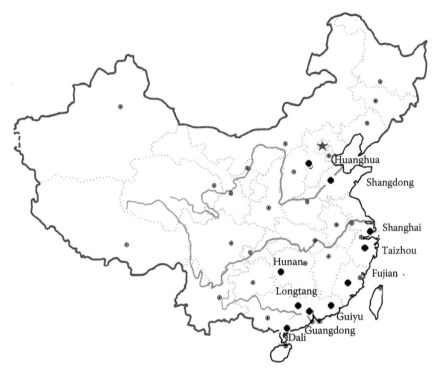

FIGURE 7.2
Location of e-waste processing sites in China. (From Ni, H.G. and Zeng, E.Y., A big picture view of halogenated chemical exposures in China—A report, pp. 1–45, 2009.)

Taizhou, and north of Qian Tang Jiang in Zhejiang Province, 1 site south of Shanghai, 2 sites in Hebei Province, and at least 1 site in each of the following provinces and cities: Shangdong, Jiangxi, Fujian, Sichuang, Hebei, and Hunan provinces (Figure 7.2). Similar to Guiyu, it is envisaged that all the sites are hotspots as sources of PBDEs emitted into the environment. The domestic production of brominated flame retardants in China in 2006 was 80,000 t, and the amount of PBDEs accumulated from all brominated flame retardants containing commercial products reached 200,000 ton in 2009 (Ni and Zeng 2009).

In addition to these 12 major sites, there are several hundred small-scale e-waste recycling sites distributed in the countryside, with waste materials imported from abroad as well as generated domestically, emitting a large amount of PBDEs into the environment. In general, all consumer products, including electrical and electronic equipment, polymer materials, and textiles, are the main sources of PBDEs emitted daily everywhere in China, both in cities and in the countryside, especially at e-waste recycling sites that are hotspots for PBDEs contamination.

Current Contamination Status of PBDEs

Environmental Matrices

Air

PBDEs are emitted into air and deposited into water and soil during their manufacture and usage in consumer products. When PBDEs are suspended in air, they may be present with particles and can deposit onto land or water column as the dust settles and are eventually washed out by snow and rainwater. It is difficult to estimate how long PBDEs remain in the air or how far they may travel. PBDEs do not dissolve easily in water; therefore, high levels of PBDEs are not usually detected in the water column (Liu et al. 2005a). When PBDEs settle into water columns, they tend to stick to particles and eventually settle into the sediments that act as reservoirs for PBDEs, especially for deca-BDE. Lower brominated PBDE congeners such as tetra- and penta- BDEs in the water column may enter aquatic organisms due to their lipophilic property; however, PBDEs do not enter groundwater since they tend to bind strongly to soil particles (Liu et al. 2005a).

When comparing data presented in Tables 7.1 and 7.2, PBDEs in the atmosphere of coastal areas were relatively lower among industrial areas and densely populated cities; for example, the average concentration of PBDEs in Hsinchu County of Taiwan was $48.5 \, pg/m^3$ (Wang et al. 2011), a rate similar to Yuen Long (urban site) and Hok Tsui (rural site) in Hong Kong (69.0 and $33.8 \, pg/m^3$ respectively), whereas in Tsuen Wan (Hong Kong), a residential area mixed with light industry, the concentration was $358 \, pg/m^3$. The concentrations of atmospheric PBDEs in Li Wan and Tian He, two inland sites within Guangzhou City, with heavy traffic, reached 204 and $372 \, pg/m^3$, respectively, which is about 4.5–8 times higher than the coastal residential areas. The atmosphere over the industrial areas of Dongguan (mainly textiles) and Shenzhen had rather high concentrations of PBDEs, 204 and $372 \, pg/m$, respectively. At the e-waste recycling site of Guiyu, total PBDEs in air reached $21,474 \, pg/m^3$ (Deng et al. 2007), which was 430 times higher than that found in the coastal areas (Chen et al. 2008a).

Among all the PBDE congeners, BDE-47, 99, and 209 were the main congeners in the air samples of Dali (e-waste recycling areas) and Dongguan (mainly textiles); both are located in Guangdong Province. The low brominated congeners such as BDE-28 mainly exist in aerosol. BDE-209 is only associated with particles, but penta-, hexa-, and octa-BDEs exist in both particles and aerosols, which was deduced from the air samples collected from the e-waste recycling site in Southern China (Dali and Guiyu) (Chen et al. 2008a,b). The air of industrial and urban areas of Guangzhou City mainly contained penta-BDE and deca-BDE, reflecting that the two congeners were derived from manufacturing industries.

TABLE 7.2

Total PBDEs and BDE-47 and BDE-99 in the Atmosphere of Hong Kong, Guangzhou (South China), and Taiwan (pg/m³)

Region	Hong Kong			Guangzhou		Guiyu	South China		Taiwan
Sites	YL	TW	HT	LW	TH	Burning	DG	DL	HC
Particles with aerodynamic diameter smaller than 2.5 μm (PM$_{2.5}$)									
BDE-47	3.8	49.5	3.2	26.8	52.9	5,004	/	/	/
BDE-99	3.6	93.3	2.1	25.2	66.1	5,489	/	/	/
ΣPBDEs	45.0	196	24.0	118	200	16,575	/	/	/
Total suspended particles (TSP)									
BDE-47	4.8	79.7	3.5	71.1	122	6,456	95.1	2,685	/
BDE-99	7.3	129	2.6	61.0	122	5,519	36.6	1,656	/
BDE-209	/	/	/	/	/	/	98.9	2,174	/
ΣPBDEs	69.0	358	33.8	204	372	21,474	4,316	9,133	48
Reference	Deng et al. (2007)						Chen et al. (2008a)		Wang et al. (2011)

Hong Kong: YL (Yuen Long), TW (Tsuen Wan), HT (Hok Tsui); Guangzhou: LW (Li Wan) and TH (Tian He); Dongguan: DG (Dongguan clothing manufacturing areas); Dali: DL (Dali small e-waste recycling area) in Guangdong Province; Taiwan: HC (Hsinchu).

Table 7.3 further compares atmospheric PBDEs of major cities in China with those of other countries. The air samples taken from the 325 m meteorological tower in Beijing contained 2.3–18 pg/m³ PBDEs (Li et al. 2009), which were similar to the data obtained in ambient areas of United Kingdom; Lake Michigan, Louisiana, the United States; and Gotska Sando, Sweden (Agrella et al. 2004), but these were much lower than the average concentrations in the atmosphere over Europe and Asia (Harrad and Hunter 2004), for example, Osaka and Kyoto, Japan (Li et al. 2009). In general, the concentrations of PBDEs in the atmosphere over Asia were found to be much higher than that over Europe. The highest concentrations of total PBDEs were detected in the e-waste areas in China and in other parts of the world. The atmosphere over the Dali and Guiyu e-waste recycling sites had the highest concentrations (9,133 and 16,575 pg/m³, respectively). In particular, the concentrations of BDE-47, 99, and 209 were greater than those of other parts of the world (Chen et al. 2006a, Deng et al. 2007, BAN and SVTC 2002). The concentration of penta-BDE (475,004 pg/m³) was even higher than that found in the workshop of an e-waste recycling site in Sweden (1,200 pg/m³) (Chen et al. 2008a). The concentrations of BDE-47 and 99 from a textile manufacturing site in Dongguan (Guangdong Province) were higher than that detected in Guangzhou, and Chicago, the United States, due to the fact that BDE-47 and 99 mixtures were blended into the foams that make clothing thick and warm (Chen et al. 2008a, D'Silvia et al. 2004). The findings from an e-waste workshop in Sweden (see Table 7.3) showed that the concentrations of BDE-47, 99, 209, and ΣPBDEs reached 1,200, 2,600, 36,000, and 64,000 pg/m³ in indoor air, respectively (Sjödin et al. 2007).

Water

PBDEs possess a very low water solubility, which reduces according to the increase in the number of bromine atoms (Liu et al. 2005a); therefore, high-molecular-weight BDE congeners such as BDE-196, 197, 206, 207, and 209 are not normally dissolved in water. This leads to very low concentrations in both fresh and marine waters, with PBDEs mainly attached to suspended particles, floating in the water column. A study conducted by the Ministry of Environment in Japan during 1988 and 1997, indicated that no hexa-, octa-, and deca-BDEs were detected in the water column of all rivers, estuaries, and coastal seawaters in Japan (Wu et al. 2008). Moreover, only 0.1–1.0 pg/L of BDE-47, 99, and 153 in the water column and 0.1–4 pg/L of BDE-209 associated with particles were detected along the coastal waters of The Netherlands (Booij et al. 2000).

In China, there is a lack of information on PBDEs in aquatic systems, especially the spatial, temporal, and vertical distributions of PBDEs in the water column of the Pearl River Estuary. Hence, the partition behavior of PBDEs between particle and dissolved phases was investigated. Table 7.4 shows that the concentrations of PBDEs in water were lower during late spring and early

TABLE 7.3

Concentrations of ΣPBDEs and Main Congeners in the Atmosphere of China in Comparison with Other Parts of the World (pg/m^3)

Sites	BDE-47	BDE-99	BDE-209	ΣPBDEs	References
China					
Beijing air (top air of Beijing 325 m Meteorological Tower)	/	/	/	2.3–18.0	Li et al. (2009)
Guangdong Province:					
Guiyu e-waste site	6456	5519	/	21,474	Deng et al. (2007)
Guiyu burning air	5180	5889	/	16,822	Wong et al. (2007)
Dali, e-waste site	2685	1656	2,174	9,133	Chen et al. (2008a)
Dongguan, textile	95.1	36.6	98.9	316	Chen et al. (2008a)
Guangzhou urban 1	26.8	31.8	478	577	Chen et al. (2008a)
Guangzhou urban 2	26.9	17.5	264	346	Chen et al. (2008a)
South China:					
Industrial area 1	2182	1058	4,192	7,859	Chen et al. (2008a)
Industrial area 2	74.7	66.1	750	973	Chen et al. (2008a)
Nanjing urban air	5.6	5.6	65.0	93.1	Zhao et al. (2008)
Taiwan	/	/	/	48.5	Wang et al. (2011)
Taiwan, e-waste site	/	/	/	100–190	Watanabe et al. (1992)
Other parts of the world					
Osaka (1994), Japan	/	/	83–3,060	90–3,081	Watanabe et al. (1995)
Osaka (2001), Japan	/	/	100–340	104–347	Ohta et al. (2001)
Kyoto, Japan	/	/	0–48	4.5–113	Hawakawa et al. (2004)
Michigan Lake	8.4	5.3	<0.1	15.1	Hoh and Hites (2005)
Michigan, the United States	6.2	5.1	1.5	16.2	Hoh and Hites (2005)
Chicago (1999), the United States	33.0	16.0	0.3	52.2	Chen et al. (2008a)
Chicago (2003), the United States	17.0	7.4	60.1	100	Chen et al. (2008a)
Louisiana, the United States	6.9	3.0	2.8	16.4	Hoh and Hites (2005)
e-waste workshop (Sweden)	1200	2600	36,000	64,000	Sjödin et al. (2007)

(continued)

TABLE 7.3 (continued)

Concentrations of ΣPBDEs and Main Congeners in the Atmosphere of China in Comparison with Other Parts of the World (pg/m³)

Sites	BDE-47	BDE-99	BDE-209	ΣPBDEs	References
Gotska Sando (Sweden)	1.70	0.90	6.50	10.0	Agrella et al. (2004)
Ambient United Kingdom	/	/	/	6.7	Harrad and Hunter (2004)
Over Europe	/	/	/	250	Wurl et al. (2006b)
Over Asia	/	/	/	340	Wurl et al. (2006b)
Over Indian Ocean	/	/	/	2.5	Wurl et al. (2006b)

TABLE 7.4

Concentrations of PBDEs in Particulate and Dissolved Phases at the Mouth of the Pearl River (Average Values of pg/L from Surface, Middle, and Bottom of the Water Column at Each Sampling Location)

			May 2005			
Dissolved	BDE-28	BDE-47	BDE-99	BDE-183	ΣPBDEs	BDE-209
No. 1	nd	4.4	4.5	nd	11.7	nd
No. 2	2.3	5.2	4.6	nd	12.6	nd
No. 3	nd	6.0	8.3	nd	16.4	nd
Particles	BDE-28	BDE-47	BDE-99	BDE-183	ΣPBDEs	BDE-209
No. 1	0.6	5.1	5.1	1.7	15.3	588
No. 2	0.9	7.4	7.1	1.9	21.9	507
No. 3	1.8	18.2	14.9	4.5	49.2	4001
			October 2005			
Dissolved	BDE-28	BDE-47	BDE-99	BDE-183	ΣPBDEs	BDE-209
No. 1	15.3	49.4	18.5	4.6	99.3	nd
No. 2	nd	33.7	13.9	3.3	59.1	nd
No. 3	4.3	24.9	12.4	2.2	49.8	nd
Particles	BDE-28	BDE-47	BDE-99	BDE-183	ΣPBDEs	BDE-209
No. 1	1.1	14.9	9.3	nd	32.6	838
No. 2	0.3	12.2	7.9	nd	27.4	1193
No. 3	nd	12.1	7.8	nd	26.4	2121

Source: Luo, X.J. et al., *Chin. Sci. Bull.*, 53, 93, 2008.

summer, which is due to the influence of the open seawater, but were higher in October as a result of river runoff, accompanied by a higher amount of particulate matter (Guan et al. 2007, Luo et al. 2008).

Both BDE-47 and 99 were detected in all dissolved and particulate samples for both wet (May) and dry (October) seasons of 2005 in the Pearl River Delta,

TABLE 7.5

Mean Concentrations (pg/L) of PBDEs in the Surface Microlayer and Subsurface Water Column, at 1 m Depth, Hong Kong Coastal Waters

Locations	Victoria Harbor	Tolo Harbor	Aberdeen Typhoon Shelter
ΣPBDEs in SML	297.3	77.6	71.1
ΣPBDEs in SWC	94.8	57.3	11.3
Ratio SML/SWC	3.14	1.35	6.29

Sources: Wurl, O. and Obbard, J.P., *Mar. Pollut. Bull.*, 48, 1016, 2004; Wurl, O. and Obbard, J.P., *Chemosphere*, 58, 925, 2005.

as they are the most common and abundant congeners found in the environment. On the contrary, BDE-209 did not dissolve in water, due to its high log K_{ow} (close to 10). It was mostly associated with particles and had high concentrations (76–5693 pg/L) for both wet and dry seasons. Particularly high concentrations were observed during the wet season. In general, total PBDEs were much higher in the aqueous phase in the Pearl River Estuary due to river runoff (Table 7.4, Luo et al. 2008).

The surface micro layer (SML) of water represents the boundary layer between the atmosphere and the subsurface water column (SWC) and has a typical thickness of 40–100 μm consisting of enriched organic compounds (such as lipids, proteins, and deposit particles from the atmosphere). A study related to Hong Kong coastal waters revealed that PBDEs attached to particles and in aerosols, accumulated in the SML before being transferred into the water body. Table 7.5 shows that the proportion of total PBDEs in the SML over the SWC ranged from 1.35 to 6.29, indicating that the concentrations of ΣPBDEs in the sea surface microlayer (SML) were much higher than that found in the SWC in Hong Kong (Wurl and Obbard 2004, 2005, Wurl et al. 2006a).

Soil and Sediment

Since the 1980s, more than 20 companies and factories have been established along the southeastern coastal areas of China, for the commercial production of Deca-BDE, with 4000–7000 ton produced each year. Penta-BDE and Octa-BDE made up approximately 12% and 6%, respectively, of the PBDE production market in China. In general, more than 80% of the PBDEs produced in China were used to manufacture electrical and electronic equipment, polymer materials, and textiles (Chen and Zhang 1998). Due to leaching from the surface of various consumer products, together with emissions from the open burning of e-waste, PBDEs are now widely distributed in different ecological compartments, where they then become associated with particulates and are eventually deposited in soil and sediments.

Guiyu, a small town (52.4 km^2) located in the southeastern part of Guangdong Province, with a population of 150,000, has been involved in e-waste recycling for more than a decade. It was found that combusted residue from open burning contained extremely high concentrations of ΣPBDEs (excluding BDE-209; 14,700 ng/g dry wt.) and BDE-209 (48,600 ng/g dry wt.). Other areas were also affected, for example, soil from an acid leaching site, printer roller dump site, and a duck pond, all having high concentrations of ΣPBDEs (excluding BDE-209) (70–2300 ng/g) and BDE-209 (328–1270 ng/g). Primitive recycling operations consisted of scrap sorting; heating printed circuit boards to recover lead solder and electronic components; using concentrated acid solutions to leach and recover aluminum, gold, copper, platinum, and other metals; toner sweeping; dismantling electronic equipment; selling computer monitor yokes to copper recovery operations; plastic chipping and melting; and burning cables to recover copper (Table 7.6) (Leung et al. 2007).

Table 7.6 shows different hotspots of PBDEs in China. The sediments from a river in Guiyu contained extremely high concentrations of ΣPBDEs (16,088 ng/g) and BDE-209 (62.2 ng/g) (dry weight). There were a total of five hotspots with high concentrations of BDE-209, which included the mouth of the Bohai Sea, 2,776; Pearl River Estuary, 7,340; the mouth of the Pearl River, 3,575; Laizhou Bay, 6,189; and Guiyu, 48,600 ng/g (combusted residue).

The mouth of the Yangtze River and sediments from the Nanjing Yangtze River both had lower ΣPBDEs and BDE-209 concentrations of 3.89 and 0.34 ng/g, respectively (Shen et al. 2006). Therefore, PBDE contamination seemed to be more serious in the northern coastal waters (such as Bohai Sea) and southern coastal waters (Pearl River Delta) of China. This is especially true for the Pearl River Delta, which has been transformed into one of the fastest developing industrial manufacturing areas in the world (Liu et al. 2005b, Zheng et al. 2004), but even higher ΣPBDEs and BDE-209 (10.2 and 43.8 ng/g, respectively) were detected in the coastal sediment of Macao, an entertainment/commercial city located at the mouth of the Pearl River Delta (Ni and Zeng 2009). Although Xiamen (a mid-sized city located in Southeastern China) has been established as a "Special Economic Developing Region" (Zhou et al. 2009), the pace of development is comparatively slower, resulting in lower PBDE contamination in the coastal areas of the Xiamen Harbor (0.1–2.06 ng/g dry wt.) (Ou 2006).

Table 7.6 shows that the Deca-BDE manufacturing factory in Laizhou Bay (Shangdong Province) contained high ΣPBDEs and BDE-209 levels, with up to 1860 and 1814 ng/g dry wt. in the residential area, respectively, and 7191 and 7120 ng/g dry wt. at the factory entrance, respectively (Jin et al. 2008).

The concentrations of PBDEs in sediments collected from estuaries, coastal waters, and mouths of rivers in China matched those of other parts of the world. This was demonstrated using the results of the analysis of sediments collected from the mouth of the Yangtze River and the Dagu waste discharge point of Tianjin, which contained 0.5, 14.9 and 0.55, 94.6 ng/g dry wt. of

TABLE 7.6

Concentrations of PBDEs in Soils and Sediments of China (ng/g dry wt.)

Location	BDE-47	BDE-99	ΣPBDEs[a]	BDE-209	References
In sediments					
River sediments Beijing	0.04	0.03	0.91	11.6	Chen et al. (2009)
Dagu River mouth	/	/	0.1–0.5	14.9	Zou et al. (2009)
Bohai sea, China	/	/	5.40	15.55	Lin et al. (2008)
Beihe mouth, China	/	/	5.24	2,776.5	Lin et al. (2008)
Laizhou Wan	/	/	1.3–17.6	223.8	Jin et al. (2008)
Qingdao coastal	/	/	0.12–5.51	/	Yang et al. (2003)
Yangtze River Estuary	/	/	nd–0.55	94.6	Chen et al. (2006)
Yangtze River, Nanjing	0.26	0.27	2.08–3.89	0.34	Shen et al. (2006)
Xiamen Harbor, Fujian	0.54	0.11	0.10–2.06	2.06	Ou (2006)
Pearl River Estuary	/	/	0.4–94.7	7,340	Mai et al. (2005)
Pearl River, China	/	/	0.78–49.28	3,575	Huang et al. (2009a)
Guiyu River, South China	/	/	51.3–16,088	62.2	Luo et al. (2007)
Hong Kong Coastal	3.68		1.70–53.6	2.71	Liu et al. (2005b)
Macao, China	/		10.2	43.8	Lin et al. (2008)
In soils					
Dongquan, South China	0.96	0.39	0.33–1.79	66.6	Zhou et al. (2009)
Guangzhou city	2.17	0.50	0.13–2.59	34.5	Zhou et al. (2009)
Qingyuan, South China	1.18	0.89	0.98–3.81	36.8	Zhou et al. (2009)
Guiyu 1, South China[b]	73.6	149	2,300	1,270	Leung et al. (2007)
Guiyu 2, South China	129	333	930	510	Leung et al. (2007)
Guiyu 3, South China	1.78	1.70	70	398	Leung et al. (2007)
Guiyu 4, South China	0.64	0.42	10.9	37.3	Leung et al. (2007)
Guiyu 5, South China	15.4	24.0	14,700	48,600	Leung et al. (2007)
Laizhou Bay 1, China[c]	2.6	3.1	1,436.2	1,481.1	Jin et al. (2008)
Laizhou Bay 2, China	1.6	3.0	198.8	587.8	Jin et al. (2008)
Laizhou Bay 3, China	9.9	29.2	6,342	6,189	Jin et al. (2008)
Laizhou Bay 4, China	0.5	1.3	77.2	96.6	Jin et al. (2008)
Laizhou Bay 5, China	1.3	1.1	223.8	241.4	Jin et al. (2008)

The concentrations of ΣPBDEs excluded BDE-209.

[a] ΣPBDEs: excluding BDE-209.

[b] Guiyu 1, 2, 3, 4, and 5 e-waste samples were taken from the following sites: (1) acid leaching, (2) printer roller dump site, (3) duck pond, (4) rice field, and (5) combusted residue site, Guiyu, Guangdong Province, China.

[c] Laizhou 1, 2, 3, 4, and 5 soil samples were taken from the following sites: (1) A Deca-BDE production factory worker living area, (2) east side of the factory, (3) factory entrance, (4) Laizhou Bay sediment, and (5) local river sediment.

ΣPBDEs and BDE-209, respectively. These results were close to the readings detected in Tokyo Bay, Osaka Bay, Japan; the mouth of the Scheld River, the Netherlands; the mouth of the Cinca River, Spain; and Hadley Lake, Indiana, the United States, where the levels of ΣPBDEs and BDE-209 were 3.6, 85; 1.9, 350; 24, 22; 41.7, 39.9; and 13.9, 28.8 ng/g dry wt., respectively.

However, there were several exceptions; for example, Laizhou Bay and a river in Guiyu had very high ΣPBDEs and BDE-209 (7,191 and 7,120 ng/g dry wt. at Laizhou Bay, and 16,088 and 62.2 ng/g [dry wt.] at Guiyu, respectively), because of two industries operating in these two regions: a Deca-BDE factory located at Laizhou Bay and an e-waste recycling unit located at Guiyu (Hong Kong Greenpeace Organization 2003, Jin et al. 2008, Leung et al. 2007). The concentrations found in these two regions were higher than other industrial sites in the world, for example, the industrial harbor in South Korea (ΣPBDEs and BDE-209 at 2253 and 2248 ng/g, respectively), the mouth of the Tees River, United Kingdom (ΣPBDEs: 1271 ng/g); and the mouth of the Viskan River, Sweden, which is located next to a Deca-BDE production factory (BDE-209: 7100 ng/g) (Allchin et al. 1999, Moon et al. 2002, Sellström et al. 1998) (Table 7.7).

Biota

PBDEs are hydrophobic organic compounds, which can be easily attached to lipids inside the bodies of biota through two pathways: (1) by bioconcentration directly through the water column and (2) biomagnifications through the food web (Kiriluk et al. 1997). De Carlo and Ann (1979) first detected PBDEs in soil and sludge close to a PBDE production factory in the United States, while Andersson and Blomkvist (1981) detected PBDEs in fish in Sweden (Table 7.8).

When comparing bivalve samples collected along China's coastal waters (from the north such as Dalian, Qingdao, Shanghai, and Ninbo to the south of China), the highest ΣPBDEs levels were found in mussel tissue (Perna viridis) from Hong Kong coastal areas, at 27.0–83.7 ng/g dry wt. (about 300–8500 ng/g lipid). Within the same areas, ΣPBDEs in surface sediments ranged from 1.7 to 53.6 ng/g dry wt., and the bioaccumulation ratio (mussel to sediment) of Σ_{15}PBDEs was 1.6–4.7 (Liu et al. 2005b). ΣPBDEs in dolphin blubber reached 980 ng/g dry wt. in 2005 and 4030 ng/g dry wt. in 2008 (Ramu et al. 2005, Kajiwara et al. 2006, Lau 2008, Wang et al. 2007). There seems to be an indication that the concentration increased with time, although the increase may be also due to variation in each individual dolphin.

In areas with extensive e-waste activities, fish contaminated by PBDEs was commonly observed, for example, in muscle and liver of common carp collected from local rivers in Guiyu (ΣPBDEs 115–1088 and 2687 ng/g wet wt., respectively) (Liu et al. 2008) and common carp from the Yangtze River and perch from the Bohai Sea (ΣPBDEs 1100 and 3230 ng/g wet wt., respectively) (Xian et al. 2008). Like the Pearl River Delta, PBDEs from the surrounding environment appeared to be concentrated at the Yangtze River Delta where

TABLE 7.7

Comparison of Concentrations of PBDEs in Soils and Sediments of China with Those of Other Parts of the World (ng/g dry wt.)

Location	Congener No.	ΣPBDEs	BDE-209	References
Bohai Bay, China	/	5.24	2777	Lin et al. (2008)
Dagu waste discharge mouth, Tianjin, China	13	0.5	14.9	Lu et al. (2007)
Qingdao coastal, Shangdong	21	5.51	/	Yang et al. (2003)
Laizhou Bay, Shangdong	11	7,190	7119	Jin et al. (2008)
Yangtze River mouth	40	0.55	94.6	Chen et al. (2006)
Xiamen Harbor	14	2.06	2.06	Ou (2006)
Guiyu river, South China	14	16,088	62.2	Luo et al. (2007)
Hong Kong coastal waters	14	52.6	2.92	Liu et al. (2005b)
The Pearl River Delta	18	7,340	/	Zeng (2005)
Cinca River Spanish	7	41.7	39.9	Eljattat et al. (2004)
Industrial Harbor, South Korea	20	2,253	2248	Moon et al. (2002)
Niagara River, the United States	9	148	/	Samara et al. (2006)
San Francisco Bay, the United States	22	212	/	Oros et al. (2005)
Hadley lake, the United States	/	13.9	28.8	Song et al. (2004)
Tees River mouth, United Kingdom	/	1,271	339	Allchin et al. (1999)
Scheld River mouth, Holland	/	24	22	De Boer et al. (2003)
Viskan River mouth, Sweden	/	50	7100	Sellström et al. (1998)
Coastal, Portugal	/	34.1	/	Lacorte et al. (2003)
Tokyo Bay, Japan	/	3.6	85	Lin et al. (2008)
Osaka Bay, Japan	/	1.9	350	Luo et al. (2009)

The concentrations of ΣPBDEs excluded BDE-209.

high concentrations of ΣPBDEs in different aquatic organisms (mussel, oyster, clam, shrimp, and crab, 40.2–83.5 ng/g dry wt.) (Liu et al. 2005b) were detected.

At a small e-waste recycling site at Qingyang County (northern part of Guangdong Province), the concentrations of ΣPBDEs in the muscle and liver of field forage chicken reached 30.5 and 50.7 ng/g lipid, respectively, with the highest ΣPBDE concentration (7917 ng/g lipid) obtained from individual chickens (Liu et al. 2009).

In general, the levels of PBDEs obtained in different aquatic biota samples (collected from different locations) were within ng/g–μg/g lipid. The PBDE distribution patterns were similar, with BDE-47, 99, 100, 153, and 154 being the most common congeners observed, and BDE-47 contributing more than 50% of ΣPBDEs (Kazuhiko et al. 2001). However, for terrestrial animals, levels of ΣPBDEs were lower than those of aquatic biota samples; for example, Myna birds from South China had a ΣPBDEs of 5.7–13 ng/g lipid, and

TABLE 7.8

Concentrations of PBDEs (ng/g Lipid) in Biota Samples in China Compared with Those of Other Parts of the World

Location and Species	BDE-47	ΣPBDEs	References
China			
Mussels, Pearl River Delta, Guangdong	/	40.2–83.5 ng/g dry wt.	Liu et al. (2005b)
Shrimps, Pearl River Mouth	/	266.4	Xiang et al. (2007)
Fish market, Guangzhou	/	3850	Meng et al. (2007)
Fish liver of Guiyu, Guangdong	/	2687 ng/g wet wt.	Luo et al. (2007)
Carp of Guiyu, Guangdong	/	1088 ng/g wet wt.	Luo et al. (2007)
Mussel, Hong Kong coastal waters	/	40.2.0–83.5	Liu et al. (2005b)
Indo-Pacific humpback dolphin, Hong Kong	/	6000	Kajiwara et al. (2006)
Common carps, Yangtze River	/	1100	Xian et al. (2008)
Common carp muscle, Nongkang River	14.0	50.0	Xu et al. (2009)
Common carp egg, Nongkang River (Jiansu Province)	8.9	29.0	Xu et al. (2009)
Mussel, Qingdao coastal	/	0.75 ng/g dry wt.	Yang et al. (2003)
Shell fish, Laizhou Bay, Shangdong	/	720	Jin et al. (2008)
Molluscs, Bahai Bay	/	238.6 ng/ dry wt.	Wan et al. (2008)
Perch fish, Bohai Bay	/	3230 ng/g wet wt.	Wan et al. (2008)
Fishes rivers and estuaries, Taiwan	/	30.6–281	Peng et al. (2007)
Other parts of the world			
Salamander, M Psa Lakes, Norway	/	1120	Mariumssen et al. (2008)
Polar bear, Norway	/	70	Muir et al. (2006)
Feral carp of Lobregat River, Spanish	/	744	Labandeira et al. (2007)
Dolphin, Seto Inland Sea, Japan	/	1300	Kajiwara et al. (2006)
Salamander, Baltio Sea	/	180	Asplund et al. (1999)
Barracuda, Viskan River, Sweden	/	4600	Sellström et al. (1998)

All units are in ng/g lipid unless otherwise stated.

BDE-153 and 99 were the most dominant congeners commonly found in terrestrial animals (Luo et al. 2009).

BDE-209 was the main congener detected in the tissue of vulture, whereas BDE-153 was commonly found in Chinese ibis and small eagle, both of which are birds of prey. The eggs of silver gill sampled from Taihu Lake (located in the mid-eastern part of China) contained mainly BDE-209, 206, and 207

(Luo et al. 2009). It seems that terrestrial animals (including birds) may accumulate high-molecular-weight PBDE congeners, by feeding on food contaminated with them and breathing in dust containing BDE-196, 197, 206, 207, and 209, which are attached to particles drifting in the air, whereas aquatic organisms are unlikely to take up BDE-209 and other high-molecular-weight congeners because they are commonly absent in the water column (Law et al. 2006, Luo et al. 2009).

It has been commonly observed that the concentrations of ΣPBDEs in human milk have increased with time, doubling every 5 years within the past 30 years in Europe, Asia, and America (Schecter et al. 2003, Hites 2004). A study showed that ΣPBDEs in Swedish human milk increased 60 times from 0.07 ng/g lipid in 1972 to 4.01 in 1997 (Meironyté et al. 1999). It has been estimated that the concentrations of ΣPBDEs in human bodies increased almost 100 times during the past 30 years (Wang and Jiang 2008). This has become a public health concern in China recently due to the rather high concentrations of PBDEs detected in different ecological compartments, especially in several hotspots throughout China.

Table 7.9 shows ΣPBDEs in human milk samples from both urban and rural areas in Beijing were lower (0.97 and 1.22 ng/g lipid, respectively) than most other areas in China, for example, Nanjing, Central Taiwan, Zhoushan Islands, Zhejiang Province, and South China, ranging from 3.15 to 7.7 ng/g lipid (Bi et al. 2007, Sudaryanto et al. 2008). This is possibly due to different geographical and meteorological conditions and lifestyle choices, including dietary habits. There is also a potential for PBDEs to transfer from mothers to babies, with median ΣPBDE concentrations found in fetal serum at 3.9 ng/g lipid, whereas the amount of ΣPBDEs present in maternal serum is measured at 4.40 ng/g lipid (Bi et al. 2006).

Human tissues (kidney, liver, and lung) from residents of the e-waste recycling sites in Zhejiang Province contained ΣPBDEs of 118–270 ng/g lipid, with BDE-47, 99, 100, 153, and 154 being the most common congeners. In particular, BDE-47 contributed to more than half of the total PBDEs. These PBDE congeners are also commonly found in biota samples, especially in aquatic biota; however, residents and workers of e-waste sites always have high body loadings (in milk, serum, and blood) of BDE-209. This may be attributed to their increased exposure to BDE-209 due to the nature of their recycling jobs with e-waste (Liu et al. 2005a). The exposure pathways include inhalation, skin contact, and oral intake of contaminated water and food. Cancer patients from e-waste sites contained high concentrations of BDE-209 (191–270 ng/g lipid), with some individual patients containing body burdens ranging from 38,888 to 72,964 ng/g lipid, about 200 to 300 times higher than their normal counterparts (Zhao et al. 2009). At Guiyu, e-waste workers also had high BDE-209 in their serum samples, up to 3439 ng/g lipid (Bi et al. 2007).

An attempt has been made to compare the concentrations of PBDEs in Chinese people to populations in other parts of the world, with 1157 sets of

TABLE 7.9

Concentrations of PBDEs (ng/g Lipid) in Human Tissues of China Compared with Those from Other Parts of the World

	BDE-47	ΣPBDEs	BDE-209	References
Human body, China				
Breast milk, South China	/	3.15	/	Bi et al. (2006)
Fetal serum, South China	/	3.84	/	Bi et al. (2006)
Placenta, Shenzhen, South China	/	25.99	/	Zhang et al. (2008)
Cord blood, pregnant woman Shenzhen, South China	/	23.4	/	Zhang et al. (2010)
Cancer patients in e-waste disassembly Zhejiang, East China				
Kidney	53.6	182.3	191.3 (72,964)*	Zhao et al. (2009)
Liver	50.8	192.6	118.1 (61,625)*	Zhao et al. (2009)
Lung	38.7	172.0	270.0 (38,888)*	Zhao et al. (2009)
Serum, e-waste workers, Guiyu, China	/	/	3,439	Bi et al. (2007)
Patients with uterine Leiomyomas of Hong Kong				
Subcutaneous fat	2.65	10.8	/	Qin et al. (2009)
Visceral fat	3.50	13.8	/	Qin et al. (2009)
Normal subcutaneous fat	2.19	4.44	/	Qin et al. (2009)
Human breast milk, Taiwan	/	3.93	/	Chao et al. (2007)
Human breast milk, urban, Beijing	/	1.22	/	Li et al. (2008)
Human breast milk, rural, Beijing	/	0.97	/	Li et al. (2008)
Other parts of the world				
Serum, common person, the United States	/	366	/	Schecter et al. (2006)
Serum, Vitatalian, the United States	/	127	/	Schecter et al. (2006)
Human breast lipid, 1997, the United States	18.0	38.3	/	She et al. (2002)
Common lipid, 2004, the United States	132	399	/	Johnson-Restrepo et al. (2005)
Human breast milk, UK	/	150	/	Kalantzi et al. (2004)
Human breast milk, Osaka, Japan	/	291	/	Huang et al. (2005)
Human breast milk, Akita, Japan	/	5.44	/	Eslami et al. (2006)
Human lipid, 2001, Sweden	8.80	13.4	/	Haglund et al. (1997)
Placenta, 1996, Finland	0.77	1.58	/	Danial et al. (2004)

TABLE 7.9 (continued)

Concentrations of PBDEs (ng/g Lipid) in Human Tissues of China Compared with Those from Other Parts of the World

	BDE-47	ΣPBDEs	BDE-209	References
Blood, computer workers, Spanish	/	15.0		Takumi et al. (2004)
Serum of computer workers, Sweden	/	26	/	Sjödin et al. (2007)

Cancer patients: Kidney (72,964)* Indicated the highest concentrations of BDE-209 at 72,964 ng/g lipid in 1 kidney tissue among 55 patients. Liver (61,625)* and lung (38,888)* also showed the highest concentrations of BDE-209 in the liver and lung, respectively, among the 55 patients studied.

data of ΣPBDEs in blood and serum samples between 1977 and 2007, from 10 countries (Wang and Jiang 2008). Results showed that ΣPBDEs were the highest in blood serum from the United States, about 7–35 times higher than those from Europe (Table 7.9), followed by samples from Europe and Asia (Japan and South Korea). In general, ΣPBDEs from the blood serum of pregnant mothers and their newborn babies from South China were relatively lower when compared with data from other countries (Luo et al. 2009, Wang and Jiang 2008); for example, milk samples of Japanese women reached 291 ng/g lipid, which was the highest among Asian countries (Huang et al. 2005). However, milk samples of e-waste workers in China had extremely high ΣPBDEs (192 ng/g lipid), especially BDE-209 (270 ng/g lipid), with one individual reaching 72,964 ng/g lipid (Zhao et al. 2009). The data show that the levels of ΣPBDEs were much higher than those of computer workers in Spain (Takumi et al. 2004).

Impacts on Human Health

Because PBDEs are structurally similar to PCBs, concerns have been raised on their possible adverse health effects, particularly neurotoxic effects in newborns and children (Chao et al. 2007). In fact, PBDEs have been associated with endocrine disruption, reproductive-developmental toxicity, including neurotoxicity and even cancer, due to their common presence in abiotic and biotic environments and in human tissues. A study that used animals tested the effects of Deca-BDE exposure on maternal rats indicated that Deca-BDE and its metabolites were passed onto their offspring through milk, which affected their learning and memorizing ability (Chen et al. 2006b, Jiang et al. 2009).

Emerging studies on long-term neurological effects of acute administration of PBDEs during development suggests that these compounds can act as potent thyroid hormone mimetics. A hypothesis that PBDEs may also serve as risk factors for autism emerges, when these data are considered in combination with the extensive literature on stage-dependent effects of thyroid hormone on aspects of brain development, an implication specifically related to autistic brains (Messer 2010).

Based on animal models, it has been revealed that the large-scale commercial Deca-BDE mixture used in China is less toxic than lower-brominated PBDEs. However, Deca-BDE decomposes to lower BDE congeners, and rats and mice fed with food contaminated by the PBDE mixture have been shown to develop thyroid problems. It is speculated that many of the thyroid effects of PBDEs are specific to the species of test animals (Chen et al. 2006b). Subtle behavioral changes have been observed in animals exposed to PBDEs as infants. One possible explanation for the behavioral effects may be related to changes in the thyroid because the development of the nervous system is dependent on thyroid hormones. Results from animal testing also showed that PBDEs may impair the immune system. In addition, it has been observed that rats and mice that were exposed to PBDEs developed liver tumors. Although it has not been proved whether PBDEs will cause cancer in humans, the fact that cancer patients have high concentrations of ΣPBDEs, BDE-47, and BDE-209 should not be overlooked (Qin et al. 2009).

BDE-209 possesses the toxic activity of antiproliferation and induction of apoptosis in tumor cells *in vitro*. BDE-47 causes a concentration-dependent decrease in cell proliferation and is cytotoxic and genotoxic to SH-SY5Y cells *in vitro*, which results in the loss of neuronal characteristics and adversely affects learning ability (He et al. 2007, Zhang et al. 2007). The toxicological end points of concern for environmental levels of ΣPBDEs are likely to be thyroid hormone disruption, neurodevelopmental deficits, and cancer (McDonald 2002).

According to Table 7.9, it could not be confirmed whether cancer patients who had the disease had high concentrations of several BDE congeners, such as BDE-47, 99, 100, 153, 154, BDE-209, and total PBDEs, based on the data collected from the e-waste sites in Zhejiang Province. However, the patients contained a rather high concentration of PBDEs in their body (kidney, liver, and lung), with BDE 47, BDE-209, and ΣPBDEs reaching 38.7–53.6, 172–192, and 118–270 ng/g lipid, respectively, when compared with healthy people. In some human samples, BDE-209 concentrations were up to 38,888–72,964 ng/g lipid in cancer patients living around the e-waste site in Zhejiang Province (Zhao et al. 2009). Our recent study conducted in Hong Kong showed that BDE-47 and ΣPBDEs reached 2.65–3.50 and 10.8–13.8 ng/g lipid, respectively, from the subcutaneous fat and visceral fat of patients with uterine leiomyomas, which were two to three times higher than the levels found in healthy people (Qin et al. 2009). Chao et al. indicated that pregnant women exposed to rather low PBDE concentrations (about 2.0–5.0 ng/g lipid) in central Taiwan had babies with lower birth weights and shorter birth lengths (length of the baby at birth) (Chao et al. 2007).

Zhang et al. conducted an investigation on the relationship between BDE-47 and oxidative stress and apoptosis in human neuroblastoma SH-SY5Y cells, in order to determine the role of reactive oxygen species (ROS) on apoptosis. Results showed that the rate of cellular survivors in the high-dose PBDE-47-treated group was significantly higher than that found in the control group; therefore, it can be concluded that BDE-47 can induce oxidative stress and apoptosis in SH-SY5Y cells. ROS may play an important role in apoptosis induced by BDE-47 (Zhang et al. 2007).

Methoxyresorufan-*o*-demethylase (MROD) was used to analyze competitive inhibition from BDE-47, 77, 85, 119, and 126 to Cytochrome P450 1A2, which is a monooxidizing enzyme that can speed up the metabolism of ions of pollutants and carcinogenic chemicals in rat hepatocytic microsomes. It is known that PBDE congeners have similar, but somewhat lower, inhibition to Cytochrome P450 1A2, as dioxins (PCDD/Fs) (Nie et al. 2005). Therefore, the presence of PBDE congeners may reduce the human body's ability to metabolize ions of organic pollutants, leading to a high accumulation of these pollutants in the body (Yang et al. 2009).

Furthermore, PBDEs may combine with aromatic hydrocarbons and form dioxin-like chemicals; when waste materials containing PBDEs are burnt at e-waste sites, polybrominated dibenzo-*p*-dioxin (PBDDs) and polybrominated dibenzofuran (PBDFs) are produced, which possess strong acute toxicity and pose long-term adverse effects on human health (Chen et al. 2001, Wang and Jiang 2008).

Conclusion

There were seven hotspots of PBDEs throughout China as shown in Figure 7.3. Four of these hotspots were from sediments and soil as follows: (1) sediments from the mouth of the Beihe River of the Bohai Sea, 2,777 ng/g dry wt; (2) sediments of the Pearl River Estuary, 7,340 ng/g dry wt; (3) sediments from Laizhou Wan, 6,189 ng/g dry wt; and (4) the combusted residues at Guiyu e-waste sites, 48,600 ng/g dry wt. There was one hotspot with regard to air: (5) the air of Guangzhou and Dongguan, with PBDEs found to range between 7,859 and 9,133 pg/m. There were several biota that had high concentrations of PBDEs (6) dolphins, Hong Kong, 6,000 ng/g lipid; fish, Guangzhou markets, 3,850 ng/g lipid; fish, Bohai Sea, 230 ng/g lipid; and fish, Guiyu, 2,687 ng/g lipid. There were also high concentrations of PBDEs in human tissue (7) from Guiyu, 3,499 ng/g lipid, and from cancer patients in Zhejiang Province, 72,964 ng/g lipid.

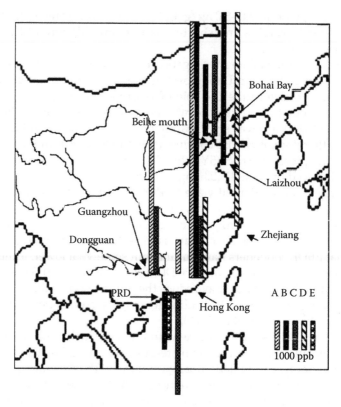

FIGURE 7.3
PBDE contamination hotspots in China: (A) air, (B) sediment, (C) fish tissue, (D) human tissue, and (E) seawater column.

PBDEs are known to be endocrine disruptors, which destroy thyroid hormone balance, impose neurodevelopmental deficits and reproductive developmental toxicity, and even cause cancer. The concentrations of ΣPBDEs in air, sediment, soil, fish, and human tissues from Guiyu, the most intensive e-waste site, were the highest among the 12 large-scale e-waste recycling sites in China. There seems to be sufficient evidence showing bioaccumulation and biomagnifications of PBDEs in some of these sites, leading to very high concentrations in human tissues, particularly in cancer patients.

Due to rapid industrialization and modernization, large amounts of PBDEs have been produced and used in China. Apart from those industrial hotspots, it is envisaged that emission of PBDEs into air from electrical and electronic equipment used in offices and homes is fast becoming a serious problem in China, especially because these products are now gaining popularity in villages and more remote areas. Therefore, there is an urgent need to address these pressing issues related to PBDEs.

Acknowledgments

This research is supported by The Research Grants Council of the University Grants Committee of Hong Kong (Central Allocation Group Research Project HKBU 1/03C). We would also like to thank for the support from the mini AoE of the Hong Kong Baptist University and Special Equipment Grant of RGC RC/AoE/08-09/01 and SEG HKBU 09.

References

Agrella C, ter Schurea AFH, Svedera J. 2004. Polybrominated diphenyl ethers (PBDEs) at a solid waste incineration plant 1: Atmospheric concentrations. *Atmospheric Environment* 38: 5139–5148.

Allchin CR, Law RJ, Morris S. 1999. Levels and trends of brominated flame retardants in the European environment. *Environmental Pollution* 105: 197–207.

Andersson I, Blomkvist G. 1981. Polybrominated diphenyl ethers (PBDEs) in fish samples Sweden. *Chemosphere* 10: 1051–1060.

Asplund L, Athanasiadou M, Sjodin AA, Borjeson H. 1999. Organohalogen substances in muscle, egg and blood from healthy Baltic salmon (*Salmo salar*) and Baltic salmon that produced offspring with M74 syndrome. *Ambio* 28: 67–76.

BAN and SVTC (The Basel Action Network and Silicon Valley Toxics Coalition). 2002. Exporting harm: The high-tech trashing of Asia. February 25, 2002. http://www.ban.org/E-wastetechnotrashfinalcomp.pdf

Bi XH, Qu WY, Sheng GY, Zhang WB, Mai BX, Chen DJ, Yu L, Fu JM. 2006. Polybrominated diphenyl ethers in South China maternal and fetal blood and breast milk. *Environmental Pollution* 144: 1024–1030.

Bi XH, Thomas GO, Jones KC, Qu WY, Sheng GY, Martin FL, Fu JM. 2007. Exposure of electronics dismantling workers to polybrominated diphenyl ethers, polychlorinated biphenyls, and organic chlorinated pesticides in south China. *Environmental Science and Technology* 41: 5647–5653.

Booij K, Zegers B, Boon J. 2000. Levels of some brominated diphenyl ethers (PBDEs) flame retardants along the Dutch coast as derived from their accumulation in SPMDs and blue mussels (*Mytilus edulis*). *Organohalogen Compounds* 47: 89–92.

Chao HR, Wang SL, Lee WJ, Wang YF, Papke O. 2007. Level of polybrominated diphenyl ethers (PBDEs) in breast milk from central Taiwan and their relation to infant birth outcome and maternal menstruation effects. *Environmental International* 32: 239–245.

Chen LG, Huang YM, Peng XC, Xu ZC, Zhang SK, Ren MZ, Ye ZX, Wang XH. 2009. PBDEs in sediments of Beijing River China: Level, distribution, and influence of total organic carbon. *Chemosphere* 76: 226–231.

Chen G, Konstantinov AD, Chittim BG. 2001. Synthesis of polybrominated diphenyl ethers and their capacity to induce CYPIA by the Ah receptor mediated pathway. *Environmental Science and Technology* 35: 3749–3756.

Chen DH, Li LP, Bi XH, Zhao JP, Sheng GY, Fu JM. 2008a. PBDEs pollution in the atmosphere of typical e-waste dismantling region. *Environmental Science* 29: 2105–2110 (in Chinese).

Chen L, Mai B, Bi X. 2006a. Concentration levels, compositional profiles, and gas-particle portioning of polybrominated diphenyl ethers (PBDEs) in the atmosphere of an urban city in South China. *Environmental Science and Technology* 40: 1190–1196.

Chen LG, Mai BX, Xu ZC, Hang JL, Peng XC, Sheng GY, Fu JM. 2008b. Comparison of PCBs and PBDEs concentrations and compositions in Guangzhou atmosphere in summer. *Acta Scientiae Circumstantiae* 28: 150–155.

Chen DJ, Yu L, Liao QP, Qie WY. 2006b. Effect of maternal exposure to BDE-209 on offspring learning and memory ability and detection of the concentration of BDE-209 in serum. *Chinese Journal of Medicine* 9: 412–415.

Chen YQ, Zhang ZD. 1998. Situation and development prospect of brominated flame retardants. *Chemical Engineer* 23: 33–35.

Danial L, Maria A, Ioannis A. 2004. Preliminary study on PBDEs and HBCD in blood and milk from Mexican women. In *The Third International Workshop on Brominated Flame Retardants*, Toronto, Ontario, Canada, August 15–18, 2004, pp. 483–488.

De Boer J, Wester PG, Van HA. 2003. Polybrominated diphenyl ethers in influents, suspended particulate matter, sediments, sewage treatment plant and effluents and biota from the Netherlands. *Environmental Pollution* 122: 63–74.

De Carlo J, Ann Y. 1979. Studies on brominated chemicals in the environment. *Environmental Science and Technology* 320: 678–681.

Decision Etudes & Conseil. 2004. China: Tomorrow's leader in electronics? www.decision.eu/ang/prod_chi_a.htm

Deng, WJ, Zheng JS, Bi XH, Fu JM, Wong MH. 2007. Distribution of PBDEs in air particles from electronic waste recycling sites compared with Guangzhou and Hong Kong, South China. *Environmental International* 33: 1063–1069.

D'Silvia K, Fernandes A, Rose M. 2004. Brominated organic micropollutants igniting the flame retardant issue. *Critical Reviews in Environmental Science and Technology* 34: 141–207.

Eljattat E, De Ia Cal A, Raldua D, Duran C, Bareelo D. 2004. Occurrence and bio-availability of polybrominated diphenyl ethers and hexabromocyclododecane in sediment and fish from the Cinca River. *Environmental Science and Technology* 38: 2603–2608.

Eslami B, Koizumi A, Olata S. 2006 Concentrations of polybrominated diphenyl ethers (PBDEs) in breast milk of Japanese. *Chemosphere* 63: 554–561.

Guan YF, Wang JZ, Ni HG, Luo XJ, Mai BX, Zeng EY. 2007. Riverine inputs of polybrominated diphenyl ethers from the Pearl River Delta (China) to the coastal Ocean. *Environmental Science and Technology* 41: 6007–6013.

Haglund PS, Zook DR, Buser HR. 1997. Identification and quantification of polybrominated diphenyl ethers and methoxy-polybrominated diphenyl ethers in Baltic biota. *Environmental Science and Technology* 31: 3281–3287.

Harrad S, Hunter S. 2004. Spatial variation in atmospheric levels of PBDEs in passive air samples on an urban-rural transect. *Organohalogen Compounds* 66: 3786–3792.

Hawakawa K, Takatsuki H, Watanabe I, Sakai S. 2004. PBDEs in atmosphere over Osaka, Kyoto, Japan. *Organohalogen Compounds* 59: 299–302.

He WH, He P, Wang AG, Xia T, Xu BY, Chen XM. 2007. Effects of PBDE-47 on cytotoxicity and genotoxicity in human neuroblastoma cell in vitro. *Mutation Research/ Genetic Toxicology and Environmental Mutagenesis* 649: 62–70.

Hites R. 2004. Polybrominated diphenyl ethers in environment and people. *Environmental Science and Technology* 38: 945–1056.

Hoh E, Hites RA. 2005. Brominated flame retardants in the atmosphere of the east central United States. *Environmental Science and Technology* 39: 7794–7782.

Hong Kong Greenpeace Organization. Anthropological survey report on e-waste dismantling in Guiyu, Shantou, 2003.

Huang YM, Chen LG, Ye ZX, Xu ZC, Peng XC, Zhang SQ, Li H. 2009a. Level and distribution pattern of polybrominated diphenyl ethers (PBDEs) in sediment of Pearl River, Guangdong. *Environmental Chemistry* 28: 140–142.

Huang MY, Peng XC, Xu ZC, Ye WZX. 2009b. Polybrominated diphenyl ethers in room dust of South China: Specialty and effect. *Scientific Guidance* 78: 169–174.

Huang JL, Zhu JS, Hong ZD, Zhang XP, Hu Z, Pan J, Huang LC. 2005. Harmfulness of plastic flame-retardants PBDEs used on electronic and electrical equipment. *Research and Development* 2(5): 40–43.

Jiang HP, Yu YH, Chen GJ. 2009. Biological effect and toxicity of polybrominated diphenyl ethers. *Guangdong Medical Journal* 30: 144–146.

Jin J, Wang Y, Liu WZ, Tang XY. 2008. Level and distribution of polybrominated diphenyl ethers in soil from Laizhou Bay. *Acta Scientiae Circum Stantiae* 28: 1463–1468.

Johnson-Restrepo B, Kannan K, Rapaport DP, Rodan BD. 2005. Polybrominated diphenyl ethers and polychlorinated biphenyls in human adipose tissue from New York. *Environmental Science and Technology* 39: 5177–5182.

Kajiwara N, Kamikawa S, Ramu K, Ueno D, Yamada TK, Subramanian A, Lam PKS et al. 2006. Geographical distribution of polybrominated diphenyl ethers (PBDEs) and organochlorines in small cetaceans from Asia waters. *Chemosphere* 64: 287–295.

Kalantzi OI, Martin FL, Thomas GO, Alcock RE, Tang HR, Drury SC, Carmichael PL, Nicholson JK, Jones KC. 2004. Different levels of polybrominated diphenyl ethers (PBDEs) and chlorinated compounds in breast milk from two UK regions. *Environmental Health Perspectives* 112: 1085–1091.

Kazuhiko A, Firotaka O, Masahiro O. 2001. GC/MS analysis of polybrominated diphenyl ethers in fish collected from the inland sea of Seto, Japan. *Chemosphere* 44: 1325–1333.

Kiriluk RM, Whittle DM, Keir MJ, Carswell AA, Huestis SY. 1997. The Great Lakes Fisheries Specimen Bank: A Canadian perspective in environmental specimen banking. *Chemosphere* 34: 1921–1932.

Labandeira A, Eljarrat E, Barrel D. 2007. Congener distribution of polybrominated diphenyl ethers (PBDEs) in feral carp (Cypriuscarpio) from L Lobregat River, Spanish. *Environmental Pollution* 146: 188–195.

Lacorte S, Guillamon M, Martinez E. 2003. Occurrence and specific congener profile of 40 polybrominated diphenyl ethers in river and coastal sediments from Portugal. *Environmental Science and Technology* 37: 892–898.

La Guardia M, Haler C, Harvey E. 2006. Detailed polybrominated diphenyl ethers (PBDEs) congener composition of the widely used penta- octa- and deca-PBDE technical flame-retardant mixtures. *Environmental Science and Technology* 40: 6247–6254.

Lau KF. 2008. Risk assessment of polybrominated diphenyl ethers, coplanar poly-chlorinated biphenyls, polychlorinated dibenzo-p-dioxins and polychlorinated dibenzofurans in cetaceans in Hong Kong waters. MPhil thesis, City University of Hong Kong, Kowloon, Hong Kong.

Law RJ, Allchin CR, de Boer J, Covaci A, Herzke D, Lepom P, Morris S, Tronczynski J, de Wit CA. 2006. Levels and trends of brominated flame retardants in the European environment. *Chemosphere* 64: 187–208.

Leung AOW, Luksemburg WJ, Wong AS, Wong MH. 2007. Spatial distribution of polybrominated diphenyl ethers and polychlorinated dibenzo-p-dioxins and dibenzofurans in soil and combusted residue at Guiyu, an electronic waste recy-cling site in South China. *Environmental Science and Technology* 41: 2730–2737.

Li JG, Yu HF, Zhao YF, Zhang G, Wu YN. 2008. Levels of polybrominated diphenyl ethers (PBDEs) in breast milk from Beijing, China. *Chemosphere* 73: 182–186.

Li YM, Zhang QH, Ji DS, Wang T, Wang YW, Wang P, Ding L, Jiang GB. 2009. Levels and vertical distributions of PCBs, PBDEs, and OCPs in the atmospheric boundary layer: Observation from the Beijing 325-m meteorological tower. *Environmental Science and Technology* 43: 1030–1035.

Lin ZS, Ma XD, Zhang QH, Yao ZW, Ma YA. 2008. Study on polybrominated diphe-nyl ethers (PBDEs) in sediments surrounding Bohai Sea. *Marine Environmental Science* 27: 24–27.

Liu ZH, Lian YH, Chau ZM, Ma QM. 2007. Advance of the distribution and pattern of polybrominated diphenyl ethers (PBDEs) in the environment. *Chinese Journal of Soil Science* 38: 1227–1233.

Liu J, Luo XJ, Chen SJ, Luo Y, Zhen L, Mai BX. 2009. Concentration and distribution pat-tern of polybrominated diphenyl ethers (PBDEs) in scatterfet chickens from elec-tronic waste recycling site. *Modern Agriculture Science and Technology* 24: 293–297.

Liu JX, Xu SJ, Wu KS, Li Y, Fu X. 2008. Contamination of polybrominated diphenyl ethers (PBDEs) and their measure. *Journal of Shantou University Medical College* 21: 126–129.

Liu HX, Zhang QH, Jian GB, Cai ZW. 2005a. Polybrominated diphenyl ethers and its environmental problems. *Progress in Chemistry* 17: 554–562.

Liu Y, Zheng GJ, Yu HX, Martin M, Richardson BJ, Lam MHW, Lam PKS. 2005b. Polybrominated diphenyl ethers (PBDEs) in sediments and mussel tissues from Hong Kong marine waters. *Marine Pollution Bulletin* 50: 1173–1184.

Lu Y, Wang LN, Huang J, Wang T. 2007. Level and distribution of polybrominated diphenyl ethers in sediments and fish tissue of Haihe River and Bohai Bay. *Environmental Pollution and Protection* 29: 652–660.

Luo Q, Cai ZW, Wong MH. 2007. Polybrominated diphenyl ethers (PBDEs) in fish and sediment from river polluted by electronic waste. *Science of the Total Environment* 383: 115–127.

Luo XJ, Mai BX, Chen SJ. 2009. Advances on study of polybrominated diphenyl ethers. *Progress in Chemistry* 21: 359–368.

Luo XJ, Yu M, Mai BX, Chen SJ. 2008. Distribution and partition of polybrominated diphenyl ethers (PBDEs) in water of Zhejiang River Estuary. *Chinese Science Bulletin* 53: 93–500.

Mai BX, Chen SJ, Luo XJ, Chen LG, Yang QS, Sheng GY, Peng PG, Fu JM, Zeng EY. 2005. Distribution of polybrominated diphenyl ethers in sediments of the Pearl River Delta and adjacent South China Sea. *Environmental Science and Technology* 39: 3521–3527.

Manchester NJ, Valters K, Sonzogn WC. 2001. Comparison polybrominated diphenyl ethers (PBDEs) and polychlorinated biphenyls (PCBs) in Lake Michigan sediments. *Environmental Science and Technology* 35: 1072–1077.

Mariumssen E, Fjield E, Breivik K. 2008. Elevated levels of polybrominated diphenyl ethers (PBDEs) in fish from Lake Josa Norway. *Science of the Total Environment* 390: 132–141.

McDonald TA. 2002. A perspective on the potential health risks of PBDEs. *Chemosphere* 46: 745–755.

Meironyté D, Norén K, Bergman Å. 1999. Analysis of polybrominated diphenyl ethers in Swedish human milk. A time-related trend study, 1972–1997. *Journal of Toxicology and Environmental Health. Part A: Current Issues* 58: 329–341.

Meng X, Zeng EY, Yu L. 2007. Persistent halogenated hydrocarbons in consumer fish of China regional and global implications for human exposure. *Environmental Science and Technology* 41: 1821–1827.

Messer A. 2010. Mini-review: Polybrominated diphenyl ether (PBDE) flame retardants as potential autism risk factors. *Physiology Behavior* 100: 245–249.

Moon HB, Choi HG, Kim SS, Jeong SR, Lee PY. 2002. Contaminations of polybrominated diphenyl ethers in marine sediments from the south-eastern coastal areas of Korea. *Organohalogen Compound* 58: 217–220.

Muir DCG, Backus S, Derocher AE. 2006. Brominated flame retardants in polar bear (*Ursus maritimus*) from Alaska, the Canadian Arctic, east Greenland and Svalbard. *Environmental Science and Technology* 40: 449–455.

Ni HG, Zeng EY. 2009. A big picture view of halogenated chemical exposures in China—A Report on China E-Waste Recycling Contamination and Response Policy Workshop, Guangzhou, 2009. pp. 1–45.

Nie FH, Chen JJ, Bunce N. 2005. MROD analysis of competitive inhibition from polybrominated diphenyl ethers to P450 1A2 (CYP 1A2) in rat hepatocytic microsomes. *Chinese Agriculture Bulletin* 21: 12.

Ohta S, Ishizaki D, Nishimura H, Nakao T, Aozasa O, Okumura T, Miyata H. 2001. Polybrominated diphenyl ethers (PBDEs) in atmosphere, Japan. *Organohalogen Compounds* 52: 210–213.

Oros DR, Horver D, Rodigari F. 2005. Levels and distribution of polybrominated diphenyl ethers (PBDEs) in water surface sediments and bivalves from San Francisco Estuary. *Environmental Science and Technology* 39: 33–41.

Ou SM. 2006. Determination of flame retardant-polybrominated diphenyl ethers (PBDEs) in the sediments from Xiamen coastal area and Yuan Dan Lake by GC/MS isotope dilution method. *Fujian Analysis and Testing* 15: 1–3.

Peng, JH, Huang CW, Weng YM, Yak HK. 2007. Determination of polybrominated diphenyl ethers (PBDEs) in fish samples from rivers and estuaries in Taiwan. *Chemosphere* 66: 1990–1997.

Qin YY, Leung CKM, Leung AOW, Wu SC, Zheng JS, Wong MH. 2009. Persistent organic pollutants and heavy metals in adipose tissues of patients with uterine leiomyomas and the association of pollutants with seafood diet, BMI, and age. *Environmental Science and Pollution Research* 17: 229–240.

Ramu K, Kajiwara N, Tanabe S, Lam PKS, Jefferson TA. 2005. Polybrominated diphenyl ethers (PBDEs) and organochlorines in small cetaceans from Hong Kong waters: Levels, profile and distribution. *Marine Pollution Bulletin* 51: 669–676.

Robinson BH. 2009. E-waste: An assessment of global production and environmental impacts. *Science of the Total Environment* 408: 183–191.

Samara F, Tsai CW, Aga DS. 2006. Determination of potential sources of PCBs and PBDEs in sediments of the Niagara River. *Environmental Pollution* 139: 489–497.

Schecter A, Harris TR, Papke O. 2006. Polybrominated diphenyl ether (PBDE) levels in blood of pure vegetarians. *Toxical Environmental Chemistry* 88: 107–112.

Schecter A, Pavuk M, Papke O. 2003. Concentrations of polybrominated diphenyl ethers (PBDEs) in breast milk of American women. *Environmental Health Perspectives* 111: 1723–1729.

Sellström U, Kierkegaard A, de Wit C, Jansson B. 1998. Polybrominated diphenyl ethers and hexabromocyclododecane in sediment and fish from a Swedish river. *Environmental Toxicology and Chemistry* 17: 1065–1072.

She, JW, Perreas M, Winkler J, Visita P, Mckinney M, Kope D. 2002. PBDEs in the San Francisco Bay area: Measurements in harbour seal and human breast adipose tissue. *Chemosphere* 46: 697–707.

Shen M, Yu YJ, Zheng GJ, Yu HX, Lam PKS, Feng JF, Wei ZB. 2006. Polybrominated biphenyls and polybrominated diphenyl ethers in surface sediments from Yangtze River Delta. *Marine Pollution Bulletin* 52: 1299–1304.

Sjödin L, Hagmar E, Klasson WK. 2007. Polybrominated diphenyl ethers (PBDEs) in blood from Swedish workers. *Environmental Health Perspectives* 107: 643–648.

Song WL, Ford JC, Li AN. 2004. Polybrominated diphenyl ethers in sediments of the Great Lakes. *Environmental Science and Technology* 38: 3286–3293.

Sudaryanto A, Kajiwara N, Tsydenova O, Isobe T, Yu HX, Takahashi S, Tanabe S. 2008. Levels and congener specific profiles of PBDEs in human breast milk from China: Implication on exposure sources and pathways. *Chemosphere* 73: 1661–1668.

Takumi T, Kurunthachalam SR, Hiroaki T. 2004. Impact of fermented brown rice with aspergillus oryzae (FEBRA) intake and concentrations of polybrominated diphenyl ethers (PBDEs) in blood of humans from Japan. *Chemosphere* 57: 795–811.

Tamade Y, Shibukawa S, Osaki H, Kashimoto S, Yagi Y, Sakai S, Takasuga T. 2002. A study of brominated compounds release from appliance-recycling facility. *Organohalogen Compounds* 56: 189–192.

UNEP DEWA/GRID-Europe. 2005. E-waste, the hidden side of the equipment's manufacturing and use, Environment Alert Bulletin 5, UNEP: Nairobi, Kenya. http://www.grid.unep.ch/product/publication/download/ew_ewaste.en.pdf

Wan Y, Hu JY, Zhang K, An LH 2008. Trophodynamics of polybrominated diphenyl ethers in the marine food web of Bohai Bay, North China. *Environmental Science and Technology* 42: 1078–1083.

Wang LC, Lee WJ, Lee WS, Chang-Chien GP. 2011. Polybrominated diphenyl ethers in various atmospheric environments of Taiwan: Their levels, source identification and influence of combustion sources. *Chemosphere* 84: 936–942.

Wang YW, Jiang GB, Lam PKS, Li A. 2007. Polybrominated diphenyl ethers in the East Asian environment: A critical review. *Environment International* 33: 963–973.

Wang YW, Jiang GB. 2008. The research of human exposure to polybrominated diphenyl ethers and perfluorooctane sulfonate. *Chinese Scientific Bulletin* 53: 481–492.

Watanabe I, Kawano M, Tatsukawa R. 1995. Polybrominated and mixed polybromino/chlorinated benzo-p-dioxins and dibenzofurans in the Japanese environment. *Organohalogen Compounds* 24: 337–340.

Watanabe I, Kawano M, Wang Y, Chen Y. 1992. Concentrations of PBDEs in air over e-waste recycling for metals at North of Taiwan. *Organohalogen Compounds* 9: 309–312.

Wong MH, Wu SC, Deng WJ, Yu XZ, Luo Q, Leung AOW, Wong CSC, Luksemburg WI, Wong AS. 2007. Export of toxic chemicals—A review of the case of uncontrolled electronic-waste recycling. *Environmental Pollution* 149: 131–140.

Wu KS, Liu JX, Li Y, Fu X. 2008. Environmental distribution of polybrominated diphenyl ethers (PBDEs). *Occupation and Health* 24: 2467–2469.

Wurl O, Lam PKS, Obbard JP. 2006a. Occurrence and distribution of polybrominated diphenyl ethers (PBDEs) in the dissolved and suspended phases of the sea-surface microlayer and seawater in Hong Kong, China. *Chemosphere* 65: 1660–1666.

Wurl O, Obbard JP. 2004. A review of pollutants in the sea-surface microlayer (SML): A unique habitat for marine organisms. *Marine Pollution Bulletin* 48: 1016–1030.

Wurl O, Obbard JP. 2005. Organochlorine pesticides, polychlorinated biphenyls and polybrominated diphenyl ethers in Singapore's coastal marine sediments. *Chemosphere* 58: 925–933.

Wurl O, Potter JR, Durville C, Obbard JP. 2006b. Polybrominated diphenyl ethers (PBDEs) over the open Indian Ocean. *Atmospheric Environment* 40: 5558–5565.

Xian QM, Ramu K, Isobe T, Sudaryanto A, Liu XH, Gao ZS, Takahashi, Yu HX, Tanabe S. 2008. Levels and body distribution of polybrominated diphenyl ethers (PBDEs) and hexabromocyclododecanes (HBCD) in freshwater from the Yangtze river, China. *Chemosphere* 71: 268–276.

Xiang CH, Luo XJ, Chen SJ. 2007. Polybrominated diphenyl ethers in biota and sediments of the Pearl River estuary, South China. *Environmental Chemistry and Technology* 26: 616–623.

Xu J, Dao ZS, Xian QM, Yu HX, Feng JF. 2009. Levels and distribution of polybrominated diphenyl ethers (PBDEs) in the freshwater environment surrounding a PBDE manufacturing plant in China. *Environmental Pollution* 157: 1911–1916.

Yang WH, Hu W, Feng Z, Liu HL, Yu HX. 2009. Study on endocrine disruption and structure–activity relationship of PBDEs and their metabolites. *Asian Journal of Ecotoxicology* 4: 164–173.

Yang YL, Pan J, Li Y, Yin XC, Shi L. 2003. Persistent organic pollutants polychlorinated naphthalenes (PCN) and polybrominated diphenyl ethers (PBDEs) in coastal sediments of Qingdao, China. *Chinese Science Bulletin* 21: 2244–2251.

Zeng EY. 2005. Distribution of polybrominated diphenyl ethers in sediments of the Pearl River Delta and adjacent South China Sea. *Environmental Science and Technology* 39: 3521–3527.

Zhang M, He WH, He P, Xia T, Chen XM, Wang AG. 2007. Effects of PBDE-47 on oxidative stress and apoptosis in SH SY5Y cell. *China Journal of Physical and Hygienic Occupation Disease* 25: 145–147.

Zhang JQ, Jiang YS, Zhou J, Wu B, Liang Y, Peng ZQ, Fang DK et al. 2010. Elevated body burdens of PBDEs, dioxins and PCBs on thyroid hormone homeostasis at an electronic waste recycling site in China. *Environmental Science and Technology* 44: 3956–3962.

Zhang JQ, Sun XK, Jiang YS, Zhou J, Wang LB, Ye ZY, Fang DK, Wang GB. 2008. Levels of PCDD/Fs, PCBs and PBDEs compounds in human placenta tissue. *Chinese Medical Magazine* 42: 911–918 (in Chinese).

Zhao X, Wang GH, Liu SS, Gao SM, Shen GF, Gao SX, Wang XD, Feng JF, Wang LS. 2008. Gas chromatography-ion trap tandem mass spectrometry for determination of polybrominated biphenyl ethers in air of Nanjing. *Chinese Journal of Analytical Chemistry* 36: 137–142.

Zhao GF, Wang ZJ, Zhou HD, Zhao Q. 2009. Burdens of PBBs, PBDEs and PCBs in tissues of cancer patients in the e-waste disassembly sites in Zhejiang, China. *Science of the Total Environment* 407: 4831–4837.

Zheng GJ, Martin M, Richardson BJ, Yu HX, Liu Y, Zhou CH, Li J, Hu GJ, Lam MHW, Lam PKS. 2004. Concentrations of polybrominated diphenyl ethers (PBDEs) in Pearl River Delta sediments. *Marine Pollution Bulletin* 40: 520–524.

Zhou MY, Xia B, Ma SS, Xin FY, Sun WH. 2009. Study on characteristics and marine environmental pollution of polybrominated diphenyl ethers (PBDEs). *Progress in Fishery Sciences* 30: 142–146.

Zou MY, Gong J, Ran Y. 2009. The distribution and the environmental fate of polybrominated diphenyl ethers in watershed soils of Pearl River Delta. *Ecology and Environment Sciences* 18: 122–127.

8

Removal of Persistent Organic Pollutants and Compounds of Emerging Concern in Public-Owned Sewage Treatment Works: A Review

Ming Man and Ming Hung Wong

CONTENTS

Introduction

Persistent organic pollutants (POPs) are organic compounds that are resistant to environmental degradation through chemical, biological, and sunlight degradation processes. POPs are known to persist in the environment for long periods due to such properties and can thus be capable of long-range transportation between lands, rivers, and oceans. Such compounds will bioaccumulate in seafood and animal tissues and subsequently become biomagnified in food chains and ultimately impact human health and the environment. Many POPs are pesticides such as DDT, whereas other compounds such as PCBs are used in industrial applications and in the production of a wide range of products. In the past few years, pharmaceutically active compounds (PhACs) and personal care products (PCPs) have been discovered in various environments and water sources. Some of those compounds have been linked to ecological impacts at concentrations of less than 1 ng/L. These PhACs and PCPs were eventually grouped with other emerging contaminants, such as nitrosamines from disinfection by-products (DBPs), metals, and other new additions of compounds, such as endocrine-disrupting compounds (EDCs), and are collectively referred to as compounds of emerging concern (CECs) (Drewes et al. 2006, WERF 2007). It has been reported that certain synthetic compounds, such as detergents and pesticides, and natural compounds, such as sex hormones, can affect the balance of normal hormonal function in animals. These substances are classified as EDCs and have been linked to a variety of adverse effects in humans and wildlife (Snyder et al. 2007). The U.S. Environmental Protection Agency (U.S. EPA) refer to EDCs as exogenous agents that interfere with the "synthesis, secretion, transport, binding, action, or elimination of natural hormones in the body which are responsible for the maintenance of homeostasis, reproduction, development, and/or behavior" (U.S. EPA 1996). In humans, studies have linked the exposure of mothers to phthalates such as diethylhexyl phthalate to adverse impacts in their children, including lowered testosterone levels in boys (U.S. Department of Health and Human Services 2008). EDCs are further classified as thyroidal (compounds with direct or indirect impacts to the thyroid glands), androgenic (compounds that mimic or block natural testosterone), and estrogenic (compounds that mimic or block natural estrogen) (Snyder et al. 2007). Recently, the illegal replacement of palm oil with diethylhexyl phthalate and diisononyl phthalate (plasticizers) by some food-grade manufacturers in Taiwan has caused huge product recalls in numerous food, drink, and medicinal formulations in Taiwan and surrounding trading markets in June of 2011.

POPs and CECs cover very diverse structures and physical/chemical characteristics such as molecular weight, water solubility, Kow, pKa, and charge density. A single treatment process cannot, therefore, be designed to remove all POPs and CECs due to the diverse nature of their characteristics. It is not possible to use effluent quality contamination surrogates, such as biochemical oxygen demand (BOD) and chemical oxygen demand (COD), to predict the removal efficiencies of such POPs and CECs. Currently, most wastewater treatment facilities are not designed to remove individual POPs and CECs but are designed to meet effluent quality parameters such as total suspended solids (TSS), BOD, and COD. Only limited information was found in other literature relating to the waste sewage treatment efficiencies of POPs and CECs through the treatment processes, where the majority of the research and information was based on drinking-water treatment processes. In the USDI (2009) study, researchers have focused on understanding the fate of approximately 80 CECs primarily during potable-water treatment. They selected those compounds because there are analytical techniques available in detecting them at concentrations as low as 1 ng/L. Literature reviews on wastewater treatments showed that many POPs and CECs are only partially removed by typical sewage treatment processes (Pickering and Sumpter 2003). The main removal pathways for the removal of POPs and CECs during wastewater treatment processes are the following (Birkett and Lester 2003):

- Compounds being adsorbed onto suspended solids
- Compounds going through aerobic and anaerobic biodegradation
- Compounds going through hydrolysis
- Compounds going through volatilization into the air space

A completed public-owned sewage treatment works (POSTW) would include physical, biological, and chemical treatment processes in order to remove any corresponding contaminants, where each sewage treatment generally involves three stages called primary, secondary, and tertiary treatments. More than one tertiary treatment process may be used at any treatment plant in which its purpose is to provide a final treatment stage to raise the effluent quality, before it is discharged into the receiving environment such as seas, rivers, lakes, and ground. Sometimes, the conversion of toxic ammonia to nitrate alone is referred to as a tertiary treatment. "Effluent polishing" is the term used when chemical disinfection—such as with NaOCl, ozone, or physical disinfection, or with UV radiation, microfiltration, and so on—is involved in treatment and is always the final process if present. Only advanced sewage treatment facilities include three stages of treatment, whereas most other facilities will only include the primary stage or a combination of primary and secondary treatments.

The major objectives for this review came about in light of the findings from the literature, which examined the removal of contaminants by drinking-water and sewage treatment processes. The objectives were to evaluate the removal efficiencies of groups of POPs and CECs in POSTW and to suggest other alternatives if removal efficiencies were insufficient.

Background on POPs and CECs

The United Nations Environment Programme Governing Council in May of 1995 decided to begin investigating POPs, with a short list of 12 compounds, known as "the dirty dozen": aldrin, chlordane, DDT, dieldrin, endrin, heptachlor, hexachlorobenzene, mirex, PCB, polychlorinated dibenzo-p-dioxins, polychlorinated dibenzofurans, and toxaphene. Since then, this list has been extended to include PAHs and certain brominated flame retardants.

Many PCPs and PhACs have been discovered in surface water and ground water; some of these compounds could cause environmental and ecological impacts at trace levels. These PhACs and PCPs were eventually grouped with other emerging contaminants and are collectively referred to as CECs (Drewes et al. 2006, WERF 2007). CECs can be classified into different categories based on their impacts to humans or the environment, which include the following groups:

- Agricultural and Industrial compounds
- Hormones and antibiotics
- Other pharmaceutical compounds

Agricultural and Industrial Compounds

CECs in agriculture applications include pesticides, herbicides, and insecticides such as Atrazine, Lindane, DDT, and DEET. They are used for preventing, destroying, repelling, or mitigating pests. The residues from agricultural applications and runoff at environmentally relevant concentrations can affect the reproductive systems of animals (van Vuuren 2008). CECs used as industrial and household products such as coolants, plasticizers, wood preservatives, fire retardants, and detergents cover a broad range of compounds. PCBs are one of the most well-known pollutants within the group of industrial and household products. They were commercially produced as complex mixtures, containing multiple isomers with different degrees of chlorination and used as coolants and insulating fluids for transformers and capacitors. The toxicity associated with PCBs and other chlorinated hydrocarbons, including polychlorinated naphthalenes, were noticed very early on due to a number of industrial incidents (Drinker et al. 1937). Once PCBs enter into the influent of the sewage system, adsorption and subsequent sedimentation may immobilize them into the aquatic system for a long duration. PCBs that are redistributed into the water column can vaporize from the water surface into the air and into the environment. PCBs contained in aquatic sediments can act as environmental reservoirs and will release slowly over a long period of time.

Hormones and Antibiotics

Hormones include the naturally occurring hormones estrone (E1), 17β-estradiol (E2), and estriol (E3), and synthetic hormone such as 17α-ethinylestradiol (EE2). They are in birth control pills and are discharged into the environment mainly through wastewater effluent. The agricultural runoff introduces hormones from animal source. Antibiotics are commonly overprescribed in the industrial world. The practice of adding antibiotics to the feed of livestock and in the fish farms leads to the creation of antibiotic-resistant strains of bacteria, a harmful side effect (WHO 2002).

Other PhACs and PCPs

PhACs include prescription drugs, over-the-counter medications, drugs used in hospitals, and veterinary drugs. After intake, drugs are generally absorbed and subjected to further metabolic reactions, including hydrolysis, alkylation and dealkylation, and oxidation and reduction of the parent compound. Another major route of interaction is forming conjugates such as glucuronides and sulfonates to enhance excretion (Cunningham 2004). Significant portion of PhAC remains unmetabolized and is excreted out of the body in urine or feces. In addition, other sources of PhACs that enter the environment are ones that are directly discarded. Concentrations of PhACs in wastewater treatment influent were estimated to range from less than 1 ng/L to approximately 130,000 ng/L, but the majority were between 100 and 1,000 ng/L (AWWARF 2007).

PCPs include products for personal health and cosmetic purposes such as fragrances, lotions, and cosmetics. Currently, the most commonly detected compound from PCPs is triclosan, a potent wide-spectrum antibacterial and antifungal agent. Triclosan is used primarily in soaps, toothpastes, and detergents.

Current Information on the Removal of POPs and CECs in POSTW

Typical wastewater treatment facilities are designed to meet effluent quality parameters such as TSS, BOD, and COD and would involve one to three stages of sewage treatment referred as primary, secondary, and tertiary treatment. Some POSTW utilize only the primary treatment stage, whereas modernized types have adopted both the primary and secondary treatment stages and more advanced POSTW have incorporated all three stages of sewage treatment. POPs and CECs cover a broad range of chemical structures with diverse physical properties such as molecular weight, water solubility and dissociation constant (pKa), octanol-water partition coefficient (Kow), and charge density. A single treatment process cannot be designed to remove all POPs and CECs from sewage influent.

The primary treatment process in a POSTW involves coagulation, flocculation, and chemical softening in order to separate suspended solids during water treatment. Coagulation uses iron or aluminum salts to precipitate metal hydroxides. Softening involves the use of lime or soda ash to remove calcium (at $pH \geq 9$) and magnesium (at $pH \geq 11$). Both coagulation and softening remove suspended solids and colloidal material; meanwhile organic compounds are absorbed into the suspended particles and removed with the solid particles. Synthetic polymers with positively charged groups, such as a quaternary amino (– NR$_4$)$^+$, are now increasingly being used as coagulants for water treatment to neutralize the negatively charged particulate solids. Once suspended particles are flocculated into larger particles, they can usually be removed from the liquid by sedimentation. The removal of POPs and CECs during the primary sewage treatment process for a compound takes into account the two main sorption mechanisms (Ternes et al. 2004):

- Absorption: It is the interaction of the organic groups of the POPs and CECs with the cell membrane of the microorganisms and lipid fractions of the sludge.
- Adsorption: It is the physical binding of ions and molecules onto the surface of another molecule through electrostatic interactions of positively charged groups of a compound with the negatively charged surfaces of the microorganisms.

Any chemical that binds to particles and compounds with high lipid solubility and positively charged species can be removed by the primary treatment process. In contrast, compounds that are highly dissolvable in water or have negatively charged compounds tend not to be removed by this process. Highly hydrophobic contaminants are most likely to enter POSTW already bound to particulate matter and remain undetected in the purely liquid portion of the influent. A detailed study was conducted on the effectiveness of the primary treatment process (Snyder et al. 2007) involving the investigation of 36 targeted compounds. The findings indicated that less than 20% was removed for 34 of these targeted compounds by coagulation with alum or ferric chloride. DDT and Benzo(a)pyrene were the exceptions. The removal rates were in the range of 20%–50% for DDT and 50%–80% for Benzo(a)pyrene. Previous studies have also indicated that coagulation and softening are generally ineffective for the removal of CECs (AWWRF 2007).

The secondary treatment process was designed to substantially degrade the biological contents of the sewage that originates from human waste, food waste, soaps, and detergent. This process relies on the metabolic activity of a mixed population of bacteria to degrade organic material in water. Microorganisms select many organic and natural compounds for food and growth in their metabolic processes. It could take place in aerobic, anoxic, or anaerobic environments. An anoxic environment is a condition where oxidized forms of nitrogen (e.g., nitrate) serve as electron acceptors in the absence of molecular oxygen, whereas in anaerobic environments neither oxygen nor an oxidized form of nitrogen is present. Microorganisms use inorganic and organic electron acceptors in anaerobic environments.

CECs are usually present in trace quantities that are insufficient in providing a primary substrate for the growth of microorganisms. The removal and transformation of POPs and CECs using biological treatment cannot be predicted theoretically; hence, the success of this process requires empirical and practical proof. The biodegradation of some POPs and CECs would generate partially degraded intermediates that could be more bioactive than the parent compound (Gröning et al. 2007).

One of the difficulties in reviewing the removal efficiencies of POPs and CECs from the literature was the large volume of data presented in some of the articles, where no detailed explanations were provided. This means it is entirely left to the reader to digest and understand the actual driving force of the removal mechanism. Table 8.1 consists of data extracted from a large project by Snyder et al. (2007), in the "Removal of EDCs and Pharmaceuticals in Drinking and Reuse Treatment Processes." Using the data from the filter with biologically active anthracite column (BAF) and from the filter with biologically active carbon column (BAC), these authors concluded that "it is likely that removal shown in the pilot-scale BAC for the less biodegradable contaminants was largely due to adsorption (by activated carbon [AC]) rather than biodegradation." The data from the experiment on "Adsorption + Biodegradation, Ambient, 2 days" stated that "the removal (of hormones) in this experiment was strictly due to adsorption (by particulate matter). Biosorption appeared to be significant for some target compounds (hormones)." The authors did not explain their observations on the experiment for the filter with biologically active sand column (BAS). But they did emphasize that their experiment on riverbank filtration with BAS correlated with the data obtained in biological degradation batch tests that were conducted and published previously (Snyder et al. 2004) by their team. Based on the BAF data and the other data as presented in Table 8.1, it can be inferred that these hormones (estradiol, estriol, estrone, and ethinylestradiol) would not go through any significant biodegradation in the POSTW. But they may be removed from the influent into sludge due to biosorption to sand, carbonic,

TABLE 8.1

Data on Hormone Removal (Reported as % Removal)

Target Compound	Adsorption + Biodegradation Ambient 2 Days	Filter with Biologically Active Anthracite Column (BAF)	Filter with Biologically Active Carbon Column (BAC)	Filter with Biologically Active Sand Column (BAS)
Estradiol	98%	1%	94%	>99
Estriol	81%	<1%	92%	99
Estrone	95%	<1%	95%	>99
Ethinylestradiol	76%	1%	91%	95
Progesterone	Not reported	52%	99%	>99
Testosterone	99%	35%	96%	>99

Source: Extracted from Snyder, S.A. et al., American Waterworks Association Research Foundation (AWWARF), 2007, p. 193, 194, and 198.

and particulate matter; however, this interpretation may not be correct. Snyder et al. (2007) presented much more data from their study, and it is up to the readers to digest the information and to draw their own conclusions from them.

The traditional POSTW removes POPs and CECs based on sorption and biodegradation. It would be desirable, therefore, to make use of removal data to show the sorption and biodegradation properties of target compounds. Snyder et al. (2007) had selected 29 target compounds for an extensive study, and the results are presented in Table 8.2 where the sorption and biodegradation properties were interpreted and rated as low (L), medium (M), and high (H). The data from other literature reviews were studied before the biodegradation property for each CEC was assigned, especially when the biosorption level was strong and the biodegradation properties could not be deduced from the original data. The data as presented in Table 8.2 were from experiments with relatively clean river water and not from sewage influent. The studies have reported that the overall removal rates of POPs and CECs in full-scale POSTWs clearly show that their elimination is often incomplete (Table 8.3). As a consequence, significant fractions of POPs and CECs are discharged with the final effluent either into the aquatic environment or into sludge discharge. The deposition of POPs and CECs on land from sludge reapplication can be another significant pathway for releasing these substances into the environment. Studies report on the removal of the POPs and CECs by comparing influent and effluent concentrations, but without distinguishing between the three major fates of a substance in POSTWs, which are as follows: (a) degradation to lower-molecular-weight inactive fractions, (b) degradation to smaller or aggregation to higher-molecular-weight but active species, and (c) physical sorption to solids and removed as sludge.

Compounds with either properties of sorption or biodegradation can be removed by sewage treatment systems effectively; however, compounds that lack such properties, such as carbamazepine and iopromide, will persist in the effluent for a long time as shown in Tables 8.2 and 8.3. The removal efficiencies of diclofenac and ibuprofen are quite similar to each other, as reported for several POSTWs that are located all over the world. The results indicated that the elimination of such compounds seems unrelated to the designs of each POSTW. The reported removal efficiencies of hormones, including estrone, 17-β-estradiol, and 17-α-ethinyl-estradiol, were found to vary greatly between studies, which can be demonstrated by the following different behaviors observed: (a) an increase along the passage of the POSTW (Baronti et al. 2000, Carballa et al. 2004), (b) no significant removal (Ternes et al. 1999a), and (c) efficiencies higher than 80% (Ternes et al. 1999b, de Mes et al. 2005, Nakada et al. 2006). However, the factors that may explain these deviations cannot be fully elucidated, since many of the cases did not report adequate operational data. Complete oxidation of 17-β-estradiol to estrone would be expected to occur in less than 3 h, whereas further oxidation of estrone should occur at a slower rate (50% after 24 h),

TABLE 8.2

Review of the Removal of Emerging Chemicals by Sorption and Biodegradation Processes

Target Compound	(A+B) 2 Days[a]	(A+B) 5 Days[b]	Adsorption 5 Days[c]	Sorption[d]	BioDegradation[e]
	Percent Removal			Rated as H, M, or L	
Acetaminophen	94	99	79	M	H
Androstenedione	99	99	96	H	H (Voishvillo et al. 2004)
Atrazine	58	55	54	L	L
Caffeine	76	92	77	M	M
Carbamazepine	57	54	55	L	L
DEET	40	57	37	L	M
Diazepam	83	82	83	M	L
Diclofenac	71	74	67	M	L
Dilantin	79	77	78	M	L
Erythromycin	84	79	83	M	L
Estradiol	98	99	85	M	H
Estriol	81	99	81	M	M
Estrone	95	99	62	M	H
Ethinylestradiol	76	79	73	M	L
Floxetine	98	99	98	H	L (Redshaw et al. 2008)
Gemfibrozil	82	99	54	L	H
Hydrocodone	48	47	49	L	L
Ibuprofen	94	99	66	L	H
Iopromide	31	33	28	L	L
Meprobamate	37	48	36	L	L
Naproxen	85	98	80	M	M
Oxybenzone	91	99	83	M	M
Pentoxifylline	92	99	91	H	M
Sulfamethoxazole	80	77	81	M	L
TCEP	54	53	54	L	L
Testosterone	99	99	92	H	H (Yang et al. 2010)

(continued)

TABLE 8.2 (continued)

Review of the Removal of Emerging Chemicals by Sorption and Biodegradation Processes

Target Compound	(A+B) 2 Days[a]	(A+B) 5 Days[b]	Adsorption 5 Days[c]	Sorption[d]	BioDegradation[e]
	Percent Removal			Rated as H, M, or L	
Triclosan	97	99	97	H	L (Heidler and Halden 2009)
Trimethoprim	19	24	55	L	L

Source: Interpreted from the original data of Snyder, S.A. et al., American Waterworks Association Research Foundation (AWWARF), 2007.

Note: Some compounds fell out of these data ranges, and their biodegradation property (in *italics*) was assigned based on other references.

H, high; M, medium; l, low.

[a] (A+B) 2 days = Adsorption + Biodegradation, Ambient, 2 days. (Data taken from Table 11.1 in Snyder et al. [2007].)

[b] (A+B) 5 days = Adsorption + Biodegradation, Ambient, 5 days. (Data taken from Table 11.1 in Snyder et al. [2007].)

[c] Adsorption = 5 days exposure data with Colorada River water, with pH adjusted to 2.0. (Data taken from Table 11.1 in Snyder et al. [2007].)

[d] Sorption property interpreted as follows: Adsorption <56 set as Low (L); Adsorption >90 set as High (H); and in between as Medium (M).

[e] Biodegradation property interpreted as follows: with (A+B) 5 days <80 set as Low (L); with [{(A+B) 2 days} – Adsorption] >12 set as High (H); and [{(A+B) 5 days} – {(A+B) 2 days}] >6 set as Medium (M).

and 17-α-ethinyl-estradiol would not be appreciably removed even after 48 h (Ternes et al. 1999a). The results suggested that to accomplish the complete removal of hormones, a minimum hydraulic retention time (HRT) is required.

As demonstrated in Table 8.3, the PCB congeners that are more highly chlorinated adsorb strongly to soil and sediment and generally persist for a long time in the environment, but would be removed from the sewage system as sludge. The various congeners in soil and sediment have half-lives that extend from months to years. Adsorption of PCBs into the sludge generally increases according to the extent of chlorination of the congener and the organic carbon contents of sediment. Volatilization and biodegradation are two very slow processes, which also happen to be the major pathways for PCB removal from water and soil (Faroon et al. 2003). The removal of antibiotics such as ciprofloxacin and trimethoprim in the sewage system demonstrated the importance of sorption property. Ciprofloxacin and trimethoprim are antibiotics, and thus, their biodegradation rates during the sewage treatment process are very low; the former has a high removal rate due to its high sorption property and also has a high affinity for soil but mainly accumulates in the topsoil (Golet et al. 2003). The concentrations of ciprofloxacin in sewage sludge and

TABLE 8.3

Review of the Removal Efficiencies of Emerging Chemicals in POSTWs

Emerging Chemicals	Removal Efficiency (%)	Sorption	Biodegradation	Reference
PFOS (perfluorooctanesulfonate)	~0	Low	Low	Kurume Laboratory (2002)
PFOA (perfluorooctanoic acid)	~0	Low	Low	Loganathan et al. (2007)
PAH (polycyclic aromatic hydrocarbons)	~30 (low MW) ~70 (high MW)	Medium High	Low None	AWWRF (2005) Snyder et al. (2007)
PCB (polychlorinated biphenyls)	~80 (mostly in the sludge)	High	Low	Verbrugge et al. (1995)
PBDE (polybrominated diphenylether)	~90 (mostly in the sludge)	High	Low	Peng et al. (2009)
DDT (pesticide)	~40	Medium	Low	Snyder et al. (2007)
DEET (pesticide)	~40	Low	Medium	Snyder et al. (2007)
Ciprofloxacin (antibiotics)	~90 (mostly in the sludge)	High	Low	Golet et al. (2002)
Trimethoprim (antibiotics)	~10	Low	Low	Hernando et al. (2006)

manure-treated soils ranged from mg to kg levels (Golet et al. 2002). The sorption of several fluoroquinolone derivatives, including ciprofloxacin, to soils and pure clay minerals has been reported, where it was proposed that cation bridging was the major mechanism responsible for the fluoroquinolone sorption (Nowara et al. 1997). The pharmaceuticals iopromide (an x-ray contrast agent) and trimethoprim (an antibacterial drug) are frequently detected in effluents of wastewater treatment works and in surface waters, due to their persistence and high usage. The removal of iopromide and trimethoprim during normal sewage treatment process is minimal due to their low sorption and biodegradation properties. The findings from a laboratory-scale experiment showed that there was a significantly higher removal rate in nitrifying activated sludge, when compared to conventional activated sludge that was observed for both iopromide and trimethoprim. This suggests that the nitrifying bacteria have an important role in the biodegradation of iopromide and trimethoprim in the activated sludge with a higher solid retention time (SRT). Results from the laboratory-scale study corroborated with the removal efficiencies observed in a full-scale municipal sewage treatment system. This showed that iopromide (ranging from 0.10 to 0.27 µg/L) and trimethoprim (ranging from 0.0.08 to 0.53 µg/L) were removed more effectively in the nitrifying activated sludge, which has a higher SRT (49 days) than the conventional activated sludge (SRT of 6 days). In nitrifying activated sludge, the percentage of removal of iopromide reached 61%, whereas average removal was negligible in conventional activated sludge. For trimethoprim, removal was limited to about 1% in the conventional activated sludge, whereas in the nitrifying activated sludge the removal increased to 50% (Batt et al. 2006).

In Table 8.4, different removal efficiencies can be seen for five antibiotics (tetracycline, cephalexin, norfloxacin, erythromycin, and trimethoprim) from four different sewage treatment works (STW) (Wan Chai, Stonecutters Island, Tai Po and Shatin sewage treatment facility) in Hong Kong (data extracted from Gulkowska et al. 2008). The existing Wan Chai district sewerage system was designed and built before 1970, which at present is mainly used as a preliminary screening facility that screens for suspended matter with a diameter of more than 6 mm. No removal efficiency of any antibiotics would be expected, and this was confirmed by the data from Wan Chai STW. On the other hand, Stonecutters Island sewage treatment facility is a chemically enhanced primary treatment facility in which chemicals, including ferric iron and polymeric coagulants for water treatment, are used to treat the sewage; however, the solid retention time (SRT) was less than 2 h. As displayed in Table 8.4, there was no observable removal efficiency for any of the antibiotics. Tetracycline, cephalexin, and norfloxacin were demonstrated to have good sorption property; hence, a reasonable removal efficiency would be expected in those three antibiotics during the primary treatment process (Huang et al. 2001, Golet et al. 2002, 2003, Karthikeyan and Meyer 2006). No observable removal efficiency could be detected for tetracycline, cephalexin, and norfloxacin from Stonecutters Island chemically enhanced primary treated effluent. Therefore, it would be beneficial if further investigations could be conducted to establish the cause of such outcomes. Both Shatin and Tai Po STW are secondary treatment facilities that also include processes for primary sedimentation of suspended matter and biological treatment of sewage. As demonstrated in Table 8.4, significant efficiencies were observed in the removal of tetracycline, cephalexin, and norfloxacin. The removal of these may be principally attributed to their sorption properties, whereas a less significant amount may be justified by biological degradation. The HRT for the Tai Po facility was reported as 16 h, but it was reported to be 21 h in the case of Shatin facility. There were better removal efficiencies for all the investigated antibiotics at the latter facility, which may be explained by the HRT being 5 h longer than the former facility. Apparently, HRT is also one of the major factors in determining the removal efficiencies of antibiotics. The HRT was longer than 15 h for secondary treatment plants (Shatin and Tai Po), but the solid retention time was only 1–2 h for the primary treatment plant (Stonecutters Island STW). The extremely low concentration found at the Tai Po facility, 96 ng/mL, for tetracycline in the influent sample may be an anomaly, as all other STWs' influent concentrations for tetracycline were

TABLE 8.4

Antibiotic Concentrations (Mean ± SD) in Influent and Effluent Samples from STW in Hong Kong

STW Location	Sample Source	Sample Concentration (ng/L)				
		TET (Tetracycline)	CLX (Cephalexin)	NOR (Norfloxacin)	ERY (Erythromycin)	TMP (Trimethoprim)
Wan Chai	Influent	660±2.8	1200±18	110±2.5	810±11	120±0.7
	Effluent	620±12	980±6.0	110±7.8	850±2.1	170±2.5
Stonecutters Island	Influent	550±10	1900±8.0	280±9.0	550±8.0	210±3.0
	Effluent	510±13	1800±3.0	320±10	510±9.0	230±7.0
Tai Po	Influent	96±1.6	670±13	110±0.4	470±2.5	120±2.1
	Effluent	180±6.7	240±5.7	85±0.1	520±8.5	140±5.7
Shatin	Influent	1300±0	2900±3.5	460±48	740±1.4	320±1.4
	Effluent	370±32	330±13	100±8.7	600±11	120±4.2

Source: Data extracted from Gulkowska, A. et al., *Water Res.*, 42, 395, 2008.

much higher, ranging from 550 to 1300 ng/mL. The high removal efficiency (more than 60%) for trimethoprim at the Shatin STW was significantly different from other studies (Snyder et al. 2007) and, thus, is also worth noting.

Other Treatment Methods to be Incorporated into POSTWs

Not much research has been conducted focusing on the removal of POPs and CECs from POSTWs. USDI 2009 had reviewed major advanced treatment technologies for the removal of POPs and CECs from river waters. The data from this review have been extracted and presented for hormones and antibiotics (Tables 8.5, 8.9, and 8.13), and other pharmaceutical (Tables 8.6, 8.10, and 8.14), agricultural, and industrial compounds (Tables 8.7, 8.11, and 8.15).

Activated Carbon Adsorption

AC is a family of carbonaceous adsorbents, with a highly crystalline form and extensively internal pore structure. By using different production techniques and sources of raw material, markedly different characteristics of AC products are available from different suppliers. Surface area, BET-N2, is a measurement using nitrogen gas (N_2), to obtain the extent of the pore surface developed within the matrix of the AC. Surface area of AC is a primary indicator of the activity level based on the findings that the greater the surface area, the higher the number of available adsorptive sites. Using 1 g of AC can have a surface area in excess of 500 m^2 due to its high degree of microporosity. Traditionally, AC is grounded into fine granules of less than 1.0 mm in size. Granular activated carbon (GAC) is defined as AC being retained on a 50-mesh sieve (0.297 mm), and the American Society for Testing and Materials (ASTM) classifies powdered activated carbon (PAC) as smaller than an 80-mesh sieve (0.177 mm). GAC are sold in different size of granules. A 20 × 40 GAC carbon is made of particles that will (more than 85%) pass through a U.S. Standard Mesh Size No. 20 sieve (0.84 mm) but will be (more than 95%) retained on a U.S. Standard Mesh Size No. 40 sieve (0.42 mm).

AC removes normal organic matter and organic compounds from water through hydrophobic interactions. AC total surface area and pore structure has a large influence on adsorption capacity and kinetics. The AC adsorption process works in three stages. First, the compound adheres to the surface of the AC, the moves into the large pores, and is finally adsorbed onto the smaller pores and inner surface of the AC. When the carbon hits its breakpoint, it is referred to as "spent" and is needed to be replaced or removed and sent off to be reactivated. PAC is too small for reactivation, and GAC is typically the only form of AC to be sent for the treatment (reactivation). The reactivated carbon is then sent back for reuse.

TABLE 8.5
Removal Efficiency by Different AC Treatments in Bench-Scale Tests (Hormones and Antibiotics)

Compound	Powder Activated Carbon, 5 mg/L[a]	Powder Activated Carbon, 5 mg/L[b]	Powder Activated Carbon, 5 mg/L[c]	Powder Activated Carbon, 5 mg/L[d]	Powder Activated Carbon, 5 mg/L[e]	Powder Activated Carbon, 5 mg/L[f]	Powder Activated Carbon, 1 mg/L[f]	Powder Activated Carbon, 20 mg/L[f]
Androstenedione	58	89	84	68	>98	>98	63	>98
Erythromycin	44	48	50	64	NA	NA	NA	NA
Estradiol	2	93	91	71	>97	>97	64	97
Estriol	55	NA	NA	61	NA	NA	NA	NA
Estrone	78	NA	NA	75	NA	NA	NA	NA
Ethinylestradiol	67	87	83	61	>98	>98	57	>98
Progesterone	89	92	87	77	>94	>94	66	>94
Sulfamethoxazole	42	6	40	43	41	46	<1	>78
Testosterone	36	87	83	67	99	98	60	99
Trimethoprim	38	92	91	79	>96	>96	81	>96

Source: Data extracted from Snyder S.A. et al., American Waterworks Association Research Foundation (AWWARF), 2007.
All bench-scale experiments were tested at 4 h contact time with 100 rpm stirring.
[a] Colorado River water and WPM-carbon.
[b] Ohio River water and AC800-carbon.
[c] Ohio River water and WMP-carbon.
[d] Passaic River water and WMP-carbon.
[e] Suwannee River water and AC800-carbon.
[f] Suwannee River water and WMP-carbon.

TABLE 8.6

Removal Efficiency by Different AC Treatments in Bench-Scale Tests (Other Pharmaceutical Compounds)

Compound	Powder Activated Carbon, 5 mg/L[a]	Powder Activated Carbon, 5 mg/L[b]	Powder Activated Carbon, 5 mg/L[c]	Powder Activated Carbon, 5 mg/L[d]	Powder Activated Carbon, 5 mg/L[e]	Powder Activated Carbon, 5 mg/L[f]	Powdered Activated Carbon, 1 mg/L[f]	Powder Activated Carbon, 20 mg/L[f]
					Percentage Removal (%)			
Acetaminophen	88	61	70	68	85	86	30	>90
Caffeine	16	74	75	64	89	83	<1	95
Carbamazepine	55	80	79	62	94	92	40	99
Diazepam	56	69	69	52	96	92	43	99
Diclofenac	64	49	38	45	36	51	-2	>93
Dilantin	<1	44	40	46	85	81	26	>93
Floxetine	93	94	93	89	>93	>93	85	>93
Gemfibrozil	<1	51	38	18	60	57	17	>96
Hydrocodone	70	NA	NA	74	NA	NA	NA	NA
Ibuprofen	48	20	2	17	12	37	3	80
Iopromide	33	6	9	30	30	53	14	>92
Meprobamate	<1	26	31	32	58	45	9	94
Naproxen	86	41	44	49	43	62	<1	>86
Oxybenzone	93	>97	>97	92	91	92	76	>95
Pentoxifylline	65	82	82	66	>93	>93	45	>93

Source: Data extracted from Snyder S.A. et al., American Waterworks Association Research Foundation (AWWARF), 2007.
All bench-scale experiments were tested at 4 h contact time with 100 rpm stirring.

[a] Colorado River water and WPM-carbon.
[b] Ohio River water and AC800-carbon.
[c] Ohio River water and WMP-carbon.
[d] Passaic River water and WMP-carbon.
[e] Suwannee River water and AC800-carbon.
[f] Suwannee River water and WMP-carbon.

TABLE 8.7

Removal Efficiency by Different AC Treatments in Bench-Scale Tests (Agricultural and Industrial Chemicals)

Compound	Percentage Removal (%)							
	Powder Activated Carbon, 5 mg/L[a]	Powder Activated Carbon, 5 mg/L[b]	Powder Activated Carbon, 5 mg/L[c]	Powder Activated Carbon, 5 mg/L[d]	Powder Activated Carbon, 5 mg/L[e]	Powder Activated Carbon, 5 mg/L[f]	Powder Activated Carbon, 1 mg/L[f]	Powder Activated Carbon, 20 mg/L[f]
Atrazine	55	69	60	53	90	85	25	99
Benzo(a) pyrene	>94	>90	>90	>85	69	72	52	87
DDT	81	78	74	56	>85	>85	77	>85
DEET	<1	49	46	40	83	74	24	99
1,4-Dioxane	NA	NA	NA	NA	NA	NA	NA	NA
Fluorene	>95	>93	>93	94	NA	NA	NA	NA
Lindane	66	86	81	63	>86	>86	50	>86
TCEP	70	47	50	50	81	70	15	>90
Triclosan	93	>98	95	83	>96	>96	75	95

Source: Data extracted from Snyder S.A. et al., American Waterworks Association Research Foundation (AWWARF), 2007.
All bench-scale experiments were tested at 4 h contact time with 100 rpm stirring.
[a] Colorado River water and WPM-carbon.
[b] Ohio River water and AC800-carbon.
[c] Ohio River water and WMP-carbon.
[d] Passaic River water and WMP-carbon.
[e] Suwannee River water and AC800-carbon.
[f] Suwannee River water and WMP-carbon.

AC usually binds strongly with most organic matters, but it does not bind well to certain chemicals, including alcohols, glycols, strong acids and bases, and metal and most inorganics, with the exception of iodine. In fact, the iodine number is expressed in mg/g, whereby the ASTM D28 Standard Method test is used as an indication for the total surface area. AC is most effective in removing organic contaminants from water. AC filtration has the ability to remove chlorine and trihalomethanes, pesticides, halogenated hydrocarbons, PCBs, and PAHs. AC filtration does not remove bacteria, sodium, nitrates, fluoride, and Ca and Mg hydroxides and carbonates. Generally, the least water-soluble organic molecules are strongly adsorbed. AC treatment would be less effective in removing organic molecules with bulky size, such as Iopromide, due to the difficulty getting into smaller pores of AC. Lindane, a pesticide with a log Kow of 3.72, was reduced from a concentration of 10–0.1 ppb (Kouras et al. 1998) in a bench-scale study using 20 mg/L of PAC with a 1 h contact time. The removal of 17-β-estradiol by GAC was 49% removal at 50 min and 81% removal at 180 min (Fuerhacker et al. 2001).

Pilot and full-scale studies have shown that PAC and GAC are effective for the removal of EDCs and PPCPs (AWWARF 2007). The studies found that NOM competes for binding sites and can block the pores within the AC structure, reducing the efficacy of the AC treatment process in the removal of POPs and CECs. As expected, the removal efficiencies of organic contaminants were more effective at full-scale treatment facilities that routinely regenerate or replace AC. The facilities with GAC that had been in operation without the regeneration or replacement of AC performed poorly in removal of POPs and CECs (AWWARF 2007). Generally, compounds with log Kow of greater than 2 can be effectively removed using PAC or GAC. Table 8.5 (hormones and antibiotics), Table 8.6 (other pharmaceuticals), and Table 8.7 (agricultural and industrial chemicals) summarize the eight experiments performed using different sources of water (Colorado, Ohio, and Suwannee rivers), different brands of AC (WWP and AC800), and different loadings of AC (5, 1, and 20 mg/L). The possible outliners with unexpected results can be identified from this extensive study. In Table 8.5, the removal of 2% estradiol and 6% sulfamethoxazole were probable outliners because the data from each one differed significantly from the other five points, using 5 mg/L AC loading. Other possible outliners were the 16% removal of Caffeine, <1% removal of Gemfibrozil, <2% removal of Ibuprofen, <1% removal of Meprobamate, as shown in Table 8.6, and the <1% removal of DEET, as shown in Table 8.7, as they all differed significantly from the other five points for each compound, using 5 mg/L AC loading. The total number of possible outliners was 7 out of approximately 200 data points or about 3.5%, which was a reasonable and respectable outcome considering the enormity of the investigation. By examining the generated data using 1 and 20 mg/L AC loading, it can be inferred that the removal rates of the following compounds—Caffeine, Diclofenac, Ibuprofen, Naproxen, and Sulfamethoxazole—tend to be very sensitive to AC loading.

The ability of AC to remove POPs and CECs depends on compound/ carbon loading ratio, surface area of AC, contact time, and physical properties of the target compounds. As part of the Safe Drinking Water Act Amendments of 1986, the use of GAC adsorption and filtration was recommended as the best available technology for the removal of organic contaminants from drinking water. But influent in a POSTW can have highly diverse organic contaminants and natural organic matter (NOM). The balk content of NOM could compete for binding sites within the AC structure and would reduce the efficacy of AC. Pretreatment for NOM or total organic carbon (TOC) would be an important factor to improve the efficiency of the AC treatment process. It is also important to replace and regenerate the spent GAC by reactivation treatment.

The article "Removal of EDCs and Pharmaceuticals in Drinking and Reuse Treatment Processes" by Snyder et al. (2007) contains more than 331 pages with numerous tables and figures. The U.S. Department of the Interior Bureau of Reclamation (October 2009) issued a 108-page report on CECs and treatment technologies that are applicable to reducing them. Some of the original data in Snyder's article were tabulated in the USDI article. It was difficult to establish why there were differences in the performance of GAC and PAC with regard to the removal efficiency between the two AC, where some of these data have been summarized in Table 8.8. The AC as listed in Table 5.1 of the USDI article did not specify whether it was PAC or GAC. The original data from Snyder et al. (2007) were reviewed carefully and summarized in Tables 8.9 (hormones and antibiotics), 8.10 (other pharmaceuticals), and 8.11 (agricultural and industrial chemicals).

TABLE 8.8

Removal Efficiency by Different AC Treatments

Compounds	Activated Carbon[a]	Powder Activated Carbon[b]	GAC[b]
Atrazine	63	3	63
Caffeine	59	16	59
Carbamazepine	72	16	72
Erythromycin	52	8	52
Estriol	58	<1	58
Iopromide	31	72	31
Naproxen	60	6	60
Pentoxifylline	71	26	71

Source: Data extracted from U.S. Department of the Interior Bureau of Reclamation, Secondary/Emerging Constituents Report, Southern California Regional Brine-Concentrate Management, Study—Phase I, Lower Colorado Region, October 2009.

[a] Data from USDI 2009, Table 5-1.
[b] Data from USDI 2009, Table B-1.

TABLE 8.9

Removal Efficiency by Different AC Treatments (Hormones and Antibiotics, with *Heading Corrected*)

Compound	Percentage Removal (%)							
	Powder Activated Carbon[a]	GAC[b]	Powder Activated Carbon[b]	Powder Activated Carbon[c]	GAC[d]	Powder Activated Carbon[e]	Powder Activated Carbon, 5 mg/L[f]	Powder Activated Carbon, 35 mg/L[g]
Androstenedione	70	61	70	70	61	50–80	36	92
Erythromycin	52	8	52	52	8	20–50	23	86
Estradiol	55	NA	55	55	NA	50–80	58	97
Estriol	58	<1	58	58	<1	20–50	43	97
Estrone	77	NA	77	77	NA	50–80	71	>99
Ethinylestradiol	70	NA	70	70	NA	50–80	55	96
Progesterone	84	NA	84	84	NA	>80	45	>87
Sulfamethoxazole	43	84	43	43	84	20–50	<1	90
Testosterone	71	74	71	71	74	50–80	NA	NA
Trimethoprim	69	64	69	69	64	50–80	73	>99

[a] Data from USDI 2009, Table 5-1.
[b] Data from USDI 2009, Table B-1.
[c] Data from Snyder et al. 2007, Table 5.1 (5 mg/L PAC, 4 h contact time).
[d] Data from Snyder et al. 2007, Table 5.5.
[e] Data from Snyder et al. 2007, Table 5.6.
[f] Data from Snyder et al. 2007, Table D.8, with 5 mg/L PAC, 5 h contact time.
[g] Data from Snyder et al. 2007, Table D.8, with 35 mg/L PAC, 5 h contact time.

TABLE 8.10

Removal Efficiency by Different AC Treatments (Other Pharmaceutical Compounds, with *Heading Corrected*)

Compound	Powder Activated Carbon[a]	GAC[b]	Powder Activated Carbon[b]	Powder Activated Carbon[c]	GAC[d]	Powder Activated Carbon[e]	Powder Activated Carbon, 5 mg/L[f]	Powder Activated Carbon, 35 mg/L[g]
				Percentage Removal (%)				
Acetaminophen	78	99	78	78	99	50–80	58	97
Caffeine	59	16	59	59	16	50–80	45	95
Carbamazepine	72	16	72	72	16	50–80	51	96
Diazepam	67	NA	67	67	NA	50–80	NA	NA
Diclofenac	49	69	49	49	69	20–50	18	83
Dilantin	56	23	56	56	23	20–50	45	90
Floxetine	91	NA	91	91	NA	>80	56	94
Gemfibrozil	38	8	38	38	8	20–50	21	85
Hydrocodone	72	56	72	72	56	50–80	NA	NA
Ibuprofen	26	16	26	26	16	<20	<1	74
Iopromide	31	72	31	31	72	<20	<1	65
Meprobamate	36	13	36	36	13	20–50	37	>44
Naproxen	60	6	60	60	6	20–50	<1	89
Oxybenzone	92	NA	92	92	NA	>80	77	>98
Pentoxifylline	71	26	71	71	26	50–80	52	97

[a] Data from USDI 2009, Table 5-1.
[b] Data from USDI 2009, Table B-1.
[c] Data from Snyder et al. 2007, Table 5.1 (5 mg/L PAC, 4 h contact time).
[d] Data from Snyder et al. 2007, Table 5.5.
[e] Data from Snyder et al. 2007, Table 5.6.
[f] Data from Snyder et al. 2007, Table D.8, with 5 mg/L PAC, 5 h contact time.
[g] Data from Snyder et al. 2007, Table D.8, with 35 mg/L PAC, 5 h contact time.

TABLE 8.11

Removal Efficiency by Different AC Treatments (Agricultural and Industrial Chemicals, with Heading Corrected)

Compound	Percentage Removal (%)							
	Powder Activated Carbon[a]	GAC[b]	Powder Activated Carbon[b]	Powder Activated Carbon[c]	GAC[d]	Powder Activated Carbon[e]	Powder Activated Carbon, 5 mg/L[f]	Powder Activated Carbon, 35 mg/L[g]
Atrazine	63	3	63	63	3	50–80	47	94
Benzon(a) pyrene	72	NA	72	72	NA	>80	NA	NA
DDT	70	NA	70	70	NA	50–80	<1	>13
DEET	54	63	54	54	63	20–50	38	93
1,4-Dioxane	<20	NA	NA	NA	NA	NA	NA	NA
Lindane	70	NA	70	70	NA	50–80	NA	NA
TCEP	60	NA	60	60	NA	20–50	39	91
Triclosan	90	NA	90	90	<1	>80	82	84
Fluorene	94	NA	94	94	NA	>80	83	>89

a   Data from USDI 2009, Table 5-1.
b   Data from USDI 2009, Table B-1.
c   Data from Snyder et al. 2007, Table 5.1 (5mg/L PAC, 4h contact time).
d   Data from Snyder et al. 2007, Table 5.5.
e   Data from Snyder et al. 2007, Table 5.6.
f   Data from Snyder et al. 2007, Table D.8, with 5mg/L PAC, 5h contact time.
g   Data from Snyder et al. 2007, Table D.8, with 35mg/L PAC, 5h contact time.

TABLE 8.12

Removal Efficiency by Different AC Treatments[a]

Compounds	Powder Activated Carbon[a]	GAC[a]	Powder Activated Carbon[a]
Atrazine	63	3	63
Caffeine	59	16	59
Carbamazepine	72	16	72
Erythromycin	52	8	52
Estriol	58	<1	58
Iopromide	31	72	31
Naproxen	60	6	60
Pentoxifylline	71	26	71

[a] Data from Table 8.8 with *corrected headings*.

It was concluded that in USDI 2009 (Table 5.1) the heading for "Activated Carbon Adsorption" should be "Powder Activated Carbon Absorption" and for Table B-1 the heading for "PAC" should actually read "GAC" and "GAC" should actually read "PAC." The data as presented in Table 8.8 were retabulated in Table 8.12. The performance of PAC was generally better than GAC, with one exception that involved the removal of Iopromide, which probably occurred due to the bulky size of the target compound and the smaller pore sizes that exist in PAC particles.

Ozone Treatment

Ozone is a pale blue gas and typically obtained from an ozone generator. The generator uses a high-voltage dielectric discharge gap where oxygen in the air is converted into ozone, a higher-energy state with a triatomic form (O_3). Ozone is a very reactive gas that can oxidize bacteria, moulds, organic material, and other pollutants found in water. Oxidation reactions by ozone are not selective, and it reacts with a wide variety of organic and inorganic materials in water. It is a strong oxidant and disinfectant that does not maintain a residual concentration and decays very rapidly in aqueous solutions, unlike free chlorine or chloramines (NH_2Cl). Ozone through the formation of free radicals such as HO^* radical reacts directly with many EDC and PPCP compounds. The reaction reduces POPs and CECs at the level of 5 mg/L dose for typical water disinfection and is an effective means for removing target organic compounds in rivers and streams (AWWARF 2007). Ozone is effective in treatment with phenolic compounds (with electron-donating groups) such as acetaminophen, oxybenzone, and several estrogens (with phenolic group such as estradiol, estriol, estrone, and ethinyl-estradiol) can be effectively removed using ozone. A number of compounds with electron-donating capability, including gemfibrozil, hydrocodone, naproxen, and sulfamethoxazole, also were almost completely removed using ozone (Tables 8.13 through 8.15).

TABLE 8.13

Removal Efficiency by Different Advanced Treatment Processes (Hormones and Antibiotics)

Compound	Structure	Percentage Removal (%)[a]			
		Powder Activated Carbon Adsorption	Ozone	Reverse Osmosis	Biologically Active Carbon
Androstenedione (hormone)		70	>80	>61	97
Erythromycin (antibiotic)		52	>95	>98	78
Estradiol (hormone)		55	>95	NA	94

Compound				
Estriol (hormone)	58	>95	NA	92
Estrone (hormone)	77	>95	>95	95
Ethinylestradiol (hormone)	70	>95	NA	91
Progesterone (hormone)	84	>80	NA	99

(continued)

TABLE 8.13 (continued)

Removal Efficiency by Different Advanced Treatment Processes (Hormones and Antibiotics)

Compound	Structure	Percentage Removal (%)[a]			
		Powder Activated Carbon Adsorption	Ozone	Reverse Osmosis	Biologically Active Carbon
Sulfamethoxazole (antibiotic)		43	>95	>99	63
Testosterone (hormone)		71	>80	NA	96
Trimethoprim (antibiotic)		69	>95	>99	94

[a] Data extracted from USDI 2009, Table 5.1.

TABLE 8.14

Removal Efficiency by Different Advanced Treatment Processes (Other Pharmaceutical Compounds)

Compound	Structure	Percentage Removal (%)[a]			
		Powder Activated Carbon Adsorption	Ozone	Reverse Osmosis	Biologically Active Carbon
Acetaminophen		78	>95	>90	95
Caffeine		59	>80	>99	93
Carbamazepine		72	>95	>99	90
Diazepam		67	>95	NA	84

(continued)

TABLE 8.14 (continued)

Removal Efficiency by Different Advanced Treatment Processes (Other Pharmaceutical Compounds)

Compound	Structure	Percentage Removal (%)[a]			
		Powder Activated Carbon Adsorption	Ozone	Reverse Osmosis	Biologically Active Carbon
Diclofenac		49	96	>97	75
Dilantin		56	50–80	>99	80
Floxetine		91	>95	>96	>99
Gemfibrozil		38	>95	>99	74

Compound				
Hydrocodone	72	>95	>98	92
Ibuprofen	26	50–80	>99	83
Iopromide	31	20–50	>99	42
Meprobamate	36	20–50	>99	71
Naproxen	60	>95	>99	82

(continued)

TABLE 8.14 (continued)

Removal Efficiency by Different Advanced Treatment Processes (Other Pharmaceutical Compounds)[a]

Compound	Structure	Percentage Removal (%)[a]			
		Powder Activated Carbon Adsorption	Ozone	Reverse Osmosis	Biologically Active Carbon
Oxybenzone		92	>95	>93	98
Pentoxifylline		71	>80	>96	90

[a] Data extracted from USDI 2009, Table 5.1.

TABLE 8.15

Removal Efficiency by Different Advanced Treatment Processes (Agricultural and Industrial Compounds)

Compound	Structure	Percentage Removal (%)[a]				
		Powder Activated Carbon Adsorption	Ozone	Reverse Osmosis	Biologically Active Carbon	
Atrazine (pesticide)		63	20–50	NA	83	
Benzon(a) pyrene (PAH)		72	NA	>90	89	
DDT (pesticides)		70	NA	NA	85	
DEET (insecticide)		54	50–80	>95	80	

(continued)

TABLE 8.15 (continued)

Removal Efficiency by Different Advanced Treatment Processes (Agricultural and Industrial Compounds)

Compound	Structure	Percentage Removal (%)[a]			
		Powder Activated Carbon Adsorption	Ozone	Reverse Osmosis	Biologically Active Carbon
1,4-Dioxane (industrial)		<20	<35	20–50	<20
Fluorene (PAH)		94	NA	NA	>94
Lindane (pesticides)		70	NA	NA	91
TCEP (flame retardant)		60	<20	>91	80
Triclosan (anti-microbial)		90	>95	97	97

[a] Data extracted from USDI 2009, Table 5.1.

From those tables the notable exceptions of oxidation-resistant compounds are TCEP, atrazine, iopromide, and meprobamate. Generally, ozone treatment decomposes POPs and EDCs to smaller molecules and increases the opportunity for further biodegradation. TSS, reduced forms of iron and manganese, and other reducing agents decrease the effectiveness of ozone. Data from the literature on ozone treatment with regard to the removal of POPs and CECs are summarized in Tables 8.13 through 8.15.

Reverse Osmosis

Reverse osmosis (RO) is a filtration method wherein a selective membrane is used and pressure is applied to the solution from one side, forcing the removal of many types of large molecules and ions from solutions. Pure solvent (such as water) is allowed to pass through the membrane into the other side, and the large molecules and ions are retained on the pressurized side of the membrane. The membrane is created to be "selective," preventing large molecules and ions from passing through the pores but allowing the smaller molecules in the solution (such as water) to pass through the membrane freely. RO is most commonly known for its use with removing salts and minerals as part of the drinking water purification process using seawater. In normal osmosis, the water (solvent) molecules pass through a selective permeable membrane into a region of higher salt (solute) concentration. This is a normal physical process in which any solvent moves without the input of energy or pressure. In reverse osmosis, the input of energy or pressure is to reverse the normal osmosis process. RO involves a diffusive mechanism where the separation efficiency is dependent on pressure, solute concentration, and rate of water flux (Crittenden et al. 2005). RO is one of the most effective process technologies (but not efficient) that is able to remove soluble inorganic and organic compounds from water, including POPs and CECs. There are very few compounds that are not fully removed by RO. However, it is an incredibly inefficient process; typically 3 gal of water is wasted for every gallon of purified water it produces. Although RO is an inefficient process, for areas with limited resource of surface water or groundwater, it is the most preferred and commonly used method for desalination. The data from the literature on RO treatment in the removal of POPs and CECs are summarized in Tables 8.13 through 8.15.

Biological Aerated Filters with Active Carbon Coupled with Ozone Treatment (BAC)

Biological aerated filters (BAF) are submerged filters with attached biological growth with biofilm layer where biodegradation can take place. In an aerated BAF system, an air header is provided to ensure that oxygen is going throughout the entire media bed layer. It was reported that the inner part of the BAF is usually oxygen deficient and allows for nitrate formed through nitrification to be converted into nitrogen gas (Tchobanoglous et al. 2003). GAC is used in conjunction with ozone in such an advanced water treatment process known as the BAC. The basis of this treatment system involves the use of ozone as an oxidizing agent, which causes the breakdown of POPs and EDCs into smaller compounds. The formed molecules are degraded biologically on bacterial biomass that develops on the surface of the AC, whereas the adsorption of POPs and EDCs as well as the removal of biodegradation residues could be achieved through AC conventional adsorption processes. The removal rate using BAC is a combination of the capability of ozone oxidative degradation, biological degradation, and the adsorptive properties of AC (AWWARF 2007). The data from the literature investigating BAC treatment in the removal of POPs and CECs are summarized in Tables 8.13 through 8.15.

Conclusion

Wastewater treatment facilities are designed to meet effluent quality parameters such as TSS, BOD, and COD. Most conventional POSTW facilities primarily use adsorptive, precipitative, oxidative, and biological processes, to remove mostly natural organic materials. POPs and CECs are a diverse group of compounds that include PhACs, PCPs, hormones, pesticides, and industrial and household chemicals. There are growing concerns about these compounds because of the detrimental effects associated with their use, and some of them may have an adverse impact on the development and reproduction of fish and amphibians. Other evidences show that mammals could be sensitive to extremely low concentrations of POPS and CECs (AWWARF 2007). POPS and CECs have evolved into popular concerns because of heightened public perception and increasing attention from the scientific and environmental communities. However, it is difficult to establish firm evidence for the association between adverse human health effects and low-dose exposure to POPs and CECs.

The most cost-effective option for moderate POPs and CECs removal treatment technologies is GAC for waste- and ground-water treatment. Ozonation is another cost-effective option in achieving efficient removal of POPs and CECs. Sparingly soluble salts such as silica limit the reverse osmosis recovery to around 35%, which is necessary to pretreatment in softening process to increase water recovery. Despite the high cost, precipitative softening reverse osmosis (PSRO) is the only technology that removes not only POPS and CECs but also TDS.

POPs and CECs are a collective group of diverse compounds whose physical and chemical characteristics vary broadly even within the same subcategory. About 87,000 emerging compounds have been identified as possible EDCs (U.S. EPA 2008). At present, no single technology is available to treat all POPs and CECs effectively because of their diversity; hence, multiple-barrier treatments may be the most effective approach to take.

The objective of sewage treatment is to produce an environmentally safe stream (effluent) and a solid waste (sludge) suitable for disposal or reuse (such as fertilizer). It is now possible, using advanced technologies, to reuse sewage effluent for drinking water. Singapore is the only country to implement such technologies on a large manufacturing scale, in its creation of NEWater (2011). In 1972, Singapore's first water master plan was drawn up, and 2 years later, in 1974, PUB built a pilot plant to turn wastewater into potable water, which is the precursor of today's NEWater factories. But the plan was ahead of its time as the costs were astronomical and the membranes selected were unreliable. Thus, the plan was shelved and had to wait for further scientific advancements. By 1998, the necessary technologies had been developed that brought the drinking-water costs down; thus, given these developments, in May 2000, the first NEWater plant was completed. There are currently five NEWater plants in Singapore, with the fifth and the largest one located at Changi, opened in May 2010, having the capacity of 50 million gallons per day (mgd). After the expansion of the existing plants, NEWater is estimated to provide 30% of Singapore's total water demand. The water regeneration achievements in Singapore demonstrate the future alternative sewage treatment possibilities that other countries can potentially adopt.

References

AWWARE. 2007. Removal of EDCs and Pharmaceuticals in Drinking and Reuse Treatment Process. Published by American Water Works Association Research Foundation.

Baronti C, Curini R, D'Ascenzo G, Di Corcia A, Gentili A, and Samperi R. 2000. Monitoring natural and synthetic estrogens at activated sludge sewage treatment plants and in a receiving river water. *Environmental Science and Technology* 34(24): 5059–5066.

Batt AL, Kim S, and Aga DS. 2006. Enhanced biodegradation of iopromide and trimethoprim in nitrifying activated sludge. *Environmental Science and Technology* 40 (23): 7367–7373.

Birkett JW and Lester JN. 2003. *Endocrine Disruptors in Wastewater and Sludge Treatment Processes.* Lewis Publishers, Boca Raton, FL.

Carballa M, Omil F, Lema JM, Llompart M, Garcia-Jares C, Rodriguez I, Gomez M, and Ternes T. 2004. Behavior of pharmaceuticals, cosmetics and hormones in a sewage treatment plant. *Water Research* 38: 2918–2926.

Crittenden J, Trussell R, Hand D, Howe K, and Tchobanoglous G. 2005. *Water Treatment Principles and Design,* 2 edn. John Wiley & Sons, Hoboken, NJ.

Cunningham VL. 2004. Special characteristics of pharmaceuticals related to environmental fate. In: Kümmerer K (ed.) *Pharmaceuticals in the Environment: Sources, Fate, Effects and Risks.* Springer, Berlin, Germany, pp. 12–24.

de Mes T, Zeeman G, and Lettinga G. 2005. Occurrence and fate of estrone, E2 and EE2 in STPs for domestic wastewater. *Reviews in Environmental Science and Biotechnology* 4: 275–311.

Drewes JE, Hemming J, Schauer JJ, and Sonzogni W. (2006). Water Environment Research Foundation: Wastewater Treatment and Reuse, Final Report "Removal of Endocrine Disrupting Compounds in Water Reclamation Processes".

Drinker CK, Warren MF, and Bennet GA. 1937. The problem of possible systemic effects from certain chlorinated hydrocarbons. *Journal of Industrial Hygiene and Toxicology* 19 (7): 283.

Faroon OM, Keith LS, Smith-Simon C, and De Rosa CT. 2003. *Polychlorinated Biphenyls: Human Health Aspects*. World Health Organization, Geneva, Switzerland.

Fuerhacker M, Durauer A, and Jungbauer A. 2001. Adsorption isotherms on 17β-estradiolon granular activated carbon (GAC). *Chemosphere* 44: 1573–1579.

Golet EM, Strehler A, Alder AC, and Giger W. 2002. Determination of fluoroquinolone antibacterial agents in sewage sludge and sludge-treated soil using accelerated solvent extraction followed by solid-phase extraction. *Analytical Chemistry* 74: 5455–5462.

Golet EM, Xifra I, Siegrist H, Alder AC, and Giger W. 2003. Environmental exposure assessment of fluoroquinolone antibacterial agents from sewage to soil. *Environmental Science and Technology* 15: 3243–3249.

Gröning J, Held C, Garten C, Claussnitzer U, Kaschabek SR, and Schlomann M. 2007. Transformation of diclofenac by the indigenous microflora of river sediments and identification of a major intermediate. *Chemosphere* 69: 509–516.

Gulkowska A, Leung HW, So MK, Taniyasu S, Yamashita N, Yeung LW, Richardson BJ, Lei AP, Giesy JP, and Lam PKS. 2008. Removal of antibiotics from wastewater by sewage treatment facilities in Hong Kong and Shenzhen, China. *Water Research* 42: 395–403.

Heidler J and Halden RU. 2009. Fate of organohalogens in US wastewater treatment plants and estimated chemical releases to soils nationwide from biosolids recycling. *Journal of Environmental Monitoring* 11: 2207–2215.

Hernando MD, Mezcua M, Fernández-Alba AR, and Barceló, D. 2006. Environmental risk assessment of pharmaceutical residues in wastewater effluents, surface waters and sediments. *Talanta* 69: 334–342.

Huang CH, Renew JE, Smeby KL, Pinkerston K, and Sedlak DL. 2001. Assessment of potential antibiotic contaminants in water and preliminary occurrence analysis. *Water Resource* 120: 30–40.

Karthikeyan KG and Meyer MT. 2006. Occurrence of antibiotics in wastewater treatment facilities in Wisconsin. *Science of the Total Environment* 361: 196–207.

Kouras A, Zouboulis A, Samara C, and Kouimtzis Th. 1998. Removal of pesticides from aqueous solutions by combined physicochemical processes—The behaviour of lindane. *Environmental Pollution* 103: 193–202.

Kurume Laboratory. 2002. Final report, biodegradation test of salt (Na, K, Li) of perfluoroalkyl (C = 4–12) sulfonic acid, test substance number K-1520 (test number 21520). Kurume Laboratory, Chemicals Evaluation and Research Institute, Japan.

Loganathan BG, Sajwan KS, Sinclair E, Senthil Kumar K, and Kannan K. 2007. Perfluoroalkyl sulfonates and perfluorocarboxylates in two wastewater treatment facilities in Kentucky and Georgia. *Water Research* 41: 4611–4620.

Nakada N, Tanishima T, Shinohara H, Kiri K, and Takada H. 2006. Pharmaceutical chemicals and endocrine disrupters in municipal wastewater in Tokyo and their removal during activated sludge treatment. *Water Research* 40: 3297–3303.

NEWater. 2011. Singapore's national water agency (www.pub.gov.sg), History of NEWater.

Nowara A, Burhenne J, and Spiteller M. 1997. Binding of fluoroquinolone carboxylic acid derivatives to clay minerals. *Journal of Agricultural and Food Chemistry* 45: 1459–1463.

Peng X, Tang C, Yu Y, Tan J, Huang Q, Wu J, Chen S, and Mai B. 2009. Concentrations, transport, fate, and releases of polybrominated diphenyl ethers in sewage treatment plants in the Pearl River delta, South China. *Environment International* 35: 303–309.

Pickering AD and Sumpter JP. 2003. Comprehending endocrine disruptors in aquatic environments. *Environmental Science and Technology* 37: 331A–336A.

Redshaw CH, Cooke MP, Talbot HM, McGrath S, and Rowland SJ. 2008. Low biodegradability of fluoxetine HCl, diazepam and their human metabolites in sewage sludge-amended soil. *Journal of Soils and Sediments* 8: 217–230.

Snyder S, Vanderford B, Adham S, Oppenheimer J, Gillogly T, Bond R, and Veerapaneni V. 2004. Pilot scale evaluations of membranes for the removal of endocrine disruptors and pharmaceuticals. In *American Waterworks Association (AWWA) Water Quality and Technology Conference*, San Antonio, TXNovember 14–18.

Snyder SA, Wert EC, (Dawn) Lei H, Westerhoff P, and Yoon Y. 2007. Removal of EDCs and Pharmaceuticals in Drinking and Reuse Treatment Process. p. 193, 194, and 198. American Water Works Association Research Foundation (AWWARF).

Ternes TA, Herrmann N, Bonerz M, Knacker T, Siegrist H, and Joss A. 2004. A rapid method to measure the solid–water distribution coefficient (Kd) for pharmaceuticals and musk fragrances in sewage sludge. *Water Research* 38: 4075–4084.

Ternes TA, Kreckel P, and Mueller J. 1999a. Behaviour and occurrence of estrogens in municipal sewage treatment plants—II. Aerobic batch experiments with activated sludge. *Science of the Total Environment* 225: 91–99.

Ternes TA, Stumpf M, Mueller J, Heberer K, Wilken RD, and Servos M. 1999b. Behavior and occurrence of estrogens in municipal sewage treatment plants—I. Investigations in Germany, Canada and Brazil. *Science of the Total Environment* 225: 81–90.

United States Department of Health and Human Services. 2008. Effects of Bisphenol A and Phthalates. Testimony of Dr. John Bucher on June 10, 2008. www.hhs.gov/asl/testify/2008/06/t20080610a.html

United States Environmental Protection Agency (U.S. EPA). 1997. Endocrine Disruptors Screening and Testing Advisory Committee (EDSTAC) Organizational Meeting—Final Meeting Summary. www.epa.gov/endo/pubs/edsparchive/006crs.htm

U.S. Department of the Interior Bureau of Reclamation. October 2009. Secondary/Emerging Constituents Report. Southern California Regional Brine-Concentrate Management. Study—Phase I, Lower Colorado Region.

U.S. EPA. 1988. *Drinking Water Criteria Document for Polychlorinated Biphenyls (PCBs). Final.* Cincinnati, OH, US Environmental Protection Agency, Office of Health and Environmental Assessment, Environmental Criteria and Assessment Office (ECAO-CIN-414).

U.S. EPA. 2008. Drinking water contaminants. http://www.epa.gov/safewater/contaminants/index.html

van Vuuren L. March/April 2008. The ABCs of EDCs—Global concern spurs researchers into action. *The Water Wheel* 7(2): 24–27.

Verbrugge DA, Giesy JP, Mora MA, Williams LL, Rossmann R, Moll RA, and Tuchman M. 1995. Concentrations of dissolved and particulate polychlorinated biphenyls in water from the Saginaw river, Michigan. *Journal of Great Lakes Research* 21: 219–233.

Voishvillo NE, Andriushina VA, Savinova TS, and Stytsenko TS. 2004. Conversion of androstenedione and androstadienedione by sterol-degrading bacteria. *Prikl Bioklim Mikrobiology* 40: 536–543.

Water Environment Research Foundation (WERF). 2007. "Fate of pharmaceuticals and personal care products through municipal wastewater treatment Processes" by Stephenson, Roger and Joan Oppenheimer.

World Health Organization (WHO). 2002. Global assessment of the state of the science of endocrine disruptors. International Programme on Chemical Safety. http://www.who.int/ipcs/publications/new_issues/endocrine_disruptors/en

Yang YY, Borch T, Yonag RB, Goodridge LD, and Davis JG. 2010. Degradation kinetics of testosterone by manure-borne bacteria: Influence of temperature, pH, glucose amendments, and dissolved oxygen. *Journal of Environmental Quality* 39: 1153–1160.

9

Inverse Relationship between Urinary Retinol-Binding Protein, Beta-2-Microglobulin, and Blood Manganese Levels in School-Age Children from an E-Waste Recycling Town

Jinrong Huang, Xijin Xu, Guina Zheng, Kusheng Wu,
Yongyong Guo, Junxiao Liu, Hao Ban, Weitang Liao, Wei Liu,
Hui Yang, Qiongna Xiao, Yuanping Wang, and Xia Huo

CONTENTS

Introduction

Electronic waste or e-waste refers to end-of-use electronic products, including computers, printers, television sets, refrigerators, mobile phones, and toys; these common consumer products contain a variety of potential environment contaminants such as heavy metals, polychlorinated biphenyls (PCBs), and polybrominated diphenyl ethers (PBDEs). Disposal of e-waste is an increasing global environmental problem. The current global production of e-waste is estimated to be 20–25 million tons per year, and it has become the fastest growing waste stream in the world (Robinson 2009). The techniques used in recycling of e-waste are often primitive. Guiyu, which is located in Shantou City, South China, is infamous for its involvement in primitive e-waste processing and recycling activities that cause severe damage to the environment and the health of local residents. Several studies reported elevated levels of toxic heavy metals in air, road dust, soil, and sediment of Guiyu (Deng et al. 2006, Wong et al. 2007b,c, Leung et al. 2008). In our previous investigations, we found that children and neonates from Guiyu had elevated levels of blood lead and other metals (Huo et al. 2007, Li et al. 2008, Zheng et al. 2008). It was reported that dissolved Mn concentrations were higher in Lianjiang (206–246 µg/L) and Nanyang River (188–217 µg/L) within Guiyu than in the reservoir (2–6 µg/L) outside of Guiyu (Wong et al. 2007a). Elevated levels of Mn in human hair were found in Taizhou, another e-waste recycling area in Zhejiang province in China (Wang et al. 2009).

Mn is an essential element. Deficiencies of Mn have been reported to cause weight loss, dermatitis, nausea, slow growth, skeletal deformities, sterility, and neonatal deaths in animal (Wedler 1993). The biological actions of Mn are quite diverse as confirmed by its wide-ranging role as an essential component for a variety of enzymes involved in the basic metabolic regulation of the cell. It is required for normal immune function, regulation of blood sugar levels, cellular energy, reproduction, digestion, bone growth, and defense mechanisms against free radicals (Aschner and Aschner 2005). Low levels of Mn can have a protective effect. However, excess Mn is toxic, and its effects can lead to physiological damage. Exposure to high oral or ambient air concentrations of Mn can result in elevations of tissue Mn levels and neurological effects (Santamaria 2008).

In most living environments, people are exposed to a variety of chemicals, including both organic and inorganic agents. Among possible target organs of environmental chemicals, the kidney appears to be the most sensitive one. The release of cytoplasmic proteins and other constituents into the urine following injury to tubular cells has been clinically exploited to assess the site and severity of renal tubular damage associated with disease and toxic exposures (Bernard and Lauwerys 1991, Zhou et al. 2008). One of the most sensitive measures of renal toxicity is the excretion of low-molecular-weight proteins in the urine, such as beta-2-microglobulin (β_2-MG) and retinol-binding protein (RBP). Elevated levels reflect dysfunction of the proximal tubule as RBP is a low-molecular-weight protein freely filtered at the glomerulus. The fractional tubular reabsorption of RBP is 99.97%, and increased excretion is, therefore, a sensitive marker of tubular dysfunction (Smith et al. 1994). β_2-MG, identified as a low-molecular-weight protein of 11,800 Da, is a light chain of HLA class I molecule. It also is filtered by glomerulus, and reabsorbed and catabolized by proximal tubules. Based on extensive clinical and experimental studies, elevated levels of these two biomarkers are generally accepted as valid tests for renal damage induced by environment chemicals in humans (Bernard and Hermans 1997).

Children are considered more vulnerable than adults to biochemically toxic agents because they absorb more as a proportion of body mass and because of their more rapid biological growth and development (Mielzynska et al. 2006).

We used urinary β_2-MG and RBP as markers of tubular effects. This study assessed the possibility of relationship between MnB and sensitive markers of tubular dysfunction in school-age children. To our knowledge, no previous studies have investigated the impact of Mn on the renal tubular function of school-age children from e-waste recycle site.

Materials and Methods

Study Areas

Guiyu is one of the largest e-waste recycling sites in the world, with a total area of 52 km² and a population of 150,000. Nearly 60%–80% of families in Guiyu are engaged in the business of processing e-waste. We used Liangying town, where the population, traffic density, lifestyle, socioeconomic status, and geographic location were similar to those of Guiyu, as a control. In Liangying, the local residents' work is mainly clothing manufacturing and is not related to e-waste processing.

Study Population

A total of 130 school-age children, aged 8–14 years, took part in this study. Sixty-one children (33 females and 28 males) with an average age of 9.6 years living in Nanyang village of Guiyu were selected as participants, and 69 children (28 females and 41 males) with an average age of 10.5 years from Liangying served as control.

Children were recruited on a volunteer basis with the consent of their parents. No children involved in this study had any occupational exposure to e-waste. All parents completed a self-administered questionnaire form including items concerning children's demographic variables (age, gender) and health status (past and present diseases, including chronic diseases). According to their self-reports, there were no children suffering from renal disease or diabetes. The study was approved by the Human Ethics Committee of Shantou University Medical College.

Evaluation of Physical Development Indices

Children's physical growth and development, such as height and weight, were measured before blood samples were collected. Weight and height were measured using a weight and height scale (RGZ-120-RT, Mike Balance Instrument Factory, Nanjing, China) with maximum weight of 120 kg and maximum height of 190 cm. Children were required to urinate and take off shoes. All work was carried out by trained study members.

Collection and Analysis of Samples

Biologic samples were collected by trained nurses from both groups after they had fasted for 12 h. Blood samples (2 mL) were drawn from a forearm vein into heparinized glass tubes (with polyethylene stoppers) and mixed by gentle inversion. Freshly voided, first-morning urine samples were collected in 15 mL polyethylene bottles with screw caps. The urine samples were centrifuged at 1000 rpm for 5 min. Blood and urine samples were stored at –20°C until analysis.

All the glassware and plasticware were soaked in 20% (v/v) HNO_3 for 24 h, rinsed with ultrapure water, dried and stored in a closed polypropylene container. High-quality water, obtained using a Milli-Q system (Millipore), was used exclusively. Before analysis, the frozen blood samples were slowly thawed at ambient temperature, homogenized in a vortex mixer and diluted 1:9 with 0.5% (v/v) HNO_3. Gloves were used for all biological manipulations. The concentrations of Mn in blood samples were determined by graphite furnace atomic absorption spectrometry with Zeeman background correction (Zeenit-650, Analytik Jena, Germany). The method was validated before use by standard evaluation. A correlation coefficient of at least 0.99 for the calibration curve was considered satisfactory. The main parameters used for blood Mn determination were a wavelength of 279.5 nm; a lamp current of 4 mA; a slit width of 0.2 nm; drying at 90°C, 105°C, and 120°C; ashing at 850°C; and atomization at 1700°C. The detection limit was 0.15 µg/L.

All analyses of renal biomarkers were performed under similar experimental condition in the same laboratory. Urinary RBP and β_2-MG were measured with ELISA (kit from Sun Biotech Company of Shanghai, China) methods (Cui et al. 2005). Creatinine concentrations of urine were measured using the picric acid method (kit from Jiancheng Bioengineering Institute, Nanjing, China). All urinary parameter assays were adjusted for creatinine in urine. Urine samples with creatinine level <0.3 or >3.0 g/L were excluded from the analyses (N = 6).

Statistical Analysis

Nonparametric analyses were used for data with skew distributions, and chi-square tests were used for categorical data. Spearman correlation analysis was conducted to evaluate the relationship between RBP, β_2-MG, and MnB. The level of significance was set at $P < 0.05$ (two tailed). All statistical analyses were performed using SPSS version 13.0 software (SPSS, Chicago, IL).

Results and Discussion

Participant's Characteristics

In total, 61 children from Guiyu and 69 children of control group were recruited in our investigation. The demographic characteristics of the participants are summarized in Table 9.1. The distributions of gender were not statistically different between the school-age children from Guiyu and those from the control group ($P > 0.05$). Weight, height, and body mass index of children in Guiyu were significantly lower than those in the control group ($P < 0.01$).

Manganese Levels in Blood

MnB of children in Guiyu ranged from 10.5 to 36.6 µg/L, and in the control-group children MnB ranged from 6.7 to 30.8 µg/L. Median values of MnB are shown in Table 9.2. MnB in children from Guiyu were higher than that found in the control group ($P < 0.01$). As expected, both boys and girls in Guiyu had higher MnB compared to those in control group ($P < 0.01$). MnB were not significantly different between genders.

Parameters of Renal Dysfunction

Numerous studies have characterized the association of urinary proteins with changes in renal tubular integrity associated with disease or nephrotoxic exposure in both adults and children (de Burbure et al. 2003, Trachtenberg and Barregard 2007). The most frequently observed health effects include tubular proteinuria that results from impaired renal reabsorption of low-molecular-weight proteins such as β_2-MG and RBP. The urinary indicators for renal dysfunction are shown in Table 9.2. The median urinary RBP of children was 63.0 µg/g creatinine in Guiyu and 48.9 µg/g creatinine in

TABLE 9.1

Characteristics of Children

Group	N	Age (Years)	Gender Boys	Gender Girls	Weight (kg)	Height (cm)	BMI (kg/m²)
Guiyu	61	9.6 ± 1.4	45.9%	54.1%	25.4 ± 5.6	127.8 ± 9.2	15.4 ± 1.6
Control	69	10.5 ± 1.4	59.4%	40.6%	30.5 ± 5.6	135.4 ± 8.4	16.5 ± 2.0
P		0.001	0.123	0.000	0.000	0.003	

Characteristics of the participating children are expressed either as number (N), mean value ± SD, or percentage (%).

P-value is calculated using Chi-square test for categorical data, and Mann–Whitney U test for quantitative data with skewed distributions.

BMI, body mass index.

TABLE 9.2

Median Values of RBP, β_2-MG, and MnB by Group and Gender

	Control Group[a]			Guiyu Group		
	Boys	**Girls**	**Total**	**Boys**	**Girls**	**Total**
N	41	28	69	28	33	61
RBP μg/g Cre	47.0	57.0	48.9	52.7	71.5	63.0*
β_2-MG μg/g Cre	89.2	92.6	92.1	62.9	73.8	70.7
MnB μg/L	14.0	15.7	14.9	20.8[b]**	20.9[b]**	20.9**

Cre, Creatinine.
Mann–Whitney U test compared with control group; *$P<0.05$, **$P<0.01$.
[a] Control group from Liangying and exposed group from Guiyu.
[b] Boys and girls from the Guiyu group were compared to those in control.

control. There was a significant increase in RBP in Guiyu compared to controls ($P<0.05$). When β_2-MG and RBP concentrations in boys and girls from Guiyu were separately compared to those in control, and the results showed no significant differences. The urinary RBP in girls were significantly higher than in boys for the full study population ($P<0.01$). No significant differences in urinary β_2-MG were observed between the children in Guiyu and those in the control group.

Correlation Analysis of β_2-MG, RBP, and MnB

Spearman correlation tests were performed to investigate the relationship between RBP, β_2-MG, and MnB in the study groups (Table 9.3). The correlation between β_2-MG and RBP was 0.803 ($P<0.001$) (not shown in Table 9.3). When the test was performed for boys and girls separately, it was found that there was a closer correlation between MnB and β_2-MG in boys ($r=-0.310$, $P<0.01$), whereas no significant relationships were found in girls. When the test was performed for MnB in Guiyu and control group separately, significant negative correlation with RBP and β_2-MG ($P<0.05$) were found in Guiyu group. The correlation between β_2-MG and MnB was -0.231 for the full study population ($P<0.01$).

TABLE 9.3

Correlation between RBP, β_2-MG, and MnB

	Group		Gender		
	Control	**Guiyu**	**Boys**	**Girls**	**Total**
RBP	−0.167	−0.285*	−0.185	−0.096	−0.114
B$_2$-MG	−0.095	−0.261*	−0.310**	−0.194	−0.231**

Sperman correlation *$P<0.05$, **$P\leq0.01$.

An unexpected finding in this study in children was an overall inverse relationship between β_2-MG and RBP and observed Mn levels, similar to effects described in some experimental animal models. To our knowledge, there are no reports showing evidence of tubular effects at such levels of Mn. Manganese porphyrin can reduce renal injury and mitochondrial damage during ischemia/reperfusion in rats (Saba et al. 2007). Such an association between MnB and urinary β_2-MG and RBP may be biologically plausible. Environmental exposure is a complex mixture of hazardous compounds with different mechanisms of toxicity. Oxidative stress has been suggested to be an important molecular mechanism of toxic effects of environmental pollutants such as lead and cadmium in the kidney (Jurczuk et al. 2006, Conterato et al. 2007). Mn is a constituent of mitochondrial manganese superoxide dismutase (Mn-SOD), which plays an important role in protection against oxidative stress (Luk et al. 2005). It was reported that Mn-SOD attenuates Cisplatin-induced renal injury (Davis et al. 2001). The *in vitro* antioxidant action of Mn in differently mediated lipid peroxidation conditions and its ability to scavenge superoxide anions and oxygen-free radicals have been described (Coassin 1992, Hussain and Ali 1999, Anand and Kanwar 2001). Low dose of Mn may have an antioxidant effect in kidneys of gentamicin-administrated rats, but its high doses had no beneficial effect (Atessahin et al. 2003). It seems that currently observed Mn levels in themselves have no significantly harmful effect on renal tubular function in children. Whether the higher levels of Mn in children may result in adverse effects on other parameters of renal tubule function needs further investigation.

There are several limitations to this study. This was a cross-sectional study, which limits interpretation of the results for a causal relationship and the time course of exposures and effects. However, this is the first study concerning renal tubular effects and MnB in school-age children in an e-waste-polluted area. The present study demonstrates that MnB are inversely associated with β_2-MG and RBP (sensitive markers of renal tubular damage) of school-age children in an e-waste recycling town. RBP in children from Guiyu were higher than those of control, similar to MnB. It seems that elevated levels of Mn may result in adverse effects on renal tubular function.

Acknowledgment

This work was supported by the Scientific Research Foundation for the Returned Overseas Chinese Scholars, State Education Ministry; and 211 project of Guang Dong Province. The authors would like to thank Dr. Harold W. Cook for English language editing.

References

Anand RK and Kanwar U. 2001. Role of some trace metal ions in placental membrane lipid peroxidation. *Biological Trace Element Research* 82: 61–75.

Aschner JL and Aschner M. 2005. Nutritional aspects of manganese homeostasis. *Molecular Aspects of Medicine* 26: 353–362.

Atessahin A, Karahan I, Yilmaz S, Ceribasi AO, and Princci I. 2003. The effect of manganese chloride on gentamicin-induced nephrotoxicity in rats. *Pharmacological Research* 48: 637–642.

Bernard A and Hermans C. 1997. Biomonitoring of early effects on the kidney or the lung. *Science of the Total Environment* 199: 205–211.

Bernard A and Lauwerys RR. 1991. Proteinuria: Changes and mechanisms in toxic nephropathies. *Critical Reviews in Toxicology* 21: 373–405.

Coassin M, Ursini F, and Bindoli A. 1992. Antioxidant effect of manganese. *Archives of Biochemistry and Biophysics* 299: 330–333.

Conterato GM, Augusti PR, Somacal S, Einsfeld L, Sobieski R, Torres JR, and Emanuelli T. 2007. Effect of lead acetate on cytosolic thioredoxin reductase activity and oxidative stress parameters in rat kidneys. *Basic & Clinical Pharmacology & Toxicology* 101: 96–100.

Cui Y, Zhu YG, Zhai R, Huang Y, Qiu Y, and Liang J. 2005. Exposure to metal mixtures and human health impacts in a contaminated area in Nanning, China. *Environment International* 31: 784–790.

Davis CA, Nick HS, and Agarwal A. 2001. Manganese superoxide dismutase attenuates Cisplatin-induced renal injury: Importance of superoxide. *Journal of the American Society of Nephrology* 12: 2683–2690.

de Burbure C, Buchet JP, Bernard A, Leroyer A, Nisse C, Haguenoer JM, Bergamaschi E, and Mutti A. 2003. Biomarkers of renal effects in children and adults with low environmental exposure to heavy metals. *Journal of Toxicology and Environmental Health A* 66: 783–798.

Deng WJ, Louie PKK, Liu WK, Bi XH, Fu JM, and Wong MH. 2006. Atmospheric levels and cytotoxicity of PAHs and heavy metals in TSP and PM2.5 at an electronic waste recycling site in southeast China. *Atmospheric Environment* 40: 6945–6955.

Huo X, Peng L, Xu X, Zheng L, Qiu B, Qi Z, Zhang B, Han D, and Piao Z. 2007. Elevated blood lead levels of children in Guiyu, an electronic waste recycling town in China. *Environmental Health Perspectives* 115: 1113–1117.

Hussain S and Ali SF. 1999. Manganese scavenges superoxide and hydroxyl radicals: An in vitro study in rats. *Neuroscience Letters* 261: 21–24.

Jurczuk M, Moniuszko-Jakoniuk J, and Brzoska MM. 2006. Involvement of some low-molecular thiols in the peroxidative mechanisms of lead and ethanol action on rat liver and kidney. *Toxicology* 219: 11–21.

Leung AO, Duzgoren-Aydin NS, Cheung KC, and Wong MH. 2008. Heavy metals concentrations of surface dust from e-waste recycling and its human health implications in southeast China. *Environmental Science and Technology* 42: 2674–2680.

Li Y, Xu X, Wu K, Chen G, Liu J, Chen S, Gu C, Zhang B, Zheng L, Zheng M, and Huo X. 2008. Monitoring of lead load and its effect on neonatal behavioral neurological assessment scores in Guiyu, an electronic waste recycling town in China. *Journal of Environmental Monitoring* 10: 1233–1238.

Luk E, Yang M, Jensen LT, Bourbonnais Y, and Culotta VC. 2005. Manganese activation of superoxide dismutase 2 in the mitochondria of *Saccharomyces cerevisiae*. *The Journal of Biological Chemistry* 280: 22715–22720.

Mielzynska D, Siwinska E, Kapka L, Szyfter K, Knudsen LE, and Merlo DF. 2006. The influence of environmental exposure to complex mixtures including PAHs and lead on genotoxic effects in children living in Upper Silesia, Poland. *Mutagenesis* 21: 295–304.

Robinson BH. 2009. E-waste: An assessment of global production and environmental impacts. *Science of the Total Environment* 408: 183–191.

Saba H, Batinic-Haberle I, Munusamy S, Mitchell T, Lichti C, Megyesi J, and MacMillan-Crow LA. 2007. Manganese porphyrin reduces renal injury and mitochondrial damage during ischemia/reperfusion. *Free Radical Biology and Medicine* 42: 1571–1578.

Santamaria AB. 2008. Manganese exposure, essentiality & toxicity. *Indian Journal of Medical Research* 128: 484–500.

Smith GC, Winterborn MH, Taylor CM, Lawson N, and Guy M. 1994. Assessment of retinol-binding protein excretion in normal children. *Pediatric Nephrology* 8: 148–150.

Trachtenberg F, Barregard L. 2007. The effect of age, sex, and race on urinary markers of kidney damage in children. *American Journal of Kidney Diseases* 50: 938–945.

Wang T, Fu J, Wang Y, Liao C, Tao Y, Jiang G. 2009. Use of scalp hair as indicator of human exposure to heavy metals in an electronic waste recycling area. *Environmental Pollution* 157: 2445–2451.

Wedler FC. 1993. Biological significance of manganese in mammalian systems. *Progress in Medicinal Chemistry* 30: 89–133.

Wong CS, Duzgoren-Aydin NS, Aydin A, and Wong MH. 2007a. Evidence of excessive releases of metals from primitive e-waste processing in Guiyu, China. *Environmental Pollution* 148: 62–72.

Wong MH, Wu SC, Deng WJ, Yu XZ, Luo Q, Leung AO, Wong CS, Luksemburg WJ, and Wong AS. 2007b. Export of toxic chemicals—A review of the case of uncontrolled electronic-waste recycling. *Environmental Pollution* 149: 131–140.

Wong CS, Wu SC, Duzgoren-Aydin NS, Aydin A, and Wong MH. 2007c. Trace metal contamination of sediments in an e-waste processing village in China. *Environmental Pollution* 145: 434–442.

Zheng L, Wu K, Li Y, Qi Z, Han D, Zhang B, Gu C et al. 2008. Blood lead and cadmium levels and relevant factors among children from an e-waste recycling town in China. *Environment Research* 108: 15–20.

Zhou Y, Vaidya VS, Brown RP, Zhang J, Rosenzweig BA, Thompson KL, Miller TJ, Bonventre JV, and Goering PL. 2008. Comparison of kidney injury molecule-1 and other nephrotoxicity biomarkers in urine and kidney following acute exposure to gentamicin, mercury, and chromium. *Toxicological Sciences* 101: 159–170.

10

Mitigating Environmental and Health Risks Associated with Uncontrolled Recycling of Electronic Waste: Are International and National Regulations Effective?

Ming Hung Wong, Anna O.W. Leung, Shengchun Wu, Clement K.M. Leung, and Ravi Naidu

CONTENTS

General Background

The generation of end-of-life electrical and electronic equipment (EEE) (referred to as waste electrical and electronic equipment, WEEE, or simply e-waste), which includes personal computers, mobile phones, and entertainment electronics, has soared in recent decades, with 6 million generated in 1998 in Western Europe and 500 million between 1997 and 2007 in the United States (SVTC 2008). It has been estimated that the current global e-waste production is between 20 and 25 million tonnes per year (Robinson 2009) and forecast to reach 50 million tonnes per year (UNEP 2009). With today's technologically advancing societies and the demand for newer, more efficient and effective technology, older EEE are becoming obsolete at increasingly faster rates in more affluent countries. For example, the average lifespan of computers in developed countries has decreased from 6 years in 1997 to 2 years in 2005 (SVTC 2007). This has resulted in the rapid rise of e-waste, by at least 3%–5% per annum, and has, in fact, become the fastest growing waste stream in the industrialized countries (UNEP 2005).

On average, WEEE contains about 38% ferrous, 28% nonferrous, 19% plastic, 4% glass, 1% wood, and 10% other materials (APME 1995), which could serve as secondary raw materials. Recycling of e-waste is a prosperous business in some developing countries, such as China, India, Pakistan, and Africa. Significant amounts of valuable materials, such as gold, platinum, silver, copper, steel, aluminum, wires, and cables can be extracted from WEEE. More than 80% of the world's hazardous waste come from the United States (O'Connell 2002), of which 80% are exported to Asian countries, in particular China, due to lower environmental standards and cheaper labor (BAN and SVTC 2002).

Two Major E-Waste Recycling Sites in China

China has been the predominant recipient of the world's e-waste, with about 70% of the worldwide e-waste illegally imported from developed countries (BAN and SVTC 2002). Guiyu town, located in southeastern Guangdong Province, with a population of 150,000 and a land area of 52.4 km^2, has been the major e-waste dumping site for e-waste exported from North America since the 1980s (BAN and SVTC 2002). Taizhou, Zhejiang Province, with a population of 400,000 and a land area of 274 km^2, is another intensive e-waste hotspot. The city handles e-waste generated domestically and is the largest center for dismantling obsolete transformers and capacitors (which contain polychlorinated biphenyls, PCBs, as dielectric fluid) in China since the 1970s (Zhao et al. 2007). From the 1990s, different kinds of e-waste materials imported from other countries are being processed (The World Bank 2005).

Primitive Techniques Used in Recycling E-Waste

Primitive techniques are used to recycle WEEE, which include (1) baking electronic boards on top of an open fire (using homemade honeycomb-shaped coals, a low-grade fuel often made by mixing coal with local contaminated sediment), in order to separate the components and chips more effectively. This is usually carried out in houses partially converted into small-sized workshops, with workers equipped with only minimal or no health protection measures, where inhalation of noxious acid fumes is a severe health hazard; (2) stripping of gold from chips and other components using strong acids, with the treated waste haphazardly dumped along river banks, leading to high acidity and dissolved metals in the sediment and river water; (3) open burning of PVC-insulated cable wires (to extract copper), computer casings, and all materials that no longer possess any value, resulting in the emission of flame retardants (such as polybrominated diphenyl ethers, PBDEs), chlorinated and brominated dibenzo-p-dioxins and dibenzofurans (PCDD/Fs, PBDD/Fs), and polycyclic aromatic hydrocarbons (PAHs). Many of these toxic chemicals are listed in the Stockholm Convention on Persistent Organic Pollutants because they are toxic, persistent, undergo long-range transport, and can bioaccumulate in the body (Secretariat of the Stockholm Convention-United Nations Environment Programme 2009). There has been substantial evidence showing that the export of toxic chemicals from industrialized countries to developing countries exerts harmful effects on the environment and human health of those countries receiving e-waste (Wong et al. 2007).

The so-called recycling of e-waste in China and in other countries such as India, Africa, and Pakistan is largely unregulated and conducted using rudimentary techniques. The illegal transport of e-waste from developed countries to developing countries, thereby transcending international boundaries, has made e-waste recycling a global, twenty-first-century problem, which is caused by the use of highly inadequate, rudimentary nineteenth-century methods at recycling sites. The majority of the workers do not know the extent of the harm that e-waste recycling has on their health and that of the younger generation and they have no means of protecting themselves. The informal recycling of WEEE has deep-rooted environmental, social, and economic implications (Williams et al. 2008).

Major Objectives

This chapter reviews (1) the effects of different toxic chemicals emitted during uncontrolled e-waste recycling and their potential health effects, by studying the relationships between concentrations of different toxic chemicals in different ecological compartments, and food, linking with body burden and epidemiological data, and (2) the effectiveness of international and national regulations in controlling e-waste recycling practices that use primitive techniques and emit toxic chemicals.

Toxic Chemicals Emitted through Uncontrolled Open Burning of e-Waste

Toxic chemicals, including persistent organic pollutants (POPs) (PCDD/Fs, PBDEs, PAHs, PCBs) and heavy metals (Cd, Cr, Cu, Hg, especially Pb) released through uncontrolled open burning (incomplete combustion) of e-waste (containing plastic chips, wire insulations, PVC materials, and metal scraps), have resulted in severe contamination of different environmental media (i.e., air, soil, sediment, water) (Wong et al. 2007). Flame retardants are commonly added in EEE. Commercial penta-BDE and octa-BDE were included in the list of POPs for elimination under the Stockholm Convention on Persistent Organic Pollutants in May 2009 (Stockholm Convention 2008).

Sampling of the air in Guiyu revealed that $PM_{2.5}$ was dominated by PBDE congeners, BDE-28, 47, 66, 100, 99, 154, 153, 183, and 191 which accounted for 84.3% of the total PBDE concentration (22 congeners). The more toxic mono- to penta-brominated congeners contributed to 92.8%. The concentrations of the PBDE congeners were 100 times higher than other published data and were attributed to uncontrolled recycling activities such as the baking of printed circuit boards inside recycling workshops and open burning of e-waste containing PBDEs (Deng et al. 2007). The combusted residue and ash of plastic cables and plastic chips were also found to contain extremely high levels of PBDEs, with BDE-209, the main component of the commercial flame retardant, deca-BDE, having the highest abundance (up to 81%) (Leung et al. 2007). Sediment collected from the rivers in Guiyu had a higher level of contamination than other polluted sites reported in the world due to dumping of combusted residue and ash (Wang et al. 2009).

Toxic Chemicals Are Bioaccumulated and Biomagnified in Food Chains

POPs are generally lipophilic (fat-loving) and can be stored in the lipid of biota, and therefore, certain fish species that have high lipid content in their tissues have higher concentrations of PBDEs. The concentrations of total PBDEs in tilapia (*Oreochromis* spp.) collected from rivers in Guiyu were 600 times and 15,000 times higher than those from Canadian markets (0.18 ng/g wet wt) and U.S. markets (0.0085 ng/g wet wt), respectively (Luo et al. 2007). It is known that fish bioaccumulate POPs, which subsequently results in the accumulation of POPs in the human body when contaminated fish are consumed. People can be exposed to POPs via several exposure pathways, which include ingestion (consumption of contaminated foods, inadvertent dust ingestion), inhalation, and dermal exposure.

Body Burdens of Toxic Chemicals in Residents of E-Waste Recycling Sites

The body burdens of POPs can be magnified in different human tissues. The concentrations of different POPs in milk, placenta, and hair of local residents

TABLE 10.1

Concentrations of POPs in Human Specimens from the Study Sites

POPs	Specimen	Unit	Guiyu	Taizhou	Lin'an
ΣPCDD/Fs	Human milk	pg WHO-TEQ$_{1998}$/g fat	NA	21.0 ± 13.81a	9.35 ± 7.39a
	Placenta	pg WHO-TEQ$_{1998}$/g fat	NA	35.1 ± 15.7a	11.9 ± 7.05b
	Hair	pg WHO-TEQ$_{1998}$/g dry wt	NA	33.8 ± 17.7a	5.59 ± 4.36b
Σ_7PBDEs	Human milk	ng/g fat	94.1 ± 86.4a	70.7 ± 114b	1.43 ± 0.805c
	Placenta	ng/g fat	NA	12.9 ± 20.3a	0.529 ± 0.189b
	Hair	ng/g dry wt	NA	82.9 ± 163a	2.44 ± 1.54b
Σ_{36}PCBs[a]/ Σ_{23}PCBs[b]	Human milk	ng/g fat	9.5 ± 15.7c	363 ± 470a	116 ± 108b
	Placenta	ng/g fat	NA	224 ± 97.0a	44.6 ± 77.0b
	Hair	ng/g dry wt	NA	386 ± 465a	51.0 ± 37.2b

Sources: Chan, J.K.Y. et al., *Environ. Sci. Technol.*, 41, 7668, 2007; Chan, J.K.Y., Dietary exposure, human body loadings, and health risk assessment of persistent organic pollutants at two major electronic waste recycling sites in China, PhD thesis, Hong Kong Baptist University, Kowloon Tong, Hong Kong, 2008; Leung, A.O.W. et al., *Environ. Sci. Pollut. Res.*, 17, 1300, 2010; Xing, G.H. et al., *Environ. Int.*, 35, 76, 2009; Xing, G.H., Human exposure and health risk assessment of polychlorinated biphenyls at two major electronic waste recycling sites in China, PhD thesis, Hong Kong Baptist University, Kowloon Tong, Hong Kong, 2008.

NA, Not available.

Means having different letters (across a row) are significantly different at the 5% probability level among the three study sites for a given specimen according to the Tukey Test.

Σ_7PBDEs = sum of BDE-28, 47, 99, 100, 153, 154, 183.

[a] Σ_{36}PCBs (Guiyu) = sum of PCB-18, 37, 44, 49, 52, 70, 74, 77, 81, 87, 99, 101, 105, 114, 118, 119, 123, 126, 128/158, 138, 151, 153, 156, 157, 167, 168, 169, 170, 177, 180, 183, 187, 189, 194, 199.

[b] Σ_{23}PCBs (Taizhou, Lin'an) = sum of PCB-28, 52, 77, 81, 99, 101, 105, 114, 118, 123, 126, 128, 138, 156, 157, 167, 169, 170, 180, 183, 187, 189.

of Guiyu and Taizhou were significantly higher than those of Lin'an (a city 245 km away from Taizhou, serving as a control site), and obviously linked with the polluted environment (Table 10.1).

The average concentration of PBDEs in human milk from Guiyu residents was found to be higher than Taizhou due to the intensive processing of computers and plastic cables and wires in Guiyu. These WEEE contain PBDEs—flame retardants—that are not chemically bound to the product and, as a result, could be released into the environment under high-temperature conditions. Total PBDE concentrations in umbilical cord blood of neonates in Guiyu (median 13.84, range 1.14–504.97 ng/g lipid) have also been found to be higher than that of samples from a control site (5.23, range 0.29–363.70 ng/g lipid) (Wu et al. 2010).

The concentrations of PCBs in the three types of human samples (hair, placenta, and milk) collected from Taizhou were all significantly higher ($p < 0.05$) (7.6, 5.0, and 3.1 times, respectively) than those at the control site (Lin'an) (Xing 2008). When comparing with the results from the third WHO-coordinated study, in which all human milk samples were collected after 2000, Taizhou topped the list with regard to WHO-PCB-TEQ values. On a global basis, the PCB levels and WHO-PCB-TEQ levels in placenta and hair samples collected from both Taizhou and Lin'an ranked at the high end. However, PCB levels in human milk were not as high as expected. The average PCB concentration in human milk was 38 and 12 times lower in Guiyu than in Taizhou and Lin'an, respectively. The e-waste workers at Guiyu have a lower standard of living than the workers in the other two towns, as reflected by a lower consumption of fish and meat (Xing 2008). Furthermore, the environmental contamination attributed to PCBs at Guiyu was not as prevalent as in the other two cities. Most of the donors were not primiparous mothers; therefore, a decrease of the PCB accumulation in the body may have been a result of fore lactation (Fitzgerald et al. 1998). Moreover, median PBDE concentrations in blood serum of e-waste workers in Guiyu were found to be three times higher compared to the inhabitants of a fishing district 50 km east of Guiyu (Bi et al. 2007). Levels of Pb and Cd in the blood of Guiyu children were higher than that of children from a reference site (Zheng et al. 2008).

Potential Health Risk of Adults and Infants at the E-Waste Recycling Sites

Table 10.2 shows the estimated daily intakes (EDI) of different POPs of adults and breast-fed infants at the two e-waste recycling sites (Guiyu and Taizhou) compared with the control site (Lin'an).

Several EDIs of PCDD/Fs, PBDEs, and PCBs were found to be higher than the reference doses for adults and infants at the 5% probability level. The daily intakes of PCDD/Fs by infants via breast-feeding exceeded the WHO Tolerable Daily Intake (1–4 pg WHO-TEQ/kg bw/day) by at least 25 and 11 times in Taizhou and Lin'an, respectively (Chan et al. 2007). The recently proposed reference dose for BDE-47 by U.S. EPA, based on developmental neurotoxicity, is 100 ng/kg/day (Costa and Giordano 2007). Therefore, the estimated daily intake of BDE-47 (584 ng/kg bw/day) by Guiyu women exceeded the reference dose (100 ng/kg/day) by nearly 6 times. The maximum EDI value for BDE-47 for breast-fed Taizhou infants was 534 ng/kg bw/day, which also exceeded the reference dose for BDE-47 (Leung et al. 2010).

In comparison to published average daily intakes for adults in other countries, the EDI values for Taizhou mothers were at the higher end, exceeding the FAO/WHO tolerable intake limit (70 pg TEQ/kg bw/month) (JECFA 2001). The EDIs of PCB levels at Lin'an and Guiyu were comparable to those of the United Kingdom (0.22–0.46 ng/person/day) (Juan et al. 2002), Japan (0.166–0.523 ng/person/day) (Koizumi et al. 2005), and northern Italy (0.26 ng/person/day) (Turci et al. 2006).

TABLE 10.2

Estimated Daily Intakes of POPs for Adults and Infants at the Study Sites

Estimated Daily Intake	Population Group	Guiyu	Taizhou	Lin'an
ΣPCDD/Fs (pg-TEQ/kg body wt/day)	Adult*	1.95±1.25a	0.37±0.36b	0.03±0.03c
	Breast-fed infant	NA	103±67.7a	45.8±36.2b
Σ_7PBDEs (ng/kg bw/day)	Adult**	931±772a	44.7±26.3b	1.94±0.86c
	Breast-fed infant	461±423a	346±559a*	7.01±3.95b
Σ_{37}PCBs (ng/kg bw/day)	Adult**	5.36±4.60b	92.8±77.5a	7.31±4.73b
	Breast-fed infant	46.6±76.9c	1779±2303a	568±529b

Sources: Chan, J.K.Y. et al., *Environ. Sci. Technol.*, 41, 7668, 2007; Chan, J.K.Y., Dietary exposure, human body loadings, and health risk assessment of persistent organic pollutants at two major electronic waste recycling sites in China, PhD thesis, Hong Kong Baptist University, Kowloon Tong, Hong Kong, 2008; Xing, G.H., Human exposure and health risk assessment of polychlorinated biphenyls at two major electronic waste recycling sites in China, PhD thesis, Hong Kong Baptist University, Kowloon Tong, Hong Kong, 2008.

NA, Not available.
* Exposure from fish consumption.
** Exposure from nine food groups.

The EDIs of total TEQ values (502 pg/kg bw/day) for Taizhou infants exceeded the FAO/WHO tolerable daily intake (70 pg TEQ/kg/month) by more than 200 times (JECFA 2001). Taking into account the placental transfer of PCBs, this estimation of risk for infants in Taizhou could be underestimated. Such an acute exposure is of concern due to the long half-lives of the compounds that could impose serious effects on the thyroid hormone system and immunological functions in infants and children (Nagayama et al. 2007).

Epidemiological Data from Taizhou

According to the epidemiological data for adults provided by the Center for Disease Control and Prevention in Taizhou, from 2004 to 2006, there was an obviously increasing trend in morbidity due to incidences of malignant tumors and trauma; disorders in digestive, cardiovascular, respiratory, endocrine, and urinary systems; gynecological disease; and ophthalmologic and otolaryngological disorder ($p < 0.05$). In addition, the average rate of incidence of cardiovascular disease was significantly higher than all other disorders (79%, $p < 0.001$), followed by respiratory (53%) and digestive system (48%) disorders (Leung et al. 2010, Wong et al. 2008). Health checks on workers and children in Guiyu revealed the prevalence of digestive, neurological, and respiratory problems and high incidences of bone disease (Qiu et al. 2004).

Although only a limited number of epidemiological studies on e-waste recycling sites exist, it seems reasonable from a public health standpoint that the levels of toxic chemicals released from uncontrolled e-waste recycling operations into the environment should be drastically reduced.

Significance of the Findings

Research conducted at the two major e-waste recycling sites has adequately demonstrated the sources, fates, and effects (both environmental and health effects) of toxic chemicals generated by primitive and uncontrolled recycling of e-waste. The highly toxic chemicals (such as PCDD/Fs, PBDEs, and PCBs) are concentrated in human tissues, due to the high background levels of these toxic chemicals in different ecological compartments, for example, through inhalation and dermal contact, as well as high oral intake due to continuous consumption of contaminated food items. There is also a danger that these toxic chemicals are passed on to our next generation, through placenta transfer and lactation. The epidemiological data provided by the hospital in Taizhou have confirmed the increasing morbidity of several major diseases in the area.

There is an urgent need to reduce the negative impacts of e-waste recycling activities on the environment and humans.

Addressing the Problems of E-Waste

Laws and Regulations Concerning E-Waste

International

Basel Convention

The movement of hazardous waste across international boundaries is regulated by the Basel Convention on the Control of Transboundary Movements of Hazardous Wastes and Their Disposal, whereby e-waste is included under List A of Annex VIII of the Convention. The convention was adopted more than 20 years ago, on March 22, 1989, and entered into force on May 5, 1992. Transboundary movements of hazardous waste are restricted, and prior written notification of the exporting country to the competent authorities of the importing country is a requirement (Secretariat of the Basel Convention-United Nations Environment Programme 1992). The spirit of the Basel Convention is to protect human health and the environment from the dangers posed by hazardous wastes. It outlines a three-step strategy for minimizing the generation of wastes, treating wastes as near as possible to the location where they are generated, and reducing international movement of hazardous wastes. Despite this convention, developed countries still routinely export their e-waste to developing countries. Notably, the United States is the world's largest e-waste producer and has neither ratified the original Basel Convention nor its amendment.

In November 2008, CBS 60 Minutes exposed the illegal export of e-waste from the United States to Hong Kong and reported that environmentally conscientious citizens who return their WEEE to recycling depots could be cheated by organizations whose real intention is to illegally export the materials to developing countries. Sometimes the e-waste is unwittingly sold to wholesale buyers who then handle the e-waste illegally.

European Union Directives

Two regulations concerning e-waste have been introduced by the European Union (EU): (1) The Waste Electrical and Electronic Equipment (Amendment) Regulations 2007 (WEEE Directive 2002/96/EC), which came into force on January 1, 2008. This aims to reduce the amount of WEEE being generated; encourage stakeholders to reuse, recycle, and recover these materials; and improve the environmental performance of relevant businesses that manufacture, supply, use, recycle, and recover EEE. (2) The Restriction of the Use of Certain Hazardous Substances in EEE (RoHS) Regulations 2006 (Directive 2002/95/EC), which prohibits the use of brominated flame retardants—polybrominated diphenyl ethers (PBDEs), namely, penta-BDE and octa-BDE, and polybrominated biphenyls PBBs—lead, mercury, cadmium, and hexavalent chromium exceeding set maximum concentration values in new EEE products that are available in the EU market after July 1, 2006.

National

In recent years, China has developed several regulations, laws, and measures to deal with e-waste.

The Chinese version of the EU WEEE directive is called "The Management Regulations on the Recycling and Disposal of Waste Electrical and Electronic Products" (The Central People's Government of the Peoples' Republic of China 2009). This law was enacted by the National Development and Reform Commission (NDRC) and was implemented on January 1, 2011. It aims to regulate recycling and disposal of five types of waste related to household electrical appliances: TV sets, refrigerators, washing machines, air conditioners, and computers. It requires mandatory recycling of electrical and electronic appliances discarded by the consumer. It stipulates that recycling be carried out by operators licensed by the relevant authority; a special fund to support WEEE recycling and disposal be established; environmentally friendly designs that favor "circular use" be created; materials extracted be nontoxic, nonhazardous, and recyclable; and information to aid recycling and information on product composition, recycling, and treatment be provided (Yang 2008). The regulation stipulates relatively clear and specific legal liabilities and penalties. For example, unauthorized dismantling will result in a fine ranging from U.S.$700 to U.S.$6000.

The law "Management Measure for the Prevention of Pollution from Electronic Information Products" is a counterpart to the EU RoHS Directive and came into effect on March 1, 2007. It was developed by several governmental bodies and aims to reduce the hazardous and toxic substances and materials contained in electronic appliances and to minimize pollution caused by the production, recycling, and disposal of these products (Yang 2008). The following important points are covered: (1) restricting six hazardous and toxic substances (the same ones stipulated in the EU RoHS Directive); (2) products and their packaging should be nontoxic, nonhazardous, degradable, and recyclable; (3) information on the components and hazardous substances contained in the products and their safe use and recycling should be specified by the producers. The original "Chinese RoHS" only covers electronic information products (e.g., computers, televisions, radios); however, in 2011, a revised RoHS has been proposed, which will cover electrical products such as air conditioners and refrigerators.

In February 2008, the Administrative Measures on Prevention and Control of E-waste Pollution (SEPA Order No. 40) came into effect with regard to the prevention and control of pollution arising from the generation, storage, disassembly, utilizing, recycling, and disposal of electronic waste (Ministry of Environmental Protection, www.mep.gov.cn). It stipulates that an environmental impact assessment or an EIA screening process be undertaken for newly built, rebuilt, or expended projects for dismantling, utilization, and disposal of e-waste. The regulation specifies the responsibilities of stakeholders; for example, manufacturers, importers, and retailers of EE products should establish a take-back system to collect waste EE products and be responsible for treating and disposal of e-waste.

Although China has several comprehensive regulations and policies governing e-waste, how effective these will be depends on the monitoring and strict enforcement of their implementation. Currently, an effective framework for implementation and systematic monitoring is lacking, and the Chinese WEEE regulations still contain several loopholes (Chung and Zhang 2011). The regulatory development for the management of e-waste is carried out by individual departments within their own mandates, which in turn creates gaps and conflicts in existing legislations (Yang 2008). Thus, there needs to be better communication, coordination, and support between the departments or ministries. In spite of these shortcomings, the tightening up of the illegal transport of e-waste into China by the Chinese government has resulted in large amounts of e-waste stored in Hong Kong—the doorstep of China.

Defining E-Waste

The definition of e-waste needs to be clearly defined because it is currently loosely used as a catch-all term for end-of-life EEE, which are disposed, sold, or donated by the original owners or are destined for reuse, resale, recycling, or disposal. In some countries, EEE that are destined for direct reuse are not considered e-waste. Should a computer that is to be reused by a secondary user be considered e-waste? The fact that it is to be reused means that it has not yet reached its end of life. There should be specific guidelines dictating when an item should be considered e-waste. The question is should "end-of-life" be defined from the point of view of the original user or based on the actual end-of-life of the EEE?

Closing the Loopholes

The Basel Convention contains a legal loophole in that the export of whole products is permitted to other countries as long as they are not meant for recycling. Therefore, in essence, there is no transboundary movement of *waste*, and dishonest brokers could make use of this loophole. A considerable amount of waste is now being exported—not as waste, but as reusable equipment—some under the guise of reusable EEE, but which cannot be repaired without expert technological knowledge, and this has emerged as a point of concern on a global scale. It is difficult to trace whether "reusable" items in transit will actually be reused overseas. The export of reusable equipment is outside the control of the Basel Convention.

The U.S. "E-waste" Bill (H.R. 2595) introduced in May 2009 at the 11th Congress will open the gateway for the export of toxic waste to continue because although it bans the export of listed "restricted electronic waste" to non-OECD countries for recycling, the bill allows for the export of used electronic equipment for refurbishment and subsequent reuse (govTrack. us 2009). Unscrupulous recyclers could take advantage of this loophole by sending damaged and outdated equipment for recycling. To close this legal loophole, only the export of commodities (crushed glass, shredded plastic, aggregated metal) that could be used as raw materials for other products should be permitted.

In June 2011, Bill H.R. 2284 was introduced to prohibit the export of certain e-waste from the United States to developing nations. It allows for the export of tested and fully functional and usable electronic goods to promote reuse, if consent is received from the importing country (GovTrack.us 2011). This bill is expected to create employment in the recycling and refurbishment industry in the United States.

Global and National Cooperation

A holistic approach consisting of systematic governance and detailed guidelines on a nation-wide basis is needed to successfully address the e-waste issue. The illegal import of e-waste must be restricted by eradicating all possible regulatory loopholes. E-waste is a global challenge; therefore, international cooperation is very much necessary to tackle transboundary movement of e-waste. As stated in Article 10 of the Basel Convention, parties to the convention should cooperate with each other to improve and achieve environmentally sound management of hazardous wastes and other wastes (Secretariat of the Basel Convention-United Nations Environment Programme 1992). This can be done by sharing technological knowledge and continuously improving and harmonizing technical standards and codes of practices and technologies for safe, efficient, and effective management of e-waste. Parties should also cooperate in monitoring the effects of the management of hazardous wastes on human health and the environment and develop and employ appropriate means to assist developing countries in building capacity (technical or otherwise) for handling e-waste.

A fundamental component of successful e-waste management is full support from the government in terms of legislating extended producer responsibility, which requires manufacturers of EEE to establish a take-back system that incorporates a circular economy with minimum risks to the environment and human health, and subsidizing e-waste reuse and recycling programmes. Sophisticated centralized recycling centers should be built to process regional WEEE. A national recycling levy, such as that implemented in Japan, could be introduced, whereby all manufacturers, sellers, and consumers pay a recycling tax that will help finance the recycling of EEE. In many countries, WEEE regulations are either nonexistent or the initiation, drafting, and adoption of these regulations have been slow (Ongondo et al. 2011).

Furthermore, it is important to launch environmental education and ethics programs, epidemiological studies, environmental monitoring and risk assessment, remediation of contaminated sites, mathematical modeling to predict likely effects, and auditing to check whether all regulations are fully implemented and enforced. It seems necessary to conduct a cost-benefit analysis to provide information on the cost of hospitalization of workers and local residents due to adverse health impacts arising from e-waste recycling and for remediation of the contaminated environment at the e-waste recycling sites.

The long-term health consequences of primitive and crude recycling processes on the workers and their children are still largely unknown. The poor do not have the freedom of choice, especially because the "not in my backyard" mentality of the developed countries have made "recycling" of digital dumps in the backyards of the poor their sole means of survival.

Acknowledgments

The authors thank all the team members who took part in the e-waste project. Financial support from the Special Equipment Grant (SEG HKBU09), Group Research (HKBU/1-03C) of the Research Grants Council of Hong Kong, and the Area of Excellence Fund (Hong Kong Baptist University) is gratefully acknowledged. Part of this chapter was presented at Cleanup 09, *Third International Contaminated Site Remediation Conference*, Adelaide, South Australia (September 27–30).

References

APME (Association of Plastics Manufactures in Europe). 1995. Plastics—A material of choice for the electric and electronic industry, plastics consumption and recovery in Western Europe. APME Report Code 98-2004, Brussels, Belgium.

BAN and SVTC (The Basel Action Network and Silicon Valley Toxics Coalition). 2002. Exporting harm: The high-tech trashing of Asia. February 25, 2002. http://www.ban.org/E-wastetechnotrashfinalcomp.pdf

Bi XH, Thomas GO, Jones KC, Qu WY, Sheng GY, Martin FL, Fu JM. 2007. Exposure of electronics dismantling workers to polybrominated diphenyl ethers, polychlorinated biphenyls, and organochlorine pesticides in South China. *Environmental Science and Technology* 41: 5647–5653.

CDC (Centers for Disease Control and Prevention). 2005. *Preventing Lead Poisoning in Young Children*. Centers for Disease Control, Atlanta, GA.

Chan JKY. 2008. Dietary exposure, human body loadings, and health risk assessment of persistent organic pollutants at two major electronic waste recycling sites in China. PhD thesis, Hong Kong Baptist University, Kowloon Tong, Hong Kong.

Chan JKY, Xing GH, Xu Y, Liang YX, Chen LX, Wu SC, Wong CKC, Leung CKM, Wong MH. 2007. Body loadings and health risk assessment of polychlorinated dibenzo-p-dioxins and dibenzofurans at an intensive electronic-waste recycling site in China. *Environmental Science and Technology* 41: 7668–7674.

China Daily. 2004. City makes efforts to clean up in recycling business. http://www.china.org.cn/english/environment/98233.htm, dated June 15, 2004 (accessed June 26, 2008).

Chung SS, Zhang C. 2011. An evaluation of legislative measures on electrical and electronic waste in the People's Republic of China. *Waste Management* 31: 2638–2646.

Costa LG, Giordano G. 2007. Developmental neurotoxicity of polybrominated diphenyl ether (PBDE) flame retardants. *Neurotoxicology* 28: 1047–1067.

Deng WJ, Zheng JS, Bi XH, Fu JM, Wong MH. 2007. Distribution of PBDEs in air particles from an electronic waste recycling site compared with Guangzhou and Hong Kong, South China. *Environment International* 33: 1063–1069.

Fitzgerald EF, Hwang SA, Bush B, Cook K, Worswick P. 1998. Fish consumption and breast milk PCB concentrations among Mohawk women at Akwesasne. *American Journal of Epidemiology* 148: 164–172.

Fu JJ, Zhou QF, Liu JM, Liu W, Wang T, Zhang QH, Jiang GB. 2008. High levels of heavy metals in rice (*Oryza sativa* L.) from a typical e-waste recycling area in southeast China and its potential risk to human health. *Chemosphere* 71: 1269–1275.

Geering AC. 2007. Modeling of existing e-waste collection systems in China, a case study based on household surveys in the Taizhou prefecture. Master thesis, Swiss Federal Institute of Technology, Switzerland.

GovTrack.us. 2009. Text of H.R. 2595: To restrict certain exports of electronic waste. http://www.govtrack.us/congress/billtext.xpd?bill=h111-2595 (accessed March 3, 2011).

GovTrack.us. 2011. Text of H.R. 2284: To prohibit the export from the United States of certain electronic waste, and for other purposes. http://www.govtrack.us/congress/billtext.xpd?bill=h112-2284 (accessed October 11, 2011).

Huo X, Peng L, Xu XJ, Zheng LK, Qiu B, Qi ZL et al. 2007. Elevated blood lead levels of children in Guiyu, an electronic waste recycling town in China. *Environmental Health Perspectives* 15: 1113–1117.

IISD (International Institute for Sustainable Development). 2009. Earth Negotiations Bulletin—COP4 Final. http://www.iisd.ca/chemical/pops/cop4

JECFA (Joint FAO/WHO Expert Committee on Food Additives). 2001. *57th Meeting*, June 5–14, Rome, Italy. http://www.fao.org/es/esn/jecfa/fecfa57c.pdf

Juan CY, Thomas GO, Sweetman AJ. 2002. An input–output balance study for PCBs in humans. *Environment International* 28: 203–214.

Koizumi A, Yoshinaga T, Harada K, Inoue K, Morikawa A, Muroi J et al. 2005. Assessment of human exposure to polychlorinated biphenyls and polybrominated diphenyl ethers in Japan using archived samples from the early 1980s and mid-1990s. *Environmental Research* 99: 31–39.

Leung AOW, Chan JKY, Xing GH, Xu Y, Wu SC, Wong CKC, Leung CKM, Wong MH. 2010. Body burdens of polybrominated diphenyl ethers in childbearing-aged women at an intensive electronic-waste recycling site in China. *Environmental Science and Pollution Research* 17: 1300–1313.

Leung AOW, Luksemburg WJ, Wong AS, Wong MH. 2007. Spatial distribution of polybrominated diphenyl ethers and polychlorinated dibenzo-p-dioxins and dibenzofurans in soil and combusted residue at Guiyu, an electronic waste recycling site in southeast China. *Environmental Science and Technology* 41: 2730–2737.

Luo Q, Cai ZW, Wong MH. 2007. Polybrominated diphenyl ethers in fish and sediment from river polluted by electronic waste. *Science of the Total Environment* 383: 115–127.

Man YB, Sun XL, Zhao YG, Lopez BN, Cheung KC, Wong MH. 2009. Toxicity assessments of agricultural soil contaminated by changing land use, with emphasis on e-waste storage and recycling. *Proceedings of CLEANUP 09, 3rd International Contaminated Site Remediation Conference*, September 27–30, 2009, Adelaide, South Australia.

Ministry of Environmental Protection of the People's Republic of China. 2006. The technical policy for the prevention of pollution from waste electrical and electronic products. http://www.sepa.gov.cn/tech/hjbz/bzwb/wrfzjszc/200607/t20060720_91676.htm (accessed June 23, 2008).

Nagayama J, Kohno H, Kunisue T, Kataoka K, Shimomura H, Tanabe S, Konish S. 2007. Concentrations of organochlorine pollutants in mothers who gave birth to neonates with congenital hypothyroidism. *Chemosphere* 68: 972–976.

O'Connell KA. 2002. Computing the damage, wastage. http://wasteage.com/ar/waste_computing_damage/ (accessed June 24, 2008).

Ongondo FO, Williams ID, Cherrett TJ. 2011. How are WEEE doing? A global review of the management of electrical and electronic wastes. *Waste Management* 31: 714–730.

Qiu B, Peng L, Xu XJ, Lin XS, Hong JD, Huo X. 2004. Medical investigation of e-waste demanufacturing industry in Guiyu Town. In *Proceedings of the International Conference on Electronic Waste and Extended Producer Responsibility in China*, April 21–22, 2004, Beijing, China, Greenpeace China, pp. 79–83.

Robinson BH. 2009. E-waste: An assessment of global production and environmental impacts. *Science of the Total Environment* 408: 183–191.

Secretariat of the Basel Convention-United Nations Environment Programme. 1992. Basel convention on the control of transboundary movements of hazardous wastes and their disposal. http://www.basel.int/text/con-e-rev.pdf (accessed Nov 1, 2009).

Secretariat of the Stockholm Convention-United Nations Environment Programme, Stockholm Convention on Persistent Organic Pollutants (POPs). 2009. http://chm.pops.int/ (accessed July 27, 2009).

Stockholm Convention. 2008. http://chm.pops.int/Programmes/New%20POPs/The%209%20new%20POPs/tabid/672/language/en-US/Default.asp

SVTC (Silicon Valley Toxics Coalition). 2007. Just say no to e-waste: Background document on hazards and waste from computers. http://www.svtc.org/cleancc/pubs/sayno.htm (accessed September 7, 2007).

SVTC (Silicon Valley Toxics Coalition). 2008. Global e-waste crisis: Threatening communities around the globe. http://www.etoxics.org/site/PageServer?pagename=svtc_global_ewaste_crisis (accessed May 18, 2008).

UNEP, GRID-Europe. 2005. E-waste, the hidden side of IT equipment's manufacturing and use. *Environment Alert Bulletin* 5.

UNEP. 2009. Background information in relation to the emerging policy issue of electronic waste (SAICM/ICCM.2/INF/36), March 31, 2009. In *International Conference on Chemicals Management*, Second Session, May 11–15, 2009, Geneva, Switzerland.

The Central People's Government of the Peoples' Republic of China. 2009. The management regulations on the recycling and disposal of waste electrical and electronic products, No. 551. http://www.gov.cn/zwgk/2009-03/04/content_1250419.htm (in Chinese) (accessed March 23, 2011).

The World Bank. 2005. Environmental impact assessment of Zhejiang Province for PCB management and disposal demonstration project. http://www-wds.worldbank.org/servlet/WDSContentServer/WDSP/IB/2005/04/08/0000120 09_20050408104556/Rendered/INDEX/E10960v10rev0ZhejiangEIAApril2005.txt (accessed March 2, 2008).

Turci R, Turconi G, Comizzoli S, Roggi C, Minoia C. 2006. Assessment of dietary intake of polychlorinated biphenyls from a total diet study conducted in Pavia, Northern Italy. *Food Additives and Contaminants* 23: 919–938.

Wang F, Leung AOW, Wu SC, Yang MS, Wong MH. 2009. Chemical and ecotoxicological analyses of sediments and elutriates of contaminated rivers due to e-waste recycling activities using a diverse battery of bioassays. *Environmental Pollution* 157: 2082–2090.

Widmer R, Oswale-Krapf H, Sinha-Khetriwal D, Schnellmann M, Böni H. 2005. Global perspectives on e-waste. *Environmental Impact Assessment Review* 25: 436–458.

Williams E, Kahha, R, Allenby B, Kavazanjian E, Kim J, Xu M. 2008. Environmental, social, and economic implications of global reuse and recycling of personal computers. *Environmental Science and Technology* 42: 6446–6454.

Wong MH, Weber R, Leung A. 2008. The complex mix of toxic compounds created by e-waste recycling—A risk assessment challenge. Paper presented at *The Fifth NIES Workshop on E-waste*, November 17, 2008, Kyoto, Japan.

Wong MH, Wu SC, Deng WJ, Yu XZ, Luo Q, Leung AOW, Wong CSC, Luksemburg WJ, Wong AS. 2007. Export of toxic chemicals—A review of the case of uncontrolled electronic-waste recycling. *Environmental Pollution* 149: 131–140.

Wu KS, Xu XJ, Liu JX, Guo YY, Li Y, Huo X. 2010. Polybrominated diphenyl ethers in umbilical cord blood and relevant factors in neonates from Guiyu, China. *Environmental Science and Technology* 44: 813–819.

Xing GH. 2008. Human exposure and health risk assessment of polychlorinated biphenyls at two major electronic waste recycling sites in China. PhD thesis, Hong Kong Baptist University, Kowloon Tong, Hong Kong.

Xing GH, Yang Y, Chan JKY, Tao S, Wong MH. 2008. Bioaccessibility of polychlorinated biphenyls in different foods using an *in vitro* digestion method. *Environmental Pollution* 156: 1218–1226.

Xing GH, Chan JKY, Leung AOW, Wu SC, Wong MH. 2009. Environmental impact and human exposure to PCBs in Guiyu, an electronic waste recycling site in China. *Environment International* 35: 76–82.

Yang WH. 2008. Regulating electrical and electronic wastes in China. *Review of European Community and International Environmental Law* 17: 337–346.

Zhao GF, Xu Y, Li W, Han GG, Ling B. 2007. PCBs and OCPs in human milk and selected foods from Luqiao and Pingqiao in Zhejiang, China. *Science of the Total Environment* 378: 281–292.

Zheng LK, Wu KS, Li Y, Qi ZL, Han D, Zhang B et al. 2008. Blood lead and cadmium levels and relevant factors among children from an e-waste recycling town in China. *Environmental Research* 108: 15–20.

11

Decision-Making Support Tools for Managing Electronic Waste

Peeranart Kiddee, Ravi Naidu, and Ming Hung Wong

CONTENTS

Introduction

There are a number of definitions of electronic wastes (or e-waste), which is also known as waste electrical and electronic equipment (WEEE) (EU 2002; Puckett and Smith 2002; Electronic Recyclers International 2006; Wong et al. 2007a). For the purpose of this chapter, e-waste will be defined as consisting of old, end-of-life electronic devices such as televisions, refrigerators, washing machines, vacuum cleaners, computers, computer peripherals, and mobile phones that original users no longer want because they are obsolete or irreparable.

E-waste is one of the world's rapidly growing problems (LaCoursiere 2005). In most countries around the world, including Australia, it is projected that the volume of e-waste materials is growing at a rate of 3–5% per annum and will be three times that of other individual waste streams in the solid waste sector (Schwarzer et al. 2005). The advent of new design and technology at regular intervals in the electronic sector is causing the early obsolescence of many electronic items used around the world today. For example, the introduction of digital TVs in Australia and worldwide has posed significant challenge to the management of millions of analogue TVs that are currently being disposed in prescribed landfills despite the potential adverse impact that such an act can have on the environment. Coupled with this, it is now widely recognized that the lifespan of many electronic goods has substantially reduced due to advanced electronics, attractive design, and compatibility. For example, the average lifespan of a new model computer has decreased from 4.5 years in 1992 to an estimated 2 years in 2005 and is further decreasing (Widmer et al. 2005). A UNEP (Bushehri 2010) report estimates that more than 130 million computers, monitors, and televisions become obsolete each year and that amount is growing each year in the United States. The UNEP report further reveals that around 500 million computers will become obsolete in the United States alone between 1997 and 2007 and that 610 million computers are to be discarded in Japan by the end of December 2010. It is estimated that in China 5 million new computers and 10 million new televisions are purchased every year since 2003 (Hicks et al. 2005) and around 1.11 million tonnes e-waste is generated every year coming mainly from electrical and electronic manufacturing and production processes, end-of-life of household appliances, information technology products, and import from other countries (Xuefeng et al. 2006). In Canada a quantity of 140,000 tonnes of e-waste is expected to be managed. The same scenario applies to other developed countries as well.

E-waste problem in developing countries is in transition and is not so severe given the amount of waste generated by these countries and given much longer half-life of electronic goods, which is needed because of the financial constraints or lower affordability on the part of both the local community and the nation as a whole. However, the major e-waste problem in these countries arises from the flood of e-wastes and electronic goods imported from developed countries; 80% of all e-wastes are exported to the developing countries. Given the limited policies, safeguards, legislation, and enforcement, imported e-wastes and electronic goods have led to serious human and environmental problems in these countries. Concern arises not just from the large volume of e-wastes imported in developing countries but also from the large amounts of toxic chemical contents in such e-wastes received. Numerous researchers have demonstrated that toxic metals and polyhalogenated organics, including polychlorinated biphenyls (PCBs), can be released from e-waste during recycling, subjecting the local residents around the e-waste recycling site and the living environment to extremely harmful levels of toxicity, which is caused by deposits of impurities into the soil or emitting of other airborne pollutants into the atmosphere during the recycling process (Czuczwa and Hites 1984; Williams et al. 2008). A review of published report on e-waste problems in developing countries reveals that developing countries and countries in transition such as China, Cambodia, India, Indonesia, Pakistan, and Thailand, including African countries such as Nigeria, receive e-waste from all around the world; nevertheless, the nature and scope of problems faced on account of e-waste import differ between these countries. For instance, although African countries mainly reuse disposed electronics, Asian countries dismantle them, using mostly unsafe methods (U.S. Government Accountability Office 2008). With the recent recognition of social and human health problems experienced in some developing countries, it is worth noting that China, India, and other Asian countries have recently amended their laws to address e-waste imports (Widmer et al. 2005). Moreover, some manufacturers of electronic goods have also attempted to safely dispose of e-waste with advanced technologies (Widmer et al. 2005; U.S. Government Accountability Office 2008).

In this chapter, we present an overview of e-waste problems experienced by most countries and possible decision support tools that could be used to manage e-wastes.

E-Waste Categories

E-waste is classified into 10 different categories, including large household appliances, small household appliances, IT and telecommunications equipment, consumer equipment, lighting equipment, electrical and electronic tools (with the exception of large-scale stationary industrial tools), toys, leisure and sports equipment, medical devices (with the exception of all implanted and infected products), monitoring and control instruments, and automatic dispensers (EU 2002) (Table 11.1).

TABLE 11.1

E-Waste Categories and Examples of Products according to the EU Directive on WEEE

No.	Category	Label	Products
1	Large household appliances	Large HH	Refrigerators, freezers, washing machines, clothes dryers, dish washing, etc.
2	Small household appliances	Small HH	Vacuum cleaners, carpet sweepers, irons, toasters, fryers, grinders, coffee machines, etc.
3	IT and telecommunications equipment	ICT	Personal computers (CPU, mouse, screen, and keyboard), notebook computers, notepad computers, printers, copying equipment, etc.
4	Consumer equipment	CE	Radio sets, television sets, video cameras, video recorders, audio amplifiers, musical instruments, etc.
5	Lighting equipment	Lighting	Luminaries for fluorescent lamps, straight fluorescent lamps, compact fluorescent lamps, etc.
6	Electrical and electronic tools (with the exception of large-scale stationary industrial tools)	E & E tools	Drills; saws; sewing machines; equipment for turning, milling, sanding, grinding, sawing cutting, shearing, drilling, etc.
7	Toys, leisure, and sports equipment	Toys	Electric trains or car racing sets, hand-held video game consoles, video games, etc.
8	Medical devices (with the exception of all implanted and infected products)	Medical equipment	Radiotherapy equipment, cardiology, dialysis, pulmonary ventilators, nuclear medicine, etc.
9	Monitoring and control instruments	M & C	Smoke detector, heating regulatory, thermostats, measuring, etc.
10	Automatic dispensers	Dispensers	Automatic dispensers for hot drinks, automatic dispensers for hot or cold bottles or cans, automatic dispensers for money, etc.

Source: EU, *Off. J. Eur. Union*, L037, 0024, 2002.

Toxic and Hazardous Substances in E-Waste

E-waste consists of a large variety of materials (Zhang and Forssberg 1997), some of which contain a variety of toxic substances that can contaminate the environment and threaten human health if the end-of-life stage for such products containing those materials is not meticulously managed. E-waste disposal methods, including landfill and incineration, pose considerable contamination risks. Landfill leachates can potentially transport toxic substances into groundwater, whereas their combustion in an incinerator can emit toxic gas into the atmosphere. Moreover, even recycling e-waste can distribute hazardous substances into the environment and may affect human health. Although there are more than 1000 toxic substances associated with e-wastes, the more commonly reported substances include the following:

- Toxic metals, including barium, beryllium, cadmium, cobalt, copper, iron, lead, lithium, lanthanum, mercury, manganese, molybdenum, nickel, and so on
- Neurotoxin, carcinogens, and toxic dioxins, including hexavalent chromium and brominated flame retardants, and PVC

Impacts of E-Waste

The rapid growth of e-waste and the effectiveness of its management will have profound impacts on the environment and human health.

Environment

E-waste management by its disposal via landfills, recycling, or incineration can cause hazardous risks to the environment. E-waste disposal in landfills is seen to be a serious problem given the volume of material disposed and the toxicity of many substances released from such wastes. The pollutants have the potential to migrate through soils and groundwater within and around landfill sites (Kasassi et al. 2008). Organic and putrescible material in landfills decomposes and is transported by water to percolate through soil as landfill leachate. Where such materials have been incinerated using tools that do not capture emissions, significant levels of toxic substances have been found in the emissions. Such toxic substances have been shown to pose risk to both the environment and the people at the site and also locations in and around the site due to the movement of the airborne toxic substances.

Leachate can comprise high concentrations of dissolved and suspended organic substances, inorganic compounds, and heavy metals. However, the concentration of toxic substances from leachate depends on the waste characteristics and stages of waste decomposition in a particular landfill (Qasim and Chiang 1994). There have been a number of studies that investigated the leachability of the components that comprise e-waste. The leachability of lead from cathode ray tubes (CRTs) in televisions and computer monitors is one of a number of toxic substances that can leach to the wider ecosystem

(Musson et al. 2000). Jang and Townsend (2003) investigated 11 Florida landfills to determine lead leachability from printed circuit boards in computers and cathode ray tubes from computers and televisions. Townsend et al. (2004) studied leachability of 12 types of electronic appliances, including CPUs, computer monitors, laptops, printers, color TV, VCRs, cell phones, keyboards, mice, remote controls, smoke detectors, and flat-panel displays to examine the concentration of lead, iron, copper, and zinc from their leachate. Li et al. (2006) examined the leachate from motherboards, various expansion cards, disk drives, and power supply units to test the concentration of eight elements, namely, arsenic, barium, cadmium, chromium, lead, mercury, selenium, and silver. Spalvins et al. (2008) studied the impact of lead leachability from electronic equipment, including computers, keyboards, mouse devices, smoke detectors, monitors, cell phones, and cell phone batteries in the simulated landfills, and Li et al. (2009) investigated 18 heavy metals in the leachate of personal computers and cathode ray tubes (CRTs) in the simulated landfills.

Spalvins et al. (2008) studied the lead concentration from e-waste in landfill leachate. They found lead concentrations of 7–66 µg/L in the simulated landfill column containing e-waste and further found that the simulated landfill column without e-waste was <2–54 µg/L. Heavy metals accumulate in both soil and groundwater around landfill sites through vertical and horizontal migration (Rawat et al. 2008). Li et al. (2009) analyzed the soil samples in simulated landfill columns and found heavy metals, including Pb, Cr, Ni, and Sn; however, they could not detect these in the leachate.

Vast quantities of e-waste are now being exported around the world for recycling using processes that emit the pollution into the environment in developing countries. Guiyu and Taizhou are the large e-waste recycling sites in China where extensive pollution is emitted due to e-waste recycling processes. The whole ecosystem, including soil, sediment, water, and air, is being contaminated by these toxic substances. Many researchers are now investigating the distribution and contamination of pollution from the e-waste recycling process in the environment. Soil samples are collected to test the contamination of polybrominated diphenyl ethers (PBDEs) (Wang et al. 2005; Cai and Jiang 2006; Leung et al. 2007), polycyclic aromatic hydrocarbons (PAHs) (Leung et al. 2006; Yu et al. 2006; Shen et al. 2009; Tang et al. 2010), polychlorinated dibenzo-p-dioxins and dibenzofurans (PCDD/Fs), (Leung et al. 2007; Shen et al. 2009), and polychlorinated biphenyls (PCBs), (Shen et al. 2009; Tang et al. 2010) Cr, Cd, Hg, Ni, Pb, and Zn (Tang et al. 2010) either in recycling sites or around the areas. Also, water samples are examined for heavy metals, including Ag, As, Be, Cd, Cr, Co, Cu, Li, Mo, Ni, Pb, Sb, Se, and Zn (Wong et al. 2007b). Air samples are collected to examine contamination by heavy metals such as As, Cd, Cr, Cu, Mn, Ni, Pb, Zn, and PAHs (Deng et al. 2006); PBDEs (Deng et al. 2007; Chen et al. 2009); PCDD/Fs; and polybrominated dibenzo-p-dioxins and dibenzofurans (PBDD/Fs) (Li et al. 2007). As a result, the environment around recycling sites in China has been found to be extensively contaminated by heavy metals and chlorinated and brominated compounds from e-waste.

Human Health

E-waste disposals impact human health in two main ways. One is the contamination by toxic substances from e-waste disposal either in landfill or release of toxic substances into the environment due to the use of primitive recycling processes, which later enter into the food chain and are transferred to humans. The other is the direct impact on workers who work at sites where such primitive recycling processes take place. The danger of e-waste toxicity to human health, both chronic and acute, has become a serious problem in society. Qu et al. (2007) studied PBDE exposure of workers in e-waste recycling areas and found high levels of PBDE in the serum of the sample groups. Zheng et al. (2008) investigated blood lead and cadmium levels in children around e-waste recycling regions and found high levels of lead and cadmium in their blood.

Strategies for Managing E-Waste

The constant improvement in information and communications technologies (ICT) has decreased the lifespan of electronic products, leading to an increase in the volume of e-waste. Although the e-waste problem was typically associated with developed countries in the past, it is no longer restricted to developed countries and is now indeed a global problem. In response to this global challenge, the international community has come together in a collaborative research effort to reduce e-waste impacts. Moreover, researchers have investigated the volume, nature, and potential environmental and human health impacts of e-waste (Widmer et al. 2005). Such studies have contributed to the development of methodology for assessing informal e-waste management systems, for instance, a comprehensive tool developed for assessing e-waste impact as shown in Figure 11.1. This tool assists both qualitative and quantitative assessment of the impact of e-wastes. Widmer et al. (2005) developed an assessment indicator system to measure and compare e-waste management systems, which included the investigation and management of e-waste problems using three main points: policy and legislation; economy, society, and culture; and science and technology.

Policy and Legislation

There is currently an extensive research into both policy and legislative aspects of e-waste in order to mitigate problems at the national level. The current state of the national policy and legislative framework is very diverse.

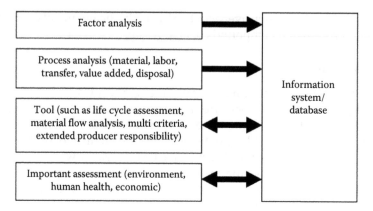

FIGURE 11.1
Overview assessment methodology. (Modified from Widmer, R. et al., *Environ. Impact Assess. Rev.*, 25, 436, 2005.)

European countries have developed e-waste regulations since the early 1990s. These regulations are based on the concept of extended producer responsibility (EPR), where producers of the product are encouraged to mitigate e-waste issues (Tojo 2001). Japan, with its large and mature ICT industry, has ongoing concerns about e-waste issues. Its home appliance recycling law was passed in May 1998 and enforced in 2001 to encourage recycling and curtail disposal costs (Kawakami 2001). In China, environmental concerns are growing appreciably after the huge importation of e-wastes from around the world for reprocessing. The Chinese government has drafted a legislation and proposed market mechanisms to encourage the recycling and safe disposal of e-waste. The objective of the draft is to improve the framework of the country's e-waste recycling system. Pilot programs are operating in Qingdao and Zhejiang to change the attitudes and practices of Chinese toward e-waste (Hicks et al. 2005). Sinha-Khetriwal et al. (2006) studied the progress of e-waste legislation within the framework of the Swiss global e-waste program in China, India, and South Africa. The authors identified five areas of concern: the definition of e-waste, producer responsibilities, orphan appliances, setting collection and recycling targets, and monitoring authority. These questions provide a broad framework for successful development of legislation. Nnorom and Osibanjo (2008) also investigated the concept of extended producer responsibility (EPR) in e-waste management and regulation with a view to progressing EPR in developing nations.

Australia has been concerned about the impact of e-waste since the 1990s. Environment Australia provided guidelines for the treatment of hazardous waste electronic scrap under the act in 1999 (Environment Australia 1999). In 2005, the State EPAs collaborated to survey the amount of household e-waste (IPSOS 2005). In addition, the Environment Protection and Heritage Council offered to investigate the willingness of consumers to pay for the recycling of e-waste, particularly televisions and computers (Rolls et al. 2009), and also supported regulation for end-of-life treatment of televisions and computers (Environment Protection and Heritage Council 2009). A growing number of researchers are concerned about e-waste issues in Australia. Herat (2007) reviewed sustainable management in e-waste problems. Davis and Herat (2008) surveyed the level of understanding and action on e-waste to assist with e-waste management efforts by the local government in Queensland, Australia.

Economy, Society, and Culture

The growth and rapid advances in electronic technology point to more consumers generating ever-larger volumes of e-waste. Some e-waste is exported for reprocessing in developing countries where it creates job opportunities, whereas some e-waste is recycled locally. The huge impact of e-waste recycling on both the economic system and society is that it provides income for poor people, but it also increases the costs for consumers and producers. Jung and Bartel (1999) investigated economic data in a computer collection and recycling company for comparison with alternative electronic device management systems. Kang and Schoenung (2006) examined many techniques for e-waste treatment and considered their costs and returns. Nixon and Saphores (2007) studied the willingness to pay for the recycling of e-waste in California. They found that the majority of respondents pay only 1% of the advanced recycling fee. Gregory and Kirchain (2008) created a framework that considered various factors such as the activities of recycling process, cash flow elements, and resources for evaluating economic performance in a recycling system. Liu et al. (2009) investigated the economic assessment of recycling systems in order to evaluate the economic feasibility of e-waste treatment in municipal China. Darby and Obara (2005) studied consumer behavior and attitudes toward discarding small items of e-waste at the household level. Manhart (2007) studied the social impacts of electronic devices and e-waste recycling. He found that e-waste recycling has a positive impact on the economy, providing jobs for approximately 700,000 people in China. However, these industries pose a risk to human health and the environment. Saphores et al. (2009) surveyed the volumes of e-waste stored in basements and attics and found 4.2 items of small e-waste and 2.4 items of large e-waste per U.S. household.

Science and Technology

There have been a number of models developed to support decision making to overcome the problems of e-waste. For example, Yu et al. (2000) developed a decision-making model for material recycling and disposition of e-waste in terms of environmental impact, recycling feasibility, and intensity of resource recovery. Nagurney and Toyasaki (2005) created the reverse supply chain management model of e-waste from a variety of components, which included the e-waste sources, the recyclers, and the processors and consumers of different devices. Life cycle assessment (LCA) is one of the tools in e-waste management that is used to assess the electronic products such as computers (Ahluwalia and Nema 2007; Babbitt et al. 2009; Duan et al. 2009), televisions (Dodbiba et al. 2007), mobile phones (Scharnhorst et al. 2005), and laptops (Lu et al. 2006). In addition, "multi-criteria," a tool that determines the appropriate options in e-waste management, is also used (Lee et al. 2001; Rousis et al. 2008).

Material flow analysis is a tool to study the route e-waste takes to flow into recycling sites or disposal areas. This tool has been used to manage e-wastes. Shinkuma and Nguyen Thi Minh (2009) used material flow analysis to investigate the flow of e-waste in Asia. They found that most of the e-waste is recycled in China. This study led to a recommendation to provide suitable recycling services in China to service the Asia region. Steubing et al. (2010) investigated e-waste generation using material flow analysis. They found that the e-waste generation will increase four to five times during 2010–2019.

Many technologies are created to recover material in recycling process or reduce material use. Mohabuth and Miles (2005) used the vertical vibration technique to recover material from e-waste. This technique splits a blend of plastic and bronze in water. Cui and Zhang (2008) reviewed metallurgical processing to recover valuable material from e-waste. The authors state that biotechnology is the best technique for recovering precious metals and copper. Jiang et al. (2008) used a roll-type corona-electrostatic separator in a recycling process, which is an efficient technique. Moreover, Platcheck et al. (2008) investigated the methodology for ecodesign of electronic equipment to be environmentally friendly.

Tools for Decision Making

E-waste is one of the most complicated solid wastes to deal with, due to a range of difficulties encountered in managing electronic components that comprise toxic material. However, numerous researchers have used life cycle assessment (LCA) and risk assessment (RA) as part of their decision making to support e-waste management.

LCA and RA are both effective tools for evaluating environmental and human health issues as well as decision making, as illustrated in Table 11.2. For instance, Olsen et al. (2001) compared similarities, differences, and relationships between life cycle assessment and risk assessment with reference to chemical substances over the entire lifespan of certain products, with a view to reducing environmental and health problems. The authors stated that neither tools can replace the other but they can benefit each other. Anex and Focth (2002) found there was a shared need for public participation in

TABLE 11.2

Life Cycle Assessment and Risk Assessment

Authors	Areas
Olsen et al. (2001)	Chemicals
Anex and Focht (2002)	Public participation
Cowell et al. (2002)	Policy
Nishioka et al. (2002)	Insulation
Sonnemann et al. (2004)	Industrial processes
Benetto et al. (2006)	Mineral waste
Mouron et al. (2006)	Agriculture
Socolof and Geibig (2006)	Solder product
Wright et al. (2008)	Economic

life cycle assessment and risk assessment. They found that social values applied equally to both tools. Benetto et al. (2006) used both life cycle assessment and risk assessment to support decision making in mineral waste reuse scenarios. These tools are more efficient for environmental assessment and for finding beneficial ways to reuse mineral wastes in cases where there are no regulations or standards. Socolof and Geibig (2006) applied LCA and RA tools to human and ecological impacts through life cycle of lead solder products: Both tools enhanced understanding of issues in the European ban on lead solder and the resulting improvement of public health. Wright et al. (2008) investigated a screening-level analysis that is assessed with an economic input–output life cycle assessment model and a risk assessment tool. They found that these tools are appropriate for conducting screening-level analysis and can be used to classify the critical life cycle stage and, in particular, emerging chemicals during the recycling of e-waste.

We believe that by integrating LCA and RA to assist decision making a system could be devised for managing the disposal of e-waste using landfills. LCA and RA are combined as conceptual tools and complement one another in the assessment of environmental issues (Olsen et al. 2001; Benetto et al. 2006; Mouron et al. 2006; Socolof and Geibig 2006) and as decision support mechanisms (Anex and Focht 2002; Cowell et al. 2002; Benetto et al. 2006; Shatkin 2008). The focus of LCA is in the product and process areas, whereas RA emphasizes the chemical area, based on toxicology and ecotoxicology (Flemström et al. 2004).

Life Cycle Assessment

Life cycle assessment (LCA) is a tool for evaluating the environmental impact of products, processes, and services during a product's entire life span and is also used for environmental decision making. The technical framework of

LCA methodology is divided into four phases, namely, goal and scope definition, inventory analysis, impact assessment, and interpretation, as shown in Figure 11.2 and summarized later.

Goal and Scope Definition

The first phase of goal and scope definition is designed to determine the aim and boundaries for the LCA study.

The system boundaries are the points where materials become waste and where waste is transformed to inert waste, leachate, and gas. The major environmental impact for landfill includes leachate treatment but does not cover waste separation, collection, and transportation.

Inventory Analysis

The second phase collects the data for a unit process that include gas emissions, water emissions, the public expectations, waste characteristics and composition, and cost management. The data are run on SimaPro.

Impact Assessment

The impact assessment focuses on e-waste impacts in landfill that affect the environment and the society. This includes investigating the effects of eco-toxicity on the ecosystem. Eco-indicator 99 methodology is used to assess environmental impacts.

Interpretation

To sum up the LCA phase, the process under investigation is assessing the environmental impact. The environmental performance can then be used

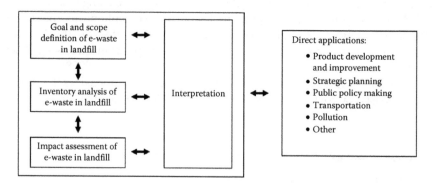

FIGURE 11.2
Life cycle assessment framework in the ISO 14041 LCA standard (ISO 14041 1998). (Modified from Baumann, H. and Tillman, A.M., *The Hitchhiker's Guide to LCA*, Studentlitteratur, Lund, Sweden, 2004.)

to support decision making about the different impacts of various types of e-waste in landfills.

Risk Assessment

Risk assessment is an analytical tool used to organize, structure, and compile scientific information to aid hazard identification, predict potential problems, determine priorities, and offer remedial action. It can also be used to inform decision making (SETAC 1997).

Framework of Environmental Risk Assessment

Environmental risk assessment (ERA) evaluates the potential harmful effects that occur as a result of pollutant release into the environment and the biota affecting both human health and the environment. There are four phases, namely, problem formulation, analysis, risk characterization, and risk management, as shown in Figure 11.3.

Problem Formulation

Problem formulation is the first phase; its aim is to define the problem and the plan for dealing with it. It investigates e-waste issues in landfill that affect the ecosystem and the human health (Figure 11.4).

Analysis

This phase consists of the characterization of exposure and the characterization of ecological effects. The analysis uses LandSim, a specialized computer model, to evaluate landfill risk (Butt et al. 2008). Leachate is monitored to assess the impact of e-waste from the landfill. The elements of concern in this investigation are lead, bromine, zinc, antimony, arsenic, chromium, cadmium, and mercury. The risk evaluation also includes geological barriers and landfill liners.

Risk Characterization

Risk characterization is the third stage of the risk assessment process and is used to estimate and describe risks. This phase is divided in two parts: risk estimation and risk description. Risk estimation models the exposure response analyses and the associated uncertainty. Risk description is used to assist the risk managers in understanding what they are dealing with.

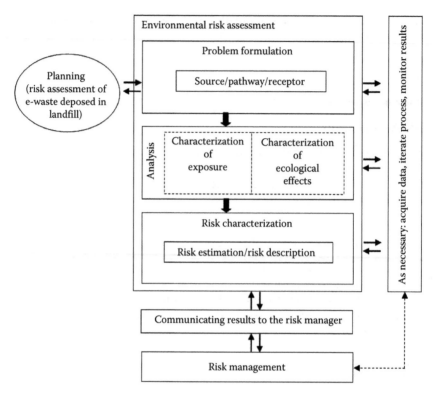

FIGURE 11.3
Framework of Environmental Risk Assessment. (Modified from U.S. EPA, *Guidelines for Ecological Risk Assessment*, EPA/630/R-95/002F, Risk assessment forum, U.S. EPA, Washington, DC, 1998.)

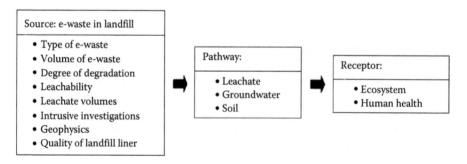

FIGURE 11.4
Problems of e-waste dispose in landfill.

Risk Management

The last stage is the decision-making process for remediation. It offers policy input to the problem formulation, indicates the results of the risk analysis, and makes a decision as to the optimum remediation strategy.

Assessment and Interpretation for Integrated Life Cycle and Risk Assessment

LCA and RA have been developed with different aims and on a conceptually different basis. However, the two tools provide varying perspectives of the same issue, which can benefit and complement each other. The relationships between life cycle assessment and risk assessment are illustrated in Figure 11.5. The results of using them in tandem can provide valuable support for decision making and help to solve e-waste disposal problems.

Conclusion

It is now well recognized that e-waste is a major problem not just in developed countries, but also in developing countries and countries in transition. The export of these wastes to developing countries have created major environmental, human health, and social hazards that, unless resolved, would lead to major calamity among the community of workers and their families who handle e-wastes as a source of income to support themselves and their families. In addition to developing effective policies for handling e-wastes, these is a need for an effective decision-making tool that not only captures the disposal of such wastes to landfills but also helps regulators manage the import and distribution of e-wastes in their countries. The decision support tool presented in this chapter should provide directions to regulators and waste handlers for the management of e-wastes.

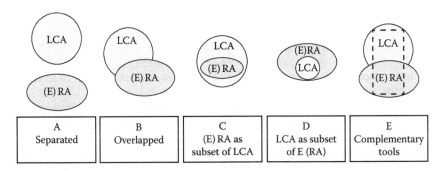

FIGURE 11.5
Relations between LCA and ERA. (From Flemström, K. et al., *Relationships between Life Cycle Assessment and Risk Assessment—Potentials and Obstacles,* Industrial Environmental Informatics (IMI), Chalmers University of Technology, Naturvårdsverket, Sweden, 82pp., 2004.)

References

Ahluwalia PK, Nema AK. 2007. A life cycle based multi-objective optimization model for the management of computer waste. *Resources, Conservation and Recycling* 51: 792–826.

Anex RP, Focht W. 2002. Public participation in life cycle assessment and risk assessment: A shared need. *Risk Analysis* 22: 861–877.

Babbitt CW, Kahhat R et al. 2009. Evolution of product lifespan and implications for environmental assessment and management: A case study of personal computers in higher education. *Environmental Science and Technology* 43: 5106–5112.

Baumann H, Tillman AM. 2004. *The Hitchhiker's Guide to LCA*. Studentlitteratur, Lund, Sweden.

Benetto E, Tiruta-Barna L et al. 2006. Combining lifecycle and risk assessments of mineral waste reuse scenarios for decision making support. *Environmental Impact Assessment Review* 27: 266–285.

Bushehri FI. 2010. UNEP's role in promoting environmentally sound management of e-waste. *5th ITU Symposium on ICTs, the Environment and Climate Change*, 2–3 November. Cairo, Egypt.

Butt TE, Lockley E et al. 2008. Risk assessment of landfill disposal sites—State of the art. *Waste Management* 28: 952–964.

Cai Z, Jiang G. 2006. Determination of polybrominated diphenyl ethers in soil from e-waste recycling site. *Talanta* 70: 88–90.

Chen D, Bi X et al. 2009. Pollution characterization and diurnal variation of PBDEs in the atmosphere of an e-waste dismantling region. *Environmental Pollution* 157: 1051–1057.

Cowell SJ, Fairman R et al. 2002. Use of risk assessment and life cycle assessment in decision making: A common policy research agenda. *Risk Analysis* 22: 879–894.

Cui J, Zhang L. 2008. Metallurgical recovery of metals from electronic waste: A review. *Journal of Hazardous Materials* 158: 228–256.

Czuczwa JM, Hites RA. 1984. Environmental fate of combustion-generated polychlorinated dioxins and furans. *Environmental Science and Technology* 18: 444–450.

Darby L, Obara L. 2005. Household recycling behaviour and attitudes towards the disposal of small electrical and electronic equipment. *Resources, Conservation and Recycling* 44: 17–35.

Davis G, Herat S. 2008. Electronic waste: The local government perspective in Queensland, Australia. *Resources, Conservation and Recycling* 52: 1031–1039.

Deng WJ, Louie PKK et al. 2006. Atmospheric levels and cytotoxicity of PAHs and heavy metals in TSP and PM2.5 at an electronic waste recycling site in southeast China. *Atmospheric Environment* 40: 6945–6955.

Deng WJ, Zheng JS et al. 2007. Distribution of PBDEs in air particles from an electronic waste recycling site compared with Guangzhou and Hong Kong, South China. *Environment International* 33: 1063–1069.

Dodbiba G, Furuyama T et al. 2007. Life cycle assessment: A tool for evaluating and comparing different treatment options for plastic wastes from old television sets. *Data Science Journal* 6: 39–50.

Duan H, Eugster M et al. 2009. Life cycle assessment study of a Chinese desktop personal computer. *Science of the Total Environment* 407: 1755–1764.

Electronic Recyclers International. 2006. The history of e-waste: E-waste defined. Retrieved December 1, 2009, from www.electronicrecyclers.com/historyofe-waste.aspx

Environment Australia. 1999. Hazard status of waste electrical and electronic assemblies or scrap, Environment Australia, Department of the Environment and Heritage, Canberra, ACT, Australia, 24pp.

Environment Protection and Heritage Council. Decision regulatory impact statement: Televisions and computers, 2009, pp. 257. PricewaterhouseCoopers Australia. http://www.ephc.gov.au/sites/default/files/PS_TV_Comp_Decision_RIS_Televisions_and_Computers_200911_0.pdf

EU. 2002. Directive 2002/96/EC of the European parliament and of the council of 27 January 2003 on waste electrical and electronic equipment (WEEE). *Official Journal of the European Union* L037: 0024–0039.

Flemström K, Carlson R et al. 2004. *Relationships between Life Cycle Assessment and Risk Assessment—Potentials and Obstacles*. Industrial Environmental Informatics (IMI), Chalmers University of Technology, Naturvårdsverket, Sweden, 82pp.

Gregory JR, Kirchain RE. 2008. A framework for evaluating the economic performance of recycling systems: A case study of North American electronics recycling systems. *Environmental Science and Technology* 42: 6800–6808.

Herat S. 2007. Sustainable management of electronic waste (e-waste). *Clean* 35: 305–310.

Hicks C, Dietmar R et al. 2005. The recycling and disposal of electrical and electronic waste in China-legislative and market responses. *Environmental Impact Assessment Review* 25(5): 459–471.

IPSOS. 2005. Household electrical & electronic waste survey 2005 report of findings, Prepared for Department of Environment and Conservation, Sydney, NSW, Australia.

Jang YC, Townsend TG. 2003. Leaching of lead from computer printed wire boards and cathode ray tubes by municipal solid waste landfill leachates. *Environmental Science and Technology* 37: 4778–4784.

Jiang W, Jia L et al. 2008. Optimization of key factors of the electrostatic separation for crushed PCB wastes using roll-type separator. *Journal of Hazardous Materials* 154: 161–167.

Jung LB, Bartel TJ. 1999. Computer take-back and recycling: An economic analysis for used consumer equipment. *Electronics Manufacturing* 9: 67–77.

Kang HY, Schoenung JM. 2006. Economic analysis of electronic waste recycling: Modeling the cost and revenue of a materials recovery facility in California. *Environmental Science and Technology* 40: 1672–1680.

Kasassi A, Rakimbei P et al. 2008. Soil contamination by heavy metals: Measurements from a closed unlined landfill. *Bioresource Technology* 99: 8578–8584.

Kawakami K. 2001. Outline of law for recycling specified kinds of home appliances in Japan. *Environmental Economics and Policy Studies* 4: 211–218.

LaCoursiere C. 2005. Electronic waste recovery business. Report Code: MST0374 Norwalk, CT.

Lee SG, Lye SW et al. 2001. A multi-objective methodology for evaluating product end-of-life options and disassembly. *Advanced Manufacturing Technology* 18: 148–156.

Leung A, Cai ZW et al. 2006. Environmental contamination from electronic waste recycling at Guiyu, southeast China. *Material Cycles and Waste Management* 8: 21–33.

Leung AOW, Luksemburg WJ et al. 2007. Spatial distribution of polybrominated diphenyl ethers and polychlorinated dibenzo-*p*-dioxins and dibenzofurans in soil and combusted residue at Guiyu, an electronic waste recycling site in Southeast China. *Environmental Science and Technology* 41: 2730–2737.

Li Y, Richardson JB et al. 2006. TCLP heavy metal leaching of personal computer components. *Environmental Engineering* 132: 497–504.

Li Y, Richardson JB et al. 2009. Leaching of heavy metals from e-waste in simulated landfill columns. *Waste Management* 29: 2147–2150.

Li H, Yu L et al. 2007. Severe PCDD/F and PBDD/F pollution in air around an electronic waste dismantling area in China. *Environmental Science and Technology* 41: 5641–5646.

Liu X, Tanaka M et al. 2009. Economic evaluation of optional recycling processes for waste electronic home appliances. *Journal of Cleaner Production* 17: 53–60.

Lu LT, Wernick IK et al. 2006. Balancing the life cycle impacts of notebook computers: Taiwan's experience. *Resources, Conservation and Recycling* 48: 13–25.

Manhart A. 2007. *Key Social Impacts of Electronics Production and WEEE-Recycling in China*. Institute for Applied Ecology, Corvallis, OR, 29pp.

Mohabuth N, Miles N. 2005. The recovery of recyclable materials from waste electrical and electronic equipment (WEEE) by using vertical vibration separation. *Resources, Conservation and Recycling* 45: 60–69.

Mouron P, Nemecek T et al. 2006. Management influence on environmental impacts in an apple production system on Swiss fruit farms: Combining life cycle assessment with statistical risk assessment. *Agriculture, Ecosystems & Environment* 114: 311–322.

Musson SE, Jang YC et al. 2000. Characterization of lead leachability from cathode ray tubes using the toxicity characteristic leaching procedure. *Environmental Science and Technology* 34: 4376–4381.

Nagurney A, Toyasaki F. 2005. Reverse supply chain management and electronic waste recycling: A multitiered network equilibrium framework for e-cycling. *Transportation Research Part E: Logistics and Transportation Review* 41: 1–28.

Nishioka Y, Levy JI et al. 2002. Integrating risk assessment and life cycle assessment: A case study of insulation. *Risk Analysis* 22: 1003–1017.

Nixon H, Saphores JDM. 2007. Financing electronic waste recycling Californian households' willingness to pay advanced recycling fees. *Journal of Environmental Management* 84: 547–559.

Nnorom IC, Osibanjo O. 2008. Overview of electronic waste (e-waste) management practices and legislations, and their poor applications in the developing countries. *Resources, Conservation and Recycling* 52: 843–858.

Olsen SI, Christensen FM et al. 2001. Life cycle impact assessment and risk assessment of chemicals—A methodological comparison. *Environmental Impact Assessment Review* 21: 385–404.

Platcheck ER, Schaeffer L et al. 2008. Methodology of ecodesign for the development of more sustainable electro-electronic equipments. *Journal of Cleaner Production* 16(1): 75–86.

Puckett J, Smith T. 2002. Exporting harm the high-tech trashing of Asia. In: Coalition SVT (Ed.), Silicon Valley Toxics Coalition, San Jose, CA, pp. 51.

Qasim SR, Chiang W. 1994. *Sanitary Landfill Leachate: Generation, Control and Treatment*. CRC Press, Boca Raton, FL.

Qu W, Bi X et al. 2007. Exposure to polybrominated diphenyl ethers among workers at an electronic waste dismantling region in Guangdong, China. *Environment International* 33: 1029–1034.

Rawat M, Singh UK et al. 2008. Methane emission and heavy metals quantification from selected landfill areas in India. *Environmental Monitoring and Assessment* 137: 67–74.

Rolls SC, Brulliard et al. 2009. Willingness to pay for e-waste recycling, Prepared for Environment Protection and Heritage Council, Adelaide, SA, Australia, 110 p.

Rousis K, Moustakas K et al. 2008. Multi-criteria analysis for the determination of the best WEEE management scenario in Cyprus. *Waste Management* 28: 1941–1954.

Saphores JD, Nixon MH et al. 2009. How much e-waste is there in US basements and attics? Results from a national survey. *Journal of Environmental Management* 90: 3322–3331.

Scharnhorst W, Althaus HJ et al. 2005. The end of life treatment of second generation mobile phone networks: Strategies to reduce the environmental impact. *Environmental Impact Assessment Review* 25: 540–566.

Schwarzer S, Bono AD et al. 2005. E-waste, the hidden side of IT equipment's manufacturing and use, UNEP Early Warning on Emerging Environmental Threats No.5., Geneva, Switzerland.

SETAC. 1997. Ecological Risk Assessment Technical Issue Paper. Pensacola, FL.

Shatkin JA. 2008. Informing environmental decision making by combining life cycle assessment and risk analysis. *Journal of Industrial Ecology* 12: 278–281.

Shen C, Chen Y et al. 2009. Dioxin-like compounds in agricultural soils near e-waste recycling sites from Taizhou area, China: Chemical and bioanalytical characterization. *Environment International* 35: 50–55.

Shinkuma T, Nguyen Thi Minh H. 2009. The flow of e-waste material in the Asian region and a reconsideration of international trade policies on e-waste. *Environmental Impact Assessment Review* 29: 25–31.

Sinha-Khetriwal D, Widmer R et al. 2006. Legislating e-waste management: Progress from various countries. *Elni Review* 1+2/06: 27–36.

Socolof M, Geibig J. 2006. Evaluating human and ecological impacts of a product life cycle: The complementary roles of life cycle assessment and risk assessment. *Human and Ecological Risk Assessment* 12: 510–527.

Sonnemann G, Castells F et al. 2004. *Integrated Life-Cycle and Risk Assessment for Industrial Processes.* Lewis Publishers, Boca Raton, FL.

Spalvins E, Dubey B et al. 2008. Impact of electronic waste disposal on lead concentrations in landfill leachate. *Environmental Science and Technology* 42: 7452–7458.

Steubing B, Böni H et al. 2010. Assessing computer waste generation in Chile using material flow analysis. *Waste Management* 30: 473–482.

Tang X, Shen C et al. 2010. Heavy metal and persistent organic compound contamination in soil from Wenling: An emerging e-waste recycling city in Taizhou area, China. *Journal of Hazardous Materials* 173: 653–660.

Tojo N. 2001. Extended producer responsibility legislation for electrical and electronic equipment -approaches in Asia and Europe, pp. 1–11, International Institute for Industrial Environmental Economics at Lund University, Lund, Sweden. http://sdsap.org/data/epr-electrical.pdf

Townsend TG, Vann K et al. 2004. RCRA toxicity characterization of computer CPUs and other discarded electronic devices. Department of Environmental Engineering Sciences, University of Florida, Gainesville, FL, 79pp.

U.S. EPA. 1998. Guidelines for ecological risk assessment. EPA/630/R-95/002F. Risk assessment forum. U.S. EPA, Washington, DC.

U.S. Government Accountability Office. 2008. Electronic waste EPA needs to better control harmful U.S. exports through stronger enforcement and more comprehensive regulation. United States Government Accountability Office, Washington, DC.

Wang D, Cai Z et al. 2005. Determination of polybrominated diphenyl ethers in soil and sediment from an electronic waste recycling facility. *Chemosphere* 60: 810–816.

Widmer R, Oswald-Krapf H et al. 2005. Global perspectives on e-waste. *Environmental Impact Assessment Review* 25: 436–458.

Williams E, Kahhat R et al. 2008. Environmental, social and economic implications of global reuse and recycling of personal computers. *Environmental Science and Technology* 42: 6446–6454.

Wong CSC, Duzgoren-Aydin NS et al. 2007a. Evidence of excessive releases of metals from primitive e-waste processing in Guiyu, China. *Environmental Pollution* 148: 62–72.

Wong MH, Wu SC et al. 2007b. Export of toxic chemicals—A review of the case of uncontrolled electronic-waste recycling. *Environmental Pollution* 149: 131–140.

Wright HE, Zhang Q et al. 2008. Integrating economic input–output life cycle assessment with risk assessment for a screening-level analysis. *The International Journal of Life Cycle Assessment* 13: 412–420.

Xuefeng W, Jinhui L et al. 2006. An agenda to move forward e-waste recycling and challenges in China. In *Proceedings of the 2006 IEEE International Symposium on Electronics and the Environment*, May 8–11, San Francisco, CA, 2006 pp. 315–320.

Yu XZ, Gao Y et al. 2006. Distribution of polycyclic aromatic hydrocarbons in soils at Guiyu area of China, affected by recycling of electronic waste using primitive technologies. *Chemosphere* 65: 1500–1509.

Yu Y, Jin K et al. 2000. A decision making model for materials management of end of life electronic products. *Manufacturing Systems* 19: 94–107.

Zhang S, Forssberg E. 1997. Mechanical separation-oriented characterization of electronic scrap. *Resources, Conservation and Recycling* 21: 247–269.

Zheng L, Wu K et al. 2008. Blood lead and cadmium levels and relevant factors among children from an e-waste recycling town in China. *Environmental Research* 108: 15–20.

Part III

Ecological Restoration of Contaminated Sites: Bioremediation

12

Biomethylation of Arsenic in Contaminated Soils

Richard P. Dick, Qin Wu, and Nicholas T. Basta

CONTENTS

Introduction

Sources and Toxicology

Arsenic (As) is a naturally occurring metalloid found in organic and inorganic compounds which ultimately are of geological origin and so is found in soil and underground water. Arsenic occurs naturally in oxide, hydrous oxide, sulfide, phosphate, and other minerals (Garelick et al. 2009). In soil the concentration of As ranges from 0.1 to 40 mg/kg, with a median concentration of total As being 6.0 mg/kg worldwide (Bowen 1979). Volcanic activity, thermal springs, or anthropogenic activities such as mining and industrial production can transport or concentrate As in the environment (Nordstrom 2002).

Arsenic is carcinogenic (Rosen 1971) and has been ranked the number one hazardous substances since 1997 in the U.S. Comprehensive Environmental Response, Compensation, and Liability Act (CERCLA) (Agency for Toxic Substance and Disease 2010). Inorganic arsenic is classified by the U.S. Environmental Protection Agency as a class A human carcinogen (EPA 1998). Additionally, arsenic compounds are poisonous and classified into three broad toxicological categories: as arsine gas, inorganic compounds (solid and solution phases), and organic As.

Arsenic has four states of oxidation (V, III, 0, or –III) which controls its bioavailability and toxicity to organisms. However, in the environment the inorganic forms of arsenate (AsO_4^{-3}) (V) and arsenite (AsO_2^{-2}) (III) dominate (Cullen and Reimer 1989), with arsenite being more toxic (Lerman et al. 1983, Vahidnia et al. 2007). Arsenate is toxic because of its similarity to phosphate, causing phosphorylation inhibition (Moore et al. 1983, Coddington 1986) and forming ADP-arsenate that is unstable (Crane and Lipmann 1953, Gresser 1981, Moore et al. 1983). Arsenite causes toxicity by reacting with dithiols such as glutaredoxin and with sulfhydryl groups of proteins (Oehme 1972, Knowles and Benson 1983).

Direct contamination of soils with As can occur from several human activities. In agriculture arsenical pesticides and herbicides have been major sources of As of soils (Smith et al. 1998, WHO 2001, Liao et al. 2005). Currently, approximately 50% of arsenic production is pesticides, with organic forms being the most important forms (Matschullat 2000). Another important and extensive As source is pressure-treated wood.

From the 1970s until about 2004, 90% of the wood used in the United States for outdoor applications was treated with chromated-copper arsenate (CCA). Although this was banned in 2004 in the United States, some seven billion board feet was used widely in fences, home decks, play structures, and other uses (Raloff 2004). Arsenic in CCA-treated wood structures and poles can leach into soils or as CCA wood decomposes As is released to soils (Zagury and Pouschat 2005). Additionally, CCA-treated wood manufacturing plants have caused incidental contamination of soil with As (Zagury and Pouschat 2005).

Incidental soil ingestion by children is an important exposure pathway for assessing public health risks associated with exposure to As-contaminated soils. The importance of soil ingestion by children as a health issue has been reported by many researchers and highlights the importance of this pathway in terms of subsequent chemical exposure (Basta et al. 2001, Scheckel et al. 2009, Hale et al. 2010).

Arsenic has become a widespread hazard in India, Bangladesh, Vietnam, Chile, and Taiwan. In some cases this is due to pollution but additionally vast areas of these countries have been contaminated by As of geological origin (Nordstrom 2002). Epidemiological studies in these regions indicated that skin, liver, lung, bladder, and other cancers are highly related to the exposure to As (Abernathy et al. 1999). Of further concern, particularly in Bangladesh, is that groundwater containing high levels of geologic As was extensively used for irrigation water between the 1980s and 2000s (Meharg and Rahman 2002). This has elevated As in rice soils of Bangladesh from background levels of 10 to as much as 46 μg As/g soil and caused elevated levels of As in rice grain (Meharg and Rahman 2002).

A major concern is for children who ingest soil contaminated with As. This is based on human or animal studies that have shown that ingested inorganic As salts are absorbed from the gastro-intestinal tract and enter the blood stream. The relative bioavailability of dosed sodium arsenate is 80% (Basta et al. 2001, Scheckel et al. 2009). This is a much higher rate than most other nonessential elements and, in part, is likely due to arsenate's (AsO_4^{3-}) chemical similarity to the essential nutrient, phosphate (PO_4^{3-}), that is readily absorbed through the gastric intestinal track of animals (Caussy and Priest 2008).

Remediation Strategies

There are potentially a number of approaches for remediating As-contaminated soils. One strategy would be physical removal of these soils to contained landfill sites. Although this may be appropriate for highly contaminated and localized pollution, it would be too expensive for As contamination over large areas or where the soil would need to be replaced for agricultural or landscape vegetation.

Phytoremediation is an *in situ* approach whereby hyperaccumulating As plants are grown and As is removed by harvesting plants. Disadvantages of this approach include the following: As accumulating plants must be locally adapted; requiring plant management and repeated harvests over an extended period; and disposal of As plant biomass, all of which limits the economic feasibility of this method.

An alternative As remediation approach is *in situ* volatilization of As. This involves As biotransformation by reduction and methylation of As in As-enriched soils. Microorganisms can release volatile derivatives of As (mainly arsines, mono-, di- or tri-methylarsine, and As oxides). This approach could be done as a stand-alone method or in tandem with phytoremediation of As-contaminated soils.

A concern of this approach is the dispersion of toxic forms of As in the atmosphere. However, it first should be noted, as mentioned earlier that As volatilization is a natural process. As gases are released from natural and anthropogenic sources that includes sewage treatment plants, lake and marine sediments, landfills, volcanic activity, and mining wastes (Hirner et al. 1994, Feldmann and Hirner 1995, Craig and Jenkins 2004, Cernansky et al. 2009). It has been estimated that 45,000 and 28,000 Mg As are released annually from natural and anthropogenic sources, respectively (Chilvers and Peterson 1987). It is thought that about 35% of the atmospheric As comes from soil volatilization of As (Chilvers and Peterson 1987).

Second, once volatile methylated As is in the air, it is rapidly oxidized and demethylated, and dispersed by the flow of air (Gao and Burau 1997, Pongratz 1998, Planer-Friedrich and Merkel 2006). This is important because mono- and dimethylated arsines being very genotoxic are rapidly converted to less toxic arsenate. And furthermore the levels released on a daily basis would be so small as to not change the equilibrium levels of As in the atmosphere from a given As remediation site.

Volatilization of As could be done by stimulating the native soil microbial community or by inoculating soils with specific microorganisms that accelerate this process (Huysmans and Frankenberger 1991, Thomas and Rhue 1997, Edvantoro et al. 2004).

Recent studies and reviews on As biomethylation, especially on biomethylation in contaminated soils, are limited. In fact, most of what is known comes from studies done with soils spiked with As in the lab with very little information about biomethylation from soils contaminated with As and then aged under field conditions.

Reported biomethylation rates vary greatly as do their experimental conditions. The large discrepancy is likely due to different experimental conditions, soil conditions, forms of As in soil, and varying microbial communities. The objective of this chapter is to review biomethylation rates from key biomethylation studies and the effect of experimental, environmental, and soil factors that affect the As biomethylation process.

Arsenic Methylation

Methylation Reactions and Pathways

The term biological methylation, typically referred to as biomethylation, describes the enzymatic transfer of a previously formed methyl group from donor atoms to acceptor atoms within living organisms. This reaction forms both volatile and non-volatile compounds of metals and metalloids, and it is an enzymatic transmethylation occurring in cells (Thayer 2002, Mohapatra et al. 2008). One possible result of introducing methyl groups onto acceptor atoms is the enhancement of their volatility and solubility in lipids, which usually decreases their solubility in water (Thayer 2002).

Arsenic methylation has been observed for a long time. Reports of mysterious poisonings from moldy wallpaper in damp rooms began as early as 1815 (Thom and Raper 1932). Later in the nineteenth century, people found the poisoning was related to As in coloring substances used as Scheele's green (cupric arsenite) and Schweinfurth's green (copper acetoarsenite) in wallpaper (Challenger et al. 1933, Woolson 1983).

Gasio was the first to link fungal metabolism to the formation of As gas. He isolated several fungi strains from an As-containing potato medium that produced a garlic-like odor gas which was named "Gasio Gas" after him. He suggested this gas was dimethylarsine (DMA) (Gasio 1893, 1901), while Challenger later showed that the "Gasio Gas" was actually trimethylarsine (TMA) (Challenger et al. 1933, Challenger 1945).

Challenger established the metabolic pathways for As methylation which came to be known as the Challenger's mechanism (Challenger 1945). These pathways are a series of reactions where there is a reduction of the pentavalent form of As that is followed by the oxidative addition of a methyl group (Dombrowski et al. 2005, Wang and Mulligan 2006). This results in As chemical species that are increasingly methylated as follows: methyl arsenite (MMA), dimethyl arsenate, dimethyl arsenite, and trimethyl arsine oxide which is shown in Figure 12.1. McBride and Wolfe (1971) went on to show that under anaerobic condition, bacteria can also do As methylation as well.

The biomethylation process requires a methyl-donor or precursor (Cullen et al. 1977, Cullen and Reimer 1989, Tamaki and Frankenberger 1992), and three major co-enzymes including (1) S-adenosylmethionine (SAM), the most important methylating agent, (2) N5-methyltetrahydrofolate derivatives, and (3) vitamin B12 derivatives (Fatoki 1997, Mohapatra et al. 2008).

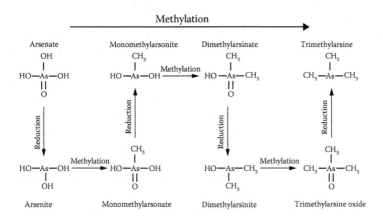

FIGURE 12.1
Revised Challenger Mechanism for arsenic methylation pathway. (Based on Wang, S. and Mulligan, C.N., *J. Hazard. Mater.*, 138, 459, 2006.)

Whereas bacteria under aerobic conditions reduction steps are mediated by glutathione and other thiol-containing compounds, anaerobic bacteria may use methylcobalamin as the electron donor (Krautler 1990, Stupperich 1993).

Biomethylation of As is thought to be a detoxification process as the methylated As products are less toxic than the inorganic As species, especially those most poisonous forms of arsenious acid (or its salts or esters) (Michalke et al. 2000, Bissen and Frimmel 2003). This is because inorganic As (III) compounds have a high affinity to sulfhydryl groups in proteins and can cause deactivation to enzymes. And inorganic As (V) can compete with phosphate in cell reactions, uncouple oxidative phosphorylation, and then impede the formation of high-energy adenosine triphosphate bonds (Bissen and Frimmel 2003).

Woolson (1983) summarized past research on As biomethylation and concluded methylation of As is a widely occurring natural process, which accounts for significant As losses from soil.

Microbial Genetic Controls and Energetics

Interestingly, bacteria have developed mechanisms to overcome the toxicity of arsenite and arsenate. This well may be why they have evolved to perform As redox reactions and volatilization.

Microorganisms can avoid the build up of intracellular arsenicals by either inhibiting uptake or actively exporting these ions (Rosenberg et al. 1977, Willsky et al. 1980, Moblev and Rosen 1982, Silver and Keacch 1982, Silver et al. 1989). The export mechanisms are probably the most important means by which microorganisms can resist As toxicity by the reduction of arsenate to arsenite which is eliminated from cells by a specific efflux pump (Silver 1996). Secondarily, they may avoid toxicity by chemical modification of As to relatively less toxic forms.

One reduction mechanism is the conversion of arsenate to arsenite which in turn is exported out of the bacterial cells. This reaction is controlled by the *ars* operon gene as a regulatory repressor, (ArsR), an arsenate reductase (ArsC), and an arsenite efflux pump (ArsB) (Xu et al. 1998). Silver (1998) and Suzuki et al. (1998) have reported that other genes in the *ars* operon extended the arrangement of this operon to be *arsRDABC*. ArsA is an ATPase that can provide ArsB energy while effluxing arsenite out of cells. ArsD is a found to be a As chaperone for the arsAB transporter. It helps transport arsenite to ArsA and thus increases the affinity of ArsA for arsenite (Lin et al. 2007).

Surprisingly, there is strong evidence that this reduction is more common under aerobic than anaerobic conditions. Macur et al. (2001) in column studies of mining tailings showed that only As (V)-reducing microorganisms could be cultivated from aerobic soils (Eh of 400 mV) and none from anaerobic columns. Analysis of 16SrDNA genes (As reducing genes) from isolates that reduced As (V) to As (III) under pure culture conditions had genes identical to that found in gene analysis of the total populations from the soils in the aerobic columns. Isolates that could reduce As (V) under aerobic conditions included *Caulobacter, Sphingomonas,* and *Rhizobium*-like isolates; and more specifically *Pseudomonas fluorescens, Pseudomonas aeruginosa,* and *Sphingomonas echinoids* (Macur et al. 2001).

The second reduction reaction is dissimilatory reduction which occurs under anaerobic conditions where arsenate is a terminal electron acceptor. In this case, As respiration reduction is done by membrane-bound arsenate reductase arrAB (Macy et al. 2000).

Arsenite can be methylated directly or oxidized to the less toxic arsenate prior to methylation (Oehme 1972, Coddington 1986, Ishinishi et al. 1986). Bacteria gain energy during the arsenite oxidation process where arsenite serves as an electron donor, together with the reduction of oxygen or nitrate (Anderson et al. 1992, Bryan et al. 2009).

Bacteria that accomplish oxidation have the *aox* gene that codes for arsenite oxidase (a DMSO reductase) (Ellis et al. 2001), which converts arsenite to arsenate (Anderson et al. 1992). AoxAB is a heterodimeric enzyme composed of a large subunit (AoxB) and small subunit (AoxA). This enzyme consists of a monomer of 86 kDa containing one molybdenum, five or six irons, and inorganic sulfide. (Anderson et al. 1992).

These results would suggest that soils potentially contain both As (V) reducing and As (III) oxidizing microorganisms. Indeed, this was shown by Macur et al. (2000) who used both culturing and molecular analysis (16SrDNA gene sequences) and found both As (III) oxidizing (*P. fluorescens, Variovorax paradoxus, Agrobacterium tumefaciens*) and As (V) reducing (*Flavobacterium heparinum, Agrobacterium tumefaciens, Microbacterium* sp., *Arthrobacter aurescens,* and *Arthbacter* sp.) bacteria in soils contaminated with As from geothermal activity.

As mentioned in the "methylation reactions and pathways" section, the general methylation pathway(s) is a sequence of reactions where the reduction of the pentavalent form of As is followed by a methyl group that is added oxidatively (Dombrowski et al. 2005). This in turn results in increasingly methylated As species (methyl arsenite, MMA; dimethyl arsenate, DMA-V; dimethyl arsenite, DMA-III; and trimethyl arsine oxide, MAO). The methylation reactions do require SAM as the source of methyl groups. All the SAM-dependent methylation steps are controlled by a distinct gene (arsM) that has been found in more than 120 prokaryotic species (and some archaea) and further characterized in *Rhodopseudomonas palustris* (Qin et al. 2006).

Factors Affecting Arsenic Biomethylation

Arsenic biomethylation is a microbial process and production of methylarsines requires a combination of enzyme activities and energy input

(Burau and Gao 1997). Studies have been conducted to find out the optimum conditions for As methylation, and both very small and very large amounts of As loss via methylation have been reported under various experimental conditions (Table 12.1). Many factors, including As sources, microorganism species, environmental conditions (soil moisture, temperature, pH, redox potential, substrates, etc.), are likely to affect As biomethylation rates.

Arsenic Sources

Forms, concentrations, and status (freshly added or residua) of arsenicals may have the potential to affect the As methylation process. Gao and Burau (1997) systemically assessed arsine evolution rates of four different As species. Under the same treatments, As loss followed the order: sodium cacodylate (dimethylarsenic acid; $(CH_3)_2AsO_2H$) (CA) > methanearsonic acid $(CH_3AsO(OH)_2)$ > sodium arsenite [As(III)] = sodium arsenate [As(V)] (Table 12.1). They suggested this was due to the different solubilities, toxicities, as well as methylation mechanisms related to these species. Their results were consistent with Woolson's finding that As loss from CA-contaminated soil was about 18%, while only 1% for As(V)-contaminated soil after 160 days aerobic incubation (Woolson 1977a, Table 12.1).

Cernansky et al. (2009) studied the effect of As valence on its biomethylation. Their results showed the average amount of volatilized As for all tested fungal strains ranged from 10% to 36% of the As added to pure cultures (Table 12.1). They concluded there was no significant difference between As (III) and As (V) on As biomethylation rates. They also claimed the initial As concentration was more important than As valence. Their experimental results showed initial As concentration of 5 or 20 mg/L resulted in As loss from 0.010 to 0.067 mg and 0.093 to 1.00 mg, respectively (Table 12.1). They suggested larger initial As concentration can lead to higher toxicity, which will result in more intensive microbial detoxification activities (Cernansky et al. 2007, 2009). However, Pearce et al. (1998) reported no correlation between initial As concentration and As loss rate under their test conditions.

Arsenic status will also affect methylation process. According to Edvantoro et al. (2004), fungal-induced arsine production was more pronounced ($\geq$2.2-fold) in soil freshly spiked with As than in long-term contaminated soil, and the authors proposed this may happen as the freshly spiked As is more readily bioavailable (Table 12.1). There is evidence that organic arsenicals are more readily volatized than inorganic forms (Braman 1975, Woolson 1979).

Microorganisms

Early work to identify As volatizing microorganisms was based on the presence of a garlic odor or known as "Gasio Gas" named after an Italian physician Bartolomeo Gosio who pioneered research on the microbiology of As

TABLE 12.1

Summary of Arsenic Biomethylation Rates under Different Methylating Conditions from Various Studies

Arsenic Source	Extent of Biomethylation[a] Yield (w)	Yield (%)	Microorganisms	Incubation Method	Environmental Conditions	Substrate	Incubation Time	Methylation Products Measured	References
100 mg As(V)/kg soil, newly spiked	0.001		No inoculation	Soil	Aerobic, 22°C, pH 6.9, 350 g H_2O/kg soil	Cellulose	70 days	Arsine	Burau and Gao (1997)
100 mg As(II)/kg soil, newly spiked	0.007		No inoculation	Soil	Aerobic, 22°C, pH 6.9, 350 g H_2O/kg soil	Cellulose	70 days	Arsine	Burau and Gao (1997)
100 mg MMAA/kg soil, newly spiked	0.021		No inoculation	Soil	Aerobic, 22°C, pH 6.9, 350 g H_2O/kg soil	Cellulose	70 days	Arsine	Burau and Gao (1997)
100 mg CA/kg soil, newly spiked	0.057		No inoculation	Soil	Aerobic, 22°C, pH 6.9, 350 g H_2O/kg soil	Cellulose	70 days	Arsine	Burau and Gao (1997)
10 mg Na_3AsO_4/kg soil, newly spiked	1		No inoculation		Aerobic, pH 5.3, 25%–35% field capacity	Ground soybean meal	160 days	DMA, TMA	Woolson (1977a)
10 mg Na_3AsO_4/kg soil, newly spiked	1.8		No inoculation		Anaerobic, pH 5.3, 25%–35% field capacity	Ground soybean meal	160 days	DMA, TMA	Woolson (1977a)

(continued)

TABLE 12.1 (continued)

Summary of Arsenic Biomethylation Rates under Different Methylating Conditions from Various Studies

Arsenic Source	Extent of Biomethylation[a]	Microorganisms	Incubation Method	Environmental Conditions	Substrate	Incubation Time	Methylation Products Measured	References
10 mg MSMA/kg soil, newly spiked	12.5	No inoculation		Aerobic, pH 5.3, 25%–35% field capacity	Ground soybean meal	160 days	DMA, TMA	Woolson (1977a)
10 mg MSMA/kg soil, newly spiked	0.8	No inoculation		Anaerobic, pH 5.3, 25%–35% field capacity	Ground soybean meal	160 days	DMA, TMA	Woolson (1977a)
10 mg CA/kg soil, newly spiked	18	No inoculation		Aerobic, pH 5.3, 25%–35% field capacity	Ground soybean meal	160 days	DMA, TMA	Woolson (1977a)
10 mg CA/kg soil, newly spiked	7.8	No inoculation		Anaerobic, pH 5.3, 25%–35% field capacity	Ground soybean meal	160 days	DMA, TMA	Woolson (1977a)
4 mg/L As(III), newly spiked	0.026–0.078 mg	Inoculated fungi		Aerobic, room temperature, pH 4		35 days	Total arsenic loss in soil	Cernansky et al. (2009)
17 mg/L As(III), newly spiked	0.143–0.257 mg	Inoculated fungi		Aerobic, room temperature, pH 4		35 days	Total arsenic loss in soil	Cernansky et al. (2009)
4 mg/L As(V), newly spiked	0.024–0.067 mg	Inoculated fungi		Aerobic, room temperature, pH 4		35 days	Total arsenic loss in soil	Cernansky et al. (2009)
17 mg/L As(V), newly spiked	0.093–0.191 mg	Inoculated fungi		Aerobic, room temperature, pH 4		35 days	Total arsenic loss in soil	Cernansky et al. (2009)

5mg/L As (V), newly spiked	0.010–0.067 mg	4–26.8	Inoculated fungi	Anaerobic, room temperature, pH 4		30 days	Total arsenic loss in soil	Cernansky et al. (2007)
20mg/L As(V), newly spiked	0.093–0.262 mg	9.3–26.2	Inoculated fungi	Anaerobic, room temperature, pH 4		30 days	Total arsenic loss in soil	Cernansky et al. (2007)
1390 mg As/kg soil, long contaminated	0.46 μg/kg soil		No inoculation	Aerobic, 25°C, pH 5–6, 75% field capacity	Cow manure	50 days	Arsine, MMA, DMA, and TMA	Edvantoro et al. (2004)
1390 mg As/kg soil, long contaminated	1.7 μg/kg soil		Inoculated fungi (*Penicillium* and *Ulocladium*)	Aerobic, 25°C, pH 5–6, 75% field capacity	Cow manure	50 days	Arsine, MMA, DMA, and TMA	Edvantoro et al. (2004)
1390 mg As/kg soil, newly spiked	3.8 μg/kg soil		Inoculated fungi (*Penicillium* and *Ulocladium*)	Aerobic, 25°C, pH 5–6, 75% field capacity	Cow manure	50 days	Arsine, MMA, DMA, and TMA	Edvantoro et al. (2004)
1390 mg As/kg soil, long contaminated		8.3	No inoculation	Anaerobic, 25°C, pH 5–6, 75% field capacity	Cow manure	5m	Total arsenic loss in soil	Edvantoro et al. (2004)
Synthetic CCA contaminated soil, 650mg As/kg soil	0	0	No inoculation	Frequently aerated, pH 7	Peat moss and poultry manure	40 days	MMAA, DMAA	Dobran and Zagury (2006)

(continued)

TABLE 12.1 (continued)

Summary of Arsenic Biomethylation Rates under Different Methylating Conditions from Various Studies

Arsenic Source	Extent of Biomethylation[a]	Microorganisms	Incubation Method	Environmental Conditions	Substrate	Incubation Time	Methylation Products Measured	References
30 mg/L Na$_2$HAsO$_4$, newly spiked	35	Inoculated methanogenic bacteria		Anaerobic, 35°C, pH 7	Cow dung and glucose	40 days	Volatilized As gas	Mohapatra et al. (2008)
CCA contaminated soils	0–0.17 mg/kg soil	Bacteria		Aerobic, room temperature, pH 7, darkness	Glucose	5 days	MMAA, DMAA	Turpeinen et al. (1999)
CCA contaminated soils	0	Bacteria		Anaerobic, room temperature, pH 7, darkness	Glucose	5 days	MMAA, DMAA	Turpeinen et al. (1999)
100 μg DSMA/g soil, newly spiked	2.2–11	No inoculation	Soil	pH 6.6, wet and moist condition	Ground corn seedlings	60 days	Total arsenic loss in soil	Atkins and Lewis (1976)

[a] Based on total volatile methylated arsenic detected over whole incubation period, unless otherwise specified. CA, cacodylic acid; MMA, monomethylarsine; DMA, dimethylarsine; TMA, dimethylarsine; MSMA, monosodium methane-arsonate; MMA, MMAA, monomethylarsonic acid; DMA, dimethylarsine; TMA, dimethylarsine; MSMA, monosodium methane-arsonate; MMA, monomethylarsine; CCA, chromated-copper arsenate; DMAA, dimethylarsinic acid; DSMA, disodium methane-arsonate.

volatilization in the late 1800s (Bentley and Chasteen 2002). Although As volatilizing organisms identified by this method were later substantiated, we only report the modern research when analytical methods allowed the measurement of individual As species evolved by microorganisms. Furthermore, we are limiting our review to As volatizing organisms found in the environment (soil, water, and sewage materials) and not those reported in mammalian intestines, feces or urine (see Bentley and Chasteen 2002).

In soils, thus far there is no evidence that there is chemical volatilization of As compounds. This was shown by Atkins and Lewis (1976) who found no As volatilization from steam-sterilized soils amended with [74]As labeled disodium methanearsonate compared to non-sterilized soils that lost up to 11% of the added As over a 60 day incubation.

A large consortium of natural, indigenous microorganisms are able to convert inorganic As to volatile alkyl As derivatives via biomethylation. Aerobic and anaerobic bacteria and archaea (Table 12.2) and microscopic fungi (Table 12.3) are responsible for this process (Cullen and Reimer 1989, Michalke et al. 2000, Bentley and Chasteen 2002, Meyer et al. 2007). These As

TABLE 12.2

Bacterial Species Shown to Volatilize As, MMAA, DMAA, MMA, DMA, TMA

Bacterial Species	Transformation	References
Methanobacterium strain MOH ATCC 33272	As (V) → As (III) → MMAA→ DMAA → DMA (anaerobic)	McBride and Wolfe (1971)
Flavobacterium, Cytophaga sp.	DMAA → TMAO (aerobic) → TMA (aerobic)	Chau and Wong (1978) Honschopp et al. (1996)
Alcaligenes sp.	As (V), As (III), MMAA DMAA → Arsine (aerobic)	Cheng and Focht (1979)
Pseudomonas sp.	As (V) → MMA, DMA, TMA (aerobic)	Shariatpanahi et al. (1981), Cheng and Focht (1979)
Corynebacterium sp., *E. coli*, *Proteus* sp.	As (V) → DMA	Shariatpanahi et al. (1981)
Nocardia sp., *Achromobacter* sp., *Aeromonas* sp.	As (V) → MMA, DMA	Shariatpanahi et al. (1983)
Strict anaerobic Gram-positive strain (ASI-1) (high similarity with *Clostridium glycolicum* based on 16S rDNA sequencing)	As (V)→ AsH$_3$, MMAs, DMAs, TMAs AsX (anaerobic)	Meyer et al. (2007)
R. palustris	As (V) → DMA, TMA	Qin et al. (2006)
Clostridium collagenovorans	As (V) → TMA	Michalke et al. (2000)
Archaea		
Methanobacterium formicicum,	As (V) → MMA, DMA, TMA	Michalke et al. (2000)
Methanosarcina barkeri, Methanobacterium thermoautotrophicum	As (V) → AsH$_3$	Michalke et al. (2000)

TABLE 12.3

Fungal Species Shown to Volatilize As

Fungal Species	Transformation	References
C. humicola ATCC 26699	CCA → TMA TMA oxide → TMA	Cullen et al. (1984), Cox and Alexander (1973b), Picket et al. (1981)
Scopulariopsis brevicaulis (*Penicillium brevicaule*) ATCC 7903	As (III), MMAA, DMAA → TMA	Huysmans and Frankenberger (1991), Frankenberger (1998), Pearce et al. (1998), Andrewes et al. (2000), Challenger (1945)
Phaeolus schweinitzii ATCC 10013	As (III) → total volatile As compounds	Pearce et al. (1998)
Gliocladium roseum ATCC 10521	MMAA, DMAA → TMA	Cox and Alexander (1973)
Neosartorya fischeri, Aspergillus clavatus, A. niger	As (III), As (V) → total volatile As compounds	Cernansky et al. (2009)
Aspergillus clavatus, A. niger, Trichoderma viride, Penicillium glabrum	As (V) → total volatile As compounds	Cernansky et al. (2007)
Penicillium sp., *Ulocladium* sp. (inoculated to soil)	As (unknown As species in As polluted soil plus DDT) → total volatile As compounds	Edvantoro et al. (2004)

biotransforming organisms have been found in diverse environments: soils (Cheng and Focht 1979, Cernansky et al. 2007, 2009, Meyer et al. 2007), sewage sludge (Cox and Alexander 1973a, Michalke et al. 2000), fresh water sediments (McBride et al. 1978, Bright et al. 1994), and composts (McBride et al. 1978).

Bacteria and Archaea

The early research of Gosio and even as late as that of Fredrick Challenger was focused on fungi. Indeed, Challenger and Bach proposed that bacteria were unable to volatilize As (Bach 1945, Challenger 1945). Earlier Puntoni (1917) detected a garlic-like odor (indicating As volatilization) from several bacteria species exposed to DMA or arsenate that had been isolated from fecal material. But in a later study, Challenger and Higginbottom (1935) were unable to show volatilization from these same species culture collection. Such outcomes are the reason why Challenger did not support the idea that bacteria could perform As volatilization. Assuming these bacteria that Challenger used were the same species as those used by Puntoni, it can be inferred that there may be only certain strains of a given bacterial species that can do As volatilization.

The most widely studied group of microorganisms relative to As volatilization are methanogenic bacteria. Under anaerobic conditions, methanogenic bacteria can convert inorganic As into DMA, which is very stable in the absence of oxygen (McBride and Wolfe 1971, Cullen et al. 1977, Woolson 1983, Bentley and Chasteen 2002) (Table 12.2). Methanogens are morphologically diverse and are found in large numbers in biosolids, composts, and freshwater sediments (McBride et al. 1978). At least one species of *Methanobacterium* can form DMA from arsenate, arsenite, and methanearsonic acid substrates (McBride et al. 1978) with biomethylation being completely inhibited by heat treatment of *Methanobacterium* cells (McBride and Wolf 1971). Michalke et al. (2000) with isolates from sewage sludge showed one bacterial species (*Clostridium collagenovorans*) and several methanogenic archaea species (Table 12.2) could volatilize arsenate to methylated As forms (AsH₃, MMA, DMA, TMA) under anaerobic conditions. They found the methanogenic archaea, *Methanobacterium formicicum*, to be the most efficient at As volatilization of the anaerobic microorganism tested.

Early work by Cheng and Focht (1979) showed that *Pseudomonas* sp. and an *Alcaligenes* sp. isolated from soil produced only arsine from arsenate or arsenite when incubated anaerobically. The bacteria *Corynebacterium* sp., *Escherichia coli*, *Flavobacterium* sp., *Proteus* sp., and *Pseudomonas* sp. isolated by Shariatpanahi et al. (1981) from the environment produced DMA from the pesticide, sodium arsenate. But only *Pseudomonas* sp. formed MMA and TMA. MMA and DMA were produced from the organic pesticide methylarsonate by *Achromobacter* sp., *Aeromonas* sp., *Alcaligenes* sp., *Flavobacterium* sp., *Nocardia* sp., and *Pseudomonas with Norcardi* sp. also producing TMA (Shariatpanahi et al. 1983). Others have been found that TMA oxide can be converted to TMA by *Fusobacterium nucleatum*, *Veillonella alcalescens*, and *Staphylococcus aureus* by Pickett et al. (1988, 1981). Honshopp et al. (1996) isolated *Flavobacterium-Cytophaga* groups from soils that only produced volatile TMA.

Michalke et al. (2000) found that *Clostridium collagenovorans*, *Desulfovibrio vulgaris*, and *Desulfovibrio gigas* formed detectable levels of TMA. They also showed that methanogenic archaea can produce arsine (*Methanobacterium formicium*) and MMA, DMA, and TMA (*Methanosarcina barkeri*, *Methanobacterium thermoautotrophicum*).

The bacteria isolated thus far that can methylate organic and inorganic forms of As aerobically have greater diversity than those isolated from anaerobic environments (Frankenberger and Huysmans 1991). A cross section of bacteria that can volatilize As under aerobic conditions are shown in Table 12.2.

The report by Islam et al. (2007) provides an estimate of the populations of bacteria that can volatize As in soils. They found that As methylating bacteria (AsMB) from ten As-contaminated sites in Bangladesh had from 0.2×10^4 to 7.8×10^4 AsMB (MPN)/kg dry soil.

It worth noting that As volatilization can proceed under anaerobic conditions. One of the first reports was by McBride and Wolfe (1971) who found *Methanobacterium bryantii* produced dimethylarsine. Isolates from various anaerobic environments that have been shown to volatilize As includes *Desulfovibrio vulgaris* (McBride and Wolfe 1971), *Methanobacterium thermoautotrophicum* (Cullen et al. 1989), *Methanobacterium formicicum* (Michalke et al. 2000), *Methanosarcina barkeri* (Michalke et al. 2000), and *Serratia marinorubra* (marine facultative anaerobe) (Vidal and Vidal 1980). Some of those identified as As volatilizers are sulfur reducers (e.g., *Methanobacterium formicicum*) (Bright et al. 1994, Michalke et al. 2000).

Meyer et al. (2007) isolated a strict anaerobe Gram-positive strain (ASI-1) from a soil with low levels of As contamination that under anaerobic and pure culture conditions produced volatile compounds as methylated As species and arsine (Table 12.2). It showed high similarity to *Clostridium glycolicum* (based on 16S rDNA sequencing) but interestingly this known species did not produce any detectable levels volatile As. From fluorescence *in situ* hybridization, it was estimated that strain ASI-1 was 2% of the total microbial flora of this soil. From this the authors suggested that this strain represented a dominant member of the metal(loid) volatilizing population in this soil. Additionally they showed ASI-1 grows fermentatively in presence of yeast extract (1%) and uses the organic compounds glucose, maltose, sucrose, mannose ribose, sorbitol, glycerol, and starch. Besides the fact that this research was on an anaerobe, it is illustrative of the type of work that should be done on a wider range of soils and under anaerobic conditions to manage soils for natural attenuation of As-contaminated soils. This approach shows the potential to develop such systems based on the knowledge of the organisms and their physiological requirements, to develop targeted substrates and optimal conditions to promote As volatilization.

An interesting observation is that bacteria adapted for biomethylation of As or resistance to As, apparently can grow better in the presence of As than without As up to some As threshold level. This was shown by Honshopp et al. (1996) who compared uptake of As by two soil bacteria (one that methylates As and the other exhibited As resistance) and found growth rates increased up to $200\mu g/mL$ in the growth medium. As expected *Flavobacterium-Cytophaga* sp. which exhibits biomethylation accumulated significantly less As in its biomass than the As resistant species.

Fungi

Fungi were the first microorganisms to be associated with As volatilization. Although there were earlier reports of fungal involvement in As volatilization, it was not until Gosio in the late 1800s did systematic studies that unraveled the mystery of As poisoning and found fungi were conclusively implicated in this phenomenon. He showed the poisonings were As gas released from wallpaper that contained arsenical pigments for coloring purposes (Agrifoglio 1954). Specifically he determined that *Penicillium brevicaule* (now called *Scopulariopsis brevicaulis* (Sacc.) could form an arsenical gas that killed rats (Gasio 1892, 1893). He implicated other fungal species but at this time the exact chemical nature of these poisonous arsenical were not known (Agrifoglio 1954).

With the advent of improved chemical methods for detecting methylated forms of As, Fredrick Challenger in the 1940s was the first to link fungal species to a specific arsenical gas, TMA. He and coworkers at the University of Leeds found low levels of TMA emitting from pure cultures of *Aspergillus glaucus* or *A. versicolor*. They further showed that these organisms and *Penicillium notatum*, *P. chrysogenum*, and two strains of *Aspergillus niger* could convert DMA to TMA (Bird et al. 1948). A number of fungal species were grown on As_2O and were indirectly associated with As volatilization by producing garlic-like odors which included *Lenzites trabea*, *Lenzites saepiaria*, *Phaaeolus schweinizii*, and *Tricohphyton rubrum* when grown in the presence of As (Zussman et al. 1975, Barret 1978, Pearce et al. 1998).

Other fungi species, associated with As biomethylation, have been isolated from various environmental samples. With the widespread use of CCA as wood preservative since the 1970s until recently, there has been interest in degradation of treated wood and release of As. Cullen et al. (1984) found that *C. humiculus* growing on preserved wood could release TMA. DMA was converted to TMA by three fungi isolated from sewage sludge (*Gliocladium roseum*, *Penicillium* sp., and *Candida humicola*, later reclassified as *Apiotrichum humicola* by Cullen et al. (1995) (Cox and Alexander 1973).

Frankenberger and coworkers have done in-depth studies of a *Penicillium* species isolated from an agricultural evaporation pond. Environmental conditions that promoted TMA release by this organism are discussed in detail later in this chapter.

A species of *Fusarium* was isolated from soil contaminated with a cattle medicinal solution dip (As trioxide) more than 30 years after the contamination (Thomas and Rhue 1997). In pure culture this organism released arsenicals of unknown chemical structure.

Cernansky et al. (2009) investigated the potential of three microscopic filamentous fungi strains isolated from soil samples collected from a mining site highly contaminated with As. In pure culture the fungi released arsenicals from As (III) or As (V) in the following order: *Neosartorya fischeri* > *Aspergillus clavatus* > *A. niger* which averaged across all fungi was 24% and 16% for As (V) addition of 4 and 17 mg/L, respectively (similar results were found for As III). Interestingly, even though As (III) is considered to be more toxic than As (V), there were minimal differences in fungal biomass or As volatilization. A more important factor was As concentration which on a percentage basis was higher at the lower concentration. However, this stands in contrast to Pearce et al. (1998) who found no effect of initial As concentration on As volatilization by fungi.

Table 12.3 presents fungi identified by Cernansky et al. (2007) that could volatilize As. An important finding of this research was that the amount of As volatilized was higher for the *A. niger* strains isolated from the uncontaminated than those from a contaminated soil. This may be due to the strain isolated from the contaminated soil having made adaptations to reduce toxicity by enhancing melanin production, which binds As (Fogarty and Tobin 1996), or by enzymatic intracellular transformation of As—all of which immobilizes As rather than allowing for biomethylation and volatile losses of As.

Of the fungi studied so far it seems *C. humicola* may be one of the most important and versatile for volatilizing As. It can use arsenate or arsenite as a substrate to produce TMA (Cullen et al. 1979). Additionally it can methylate benzenearsonic acid, methylphenylarsinic acid, and dimethylphenylarsine oxide to the volatile dimethylphenylarsine (Cullen et al. 1983). And as mentioned earlier it can deplete As from wood impregnated with the preservative CCA (Cullens et al. 1984).

Microbial Inoculation of Soils

Inoculation of soil with As methylating microorganisms can enhance biomethylation rates. Edvantoro et al. (2004) reported arsine evolution rates for inoculated (*Penicillium* sp. and *Ulocladium* sp.) to long-term As-contaminated and freshly As spiked soils were approximately 3.7 and 8.3 orders of magnitude greater than uninoculated soils (Table 12.1).

Pearce et al. (1998) found fungus isolated from uncontaminated substrate was more capable of methylating As than fungus isolated from As-contaminated substrate. They attributed this finding to the fact that fungi isolated from contaminated soil may have a larger ability than fungi from uncontaminated soil to produce melanin, a compound that may bind As and decrease As uptake (Fogarty and Tobin 1996). Similar results were found by Cernansky et al. (2007) who showed that *A. niger* (B) strains isolated from uncontaminated substrates had higher ability to volatilize As than the strain from an uncontaminated substrate. They suggested that the strain isolated from contaminated substrate may shift toward As absorbance and tolerance over volatilization.

Though conversion rates varied, many previous studies showed most extractable soil As can be transferred to the atmosphere by this microbial process (Table 12.1).

Environmental Conditions

Soil Moisture and Temperature

It stands to reason that favorable moisture and temperature regimes would enhance a biologically driven process such as As volatilization. Indeed, Woolson (1977b) concluded that the environmental factors of adequate moisture and warm temperatures that promote microbial activity in combination with high organic matter enhance biomethylation. Gao and Burau (1997) proposed that warmer temperatures and an optimum soil moisture between 250 and 350 g H_2O/kg soil can enhance arsine evolution.

Edvantoro et al. (2004) found lower rates of As loss with a soil moisture of 75% of field capacity than a dryer soil of 35%. In this case over 5 month incubation, the higher soil moisture content had only about 4% total As losses compared to 6%–8% at the lower water content of two soils amended with 30% animal manure.

pH

Frankenberger and Huysmans (1991) evaluated conditions most suitable for TMA production using a *Penicillium* species isolated from an evaporation pond in California. They suggested pH 5–6, temperatures around 20°C, and phosphate concentration of 0.1–50 mM was optimal for As volatilization.

Cox and Alexander (1973) tested several pH conditions for TMA production by *C. humicola*, indicating the optimum pH is 5.0. In sediments, Baker et al. (1983) found that As methylation was optimal at pH 3.5–5.5 which was attributed to acidification causing a release of As that was biologically available for As volatilization. This mechanism for this optimum pH was demonstrated by Signes-Pastor et al. (2007) who found that the lowest pH of their experiment (pH 5.5) had the highest soluble As which would make it more biologically available.

Redox Potential

The results on the effects of anaerobic versus aerobic on As volatilization are mixed. However, as might be expected fungi, yeasts, and a wide range of species for bacteria dominate under oxic conditions, whereas under anaerobic conditions a more narrowly focused set of bacteria, primarily methanogenic bacteria or archaea drive volatilization of As (Boyle and Jonasson 1973, Weinberg 1979).

Woolson (1977a) added isotope labeled As into soils and studied As methylation under both aerobic and anaerobic conditions for 160 days. Results indicated volatile As compounds formed most quickly under aerobic conditions, and the largest As loss (18%) was in aerobically treated soil. Similar findings were provided by Edvantoro et al. (2004) and Turpeinen et al. (1999), but all reported methylation rates in aerobic conditions were not more than 18% (Table 12.1).

Contrary to these findings, Atkins and Lewis (1976) found that 8.1% of the ^{74}As added to soil was lost over a 60 day incubation under anaerobic conditions compared to 2.2% losses under aerobic conditions. Similarly, Woolson and Kearney (1973) showed that As losses from sodium cacodylate added to soil was 35% under aerobic conditions whereas under anaerobic conditions there was a 61% loss. Interestingly, the form of As that volatilized was different due to redox potential, with organic arsenicals dominating under aerobic conditions whereas TMA dominated under aerobic conditions.

Thomas and Rhue (1997) showed there was an interaction between a readily available energy source (glucose) and redox conditions. They found that under more aerobic conditions, addition of glucose had no effect on As volatilization compared to anoxic conditions when glucose stimulated As volatilization.

These somewhat conflicting results between aerobic and anaerobic conditions might be explained by the lack of control or accounting for the exact level of redox potential in previous experiments (e.g., anaerobic vs. aerobic but no measurement of the actual redox potential across treatments). A potential insight for the role of redox potential on biomethylation may be related to the solubility or bioavailability of As as a function of Eh. Signes-Pastor et al. (2007) showed that As was most soluble under moderately reducing conditions (0 to −150 mV) due to dissolution of iron oxy-hydroxides. Interestingly upon reduction to −250 mV, As solubility decreased and was controlled by the formation of insoluble sulfides. This is important because if redox potential controls soluble As (biologically available), it therefore also controls biomethylation and rates of As volatilization.

Although most of the research on specific microbial anaerobic isolates (as described previously) has been done in relation to anaerobic sewage sludge or in pure cultures (see review by Bentley and Chasteen [2002]), anaerobic bacteria could be important in remediating As contaminated. For example, when local conditions allow, it might be possible to flood As-contaminated sites or to have pulsing systems that result in fluctuating redox potential to drive anaerobic As.

Indeed, fluctuating between low and high redox potential may be important in making As more biologically available. This was shown by Frohne et al. (2010) and others (Masscheleyn et al. 1991, Chiu et al. 1998, Mitsunobu et al. 2006) that soluble As (as III species) increased with decreasing redox potential. The mechanism is that under low redox potential, Fe (hydr)oxides are reduced which results in concurrent release of associated As (Mitsunobu et al. 2006). Thus, one could imagine subjecting soils to reducing conditions to increase As in soil solution followed by induction of aerobic conditions that would be a more favorable environment for fungi which typically do not thrive under anaerobic condition. But stimulated under aerobic conditions, fungi can be dominant players in driving As volatilization.

Non-Arsenic Substrates

Soil Amendments

Nutritional support of microorganisms has been shown to be related to As biomethylation. This was first supported by Atkins and Lewis (1976) who investigated As release from soils that had a range of soil organic matter levels. In this case the highest level of release was 11% of the isotopically labeled As soil (100 μg As/g) in the soil with the highest organic matter content (11%) after 60 days. From this it would follow that addition of organic inputs should create a larger and more active microbial community to transform soil As in to volatile forms. Indeed, Dobran and Zagury (2006) who found As methylated products were not found in soils with limited nutrients available for microbial growth.

Subsequent studies on naturally occurring substrates such as animal manure have provided evidence that these inputs can affect As volatilization. Additions of fresh manure to soil can increase the bioavailability of As species that appear to be linked to higher rates of volatilization (Walker et al. 2004). Mohapatra et al. (2008) studied the effect of cow dung on As volatilization of methanogenic bacteria in an anaerobic digester. They reported a surprisingly large methylation yield of 35% under the most suitable conditions (Table 12.1). Moreover, the volatilization rates decreased when rates of manure inputs reached a threshold, which they attributed to the formation of methane that could inhibit the formation of volatile As and led to the production of non-volatile compounds. Similarly, Edvantoro et al. (2004) added manure and urine to soils contaminated by As and DDT. Their 5 month experiments demonstrated that increasing manure applications caused more As losses.

Investigations of model compounds, most often on pure cultures, have been done to determine the mechanisms that drive As volatilization. Turpeinen et al. (2002) showed that a very simple carbon source (0.2% glucose) did not stimulate As methylation or induce the reduction of soluble As in soils. Conversely, Gao and Burau (1997) showed that cellulose, a more complex and recalcitrant compound, stimulated the soil microbial community which resulted in elevated evolution of volatile As. There was a linear relationship ($R^2 = 0.99$) between cellulose application rate (ranging from 0% to 5% wet wt.) and volatile arsine production in the soil freshly spiked with 10 mg inorganic As/kg soil over a 70 day incubation. However, in the same study, the addition of cellulose to soil containing organic As as sodium cacodylate caused a decrease in demethylation of CA. The authors suggested the microorganism preferred readily released C from cellulose over that in cacodlyate.

In pure culture, Huysmans and Frankenberger (1991) found that a *Penicillium* sp. was stimulated to convert MMAA from TMA when amino acids (phenylalanine, isoleucine, and glutamine) were added to pure culture media. Conversely, this same study showed that carbohydrates and sugar acids suppressed this process.

Incubation Time and Rates of As Volatilization

Reported incubation times for As biomethylation varies from days to months (Table 12.1). Rates of As volatilization varied considerably across various studies. This is likely to the highly variable environmental conditions, types of organic substrates that may or may not have been used, soil types, and other unknown factors.

In general, anaerobic incubation times used in published studies tend to be longer than aerobic incubation. Mohapatra et al. (2008) studied the rate of

As removal as arsenical gases via methanogenic bacteria under anaerobic conditions. The initial rate of As volatilization was large, which suggested that the microorganisms were well adapted to grow in the presence of As. At later stages of incubation, the biomethylation rate was reduced corresponding to the gradual decrease of substrate. The biomethylation rate was a first-order linear kinetics up to 40 days and thereafter biomethylation rate was negligible, likely due to the lack of substrate or auto poisoning.

Some experiments have reported relatively low rates of As volatilization. This includes Rodriguez (1998) who found only 0.0005% of total As content was lost over a 20 week period. Similarly, Turpeinen et al. (2002) under both aerobic and anaerobic conditions showed that of the arsenate added to soil, only 0.02%–0.3% As was lost as volatile TMA over a 10 day lab incubation. They reported that these results were similar to what they found in the field in soil contaminated by industrial As pollution from a wood As impregnating factory. Islam et al. (2007) found that in soil contaminated naturally with As from geologically contaminated water lost 0.4% of the total As over a 100 day incubation.

Gao and Burau (1997) found that As evolved as arsines was <1% of the incorporated As after a 70 day incubation period for all treatments in this study (<0.01%/day). The maximum fractional release over 70 days of added As as arsines (0.4%) occurred in the soil treated with 10 mg As/kg soil as sodium cacodylate (CA), plus 5% cellulose and where the system was maintained at −0.03 MPa.

The aforementioned volatilization rates are much lower than other reports in the literature. Values reported by Woolson (1977b) were in the range of 0.033%–0.386%/day for CA volatilization as alkylarsines. Woolson (1977a) observed an 18% loss from a CA-amended soil after 160 days when flushing the system with air. The soil had a pH of 5.3, a loam texture and the soil moisture was maintained at 25%–35% of field capacity. Addition of animal manure to soil (30% wet wt.) with moisture at 75% of field capacity resulted in 8.3% of the As being volatilized over a 5 month incubation (Edvantoro et al. 2004)

A factor to consider for rates of and transformations of As volatilization is the "aging" of As that occurs in soils after soils receive As compounds. Onken and Adriano (1997) showed that As volatilization rate from sodium arsenate or sodium arsenite added to soil, declined to zero after 68 days. They explained these results were due to the conversion of bioavailable forms to insoluble As compounds and this happened in both saturated and subsaturated soils.

Anions and Trace Elements

Phosphate and arsenate are chemical analogs. Because of the similarity between these two anions, competition between phosphate and arsenate uptake by microorganisms is likely to occur (Kertulis et al. 2004).

Amendment of soils with phosphate could potentially be a means of displacing As from solid phases and thus making it more biologically available for biomethylation. Indeed, Signes-Pastor et al. (2007) showed that with increased rates of P addition there were increased levels of soluble As. Since phosphate fertilizers are routinely added to soil, this amendment might be a practical means for increasing As volatilization from As-contaminated soils. Unfortunately, even though phosphate might increase the uptake of As by microorganism, because of its similar chemical properties to phosphate, phosphate interferes with the biomethylation.

This was shown by Cox and Alexander (1973) who studied phosphate effect on TMA formation by fungi, *C. humicola*. They found phosphate inhibits TMA formation from arsenite, arsenate, and monomethylarsonate, but not from dimethylarsinate. Hyperphosphite will lead to temporary reduction or even a cessation of TMA evolution and phosphite suppresses TMA production from monomethylarsonate. They postulated phosphate blocks TMA formation at the stage between the mono- and dimethylarsenic compounds, which then inhibits biomethylation. Meanwhile, their research showed that high concentrations of antimonite, selenite, selenate, and tellurate affected conversion rate of arsenate while nitrate had no effect.

Frankenberger (1998) reported novel results on the effect of trace elements on As volatilization. He tested 21 trace elements to determine their inhibition or activation on As biomethylation by *Penicillium* sp. Results showed at 10 or 100 μM, only four trace elements, Co, Fe, Al, and As (III), stimulated volatilization. At the higher concentration of 1000 μM, 17 trace elements completely inhibited biomethylation. Only As(III) As (V), Se(IV) and Se(VI) didn't show complete inhibition and As(III) surprisingly enhanced As volatilization by the magnitude of 3.48-fold. Conversely, at low concentrations, Al, Cu, Fe, Hg, Se, and Zn stimulated biomethylation (Frankenberger 1998).

Perspectives

Biomethylation may be a useful, natural attenuation method for bioremediation of soils and water contaminated by heavy metals and metalloids (Frankenberger et al. 1991). Volatilization via methylation could be used to remove As contamination from soil and water. Biomethylation as a As bioremediation technology has many advantages over other remediation approaches. First, As biomethylation is a naturally occurring process and there is no need to inoculate specific microorganisms to contaminated soils for As removal (Cernansky et al. 2007). Second, although the arsine conversion rate in soils are considered to be low to moderate, especially in those soils that have been contaminated for a long time (Edvantoro et al. 2004), this process is continuous. This is especially true when environmental factors affecting biomethylation have been optimized.

Moreover, due to the slow rate of arsine evolution, it is unlikely for the released biomethylated arsenical gases to cause health issues. To date there is not any evidence that volatile As formed by biomethylation would lead to any environmental risk (Edvantoro et al. 2004). Some argue the final As methylation products could be more toxic to human and animals because they can cause stronger DNA damages than inorganic As forms (Dopp et al. 2004). However, the concentration of methylated arsenics released to atmosphere would be much lower than the concentration that could lead to DNA damages.

In short, As biomethylation has potential as an inexpensive and feasible method for bioremediation of As-contaminated soils. However, recent studies on As biomethylation are limited to laboratory studies and most are using artificially contaminated soils rather than soils from polluted sites. Furthermore, most researchers evaluated this process with soil enriched with As as the sole contaminant (Edvantoro et al. 2004). Thus, such results may fail to represent the real situation of As biomethylation in soils that are typically contaminated with multiple inorganic and/or organic chemical contaminants. Many studies do not quantify the potential of biomethylation to remove As from polluted soil. Basic and applied research is needed to develop biomethylation as a viable bioremediation technology for As-contaminated soils.

References

Abernathy CO, Liu YP, Longfellow D, Aposhian HV, Beck B, Fowler B, Goyer R et al. 1999. Arsenic: Health effects, mechanisms of actions, and research issues. *Environmental Health Perspectives* 107: 593–597.

Agency for Toxic Substances and Diseases. 2010. The priority list of hazardous substances that will be the subject of toxicological profiles. http://www.atsdr.cdc. gov/cercla/

Agrifoglio L. 1954. Igienisti italiani degli ultimi cento anni, Milano, *Hoepli* 47: 131–135.

Anderson GL, Williams J, Hille R. 1992. The purification and characterization of arsenite oxidase from *Alcaligenes faecalis*, a molybdenum-containing hydroxylase. *The Journal of Biological Chemistry* 267: 23674–23682.

Andrewes P, Cullen WR, Polishchuk E. 2000. Antimony biomethylation by *Scopulariopsis brevicaulis*: Characterization of intermediates and the methyl donor. *Chemosphere* 41: 1717–1725.

Atkins MB, Lewis RJ. 1976. Chemical distribution and gaseous evolution of arsenic-74 added to soils as DMSA-74 Arsenic. *Soil Science Society of America* 40: 655–658.

Bach SJ. 1945. Biological methylation. *Biological Reviews of the Cambridge Philosophical Society* 20: 158–176.

Baker MD, Inniss WE, Mayfield CI, Wong PTS, Chau YK. 1983. Effect of pH on the methylation of mercury and arsenic by sediment microorganisms. *Environmental Technology Letters* 4: 89–100.

Barret DK. 1978. An improved selective medium for isolation of *Phaeolus schweinitzii. Transactions of the British Mycological Society* 71: 507–508.

Basta NT, Rodriguez RR, Casteel SW. 2001. Bioavailability and risk of arsenic exposure by the soil ingestion pathway. In W.T. Frankenberger (ed.) *Environmental Chemistry of Arsenic.* Marcel Dekker, Inc., New York.

Bentley R, Chasteen TG. 2002. Microbial methylation of metalloids: Arsenic, antimony, and bismuth. *Microbiology and Molecular Biology Reviews* 66: 250–271.

Bird ML, Challenger F, Charlton PT, Smith JO. 1948. Studies on the biological methylation. 11. The action of moulds on inorganic and organic compounds of arsenic. *Biochemical Journal* 43: 73–83.

Bissen M, Frimmel FH. 2003. Arsenic—A review. Part I: Occurrence, toxicity, speciation, mobility. *Acta hydrochimica et hydrobiologica* 31: 9–18.

Bowen GD. 1979. Integrated and experimental approaches to the study of growth of organisms around roots. In: B. Schippers, W. Gams (eds.), *Soil-borne Plant Pathogens.* Academic Press, London, pp. 207–227.

Boyle RW, Jonasson IR. 1973. The geochemistry of arsenic and its use as an indicator element in geochemical prospecting. *Journal of Geochemical Exploration* 2: 251–296.

Braman RS. 1975. Arsenic in the environment. In E.A. Woolson (ed.) *Arsenical Pesticides.* ACS Symposium Series 7, American Chemical Society, Washington, DC, pp. 108–123.

Bright DA, Brock S, Cullen WR, Hewitt GM, Jafaar J, Reimer KJ. 1994. Methylation of arsenic by anaerobic microbial consortia isolated from a lake sediment. *Applied Organometallic Chemistry* 8: 415–422.

Bryan CG, Marchal M, Battaglia-Brunet F, Kugler V, Lemaitre-Guillier C, Lievremont D, Bertin PN, Arsene-Ploetze F. 2009. Carbon and arsenic metabolism in Thiomonas strains: Differences revealed diverse adaptation processes. *BMC Microbiology* 9: 127S.

Burau RG, Gao S. 1997. Environmental factors affecting rates of arsine evolution from and mineralization of arsenicals in soil. *Journal of Environmental Quality* 26: 753–763.

Caussy H, Priest N. 2008. Introduction to arsenic contamination and health risk assessment with special reference to Bangladesh. *Reviews of Environmental Contamination and Toxicology* 197: 1–16.

Cernansky S, Kolencik M, Sevc J, Urik M, Hiller E. 2009. Fungal volatilization of trivalent and pentavalent arsenic under laboratory conditions. *Bioresource Technology* 100: 1037–1040.

Cernansky S, Urík M, Ševc J, Hiller E. 2007. Biosorption of arsenic and cadmium from aqueous solutions. *African Journal of Biotechnology* 6: 1932–1934.

Challenger FC. 1945. Biological methylation. *Chemical Reviews* 36: 315–361.

Challenger FC, Higginbottom C, Ellis L. 1933. The formation of organo-metalloidal compounds by microorganisms. Part I. Trimethylarsine and dimethylarsine. *Journal of the Chemical Society* 95–101.

Cheng CN, Focht DD. 1979. Production of arsine and methylarsines in soil and in culture. *Applied and Environmental Microbiology* 38: 494–498.

Chilvers DC, Peterson PJ. 1987. Global cycling of arsenic. In: T.C. Hutchinson, K.M. Meema (eds.), *Lead, Mercury, Cadmium and Arsenic in the Environment.* John Wiley, New York, pp. 279–301.

Chiu CH, Li GC, Young CC, Chiou RY. 1998. A study on the transformation of arsenite in soils after poultry compost and poultry manure application. *Journal of the Chinese Agricultural Chemical Society* 36: 380–393.

Coddington, JA 1986. The genera of the spider family Theridiosomatidae. *Smithsonian Contributions to Zoology* 412: 1–96.

Cox DP, Alexander M. 1973a. Production of trimethylarsine gas from various arsenic compounds by three sewage fungi. *Bulletin of Environmental Contamination and Toxicology* 9: 84–88.

Cox DP, Alexander M. 1973b. Effect of phosphate and other anions on trimethylarsine formation by *Candida humicola. Applied Microbiology* 25: 408–413.

Craig PJ, Jenkins RO. 2004. Organometallic compounds in the environment: An overview. In: A. Hirner, F. Emons (eds.), *Organic Metal and Metalloid Species in the Environment.* Springer-Verlag, Berlin, Germany, pp. 1–15.

Crane RK, Lipman R. 1953. The effect of arsenate on aerobic phosphorylation. *Journal of Biological Chemistry* 201: 235–243.

Cullen, WR, Erdman AE, McBride BC, Pickett AW. 1983. The identification of dimethylphenylarsine as a microbial metabolite using a simple method of chemofocusing. *Journal of Microbiological Methods* 1: 297–303.

Cullen WR, Froese CL, Lui A, McBride BC, Patmore DJ, Reimer M. 1977. The aerobic methylation of arsenic by microorganisms in the presence of L-methionine-methyl-d3. *Journal of Organometallic Chemistry* 139: 61–69.

Cullen, WR, Li H, Pergantis SA, Eigendorf GK, Mosi MA. 1995. Arsenic biomethylation by the microorganism *Apiotrichum humicola* in the presence of L-methionine-methyl-d3. *Applied Organometallic Chemistry* 9: 507–515.

Cullen WR, McBride BC, Manji H, Pickett AW, Reglinski J. 1989. The metabolism of methylarsine oxide and sulfide. *Applied Organometallic Chemistry* 3: 71–78.

Cullen WR, McBride BC, Pickett AW. 1979. The transformation of arsenicals by *Candida humicola. Canadian Journal of Microbiology* 25: 1201–1205.

Cullen WR, McBride BC, Reglinski J. 1984. The reduction of trimethylarsine oxide to trimethylarsine by thiols: A mechanistic model for the biological reduction of arsenicals. *Journal of Inorganic Biochemistry* 21: 45–60.

Cullen WR, Reimer KJ. 1989. Arsenic speciation in the environment. *Chemical Reviews* 89: 713–764.

Dobran S, Zagury GJ. 2006. Arsenic speciation and mobilization in CCA-contaminated soils: Influence of organic matter content. *Science of the Total Environment* 364: 239–250.

Dombrowski PM, Long W, Farley KJ, Mahony JD, Capitani JF, Di Toro DM. 2005. Thermodynamic analysis of arsenic methylation. *Environmental Science and Technology* 39: 2169–2176.

Dopp E, Hartmann LM, Florea AM et al. 2004. Uptake of inorganic and organic derivatives of arsenic associated with induced cytotoxic and genotoxic effects in Chinese hamster ovary (CHO) cells. *Toxicology and Applied Pharmacology* 201: 156–165.

Edvantoro BB, Naidu R, Megharaj M, Merrington G, Singleton I. 2004. Microbial formation of volatile arsenic in cattle dip site soils contaminated with arsenic and DDT. *Applied Soil Ecology* 25: 207–217.

Edvantoro BB, Naidu R, Megharaj M, Singleton I. 2003. Changes in microbial properties associated with long-term arsenic and DDT contaminated soils at disused cattle dipsites. *Ecotoxicology and Environmental Safety* 55: 344–351.

Èeròansk S, Urik M, Sevc J, Simonovicoya A, Littera P. 2007. Biovolatilization of arsenic by different fungal strains. *Water, Air, and Soil Pollution* 186: 337–342.

Ellis PJ, Conrads T, Hille R, Kuhn P. 2001. Crystal structure of the 100 kDa arsenite oxidase from *Alcaligenes faecalis* in two crystal forms at 1.64 angstrom and 2.03 angstrom. *Structure* 9: 125–132.

Environmental Protection Agency (EPA). 1998. Substance name—Arsenic, inorganic. CASRN 7440-38-2. Last Revised April 10, 1998. Retrieved February 28, 2007, from http://www.epa.gov/iris/subst/0278.htm

Fatoki OS. 1997. Biomethylation in the natural environment: A review. *South African Journal of Science* 93: 366.

Feldmann J, Hirner AV. 1995. Occurrence of volatile metal and metalloid species in landfill and sewage gases. *International Journal of Environmental Analytical Chemistry* 60: 339–359.

Fogarty RV, Tobin JM. 1996. Fungal melanins and their interactions with metals. *Enzyme and Microbial Technology* 19: 311–317.

Frankenberger WJ. 1998. Short communication effects of trace elements on arsenic volatilization. *Soil Biology and Biochemistry* 30: 269–274.

Frankenberger WT Jr. (ed.) 2001. *Environmental Chemistry of Arsenic*. Marcel Dekker, New York, pp. 391.

Frankenberger WJ, Huysmans KD. 1991. Evolution of trimethylarsine by a *penicillium* sp. isolated from agricultural evaporation and water. *Science of the Total Environment* 105: 13–28.

Frohne T, Rinklebe J, Diaz-Bone RA, Du Laing G. 2010. Controlled variation of redox conditions in a floodplain soil: Impact on metal mobilization and biomethylation of arsenic and antimony. *Geoderma* (in press).

Gao S, Burau RG. 1997. Environmental factors affecting rates of arsine evolution from and mineralisation of arsenicals in soil. *Journal of Environmental Quality* 26: 753–763.

Garelick MG, Chan GC, DiRocco DP, Storm DR. 2009. Overexpression of type I adenylyl cyclase in the forebrain impairs spatial memory in aged but not young mice. *Journal of Neuroscience* 29: 10835–10842.

Gasio B. 1892. Súl riconoscimento dell'arsenico per mezzo di alcune muffe. *Riv. Ig. Sanità Pubblica* 3: 261–273.

Gasio B. 1893. Action de quelques moisissures sur les composés fixes d'arsenic. *Archives Italiennes de Biologie* 18: 253–265.

Gasio B. 1901. Recherches ultérieures sur la biologie et sur le chimisme des arseniomoisissures. *Archives Italiennes de Biologie* 35: 201–211.

Gresser M.J. 1981. ADP-arsenate. Formation by submitochondrial particles under phosphorylating conditions. *Journal of Biological Chemistry* 256: 5981–5983.

Hale B, Basta N, Boreiko C, Bowers T, Locey B, Moore M, Moutiere M et al. 2010. Variation in soil quality criteria for trace elements to protect human health exposure and effects estimation. In: G. Merrington, I. Schoeters (eds.), *Soil Quality Standards for Trace Elements: Derivation, Implementation, and Interpretation*. CRC Press, Boca Raton, FL. ISBN 978-1-4398-3023-9. 184 pp., pp. 81–122.

Higginbottom W.E. 1935. Method of closing bags. US Patent 2,023,682.

Hirner AV, Feldmann J, Goguel R, Rapsomanikis S, Fischer R, Andreae M. 1994. Volatile metal and metalloid species in gases from municipal waste deposits. *Applied Organometallic Chemistry* 8: 65–69.

Honshopp S, Brunken N, Nehrkorn A, Breunig HJ. 1996. Isolation and characterization of a new arsenic methylating bacterium from soil. *Microbiology Research* 151: 37–41.

Huysmans KD, Frankenberger WT. 1991. Evolution of trimethylarsine by a *Penicillium* sp. isolated from agricultural evaporation pond water. *Science of the Total Environment* 105: 13–28.

Ishinishi N, Tsuchiya K, Vahter M, Fowler BA. 1986. Arsenic. In: L. Friberg, G.F. Nordberg, V.B. Vouk (eds.), *Handbook on the Toxicology of Metals*. Elsevier, Amsterdam, p. 43.

Islam SMA, Fukushi K, Yamamoto K, Saha GC. 2007.Estimation of biologic gasification potential of arsenic from contaminated natural soil by enumeration of arsenic methylating bacteria. *Archives of Environmental Contamination and Toxicology* 52: 332–338.

Kertulis G, Ma L, MacDonald G, Chen R, Winefordner J, Cai Y. 2004. Arsenic speciation and transport in *Pteris vittata* L. and the effects on phosphorus in the xylem sap. *Environmental and Experimental Botany* 54: 239–247.

Knowles FC, Benson AA.1983. The biochemistry of arsenic. *Trends in Biochemical Sciences* 8: 178–180.

Krautler B. 1990. Chemistry of methylcorrinoids related to their roles in bacterial C1 metabolism. *FEMS Microbiology Reviews* 7: 349–354.

Lerman SA, Clarkson TW, Gerson RJ. 1983. Arsenic uptake and metabolism by liver cells is dependent on arsenic oxidation state. *Chemico-Biological Interactions* 45: 401–406.

Liao XY, Chen TB, Xie H, Liu YR. 2005. Soil As contamination and its risk assessment in areas near the industrial districts of Chenzhou City, Southern China. *Environment International* 31: 791–798.

Lin YF, Yang J, Rosen BP. 2007. ArsD: An As(III) metallochaperone for the ArsAB As(III)-translocating ATPase. *Journal of Bioenergetics and Biomembranes* 39: 453–458.

Macur RE, McDermontt TR, Inskeep WP. 2000. Microbially mediated arsenic cycling in a contaminated soil. In *American Society of Agronomy Annual Meeting Abstracts*. Soil Science Society of America, Madison, WI, p. 235.

Macur RE, Wheeler JT, McDermontt TR, Inskeep WP. 2001. Microbial populations associated with the reduction and enhanced mobilization of arsenic in mine tailings. *Environmental Science and Technology* 35: 3676–3682.

Macy JM, Santini JM, Pauling BV, O'Neill AH, Sly LL. 2000. Two new arsenate/sulfate-reducing bacteria: mechanisms of arsenate reduction. *Archives of Microbiology* 173: 49–57.

Masscheleyn PH, DeLaune R, William HP. 1991. Effect of redox potential and pH on arsenic speciation and solubility in a contaminated soil. *Environmental Science and Technology* 25: 1414–1419.

Matschullat J. 2000. Arsenic in the geosphere-a review. *Science of the Total Environment* 249: 297–312.

McBride BC, Merilees H, Cullen WR, Pickett W. 1978. Anaerobic and aerobic alkylation of arsenic. In: F.E. Brickman, F.M. Bellama (eds.), *Organometals and Organometalloids Occurrence and Fate in the Environment*. ACS Symposium Series No. 82, American Chemical Society, Washington, DC, pp. 94–115.

McBride BC, Wolfe RS. 1971. Biosynthesis of dimethylarsine by methanobacterium. *Biochemistry* 10: 4312–4317.

Meharg AA, Rahman MM. 2002. Arsenic contamination of Bangladesh paddy field soils: Implications for rice contribution to Arsenic consumption. *Environmental Science and Technology* 37: 229–234.

Meyer J, Schmidt A, Michalke K, Hensel R. 2007. Volatilisation of metals and metalloids by the microbial population of an alluvial soil. *Systematic and Applied Microbiology* 30: 229–238.

Michalke K, Wickenheiser E, Mehring M, Hirner A, Hensel R. 2000. Production of volatile derivatives of metal(loid)s by microflora involved in anaerobic digestion of sewage sludge. *Applied and Environmental Microbiology* 66: 2791–2796.

Mitsunobu S, Harada T, Takahashi Y. 2006. Comparison of antimony behaviour with that of arsenic under various soil redox conditions. *Environmental Science and Technology* 40: 7270–7276.

Moblev HLT, Rosen BP. 1982. Energy-dependent arsenate efflux: The mechanism of plasmid-mediated resistance *Proceedings of the National academy of Sciences of the United States of America* 79: 6119–6122.

Mohapatra D, Mishra D, Chaudhury GR, Das RP. 2008. Removal of arsenic from arsenic rich sludge by volatilization using anaerobic microorganisms treated with cow dung. *Soil and Sediment Contamination* 17: 301–311.

Moore MH, Donn B., Khanna, R, A'Hearn MF. 1983. Studies of proton-irradiated cometary-type ice mixtures. *Icarus* 54: 388.

Nordstrom DK. 2002. Public health—Worldwide occurrences of arsenic in ground water. *Science* 296: 2143–2145.

Oehme FW. 1972. Mechanisms of heavy metal toxicities. *Clinical Toxicology* 5: 151.

Onken BM, Adriano DC. 1997. Arsenic availability in soil with time under saturated and subsaturated conditions. *Soil Science Society of America Journal* 61: 746–752.

Pearce RB, Callow ME, Macaskie LE. 1998. Fungal volatilization of arsenic and antimony and the sudden infant death syndrome. *FEMS Microbiology Letters* 158(2): 261–265.

Pickett AW, McBride BC, Cullen WR. 1988. Metabolism of trimethylarsine oxide. *Applied Organometallic Chemistry* 2: 479–482.

Pickett AW, McBride BC, Cullen WR, Manji H. 1981. The reduction of trimethylarsine oxide by *Candida humicola. Canadian Journal of Microbiology* 27: 773–778.

Planer-Friedrich B, Merkel BJ. 2006. Volatile metals and metalloids in hydrothermal gases. *Environmental Science and Technology* 40: 3181–3187.

Pongratz R. 1998. Arsenic speciation in environmental samples of contaminated soil. *Science of the Total Environment* 224: 133–141.

Puntoni V. 1917. Arsenioschizomiceti. *Ann. Ig.* 27: 293–303.

Qin J, Rosen BP, Zhang Y, Wang G, Franke S, Rensing C. 2006. Arsenic detoxification and evolution of trimethylarsine gas by a microbial arsenite S-adenosylmethionine methyltransferase. *Proceedings of the National Academy of Sciences USA* 103: 2075–2080.

Raloff J. 2004. Limiting dead zones. *Science News* 165: 378–380.

Rodriguez R. 1998. Bioavailability and biomethylation of arsenic in contaminated soils and solid wastes. Dissertation. Oklahoma State University, Stillwater, OK.

Rosen P. 1971. Theoretical significance of arsenic as a carcinogen. *Journal of Theoretical Biology* 32: 425.

Rosenberg H, Gerdes RG, Chegwidden K. 1977. Two systems for the uptake of phosphate in Escherichia coli. *Journal of Bacteriology* 131, 505–511.

Scheckel KG, Chaney RL, Basta NT, Ryan JA. 2009. Advances in assessing bioavailability of metal(loid)s in contaminated soils. *Advances in Agronomy* 107: 10–52.

Shariatpanahi M, Anderson AC, Abdelghani AA, Englande AJ. 1983. Microbial metabolism of an organic arsenical herbicide. In: T.A. Oxley and S. Barry (eds.), *Biodeterioration*, vol. 5. John Wiley & Sons, Chichester, U.K., pp. 268–277.

Shariatpanahi M, Anderson AC, Abdelghani AA, Englande AJ, Hughes J, Wilkinson RF. 1981. Biotransformation of the pesticide, sodium arsenate. *Journal of Environmental Science and Health Part B* 16: 35–47.

Signes-Pastor AF, Burló K Mitra, Carbonell-Barrachina AA. 2007. Arsenic biogeochemistry as affected by phosphorus fertilizer addition, redox potential and pH in a west Bengal (India) soil. *Geoderma* 137: 504–510.

Silver S. 1996. Bacterial resistances to toxic metal ions—A review. *Gene* 179: 9–19.

Silver S. 1998. Genes for all metals—A bacterial view of the periodic table. The 1996 Thom Award Lecture. *Journal of Industrial Microbiology and Biotechnology* 20: 1–12.

Silver S, Keach D. 1982. Energy-dependent arsenate efflux: The mechanism of plasmid-mediated resistance. *Proceedings of the National academy of Sciences of the United States of America* 79: 6114–6118.

Silver P, Sadler I, Osborne M. 1989. Yeast proteins that recognize nuclear localization sequences. *Journal of Cell Biology* 109: 983–989.

Smith E, Naidu R, Alston AM. 1998. Arsenic in the soil environment: A Review. *Advances in Agronomy* 64: 149–195.

Stupperich E. 1993. Recent advances in elucidation of biological corrinoid functions. *FEMS Microbiology Reviews* 12: 349–365.

Suzuki K, Wakao N, Kimura T, Sakka K, Ohmiya, K. 1998. Expression and regulation of the arsenic resistance operon of *Acidiphilium multivorum* AIU 301 plasmid pKW301 in *Escherichia coli*. *Applied and Environmental Microbiology* 64: 411–418.

Tamaki S, Frankenberger WJ. 1992. Environmental biochemistry of arsenic. *Reviews of Environmental Contamination and Toxicology* 124: 79–110.

Thayer JS. 2002. Biological methylation of less-studied elements. *Applied Organometallic Chemistry* 16: 677–691.

Thom C, Raper KB. 1932. The arsenic fungi of Gosio. *Science* 76: 548–550.

Thomas JE, Rhue RD. 1997. Volatilisation of arsenic in contaminated cattle dipping vat soil. *Bulletin of Environmental Contamination and Toxicology* 59: 882–887.

Turpeinen R, Pantsar M, Haggblom M, Kairesalo K. 1999. Influence of microbes on the mobilization, toxicity and biomethylation of arsenic in soil. *Science of the Total Environment* 236: 173–180.

Turpeinen R, Pantsar-Kallio M, Kairesalo T. 2002. Roles of microbes in controlling the speciation of arsenic and production of arsines in contaminated soils. *Science of the Total Environment* 285: 133–145.

Vahidnia A, van der Voet GB, de Wolf FA. 2007. Arsenic neurotoxicity: A review. *Human and Experimental Toxicology* 26: 823–832.

Vidal FV, Vidal VMV. 1980. Arsenic metabolism in marine bacteria. *Marine Biology* 60: 1–7.

Walker DJ, Clemente R, Bernal MP. 2004. Contrasting effects of manure and compost on soil pH, heavy metal availability and growth of *Chenopodium album* L. in a soil contaminated by pyritic mine waste. *Chemosphere* 57: 215–224.

Wang S, Mulligan CN. 2006. Natural attenuation processes for remediation of arsenic contaminated soils and groundwater. *Journal of Hazardous Materials* 138: 459–470.

Weinberg ED. 1979. *Microorganisms and Minerals*. Marcel Dekker, New York, pp. 492.

WHO. 2001. Arsenic and arsenic compounds. Environmental Health Criteria, vol. 224. World Health Organization, Geneva, Switzerland.

Willsky AS, Chow EY, Gershwin SB, Greene CS, Houpt PK, Kurkjian. AL. 1980. Dynamic model-based techniques for the detection of incidents on freeways. *IEEE Transactions on Automatic Control* AC-25: 347–360.

Woolson EA. 1977a. Generation of alkylarsines from soil. *Weed Science Society of America* 25: 412–416.

Woolson EA. 1977b. Fate of arsenicals in different environmental substrates. *Environmental Health Perspectives* 19: 73–81.

Woolson EA. 1979. Generation of alkylarsines from soil. *Weed Science Society of America* 25: 412–416.

Woolson EA. 1983. In: Fowler (ed). Biological and Environmental Effects of Arsenic. Elsevier Sci Publishers E.V., pp. 51–139.

Woolson EA, Kearney PC. 1973. Persistence and reactions of 14C-cacodylic acid in soils. *Environmental Science and Technology* 7: 47–50.

Xu C, Zhou T, Kuroda M, Rosen BP. 1998. Metalloid resistance mechanisms in prokaryotes. *The Journal of Biochemistry (Tokyo)* 123: 16–23.

Zagury GJ, Pouschat P. 2005. Comments on arsenic on the hands of children after playing in playgrounds. *Environmental Health Perspectives* 113: A2.

Zussman, JU, Zussman PP, Dalton K. 1975. Post-Pubertal Effects of Prenatal Administration of Progesterone. *Paper presented at the meeting of the Society for Research in Child Development*, Denver, Colorado, April 1975.

13

Handling Copper: A Paradigm for the Mechanisms of Metal Ion Homeostasis

Kathryn A. Gogolin and Marinus Pilon

CONTENTS

Copper in Biology

Forms and Availability of Copper

Copper is a transitional element, found between nickel and zinc on the periodic table of elements. This metal's atomic number is 29 and has an atomic weight of 63.5 (Wulfsberg 2000). Copper concentrations range from 3 to 110 ppm in soils and have an average abundance of 55 ppm in the Earth's crust (Linder and Goode 1991, Misra 2000). This is quite rare when compared to the elements aluminum (81,300 ppm), iron (50,000 ppm), and manganese (950 ppm). Other elements worth noting due to possible influence on copper transport are zinc, cobalt, lead, silver, and gold. Zinc abundance is similar to copper with amounts in soils ranging from 16 to 95 ppm and averages 70 ppm (Linder and Goode 1999, Misra 2000). Cobalt and lead are found in similar concentrations, slightly less than copper. Cobalt ranges 2–47 ppm with an average of 25 ppm, whereas lead ranges 7–48 ppm and averages 13 ppm in the Earth's crust. Gold is the least abundant averaging 0.07 ppm (Misra 2000).

In nature, the common major ores of copper are bornite (Cu_5FeS_4), chalcopyrite ($CuFeS_2$), chalcocite (Cu_2S), and malachite ($Cu_2CO_3(OH)_2$). In contrast, cuprite (Cu_2O), covellite (CuS), and native copper (Cu) are found in smaller amounts and are minor ores (Klein et al. 2002). Copper has two different oxidation states, Cu^+ (Cu(I)) and Cu^{2+} (Cu(II)) (Linder and Goode 1999). Copper (I) ions can only be found free in very acidic solutions or complexed with other molecules, whereas free Cu(II) ions are stable in neutral, aqueous solutions that are exposed to the atmosphere (Klein et al. 2002).

During early geologic time (before 2.5×10^9 years ago) both the atmosphere and the oceans were in a reduced state (Walker et al. 1983). The dominant species of carbon, nitrogen, and sulfur in this environment would have been CO, CH_4, N_2, NH_3, and H_2S (Stevenson 1983). Metal bioavailability would be limited to Fe, Mn, Zn, Co, Ni, and Mo complexes for integration into molecular cofactors (Chapman and Schopf 1983). The first oxygen producers, cyanobacteria, emerged during the Early Archean time period and many scientists believe that these photosynthesizing organisms were the cause of the oxidized atmosphere found during current time (Chapman and Schopf 1983). Even though photosynthetic organisms appeared approximately 3.5×10^9 years ago, during transitional period oxygen was only found sporadically in microclimates. It was not until 1.7×10^9 years ago that the accumulation of free oxygen resulted in widespread aerobic conditions in the Earth's atmosphere as well as the oceans (Walker et al. 1983). Within this newly oxidized environment, iron availability decreased dramatically whereas copper became available as Cu(II) (Chapman and Schopf 1983). This change in the availability of copper is arguably the second most significant event in geologic history after the evolution of oxygen, giving rise to the complex photosynthetic pathway found in modern, higher plant systems.

Copper and Nutrition

Bioavailability of copper to plants is dependent on the soil type. Copper, especially as Cu(II), has a high affinity to bind to organic matter with an estimated 98% of copper found as a complex in soil solutions (Marschner 1995). Therefore, organic soils are defined copper deficient if there is less than 20 ppm whereas inorganic soils are deficient if there is less than 4 ppm (Linder and Goode 1991). Most plants contain copper concentrations ranging from 5 to 20 μg/g (ppm) dry weight (Hemphill 1972, Marschner 1995). Symptoms of deficiency start when copper decreases below 5 μg/g dry weight in vegetative tissues, while toxicity levels can be defined as 20 μg/g dry weight or higher in the same tissue (Hemphill 1972, Marschner 1995). The highest quantity of copper can be found in seeds, nuts, and legumes of plants, with some nuts such as cashews and coconuts reaching levels of 30 µg/g or more.

Copper deficient plants can display a wide variety of symptoms depending on the plant species and developmental stage. Symptoms consist of decreased growth rate, distortion or whitening of young leaves, damage to the apical meristem, as well as a decrease in fruit formation (Childers et al. 1995, Marschner 1995). Secondary effects of copper deficiency can be a decrease in cell wall formation and lignification in several tissues, including xylem tissue which would result in insufficient water transport (Marschner 1995). Due to the elevated levels of copper found in reproductive tissue, deficiency has a severe effect on pollen development and viability, fruit and seed production, in addition to embryo development and seed viability. High nitrogen concentrations in soils can induce copper deficiency in plants by decreasing copper mobility and availability (Marschner 1995).

Copper toxicity thresholds vary greatly between species of plants and affect tissues differently depending on metabolic requirements. Excess copper concentrations in the soil tend to decrease root growth before shoot growth due to preferential copper accumulation in that organ (Marschner 1995). The most common general symptom of toxicity is chlorosis of vegetative tissue. Increased formation of free radicals resulting in oxidative stress can occur when high amounts of Cu, Cd, Pb, or Al are observed at a cellular level (Pessarakli 2005). At a molecular level, photosynthesis is affected by damage to thylakoid membranes resulting in changes to chloroplast ultrastructure, inhibition of electron transport between photosystem II (PSII) and photosystem I (PSI), and impairment of carboxylase and oxygenase activities of RUBISCO (Pessarakli 2005). Copper toxicity can also reduce iron up take, even to the point of deficiency, depending on the form of iron available in the soil (Marschner 1995).

Requirements for Metabolism

The three most abundant trace elements in biochemical systems are iron, zinc, and copper (Bhattacharya 2005). Since several proteins and enzymes require copper for proper function, copper is essential for survival of most organisms (Linder and Goode 1991). Copper containing proteins have three major functions: dioxygen transport, catalytic, or copper transport/sequestration (Bhattacharya 2005). Photosynthetic organisms do not contain copper proteins that function in dioxygen transport. However, one example of such a protein is hemocyanin that can be found in mollusks, arthropods, and annelids. Hemocyanin is an exceptionally large protein (4500–9000 kDa) that is oxygenated in the gills of mollusks and transfers oxygen to tissues throughout the organism (Bhattacharya 2005).

Copper proteins that have catalytic functions include oxidation, electron transfer, and superoxide dismutation. Laccase and ascorbate oxidase are both proteins that can oxygenate substrates by transferring four electrons to a dioxygen molecule, resulting in two water molecules (Bhattacharya 2005). Laccase is a multicopper-containing glycoprotein that has been found in arthropods, fungi, and higher plants. This protein can act as a polyphenol oxidase and has been most widely implicated in lignin synthesis (Gavnholt and Larsen 2002). In plants, laccases have a predicted N-terminal signal peptide sequence that is for the secretory pathway, which could lead to integration into the cell wall (Gavnholt and Larsen 2002). This is consistent with previous research that has isolated laccase from cell walls.

The biochemical function of ascorbate oxidase is to oxidize alpha ascorbic acid to dehydroascorbic acid (Bhattacharya 2005); however, the biological function is not well understood. In plants, ascorbate oxidase is found predominately in the apoplast because dehydroascorbic acid is more readily taken up by the cell than the charged form of the molecule (Horemans et al. 2000). Ascorbic acid has been highly studied in plants and the major biological function is in oxidative stress defense mechanisms with high levels found in chloroplasts (Smirnoff 1996). Even though ascorbate oxidase function is unknown, transcript levels are increased in the light (Pignocchi et al. 2003). More recently, it has been suggested that one role could include cell elongation through cell wall loosening mediated by the hormone auxin (Kato and Esaka 2000). Interestingly, plants contain a group of ascorbate oxidases that do not include a copper-binding site. These two groups of oxidases have approximately 25%–30% sequence similarity; however, it is expected that the proteins lacking copper have a different biological function (Nakamura and Go 2005). Both *Brassica napus* and *Arabidopsis* contain ascorbate oxidases lacking copper with functions in *Brassica* involving pollen tube growth (Hulzink et al. 2002) and directional root growth in *Arabidopsis* (Sedbrook et al. 2002).

The most notable electron transfer proteins containing copper are cytochrome *c* oxidase and plastocyanin. Cytochrome *c* oxidase is found in aerobic bacteria and mitochondria of all eukaryotes. This large, transmembrane protein is the terminal oxidase in the cellular respiration system. The biochemical function of cytochrome *c* oxidase is to use electrons from cytochrome *c* to reduce dioxygen to produce water and at the same time pump protons (Michel et al. 1998). The enzyme is located in the mitochondrial inner membrane and contributes to the electrochemical gradient of protons across the membrane. This gradient drives the synthesis of adenosine-5′-triphosphate (ATP), which is the primary energy source for living organisms. Most cytochrome *c* oxidases have three core subunits that are highly conserved across organisms. There are two copper binding sites (Cu_A and Cu_B) with a total of three copper atoms (Steffens et al. 1987). The large C-terminal domain of subunit II in the intermembrane space contains a Cu_A-center that functions as an electron conductor. In contrast, subunit I is the site of oxygen reduction, containing two atoms of copper at the Cu_B site and two heme groups (Steffens et al. 1987).

Plastocyanin is a small, blue, copper-containing protein found in photosynthetic organisms. It is defined as a blue protein because of the form of copper bound (Cu II) and the characteristics of the binding site which consists of two histidines, one cysteine, and one methionine (Bhattacharya 2005). Plastocyanin contains one copper ion, is approximately 10 kDa, and is located inside the lumen of plant chloroplasts as well as some cyanobacteria and green algae. The biochemical function is to transfer electrons between cytochrome b_6f complex and PSI in the light-mediated reaction of photosynthesis (Pessarakli 2005). The structure of plastocyanin is an eight strand, antiparallel β-barrel (Sigfridsson 1998) and the copper binding site is highly conserved across higher plant systems as well as some algae at HIS[42], CIS[92], HIS[95], and MET[100] (Pessarakli 2005). The photosynthetic machinery of photosystem II and photosystem I are located within the membrane of the thylakoid. Similar to cellular respiration in mitochondria, this machinery drives an electrochemical gradient that produces ATP as well as the reduced form of nicotinamide adenine dinucleotide phosphate (NADPH). The products from this light-dependent pathway are then used to convert carbon dioxide to sugar, for the plant to use as food (Pessarakli 2005). This carbon fixation, via the Calvin cycle, occurs in the stroma of the chloroplast.

The third type of catalytic copper protein is the superoxide dismutase enzyme which requires a metal cofactor for proper function. There are three different kinds of superoxide dismutases, manganese, iron, and copper, zinc (Linder and Goode 1991). This enzyme is responsible for reducing oxidative stress caused by superoxides (O_2^-) and highly reactive hydroxyl radials (OH) within the cell by mediating the following reaction:

$$O_2^- + O_2^- + 2H^+ \rightarrow H_2O_2 + O_2$$

Hydrogen peroxide (H_2O_2) is then converted to H_2O via catalases and peroxidases (Bowler et al. 1992, Kliebenstein et al. 1998). Manganese and iron superoxide dismutases (SOD) are structurally similar with the metal ion not having a structural role. Copper, zinc SOD has two equal subunits with each subunit containing both one copper and one zinc ion (Bhattacharya 2005). The presence of types of SOD varies between species, but many plants have a mitochondrial MnSOD, a cytosolic CuZnSOD, and a chloroplastic FeSOD and/or CuZnSOD (Bowler et al. 1992). In plants, all three kinds of SOD are nuclear encoded with targeting sequences that directed them to their subcellular locations.

The last major group of copper containing proteins, and a major focus of this review, functions in copper transport and sequestration. This group of proteins is described in detail in the following section.

Biochemistry of Copper Transport

Types of Copper Transport Proteins

Copper uptake and distribution is tightly regulated because free copper ions are very toxic, even at low levels. The presence of free copper ions (Cu(I) or Cu(II)) in the cell can cause autooxidation of proteins, lipids, and nucleic acids (O'Halloran and Culotta 2000). It is speculated that intracellular free copper concentrations are very low because of the binding capacity of proteins that can sequester and traffic these metal ions (Rae et al. 1999). The three major categories of copper homeostasis proteins are membrane transporters, copper chaperones, and metallothioneins.

Currently, it is unknown which form of copper is generally taken up by different organisms; however, experimental data in yeast suggest copper is imported as Cu(I) (Dancis et al. 1994, Knight et al. 1996, Puig et al. 2002). Some organisms such as yeast contain metalloreductases on the plasma membrane to reduce Cu(II) to Cu(I) extracellularly before transport (Hassett and Kosman 1995, Georgatsou et al. 1997). Once copper has been reduced, there are several different types of transporters that can transport Cu(I) across cell membranes. One common type of transporters is the copper transporter family (Ctr) first discovered in plants but most studied in yeast. Members of this family are predicted to have three transmembrane domains and are rich in Met and Cys/His motifs (Figure 13.1A) (Harris 2000, Puig and Thiele 2002, Dumay et al. 2006). Depending on the organism, the Met motifs can vary between MxM, MxxM, and MxMxM, but are generally found as repeat motifs at the amino terminus. Sequences that are rich in cysteine and histidine can be found at the carboxy terminus which is thought to be in the cytosol (Puig and Thiele 2002, Dumay et al. 2006). In addition to yeast, copper transporters (Ctr) are found in a wide variety of eukaryotic organisms ranging from plants to mammals, including *Arabidopsis* (COPT family) (Sancenon et al. 2003).

A second group of membrane transporters is a diverse superfamily of P-type ATPases found in virtually every kind of organism. The membrane transporters that comprise this large group are categorized by their ion specificity (Axelsen and Palmgren 2001, Kuhlbrandt 2004). The heavy-metal-associated (HMA) family of ion transporters is classified in the P_{1B}-type ATPased sub-family. This group is predicted to have eight transmembrane domains with a large cytosolic loop between the sixth and seventh transmembrane domain (Figure 13.1B) (Williams et al. 2000, Hall and Williams 2003). There is a heavy metal binding domain at the amino and/or carboxy termini as well as a CPx motif in the sixth transmembrane domain that is thought to function in ion transduction (Williams and Mills 2005, Arguello et al. 2007). Another conserved sequence is DKTGT, found in the large cytosolic loop, which is the site of phosphorylation. Once a

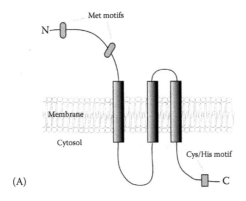

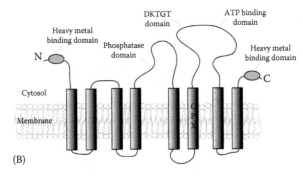

FIGURE 13.1
Schematic diagrams showing the Ctr family of transporters in yeast (Panel A) and the HMA family of P$_{1B}$-type ATPase transporters in *Arabidopsis* (Panel B).

metal binds to the transporter, ATP is utilized for phosphorylation that changes the conformation of the protein resulting in the translocation of an ion across the membrane (Arguello et al. 2007). The P-type ATPase super-family transporters are not only found on cell membranes but also within the cell in organelle membranes, specifically the Golgi complex and the chloroplast.

Lastly in Gram-negative bacteria, there is an efflux pump used to transport copper out of cells. The four-component pump found in *Escherichia coli* spans the inner membrane, periplasm, and outer membrane. Since *E. coli* is thought to diffuse Cu(I) through its cell membrane, these bacteria have most likely evolved this periplasmic export system as a defense to copper toxicity (Rensing and Grass 2003).

In addition to transporters, organisms also have proteins that shuttle Cu(I) from one place to another within the cell, called metallochaperones. Unlike other chaperones, these proteins do not aid in protein folding (Pufahl et al. 1997). Instead, metallochaperones for copper (referred to as copper

FIGURE 13.2
Ribbon model of an ATX-like copper chaperone depicting a typical βαββαβ (ferredoxin-like) fold. Model was made using SWISS-MODEL and DeepView/Swiss-PbdViewer software. (Guex, N. and Peitsch, M.C., *Electrophoresis*, 18, 2714, 1997; Peitsch, M.C., *Bio/Technology*, 13, 658, 1995; Schwede, T. et al., *Nucl. Acids Res.*, 31, 3381, 2003.)

chaperones from this point forward) are small, low molecular weight, intracellular proteins that carry Cu(I) (or other metal ions) from one target to another (Harrison and Dameron 1999). It is believed that copper chaperones are highly target-specific, always trafficking Cu(I) to and from particular proteins (O'Halloran and Culotta 2000). The crystal structure of the yeast ATX1 (Antioxidant1) chaperone shows a βαββαβ (ferredoxin-like) fold that is a conserved pattern among ATX1 homologs as well as some other copper homeostasis proteins including P_{1B}-type ATPase transporters (Figure 13.2) (Rosenzweig et al. 1999, Rosenzweig and O'Halloran 2000). Most ATX-like copper chaperones contain a putative metal binding motif, CxxC, near or in the first alpha helix (Pufahl et al. 1997, Huffman and O'Halloran 2001). Another type of copper chaperone is the CCS protein that has been found in yeast, insects, plants, and humans that functions to transport copper to cellular SOD enzymes (Culotta et al. 1997, 2006). Crystallographic structure analyses of CCS proteins indicate that it consists of three domains: an ATX-like domain (domain I), a SOD-like domain (domain II), and a unique domain (domain III) (Lamb et al. 1999, 2000, Hall et al. 2000). A CxxC metal binding sequence is present in the amino terminal, ATX-like domain of this copper trafficking protein (Lamb et al. 1999).

The third major group of copper homeostasis proteins is metallothioneins which are small, cysteine-rich proteins that have been found in a variety of different organisms. Even though these proteins are relatively small their sequence can contain up to 30% cysteine residues (Kagi et al. 1979). Some metallothioneins are able to bind approximately 12 copper ions, 6 in the β domain and 6 in the α domain (Nielson and Winge 1984). With this high binding capacity for copper, metallothioneins have the potential to sequester a large amount of excess metal in the cell. *Arabidopsis* contains four genes encoding metallothioneins that comprise two groups classified based on their structure, MT1 and MT2 (Zhou and Goldsbrough 1994). There is some evidence that increased mRNA levels of MT2 are correlated with copper tolerance in some ecotypes of *Arabidopsis* (Murphy and Taiz 1995). Experimental data, using GUS fusion analysis, suggest that metallothioneins in *Arabidopsis* have specific function. The gene promoters, *MT1a* and *MT2b*, are involved in copper transport in the phloem, while *MT2a* and *MT3* sequester excess metal ions in mesophyll cell and root tips (Guo et al. 2003).

Metal-Related Interactions and Specificity

Metal binding sequences generally include cysteine and/or histidine residues in the following motifs: CxxC, CCxSE, His-rich (including Hx and H repeats), Cys-rich, and Cys/His-rich (Arnesano et al. 2002, Arguello et al. 2007). This is especially true for P_{1B}-type ATPase transporters, but these motifs can also be found in copper chaperones and metallothioneins. Several protein–protein interactions between these transporters and copper chaperones have been observed using yeast 2-hybrid assays as well as biochemical metal transfer assays (Huffman and O'Halloran 2000, Tottey et al. 2002, van Dongen et al. 2004, Andres-Colas et al. 2006). The P_{1B}-type ATPase subgroup of transporters is further divided into two different groups based on the putative heavy metals they are hypothesized to transport. The first group is the $Zn^{2+}/Co^{2+}/Cd^{2+}/Pb^{2+}$ ion transporters and consists of HMA1, HMA2, HMA3, and HMA4 in *Arabidopsis*; while the second group includes HMA5, PAA1 (HMA6), RAN1 (HMA7), and PAA2 (HMA8) and is thought to transport Cu^{2+}/Ag^{2+} ions (Axelsen and Palmgren 1998, 2001). These two groups were originally developed based on sequence and phylogenetic analysis with the transporters in the Zn group containing histidine rich regions in the amino and/or carboxy termini areas of the proteins, whereas the transporters in the Cu group have one or two MxCxxC heavy metal binding domains in their amino terminus region. There has been a significant amount of experimental data that supports the classification of these two groups; however, more recently there has been some implication that HMA1 could be a copper transporter rather than a Zn transporter as hypothesized (Seigneurin-Berny et al. 2006). The protein sequence of HMA1 is slightly different from HMA2, 3, and 4 in that the histidine rich region is in the amino terminus and it has a rather short carboxy terminus. These differences between the HMA1 transporter and others in the Zn or Cu groups have created much dispute over the function of this protein.

Metal specificity seems to rely on both positions of conserved amino acid residues as well as the conformational structure of each protein. Recently, it has been suggested that sequences within the sixth, seventh, and eighth transmembrane domains in the P_{1B}-type ATPase transporters contribute to metal specificity (Arguello et al. 2007). Copper ions have an affinity to bind to sulfur or nitrogen ligands in histidine, methionine, and/or cysteine amino acids. This is true in plastocyanin where copper is bound to two nitrogen ligands in two histidine residues and two sulfur ligands, one in each cysteine and methionine (Figure 13.3A) (Bhattacharya 2005). In copper trafficking proteins that have a conserved MxCxxC heavy metal binding domain, copper coordination is theorized to be isolated to two sulfur ligands in the two cysteine residues (Figure 13.3B) (Rosenzweig and

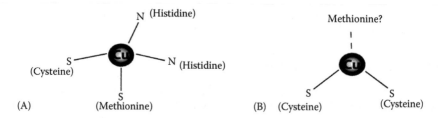

FIGURE 13.3
Diagrams depicting copper coordinating sites in the blue copper binding protein, plastocyanin (A), and possible copper binding sites in copper transporters and chaperones containing an MxCxxC heavy metal domain (B).

O'Halloran 2000). Structure analysis of the yeast ATX1 bound with Hg(II) and a domain of the Menkes disease protein (Mnk4) bound with Ag(I) indicate that the conserved methionine residue is not in close enough proximity to interact with the metal ion (Gitschier et al. 1998, Rosenzweig et al. 1999). In yeast 2-hybrid experiments it has been observed that when several conserved lysine residues located near the MxCxxC domain are mutated in the yeast ATX1 copper chaperone it loses interaction with the Ccc2 transporter (Portnoy et al. 1999). This implies that these lysines are necessary for the two proteins to interact with each other or alters the conformation of the metal binding pocket.

Copper Transport Proteins across Heterotrophic Models

Since many aspects of copper homeostasis are conserved across organisms, the following sections are a review of what is known about copper trafficking across selected model systems: bacteria, yeast, cyanobacteria, and plants. For a complete list of copper trafficking proteins and targets in these various organisms, please refer to Tables 13.1 through 13.4.

Bacterial Copper Transport Mechanisms

Many of the transporters, copper chaperones, and target enzymes for copper present in *E. coli* can be found in other prokaryotes, as well as more complex eukaryotic systems. *Enterococcus hirae* is one model organism for researching copper transport in a prokaryotic system, as a result it is vastly studied and well understood (Lu et al. 2003, Solioz and Stoyanov 2003, Magnani and Solioz 2005). These Gram-positive bacteria contain two P-type ATPase transporters, located in the plasma membrane, that function in copper transport (Figure 13.4A) (Odermatt et al. 1993, 1994). The expression of both CopA and CopB transporters are inducible under high extracellular

TABLE 13.1

List of Copper Trafficking Proteins in Bacterial Models

Copper Transport–Related Proteins and Regulatory Elements	Cellular Location	Function	Homologs	References
Bacteria				
Enterococcus hirae				
Reductase	Extracellular	Reduces Cu(II) to Cu(I) for uptake	Ndh-2 (*E. coli*)	
P-type ATPase Transporters				
CopA	Plasma membrane	Cu(I) influx transporter	CopA (*E. coli*)	Odermatt et al. (1993, 1994)
CopB	Plasma membrane	Cu(I) efflux transporter		Odermatt et al. (1993, 1994)
Cu Metallochaperone				
CopZ	Cytoplasm	Delivers Cu(I) to CopY		Odermatt and Solioz (1995), Cobine et al. (1999)
Transcriptional Regulation				
CopY	Cytoplasm	Copper-responsive repressor		Odermatt et al. (1994) Odermatt and Solioz (1995)
E. coli				
Reductase				
NDH-2	Inner membrane	Reduces Cu(II) to Cu(I) for uptake		Rapisarda et al. (1999, 2002)
P-type ATPase Transporters				
CopA	Inner membrane	Cu(I) efflux transporter	CopB (*Enterococcus hirae*)	Rensing et al. (2000)

(continued)

TABLE 13.1 (continued)

List of Copper Trafficking Proteins in Bacterial Models

Copper Transport–Related Proteins and Regulatory Elements	Cellular Location	Function	Homologs	References
Four-Component Copper Pump				
CusCFBA	Inner membrane, periplasm, outer membrane	Cu(I) efflux		Munson et al. (2000), Franke et al. (2003)
Multi-Copper Oxidase				
CueO	Periplasm	Protection of the periplasm from copper-induced damage		Grass and Rensing (2001)
Targets for Cu(I)				
Cu,Zn SOD	Periplasm	Enzyme that reduces reactive oxidative stress		
Cytochrome bo3	Periplasm	Laccase activity, detoxification?		Osborne et al. (1999)
Transcriptional Regulation				
CueR	Cytoplasm	DNA-binding regulator; member of the MerR family of metal-responsive regulators		Stoyanov et al. (2001), Outten et al. (2000)
cusRS	Cytoplasm	Chromosomal two-component regulator		Munson et al. (2000)

TABLE 13.2

List of Copper Trafficking Proteins in Yeast

Copper Transport–Related Proteins and Regulatory Elements	Cellular Location	Function	Homologs	References
Yeast				
Saccharomyces cerevisae				
Reductases				
Fre1/Fre2	Extracellular	Reduces Cu(II) to Cu(I) for uptake		Georgatsou et al. (1997), Hassett and Kosman (1995)
Copper Transporters (Ctr)				
Ctr1	Plasma membrane	High affinity Cu(I) influx	COPT1 (*Arabidopsis thaliana*)	Dancis et al. (1994)
Ctr2	Vacuole membrane?	Releases stores of copper from vacuole		Kampfenkel et al. (1995), Portnoy et al. (2001), Rees et al. (2004)
Ctr3	Plasma membrane	High affinity Cu(I) influx		Knight et al. (1996)
P-type ATPase Transporters				
Ccc2	Golgi membrane	Uptake of copper from cytosol to endomembrane system for incorporation into Fet3 for iron uptake	RAN1(*A. thaliana*)	Yamaguchi et al. (1996)
Copper Chaperones				
ATX1	Cytoplasm	Interacts with Ccc2	Atx1 (*Synechocystis*) Atx1 (*A. thaliana*) CCH (*A. thaliana*)	Lin et al. (1997)
CCS	Cytoplasm	Transports Cu(I) from cell membrane to SOD1 (Cu,Zn SOD)	CCS (*A. thaliana*)	Schmidt et al. (1999a,b)

(continued)

TABLE 13.2 (continued)

List of Copper Trafficking Proteins in Yeast

Copper Transport–Related Proteins and Regulatory Elements	Cellular Location	Function	Homologs	References
Cox17 and Cox19	Mitochondrion	Incorporation of copper into cytochrome *c* oxidase in mitochondrion	Cox17 (*A. thaliana*)	Beers et al. (1997), Nobrega et al. (2002)
Cox11	Mitochondrion	Incorporation of copper into cytochrome *c* oxidase		Horng et al. (2004), Hiser et al. (2000), Nittis et al. (2001), Winge (2003), Beers et al. (2002)
Sco1	Mitochondrion	Incorporation of copper into cytochrome *c* oxidase		
Targets for Cu(I)				
Fet3	Extracellular	Multi-copper ferroxidase/iron permease; high affinity iron uptake		Dancis et al. (1994), Askwith et al. (1996), Yuan et al. (1995), De Silva (1995)
SOD1	Cytoplasm	Enzyme that reduces reactive oxidative stress	CSD1 (*A. thaliana*)	
Cco	Mitochondrion	Terminal electron acceptor in respiration		
Transcriptional Regulation				
Mac1	Nucleus	Transcriptional activation of *FRE1*, *CTR1*, *CTR3*		Georgatsou et al. (1997), Labbe et al. (1997), Yamaguchi-Iwai et al. (1997), Zhu et al. (1998)
Ace1	Nucleus	Transcriptional activation of *SOD1*, *CRS5*, *CUP1*		Culotta et al. (1994), Thiele (1988), Gralla et al. (1991)
Metallothionein				
Crs5	Cytoplasm	Sequesters excess copper		Culotta et al. (1994)
Cup1	Cytoplasm	Sequesters excess copper		Karin et al. (1984)

TABLE 13.3

List of Copper Trafficking Proteins in Photosynthetic Cyanobacteria

Copper Transport–Related Proteins	Cellular Location	Function	Homologs	References
Cyanobacteria				
Synechocystis				
Copper Import				
FutA1	Periplasm	Aids in Cu import		Katoh et al.
FutA2	Periplasm	Aids in Cu import		(2001a,b), Waldron et al. (2007)
P-type ATPase Transporters				
CtaA	Plasma membrane	Cu(I) influx	PAA1 (*A. thaliana*)	Phung et al. (1994)
PacS	Thylakoid membrane	Cu(I) influx	PAA2 (*A. thaliana*)	Kanamaru et al. (1994)
Cu Metallochaperones				
Atx1	Cytoplasm	Transports Cu(I) from cell membrane to thylakoid membrane	Atx1 (*S. cerevisae*) Atx1 (*A. thaliana*) CCH (*A. thaliana*)	Tottey et al. (2002)
Targets for Cu(I)				
PC (Plastocyanin)	Thylakoid	Electron transport in photosynthesis		
CO (Cytochrome oxidase)	Thylakoid	Electron transport in respiration		

copper concentrations; however, CopB mutants show significantly increased cellular copper levels (Odermatt et al. 1994). This suggests that CopA is an ion importer, while CopB functions as an efflux mechanism. Intracellular levels of copper are regulated by a *cop* operon that consists of four genes: *copA, copB, copY, copZ*. The genes *copA* and *copB* encode the two copper transporters, while *copY* acts as a repressor and *copZ* functions as an activator (Odermatt and Solioz 1995). Although currently unknown, it is thought that *Enterococcus hirae* contains an extracellular reductase to convert Cu(II) to Cu(I) for cellular uptake of copper. Once copper enters the cell through the CopA transporter, a copper chaperone called CopZ shuttles Cu(I) to CopY, a transcriptional repressor (Odermatt and Solioz 1995, Strausak and Solioz 1997). The current model for copper regulation can be summarized as follows: when zinc is bound to CopY it represses transcription of the *cop* operon; however, if CopZ delivers copper to CopY, the repressor disassociates from the promoter allowing transcription to occur (Cobine et al. 1999, Magnani and

TABLE 13.4

List of Copper Trafficking Proteins in Plants

Copper Transport–Related Proteins	Cellular Location	Function	Homologs	References
Plants				
Arabidopsis thaliana				
Copper Transporters (Ctr-like)				
COPT1	Cell membrane	Cu(I) influx	Ctr1 (*S. cerevisae*)	Sancenon et al. (2003, 2004), Kampfenkel et al. (1995)
COPT2	Unknown	Unknown		Sancenon et al. (2003)
COPT3	Unknown	Unknown		Sancenon et al. (2003)
COPT4	Unknown	Unknown		Sancenon et al. (2003)
COPT5	Unknown	Unknown		Sancenon et al. (2003)
P-type ATPase Transporters				
PAA1	Chloroplast membrane	Cu(I) influx into the chloroplast	CtaA (*Synechocystis*)	Shikanai et al. (2003)
HMA1	Chloroplast membrane	Secondary copper importer?		Seigneurin-Berny et al. (2006)
PAA2	Thylakoid membrane	Cu(I) influx into the thylakoid for plastocyanin	PacS (*Synechocystis*)	Abdel-Ghany et al. (2005b)
RAN1	Golgi membrane	Cu(I) influx into Golgi complex for ethylene response	Ccc2 (*S. cerevisae*)	Andres-Colas et al. (2006), Hirayama et al. (1999)
HMA5	Plasma membrane	Copper detoxification		Andres-Colas et al. (2006)
Cu Metallochaperones				
COX17	Mitochondrion		COX17 (*S. cerevisae*)	Balandin and Castresana (2002)

	Location	Function	Homolog	References
CCH	Cytoplasm	Interacts with RAN1 and HMA5	Atx1 (*S. cerevisae*), Atx1 (*Synechocystis*)	Andres-Colas et al. (2006), Himelblau et al. (1998)
ATX1	Cytoplasm	Interacts with RAN1 and HMA5	Atx1 (*S. cerevisae*), Atx1 (*Synechocystis*)	Himelblau et al. (1998), Puig et al. (2007)
CCP	Chloroplast	Function unknown		Burkhead and Colorado State University (2003)
CCS	Chloroplast and Cytoplasm	Transports Cu(I) to CSD2 and CSD1?	CCS (*S. cerevisae*)	Abdel-Ghany et al. (2005a), Chu et al. (2005)
CutA	Chloroplast Envelope?	Unknown	CUTA (*E. coli.*)	Burkhead et al. (2003)
Targets for Cu(I)				
ETR	Endomembrane system	Ethylene signaling response		Hirayama and Alonso (2000), Hirayama et al. (1999)
Cyt-c (Cytochrome c oxidase)	Mitochondrion	Electron transport in respiration		
PC (Plastocyanin)	Thylakoid	Electron transport in photosynthesis		
CSD1(Cu,Zn SOD)	Cytoplasm	Enzyme that reduces reactive oxidative stress		Bowler et al. (1992), Kliebenstein et al. (1998)
CSD2 (Cu,Zn SOD)	Chloroplast	Enzyme that reduces reactive oxidative stress		Bowler et al. (1992), Kliebenstein et al. (1998)
CSD3 (Cu,Zn SOD)	Peroxisome	Enzyme that reduces reactive oxidative stress		Bowler et al. (1992), Kliebenstein et al. (1998)
Laccases				
Ascorbate oxidase				
Polyphenol oxidase				
Plantacyanin				

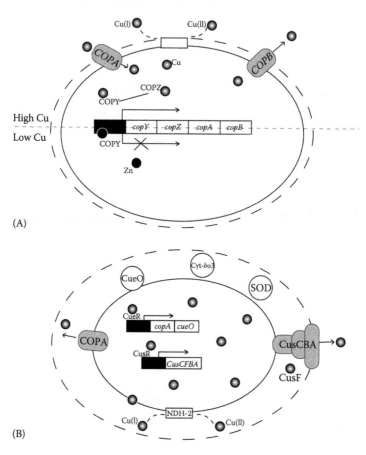

FIGURE 13.4

Illustrations of copper homeostasis in two bacterial models, *Enterococcus hirae* (Panel A) and *E. coli* (Panel B) (white rectangles symbolize copper reductases, gray rectangles are copper transporters, and white circles represent copper targets).

Solioz 2005). Structural analysis of the CopZ chaperone, via NMR spectroscopy, shows that in addition to the typical metal binding domain (MxCxxC) this protein exhibits a $\beta\alpha\beta\beta\alpha\beta$ ATX-like folding pattern (Wimmer et al. 1999).

A second prokaryotic model for studying copper transport is the bacteria *E. coli*. While copper homeostasis is not well-known in these Gram-negative bacteria, targets for copper include a Cu,Zn SOD in the periplasm (Benov and Fridovich 1994, Gort et al. 1999) as well as cytochrome bo_3 that is classified as a heme-copper oxidase (Figure 13.4B) (Osborne et al. 1999). Although there is the requirement of copper for proper function of these enzymes, most of the components that make up the known copper trafficking system in *E. coli* function in detoxification. Copper (II) ions seem to pass through the outer membrane via porins and are then reduced to Cu(I) by the copper reductase NDH-2 (Rapisarda et al. 1999, 2002). Copper, as Cu(I), can then perhaps diffuse through the cytoplasmic (inner) membrane (Beswick et al. 1976). The P-type ATPase transporter, CopA, is located in the inner membrane and functions as an efflux transporter, detoxifying the cytoplasm from copper ions (Rensing et al. 2000). Also located in the periplasm is a multi-copper oxidase, CueO, that has similarly been demonstrated to aid in copper detoxification (Grass and Rensing 2001). The expression of both the CopA and CueO proteins are regulated by a cytoplasmic copper sensing protein called CueR (Outten et al. 2000, Stoyanov et al. 2001). Another efflux mechanism for copper is a four-component pump, CusCFBA, which spans the inner membrane, periplasm, and outer membrane (Munson et al. 2000, Franke et al. 2003). Expression of this pump, in part, is regulated by a two-component signal transduction system (*cusRS*), that involves a membrane bound histidine kinase (CusS) and a cytoplasmic response regulator (CusR) (Munson et al. 2000). The CusF constituent of the efflux pump is a copper chaperone found in the periplasm that can traffic Cu(I) to the CusCBA channel for removal through the outer membrane (Franke et al. 2003).

Eukaryotic Model for Copper Regulation: *Saccharomyces Cerevisae*

With the presence of internal organelles, yeast and plants have a similar copper regulatory system which includes several direct homologs. Yeast has a copper reductase protein complex, Fre1/Fre2, that is located extracellularly to reduce Cu(II) to Cu(I) for uptake (Figure 13.5) (Hassett and Kosman 1995, Georgatsou et al. 1997). A total of three Ctr (Copper transporter) proteins have been discovered in yeast. Two of these transporters, Ctr1 and Ctr3, are located in the plasma membrane and data suggest that they are

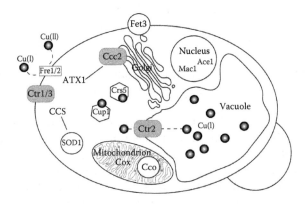

FIGURE 13.5
Copper trafficking proteins and pathways in *Saccharomyces cerevisae* (solid line depicts interactions demonstrated through experimental data, white rectangles symbolize copper reductases, gray rectangles are copper transporters, white circles represent copper targets, and white hexagons are metallothioneins).

high affinity Cu(I) importers (Dancis et al. 1994, Knight et al. 1996). The *CTR1, CTR3,* and *FRE1* genes are under the regulation of the copper-sensing MAC1 transcriptional factor (Georgatsou et al. 1997, Labbe et al. 1997, Yamaguchi-Iwai et al. 1997, Zhu et al. 1998). Copper deprivation induces a signal transduction system in which MAC1 binds to the promoter elements of the copper reductase and copper transporter genes starting transcription (Yamaguchi-Iwai et al. 1997).

The cellular localization of the Ctr2 transporter has been debated in the past; however, recent evidence indicates that it is incorporated into the vacuolar membrane where it releases copper back into the cytosol when concentrations are critically low (Kampfenkel et al. 1995, Portnoy et al. 2001, Rees et al. 2004). In addition to the Ctr family of transporters *S. cerevisae* has a P-type ATPase transporter, Ccc2, located in the Golgi (Yamaguchi et al. 1996) and is thought to translocate copper into the endomembrane system. The ATX1 (Antioxidant1) protein has been identified as the cytosolic copper chaperone for the Ccc2 transporter (Lin et al. 1997, Yuan et al. 1997). Iron metabolism has been linked to copper homeostasis in yeast through a copper requiring oxidase, Fet3, needed for ferrous iron uptake that is the target for copper in the endomembrane system (Dancis et al. 1994, De Silva et al. 1995, Yuan et al. 1995, Askwith et al. 1996).

Copper transport to the mitochondrion for integration into cytochrome *c* oxidase (Cco) in yeast has been highly studied, although it is not yet fully understood. Two putative chaperones, Cox17 and Cox19, have been implicated in cytosolic transport because of dual localization to the cytoplasm and intermitochondrial membrane space (Beers et al. 1997, Nobrega et al. 2002). However, due to biochemical and mutant analysis both proteins have been eliminated as the chaperones for mitochondrial shuttling (Cobine et al. 2006). As a result, it is currently unknown which protein is responsible for copper delivery to the mitochondrion. Once copper is in the intermitochondrial membrane space several proteins are required to work in concert to incorporate the ions into Cco. The existing model is, Cox17 donates copper to Sco1 and Cox11 (Horng et al. 2004), which then results in the integration of copper into the Cu_A site in Cox2 subunit and Cu_B site in the Cox1 subunit of cytochrome oxidase, respectively (Hiser et al. 2000, Balatri et al. 2003).

Lastly, the CCS copper chaperone is the protein for ion shuttle to a copper-requiring SOD (SOD1) (Schmidt et al. 1999a,b). In addition to the cytosol, a small fraction of both SOD1 and CCS can be found in the intermitochondrial space even though there is no presequence for targeting to the mitochondria (Sturtz et al. 2001). Yeast also contain two known metallothionein-like proteins, Cup1 and Crs5, which can sequester excess copper thereby decreasing cell toxicity (Karin et al. 1984, Culotta et al. 1994). Whereas the sequence homology between the Crs5 peptide and mammalian metallothioneins is high, there is little homology between the Crs5 and Cup1 proteins (Culotta et al. 1994). A transcriptional regulator, Ace1, has been identified that can activate transcription of Cup1 and Crs5 metallothioneins (Thiele 1988, Culotta et al. 1994), as well as the SOD1 enzyme (Gralla et al. 1991) in the presence of copper.

Copper Transport Proteins across Autotrophic Models

Copper Transport in Photosynthetic Cyanobacteria and Algae

The photosynthetic machinery and process in cyanobacteria is similar to plants, especially under sufficient copper conditions. Therefore, studying the mechanisms and regulation of copper within cyanobacteria can help to understand chloroplastic homeostasis in higher plants. Cyanobacteria contain internal thylakoid membranes, which are the site of electron transport. There are two major targets for copper within cyanobacteria, plastocyanin (PC) and cytochrome oxidase (CO), which are both found in the thylakoid (Figure 13.6) (Dworsky et al. 1995, Kerfield and Krogmann 1998). In an environment deprived of copper, some cyanobacteria can use the iron containing cytochrome c_6 protein as a terminal electron acceptor in photosynthesis instead of the copper-requiring protein plastocyanin (Zhang et al. 1992). Copper is transported through the cell membrane and the thylakoid membrane by two P-type ATPases called CtaA and PacS, respectively (Kanamaru et al. 1994, Phung et al. 1994). A copper chaperone similar to the yeast ATX protein has been identified in the model organism *Synechocystis* PCC 6803 that interacts in bacterial 2-hybrid experiments with both P-type ATPase transporters (Tottey et al. 2002). This suggests that once copper enters the cell through CtaA it is shuttled to PacS via ATX1 in the cytoplasm, for plastocyanin or cytochrome oxidase targets.

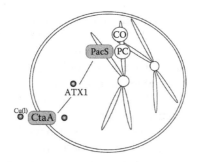

FIGURE 13.6
Copper homeostasis in the cyanobacterial model, *Synechocystis* (solid line depicts interactions demonstrated through experimental data, gray rectangles are copper transporters, and white circles represent copper targets).

Chlamydomonas is a highly studied unicellular, flagellate, green alga. Cells of *Chlamydomonas* contain one large chloroplast that has similar photosynthetic machinery as higher plants. There are three target enzymes that require copper for proper function in *Chlamydomonas*, cytochrome oxidase, plastocyanin, and multicopper oxidases involved in iron acquisition (Merchant et al. 2006). It has been proposed that copper reductases work in concert with a copper transporter to import the metal into cells. A COPT1 transporter protein has been identified in *Chlamydomonas* which is similar to the *Arabidopsis* COPT1 (Hanikenne et al. 2005). In addition, this organism contains three different HMA transporters (HMA1, HMA2, and HMA3). The HMA1 protein is similar to the *Arabidopsis* HMA1, while HMA2 and HMA3 cluster with P_{1B}-type ATPases that are copper transporters, such as RAN1 in plants (Hanikenne et al. 2005). Although copper is an essential micronutrient, ion transport in this organism is not well understood. Recently, microarray analysis has been implemented to investigate gene transcription responses to external copper concentrations in *Chlamydomonas*. In addition to genes involved in metabolic processes, stress-related and intracellular proteolysis genes were greatly affected by copper concentrations (Jamers et al. 2006).

Copper Homeostasis in Higher Plants Using Arabidopsis as a Model

Arabidopsis is a model organism for plant systems that is widely used in genetic and molecular investigations. Reasons for its use as a model include, a sequenced genome, short life cycle, and it is relatively easy to cross and mutate. Plants are complex, multicellular organisms that transport essential micronutrients, like copper, over a long distance through organs and across several membranes before they reach their final destination in target proteins.

Copper enters through the cell membrane from the apoplast by a Ctr-like transporter called COPT1 (Kampfenkel et al. 1995), and is either sequestered or trafficked to targets by copper chaperones (Figure 13.7). The COPT1 protein and its four homologs are expressed in stems, flower, leaf, as well as root tissue (Sancenon et al. 2003) and COPT1 antisense plants show abnormalities in pollen formation and increased root length (Sancenon et al. 2004). This analysis suggests that COPT1 may play a role in copper uptake in the roots from surrounding environment. Little work has been done on the other proteins in the COPT family with localizations of proteins and molecular characterizations of mutants pending.

Arabidopsis contains two different cytosolic ATX-like copper chaperones, CCH and ATX1 (Himelblau et al. 1998, Puig et al. 2007). Both proteins are functional homologs of the yeast ATX1 and have been shown to interact with the P-type ATPases, RAN1 in the Golgi membrane (Puig et al. 2007) and HMA5 in the plasma membrane (Andres-Colas et al. 2006). The RAN1 (responsive-to-antagonist1) transporter is a homolog of the yeast Ccc2 protein and functions to translocate copper into the endomembrane system were it is required for ethylene signaling (Hirayama et al. 1999, Hirayama and

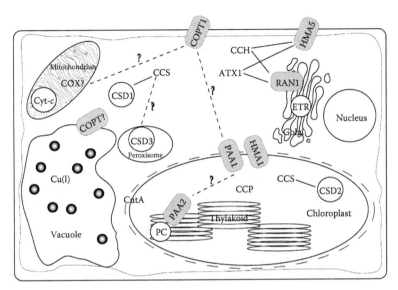

FIGURE 13.7
Copper transport pathways in *Arabidopsis thaliana* (solid lines depict interactions demonstrated through experimental data, dashed lines are hypothetical pathways, gray rectangles are copper transporters, and white circles represent copper target.

Alonso 2000). Similar to RAN1, the HMA5 membrane transporter is highly expressed in roots and flowers, but based on mutant analysis the function seems to be copper detoxification in the roots (Andres-Colas et al. 2006).

Copper chaperones that are responsible for ion transport to the mito-chondrion for cytochrome *c* oxidase are currently unknown. However, a homolog to the yeast Cox17 protein has been found in *Arabidopsis* that may play a role in mitochondrial copper delivery (Balandin and Castresana 2002). *Arabidopsis* contains three Cu,Zn SOD enzymes located in the cytosol (CSD1), stroma (CSD2), and peroxisome (CSD3). Plants possess a homolog of the yeast CCS, also called CCS that is the copper chaperone for SOD. There is only one *CCS* gene in *Arabidopsis*; however, it has been suggested that several start sites for transcription of the gene result in different subcellular locations (Abdel-Ghany et al. 2005a, Chu et al. 2005).

Within *Arabidopsis* chloroplasts there are two P-type ATPases similar to the RAN1 transporter. The outer membrane of the chloroplast is porous and many ions, including copper perhaps, can diffuse readily through it. One of the P-type ATPases, PAA1, has been localized to the membrane (inner) of the chloroplast (Shikanai et al. 2003). The second transporter, PAA2, has been localized using GFP (Green Fluorescent Protein) to the thylakoid mem-brane (Abdel-Ghany et al. 2005b). The PAA1 and PAA2 proteins are func-tional homologs of the CtaA and PacS, respectively, in cyanobacteria. Both of the PAA transporters contain a heavy metal binding motif, MxCxxC, at the N-terminal domain of the peptide along with the chloroplastic transit sequence. Although *paa1* mutants show a decrease in growth rate due to impairment of photosynthetic activity, it is not a lethal mutation (Shikanai et al. 2003). As a result of this phenotype, it is thought that copper can enter the chloroplast membrane through an alternate route. Recently, another P-type ATPase, HMA1, has been localized to the chloroplastic membrane and it has been implicated as an alternate copper transporter for SOD activity (Seigneurin-Berny et al. 2006).

In addition to the Cu, Zn SOD in the stroma, the chloroplast has a second more important target for copper, plastocyanin. Under varying copper conditions, the plant must balance regulation of copper to both of these targets within the chloroplast. Obviously, photosynthetic activity will have a higher priority for acquiring copper, but the regulators for this homeostatic mechanism are unknown. Through GFP analysis, the CCS protein has been localized to the stroma (Abdel-Ghany et al. 2005a) and additionally two other putative copper chaperones have been localized to the chloroplast (Burkhead et al. 2003, Burkhead and Colorado State University 2003). The putative chaperone, CCP (Copper Chaperone for the Plastid), is somewhat similar to the bacterial CopZ and ATX1 proteins in sequence and has an ATX-like $\beta\alpha\beta\beta\alpha\beta$ fold. However, CCP does not contain a typical metal binding domain (MxCxxC); instead it has a series of serine repeats (Burkhead and Colorado State University 2003). The function of this protein is currently unknown; however, one hypothesis is that it could be the chaperone for plastocyanin (Burkhead and Colorado State University 2003). The second putative copper chaperone, CutA, also does not have a stereotypical metal binding site; however, there is sequence similarity to a copper-related protein in *E. coli* by the same name. This protein is expressed in all major plant tissues at similar levels and purified recombinant protein of the *Arabidopsis* CutA has been shown to bind Cu(II) at a level of nearly one mole copper per one mole of protein (Burkhead et al. 2003).

The chloroplast is a complex organelle and has several membranes for copper to transverse across. Even though two of the five target proteins for copper are located in the chloroplast, transport to the organelle and within are still not well understood. No copper chaperone has been identified for transport to PAA1 in the chloroplast membrane or to PAA2 in the thylakoid membrane for plastocyanin.

References

Abdel-Ghany SE, Burkhead JL, Gogolin KA et al. 2005a. AtCCS is a functional homolog of the yeast copper chaperone Ccs1/Lys7. *FEBS Letters* 579: 2307–2312.

Abdel-Ghany SE, Muller-Moule P, Niyogi KK, Pilon M, Shikanai T. 2005b. Two P-type ATPases are required for copper delivery in *Arabidopsis thaliana* chloroplasts. *Plant Cell* 17: 1233–1251.

Andres-Colas N, Sancenon V, Rodriguez-Navarro S et al. 2006. The *Arabidopsis* heavy metal P-type ATPase HMA5 interacts with metallochaperones and functions in copper detoxification of roots. *Plant Journal* 45: 225–236.

Arguello JM, Eren E, Gonzalez-Guerrero M. 2007. The structure and function of heavy metal transport P(1B)-ATPases. *Biometals* 20: 233–248.

Arnesano F, Banci L, Bertini I, Ciofi-Baffoni S, Molteni E, Huffman DL, O'Halloran TV. 2002. Metallochaperones and metal-transporting ATPases: A comparative analysis of sequences and structures. *Genome Research* 12: 255–271.

Askwith CC, de Silva D, Kaplan J. 1996. Molecular biology of iron acquisition in *Saccharomyces cerevisiae*. *Molecular Microbiology* 20: 27–34.

Axelsen KB, Palmgren MG. 1998. Evolution of substrate specificities in the P-type ATPase superfamily. *Journal of Molecular Evolution* 46: 84–101.

Axelsen KB, Palmgren MG. 2001. Inventory of the superfamily of P-type ion pumps in *Arabidopsis*. *Plant Physiology* 126: 696–706.

Balandin T, Castresana C. 2002. AtCOX17, an *Arabidopsis* homolog of the yeast copper chaperone COX17. *Plant Physiology* 129: 1852–1857.

Balatri E, Banci L, Bertini I, Cantini F, Ciofi-Baffoni S. 2003. Solution structure of Sco1: A thioredoxin-like protein involved in cytochrome c oxidase assembly. *Structure* 11: 1431–1443.

Beers J, Glerum DM, Tzagoloff A. 1997. Purification, characterization, and localization of yeast Cox17p, a mitochondrial copper shuttle. *The Journal of Biological Chemistry* 272: 33191–33196.

Beers J, Glerum DM, Tzagoloff A. 2002. Purification and characterization of yeast Sco1p, a mitochondrial copper protein. *The Journal of Biological Chemistry* 277: 22185–22190.

Benov LT, Fridovich I. 1994. *Escherichia coli* expresses a copper- and zinc-containing superoxide dismutase. *The Journal of Biological Chemistry* 269: 25310–25314.

Beswick PH, Hall GH, Hook AJ, Little K, McBrien DC, Lott KA. 1976. Copper toxicity: Evidence for the conversion of cupric to cuprous copper in vivo under anaerobic conditions. *Chemico-Biological Interactions* 14: 347–356.

Bhattacharya PK. 2005. *Metal Ions in Biochemistry*. Alpha Science International Ltd. Harrow, U.K.

Bowler C, Montagu MV, Inze D. 1992. Superoxide dismutase and stress tolerance. *Annual Review of Plant Physiology and Plant Molecular Biology* 43: 83–116.

Burkhead JL, Abdel-Ghany SE, Morrill JM, Pilon-Smits EA, Pilon M. 2003. The *Arabidopsis thaliana* CUTA gene encodes an evolutionarily conserved copper binding chloroplast protein. *Plant Journal* 34: 856–867.

Burkhead J. 2003. *Copper Traffic in Plants: Roles for Newly Isolated Chloroplast Proteins*. Colorado State University, Department of Biology, Fort Collins, CO.

Chapman DJ, Schopf JW. 1983. Biological and biochemical effects of the development of an aerobic environment. In *Earth's Earliest Biosphere: Its Origin and Evolution* (Schopf, J.W., ed.). Princeton University Press, Princeton, NJ, pp. 302–320.

Childers NF, Morris JR, Sibbett GS. 1995 *Modern Fruit Science: Orchard and Small Fruit Culture*. Horticultural Publications, Gainesville, FL.

Chu CC, Lee WC, Guo WY, Pan SM, Chen LJ, Li HM, Jinn TL. 2005. A copper chaperone for superoxide dismutase that confers three types of copper/zinc superoxide dismutase activity in *Arabidopsis*. *Plant Physiology* 139: 425–436.

Cobine PA, Pierrel F, Bestwick ML, Winge DR. 2006. Mitochondrial matrix copper complex used in metallation of cytochrome oxidase and superoxide dismutase. *The Journal of Biological Chemistry* 281: 36552–36559.

Cobine P, Wickramasinghe WA, Harrison MD, Weber T, Solioz M, Dameron CT. 1999. The *Enterococcus hirae* copper chaperone CopZ delivers copper(I) to the CopY repressor. *FEBS Letters* 445: 27–30.

Culotta VC, Howard WR, Liu XF. 1994. CRS5 encodes a metallothionein-like protein in *Saccharomyces cerevisiae*. *The Journal of Biological Chemistry* 269: 25295–25302.

Culotta VC, Klomp LW, Strain J, Casareno RL, Krems B, Gitlin JD. 1997. The copper chaperone for superoxide dismutase. *The Journal of Biological Chemistry* 272: 23469–23472.

Culotta VC, Yang M, O'Halloran TV. 2006. Activation of superoxide dismutases: Putting the metal to the pedal. *Biochimica et Biophysica Acta* 1763: 747–758.

Dancis A, Yuan DS, Haile D et al. 1994. Molecular characterization of a copper transport protein in *S. cerevisiae*: An unexpected role for copper in iron transport. *Cell* 76: 393–402.

De Silva DM, Askwith CC, Eide D, Kaplan J. 1995. The FET3 gene product required for high affinity iron transport in yeast is a cell surface ferroxidase. *The Journal of Biological Chemistry* 270: 1098–1101.

Dumay QC, Debut AJ, Mansour NM, Saier JMH. 2006. The copper transporter (Ctr) family of Cu+ uptake systems. *Journal of Molecular Microbiology and Biotechnology* 11: 10–19.

Dworsky A, Mayer B, Regelsberger G, Fromwald S, Peschek GA. 1995. Functional and immunological characterization of both "mitochondria-like" and "chloroplast-like" electron/proton transport proteins in isolated and purified cyanobacterial membranes. *Bioelectrochemistry and Bioenergetics* 38: 35–43.

Franke S, Grass G, Rensing C, Nies DH. 2003. Molecular analysis of the copper-transporting efflux system CusCFBA of *Escherichia coli*. *The Journal of Bacteriology* 185: 3804–3812.

Gavnholt B, Larsen K. 2002. Molecular biology of plant laccases in relation to lignin formation. *Physiologia Plantarum* 116: 273–280.

Georgatsou E, Mavrogiannis LA, Fragiadakis GS, Alexandraki D. 1997. The yeast Fre1p/Fre2p cupric reductases facilitate copper uptake and are regulated by the copper-modulated Mac1p activator. *The Journal of Biological Chemistry* 272: 13786–13792.

Gitschier J, Moffat B, Reilly D, Wood WI, Fairbrother WJ. 1998. Solution structure of the fourth metal-binding domain from the Menkes copper-transporting ATPase. *Nature Structural and Molecular Biology* 5: 47–54.

Gort AS, Ferber DM, Imlay JA. 1999. The regulation and role of the periplasmic copper, zinc superoxide dismutase of *Escherichia coli*. *Molecular Microbiology* 32: 179–191.

Gralla EB, Thiele DJ, Silar P, Valentine JS. 1991. ACE1, a copper-dependent transcription factor, activates expression of the yeast copper, zinc superoxide dismutase gene. *Proceedings of the National Academy of Sciences USA* 88: 8558–8562.

Grass G, Rensing C. 2001. CueO is a multi-copper oxidase that confers copper tolerance in *Escherichia coli*. *Biochemical and Biophysical Research Communications* 286: 902–908.

Guex N, Peitsch MC. 1997. SWISS-MODEL and the Swiss-PdbViewer: An environment for comparative protein modeling. *Electrophoresis* 18: 2714–2723.

Guo WJ, Bundithya W, Goldsbrough PB. 2003. Characterization of the *Arabidopsis* metallothionein gene family: Tissue-specific expression and induction during senescence and in response to copper. *New Phytologist* 159: 369–381.

Hall LT, Sanchez RJ, Holloway SP, Zhu H, Stine JE, Lyons TJ, Demeler B, Schirf V et al. 2000. X-ray crystallographic and analytical ultracentrifugation analyses of truncated and full-length yeast copper chaperones for SOD (LYS7): A dimmer–dimer model of LYS7-SOD association and copper delivery. *Biochemistry* 39: 3611–3623.

Hall JL, Williams LE. 2003. Transition metal transporters in plants. *Journal of Experimental Botany* 54: 2601–2613.

Hanikenne M, Kramer U, Demoulin V, Baurain D. 2005. A comparative inventory of metal transporters in the green alga *Chlamydomonas reinhardtii* and the red alga *Cyanidioschizon merolae*. *Plant Physiology* 137: 428–446.

Harris ED. 2000. Cellular copper transport and metabolism. *Annual Review of Nutrition* 20: 291–310.

Harrison MD, Dameron CT. 1999. Molecular mechanisms of copper metabolism and the role of the Menkes disease protein. *Journal of Biochemical and Molecular Toxicology* 13: 93–106.

Hassett R, Kosman DJ. 1995. Evidence for Cu(II) reduction as a component of copper uptake by *Saccharomyces cerevisiae*. *The Journal of Biological Chemistry* 270: 128–134.

Hemphill DD. 1972. Availability of trace elements to plants with respect to soil–plant interaction. *Annals of the New York Academy of Sciences* 199: 46–61.

Himelblau E, Mira H, Lin SJ, Culotta VC, Penarrubia L, Amasino RM. 1998. Identification of a functional homolog of the yeast copper homeostasis gene ATX1 from *Arabidopsis*. *Plant Physiology* 117: 1227–1234.

Hirayama T et al. 1999. RESPONSIVE-TO-ANTAGONIST1, a Menkes/Wilson disease-related copper transporter, is required for ethylene signaling in *Arabidopsis*. *Cell* 97: 383–393.

Hirayama T, Alonso JM. 2000. Ethylene captures a metal! Metal ions are involved in ethylene perception and signal transduction. *Plant and Cell Physiology* 41: 548–555.

Hiser L, Di Valentin M, Hamer AG, Hosler JP. 2000. Cox11p is required for stable formation of the Cu(B) and magnesium centers of cytochrome c oxidase. *The Journal of Biological Chemistry* 275: 619–623.

Horemans N, Foyer CH, Asard H. 2000. Transport and action of ascorbate at the plant plasma membrane. *Trends in Plant Science* 5: 263–267.

Horng YC, Cobine PA, Maxfield AB, Carr HS, Winge DR. 2004. Specific copper transfer from the Cox17 metallochaperone to both Sco1 and Cox11 in the assembly of yeast cytochrome c oxidase. *The Journal of Biological Chemistry* 279: 35334–35340.

Huffman DL, O'Halloran TV. 2000. Energetics of copper trafficking between the Atx1 metallochaperone and the intracellular copper transporter, Ccc2. *The Journal of Biological Chemistry* 275: 18611–18614.

Huffman DL, O'Halloran TV. 2001. Function, structure, and mechanism of intracellular copper trafficking proteins. *Annual Review of Biochemistry* 70: 677–701.

Hulzink RJM, de Groot PFM, Croes AF et al. 2002. The 5'-untranslated region of the ntp303 gene strongly enhances translation during pollen tube growth, but not during pollen maturation. *Plant Physiology* 129: 342–353.

Jamers A, Van der Ven K, Moens L et al. 2006. Effect of copper exposure on gene expression profiles in *Chlamydomonas reinhardtii* based on microarray analysis. *Aquatic Toxicology* 80: 249–260.

Kagi JH, Kojima Y, Kissling MM, Lerch K. 1979. Metallothionein: An exceptional metal thiolate protein. *Ciba Foundation Symposium* 72: 223–237.

Kampfenkel K, Kushnir S, Babiychuk E, Inze D, Van Montagu M. 1995. Molecular characterization of a putative *Arabidopsis thaliana* copper transporter and its yeast homologue. *The Journal of Biological Chemistry* 270: 28479–28486.

Kanamaru K, Kashiwagi S, Mizuno T. 1994. A copper-transporting P-type ATPase found in the thylakoid membrane of the cyanobacterium *Synechococcus* species PCC7942. *Molecular Microbiology* 13: 369–377.

Karin M, Najarian R, Haslinger A, Valenzuela P, Welch J, Fogel S. 1984. Primary structure and transcription of an amplified genetic locus: The CUP1 locus of yeast. *Proceedings of the National Academy of Sciences USA* 81: 337–341.

Kato N, Esaka M. 2000. Expansion of transgenic tobacco protoplasts expressing pumpkin ascorbate oxidase is more rapid than that of wild-type protoplasts. *Planta* 210: 1018–1022.

Katoh H, Hagino N, Grossman AR, Ogawa T. 2001a. Genes essential to iron transport in the cyanobacterium *Synechocystis* sp. strain PCC 6803. *The Journal of Bacteriology* 183: 2779–2784.

Katoh H, Hagino N, Ogawa T. 2001b. Iron-binding activity of FutA1 subunit of an ABC-type iron transporter in the cyanobacterium *Synechocystis* sp. strain PCC 6803. *Plant Cell Physiology* 42: 823–827.

Kerfield CA, Krogmann DW. 1998. Photosynthetic cytochromes c in cyanobacteria, algae, and plants. *Annual Review of Plant Physiology and Plant Molecular Biology* 49: 397–425.

Klein C, Hurlbut CS, Dana JD, Klein C. 2002. *Manual of Mineral Science*, 22nd edn. (After James D. Dana). John Wiley & Sons, New York.

Kliebenstein DJ, Monde RA, Last RL. 1998. Superoxide dismutase in *Arabidopsis*: An eclectic enzyme family with disparate regulation and protein localization. *Plant Physiology* 118: 637–650.

Knight SA, Labbe S, Kwon LF, Kosman DJ, Thiele DJ. 1996. A widespread transposable element masks expression of a yeast copper transport gene. *Genes and Development* 10: 1917–1929.

Kuhlbrandt W. 2004. Biology, structure and mechanism of P-type ATPases. *Nature Reviews Molecular Cell Biology* 5: 282–295.

Labbe S, Zhu Z, Thiele DJ. 1997. Copper-specific transcriptional repression of yeast genes encoding critical components in the copper transport pathway. *The Journal of Biological Chemistry* 272: 15951–15958.

Lamb AL, Wernimont AK, Pufahl RA, Culotta VC, O'Halloran TV, Rosenzweig AC. 1999. Crystal structure of the copper chaperone for superoxide dismutase. *Nature Structural and Molecular Biology* 6: 724–729.

Lamb AL, Wernimont AK, Pufahl RA, O'Halloran TV, Rosenzweig AC. 2000. Crystal structure of the second domain of the human copper chaperone for superoxide dismutase. *Biochemistry* 39: 1589–1595.

Lin SJ, Pufahl RA, Dancis A, O'Halloran TV, Culotta VC. 1997. A role for the *Saccharomyces cerevisiae* ATX1 gene in copper trafficking and iron transport. *The Journal of Biological Chemistry* 272: 9215–9220.

Linder MC, Goode CA.1991. *Biochemistry of Copper*, Volume 10 of Biochemistry of the Elements. Plenum Press (ISBN: 0306436582, 9780306436581).

Lu ZH, Dameron CT, Solioz M. 2003. The *Enterococcus hirae* paradigm of copper homeostasis: Copper chaperone turnover, interactions, and transactions. *Biometals* 16: 137–143.

Magnani D, Solioz M. 2005. Copper chaperone cycling and degradation in the regulation of the cop operon of *Enterococcus hirae*. *Biometals* 18: 407–412.

Marschner H. 1995. *Mineral Nutrition of Higher Plants*. Academic Press, London, U.K.

Merchant SS, Allen MD, Kropat J et al. 2006. Between a rock and a hard place: Trace element nutrition in *Chlamydomonas*. *Biochimica et Biophysica Acta* 1763: 578–594.

Michel H, Behr J, Harrenga A, Kannt A. 1998. Cytochrome c oxidase: Structure and spectroscopy. *Annual Review of Biophysics and Biomolecular Structure* 27: 329–356.

Misra KC. 2000. *Understanding Mineral Deposits*. Kluwer Academic Publishers, Dordrecht, the Netherlands.

Munson GP, Lam DL, Outten FW, O'Halloran TV. 2000. Identification of a copper-responsive two-component system on the chromosome of *Escherichia coli* K-12. *The Journal of Bacteriology* 182: 5864–5871.

Murphy A, Taiz L. 1995. Comparison of metallothionein gene expression and nonprotein thiols in ten *Arabidopsis* ecotypes. Correlation with copper tolerance. *Plant Physiology* 109: 945–954.

Nakamura K, Go N. 2005. Function and molecular evolution of multicopper blue proteins. *Cellular and Molecular Life Sciences* 62: 2050–2066.

Nielson KB, Winge DR. 1984. Preferential binding of copper to the beta domain of metallothionein. *The Journal of Biological Chemistry* 259: 4941–4946.

Nittis T, George GN, Winge DR. 2001. Yeast Sco1, a protein essential for cytochrome c oxidase function is a Cu(I)-binding protein. *The Journal of Biological Chemistry* 276: 42520–42526.

Nobrega MP, Bandeira SC, Beers J, Tzagoloff A. 2002. Characterization of COX19, a widely distributed gene required for expression of mitochondrial cytochrome oxidase. *The Journal of Biological Chemistry* 277: 40206–40211.

Odermatt A, Krapf R, Solioz M. 1994. Induction of the putative copper ATPases, CopA and CopB, of *Enterococcus hirae* by Ag+ and Cu2+, and Ag+ extrusion by CopB. *Biochemical and Biophysical Research Communications* 202: 44–48.

Odermatt A, Solioz M. 1995. Two trans-acting metalloregulatory proteins controlling expression of the copper-ATPases of *Enterococcus hirae*. *The Journal of Biological Chemistry* 270: 4349–4354.

Odermatt A, Suter H, Krapf R, Solioz M. 1993. Primary structure of two P-type ATPases involved in copper homeostasis in *Enterococcus hirae*. *The Journal of Biological Chemistry* 268: 12775–12779.

O'Halloran TV, Culotta VC. 2000. Metallochaperones, an intracellular shuttle service for metal ions. *The Journal of Biological Chemistry* 275: 25057–25060.

Osborne JP, Cosper NJ, Stalhandske CM, Scott RA, Alben JO, Gennis RB. 1999. Cu XAS shows a change in the ligation of CuB upon reduction of cytochrome bo3 from *Escherichia coli*. *Biochemistry* 38: 4526–4532.

Outten FW, Outten CE, Hale J, O'Halloran TV. 2000. Transcriptional activation of an *Escherichia coli* copper efflux regulon by the chromosomal MerR homologue, cueR. *The Journal of Biological Chemistry* 275: 31024–31029.

Peitsch MC. 1995. Protein modeling by e-mail. *Bio/Technology* 13: 658–660.

Pessarakli M. 2005. *Handbook of Photosynthesis*. Taylor & Francis, Boca Raton, FL.

Phung LT, Ajlani G, Haselkorn R. 1994. P-type ATPase from the cyanobacterium *Synechococcus* 7942 related to the human Menkes and Wilson disease gene products. *Proceedings of the National Academy of Sciences USA* 91: 9651–9654.

Pignocchi C, Fletcher JM, Wilkinson JE, Barnes JD, Foyer CH. 2003. The function of ascorbate oxidase in tobacco. *Plant Physiology* 132: 1631–1641.

Portnoy ME, Rosenzweig AC, Rae T, Huffman DL, O'Halloran TV, Culotta VC. 1999. Structure-function analyses of the ATX1 metallochaperone. *The Journal of Biological Chemistry* 274: 15041–15045.

Portnoy ME, Schmidt PJ, Rogers RS, Culotta VC. 2001. Metal transporters that contribute copper to metallochaperones in *Saccharomyces cerevisiae*. *Molecular Genetics and Genomics* 265: 873–882.

Pufahl RA, Singer CP, Peariso KL, Lin SJ, Schmidt PJ, Fahrni CJ, Culotta VC, Penner-Hahn JE, O' Halloran TV. 1997. Metal ion chaperone function of the soluble Cu(I) receptor Atx1. *Science* 278: 853–856.

Puig S, Lee J, Lau M, Thiele DJ. 2002. Biochemical and genetic analyses of yeast and human high affinity copper transporters suggest a conserved mechanism for copper uptake. *The Journal of Biological Chemistry* 277: 26021–26030.

Puig S, Mira H, Dorcey E, Sancenón V, Andrés-Colás N, Garcia-Molina A, Burkhead JL et al. 2007. Higher plants possess two different types of ATX1-like copper chaperones. *Biochemical and Biophysical Research Communications* 354: 385–390.

Puig S, Thiele DJ. 2002. Molecular mechanisms of copper uptake and distribution. *Current Opinion in Chemical Biology* 6: 171–180.

Rae TD, Schmidt PJ, Pufahl RA, Culotta VC, O'Halloran TV. 1999. Undetectable intracellular free copper: The requirement of a copper chaperone for superoxide dismutase. *Science* 284: 805–808.

Rapisarda VA, Chehin RN, De Las Rivas J, Rodriguez-Montelongo L, Farias RN, Massa EM. 2002. Evidence for Cu(I)-thiolate ligation and prediction of a putative copper-binding site in the *Escherichia coli* NADH dehydrogenase-2. *Archives of Biochemistry and Biophysics* 405: 87–94.

Rapisarda VA, Montelongo LR, Farias RN, Massa EM. 1999. Characterization of an NADH-linked cupric reductase activity from the *Escherichia coli* respiratory chain. *Archives of Biochemistry and Biophysics* 370: 143–150.

Rees EM, Lee J, Thiele DJ. 2004. Mobilization of intracellular copper stores by the ctr2 vacuolar copper transporter. *The Journal of Biological Chemistry* 279: 54221–54229.

Rensing C, Fan B, Sharma R, Mitra B, Rosen BP. 2000. CopA: An *Escherichia coli* Cu(I)-translocating P-type ATPase. *Proceedings of the National Academy of Sciences USA* 97: 652–656.

Rensing C, Grass G. 2003. *Escherichia coli* mechanisms of copper homeostasis in a changing environment. *FEMS Microbiology Reviews* 27: 197–213.

Rosenzweig AC, Huffman DL, Hou MY, Wernimont AK, Pufahl RA, O'Halloran TV. 1999. Crystal structure of the Atx1 metallochaperone protein at 1.02 A resolution. *Structure* 7: 605–617.

Rosenzweig AC, O'Halloran TV. 2000. Structure and chemistry of the copper chaperone proteins. *Current Opinion in Chemical Biology* 4: 140–147.

Sancenon V, Puig S, Mateu-Andres I, Dorcey E, Thiele DJ, Penarrubia L. 2004. The *Arabidopsis* copper transporter COPT1 functions in root elongation and pollen development. *The Journal of Biological Chemistry* 279: 15348–15355.

Sancenon V, Puig S, Mira H, Thiele DJ, Penarrubia L. 2003. Identification of a copper transporter family in *Arabidopsis thaliana*. *Plant Molecular Biology* 51: 577–587.

Schmidt PJ, Rae TD, Pufahl RA et al. 1999a. Multiple protein domains contribute to the action of the copper chaperone for superoxide dismutase. *The Journal of Biological Chemistry* 274: 23719–23725.

Schmidt PJ, Ramos-Gomez M, Culotta VC. 1999b. A gain of superoxide dismutase (SOD) activity obtained with CCS, the copper metallochaperone for SOD1. *The Journal of Biological Chemistry* 274: 36952–36956.

Schwede T, Kopp J, Guex N, Peitsch MC. 2003. SWISS-MODEL: An automated protein homology-modeling server. *Nucleic Acids Research* 31: 3381–3385.

Sedbrook JC, Carroll KL, Hung KF, Masson PH, Somerville CR. 2002. The *Arabidopsis* SKU5 gene encodes an extracellular glycosyl phosphatidylinositol-anchored glycoprotein involved in directional root growth. *Plant Cell* 14: 1635–1648.

Seigneurin-Berny D, Gravot A, Auroy P, Mazard C, Kraut A, Finazzi G, Grunwald D et al. 2006. HMA1, a new Cu-ATPase of the chloroplast envelope, is essential for growth under adverse light conditions. *Journal of Biological Chemistry* 281: 2882–2892.

Shikanai T, Muller-Moule P, Munekage Y, Niyogi KK, Pilon M. 2003. PAA1, a P-type ATPase of *Arabidopsis*, functions in copper transport in chloroplasts. *Plant Cell* 15: 1333–1346.

Sigfridsson K. 1998. Plastocyanin, an electron-transfer protein. *Photosynthesis Research* 57: 1–28.

Smirnoff N. 1996. The function and metabolism of ascorbic acid in plants. *Annals of Botany* 78: 661–669.

Solioz M, Stoyanov JV. 2003. Copper homeostasis in *Enterococcus hirae*. *FEMS Microbiology Reviews* 27: 183–195.

Steffens GCM, Biewald R, Buse G. 1987. Cytochrome c oxidase is three-copper, two-heme-A protein. *FEBS Journal* 164: 295–300.

Stevenson J. 1983. The nature of the Earth prior to the oldest known rock record: The Hadean Earth. In *Earth's Earliest Biosphere: Its Origin and Evolution* (Schopf, J.W., ed.). Princeton University Press, Princeton, NJ, pp. 32–40.

Stoyanov JV, Hobman JL, Brown NL. 2001. CueR (YbbI) of *Escherichia coli* is a MerR family regulator controlling expression of the copper exporter CopA. *Molecular Microbiology* 39: 502–511.

Strausak D, Solioz M. 1997. CopY is a copper-inducible repressor of the *Enterococcus hirae* copper ATPases. *The Journal of Biological Chemistry* 272: 8932–8936.

Sturtz LA, Diekert K, Jensen LT, Lill R, Culotta VC. 2001. A fraction of yeast Cu, Zn-superoxide dismutase and its metallochaperone, CCS, localize to the intermembrane space of mitochondria. A physiological role for SOD1 in guarding against mitochondrial oxidative damage. *The Journal of Biological Chemistry* 276: 38084–38089.

Thiele DJ. 1988. ACE1 regulates expression of the *Saccharomyces cerevisiae* metallothionein gene. *Molecular and Cellular Biology* 8: 2745–2752.

Tottey S, Rondet SA, Borrelly GP, Robinson PJ, Rich PR, Robinson NJ. 2002. A copper metallochaperone for photosynthesis and respiration reveals metal-specific targets, interaction with an importer, and alternative sites for copper acquisition. *The Journal of Biological Chemistry* 277: 5490–5497.

van Dongen EM, Klomp LW, Merkx M. 2004. Copper-dependent protein–protein interactions studied by yeast two-hybrid analysis. *Biochemical and Biophysical Research Communications* 323: 789–795.

Waldron KJ, Tottey S, Yanagisawa S, Dennison C, Robinson NJ. 2007. A periplasmic iron-binding protein contributes toward inward copper supply. *The Journal of Biological Chemistry* 282: 3837–3846.

Walker JCG, Cornelis K, Schidlowski M, Schopf JW, Stevenson J, Walter MR. 1983. Environmental evolution of the Archean-early proterozoic earth. In *Earth's Earliest Biosphere: Its Origin and Evolution* (Schopf, J.W., ed.). Princeton University Press, Princeton, NJ, pp. 260–290.

Williams LE, Mills RF. 2005. P1B-ATPases—An ancient family of transition metal pumps with diverse functions in plants. *Trends in Plant Science* 10: 491–502.

Williams LE, Pittman JK, Hall JL. 2000. Emerging mechanisms for heavy metal transport in plants. *Biochimica et Biophysica Acta* 1465: 104–126.

Wimmer R, Herrmann T, Solioz M, Wuthrich K. 1999. NMR structure and metal interactions of the CopZ copper chaperone. *The Journal of Biological Chemistry* 274: 22597–22603.

Winge DR. 2003. Let's Sco1, Oxidase! Let's Sco! *Structure* 11: 1313–1314.

Wulfsberg G. 2000. *Inorganic Chemistry*. University Science Books, Sausalito, CA.

Yamaguchi Y, Heiny ME, Suzuki M, Gitlin JD. 1996. Biochemical characterization and intracellular localization of the Menkes disease protein. *Proceedings of the National Academy of Sciences USA* 93: 14030–14035.

Yamaguchi-Iwai Y, Serpe M, Haile D et al. 1997. Homeostatic regulation of copper uptake in yeast via direct binding of MAC1 protein to upstream regulatory sequences of FRE1 and CTR1. *The Journal of Biological Chemistry* 272: 17711–17718.

Yuan DS, Dancis A, Klausner RD. 1997. Restriction of copper export in *Saccharomyces cerevisiae* to a late Golgi or post-Golgi compartment in the secretory pathway. *The Journal of Biological Chemistry* 272: 25787–25793.

Yuan DS, Stearman R, Dancis A, Dunn T, Beeler T, Klausner RD. 1995. The Menkes/Wilson disease gene homologue in yeast provides copper to a ceruloplasmin-like oxidase required for iron uptake. *Proceedings of the National Academy of Sciences USA* 92: 2632–2636.

Zhang L, McSpadden B, Pakrasi HB, Whitmarsh J. 1992. Copper-mediated regulation of cytochrome c553 and plastocyanin in the cyanobacterium Synechocystis 6803. *The Journal of Biological Chemistry* 267: 19054–19059.

Zhou J, Goldsbrough PB. 1994. Functional homologs of fungal metallothionein genes from *Arabidopsis*. *The Plant Cell* 6: 875–884.

Zhu Z, Labbe S, Pena MM, Thiele DJ. 1998. Copper differentially regulates the activity and degradation of yeast Mac1 transcription factor. *The Journal of Biological Chemistry* 273: 1277–1280.

14

Plant Selenium Metabolism: Genetic Manipulation, Phytotechnological Applications, and Ecological Implications

Elizabeth A.H. Pilon-Smits

CONTENTS

Introduction

The element selenium (Se) is chemically similar to sulfur (S). For this reason plants and other organisms mistakenly take up and metabolize Se via S transporters and biochemical pathways. This can cause toxicity due to a combination of (1) oxidative stress caused directly by selenocompounds and (2) replacement of S by Se in proteins and other S compounds, which disrupts their function. On the other hand, Se is an essential trace element for many organisms, including mammals, many bacteria, and certain green algae (Stadtman 1990, 1996, Fu et al. 2002). For higher plants, Se has been reported to be a beneficial nutrient, but it has not been shown to be essential (Cartes et al. 2005, Djanaguiraman et al. 2005, Hartikainen 2005, Lyons et al. 2009, Pilon-Smits et al. 2009). Organisms that require Se produce essential proteins that contain selenocysteine (SeCys) in their active site. To date, no selenoproteins have been confirmed to exist in higher plants (Novoselov et al. 2002).

Selenoproteins have antioxidant or other redox functions, which is why Se deficiency often enhances the probability of developing cancers or viral infections; diseases associated with Se deficiency include Keshan disease and male infertility in humans, and white muscle disease in livestock (Whanger 1989, Ellis et al. 2004, Diwadkar-Navsariwala et al. 2006, White and Broadley 2009). There is a relatively narrow window between the amount of Se required as a nutrient and the amount that is toxic, and hence, Se deficiency and toxicity are both common problems worldwide (Terry et al. 2000). As an illustration, daily intake of 50 µg Se is recommended for humans, but long-term intake of 10 times higher levels may lead to chronic Se poisoning. The one-time ingestion of plant material containing 1000 mg/kg dry weight (dry wt.) or more Se can even lead to acute Se poisoning and death (Draize and Beath 1935, Rosenfeld and Beath 1964, Wilber 1980). Such high Se levels (1,000–10,000 mg/kg dry wt.) occur in so-called hyperaccumulator plant species that are endemic on seleniferous soils in the Western United States and parts of China where Se is naturally present. Human and livestock Se poisoning, both chronic and acute, are serious problems in these seleniferous areas (Ohlendorf et al. 1986, Harris 1991, Kabata-Pendias 1998, Terry et al. 2000).

Higher plants readily take up selenate or selenite from their environment and incorporate it into organic compounds using S transporters and S assimilation enzymes. In short, inorganic selenate is taken up and reduced to selenite, then selenide, and combined with O-acetylserine (OAS) to form SeCys. This seleno-amino acid can be nonspecifically incorporated into proteins in the place of Cys, leading to toxicity. SeCys can also be converted to selenomethionine (SeMet), which also can be misincorporated into proteins. SeMet can also be converted to volatile dimethylselenide (DMSe) (Lewis et al. 1966, Hansen et al. 1998). Furthermore, SeCys can be broken down to elemental Se and alanine (Pilon et al. 2003). Elemental Se is insoluble and relatively innocuous. SeCys can also be methylated to form methyl-SeCys, which can be safely accumulated since it is not incorporated into proteins (Neuhierl et al. 1999). Methyl-SeCys can also act as a precursor for the production of another form of volatile Se, dimethyldiselenide (DMDSe) (Terry et al. 2000, Sors et al. 2005). This is the main volatile form of Se emitted by hyperaccumulator species.

The capacity of plants to accumulate, metabolize, and volatilize Se is all useful for Se phytoremediation. Soil or water rich in Se, either naturally or due to human activities (e.g., mining, oil refining, agriculture), may be cleaned up using plants, which may either release the Se into the atmosphere in relatively nontoxic volatile form and/or accumulate it in their harvestable tissues. If Se is the only pollutant present, the use of Se-accumulating plants for phytoremediation may produce a value-added crop that may be sold, offsetting the phytoremediation costs. Some selenocompounds have particularly potent anticarcinogenic properties (Unni et al. 2005), e.g., methyl-SeCys. Examples of plant species that accumulate this compound are the crop species broccoli and garlic, and the hyperaccumulator species two-grooved milkvetch (*Astragalus bisulcatus*) and prince's plume (*Stanleya pinnata*).

Plants known to accumulate high levels of S compounds, such as many *Brassica* and *Allium* genera (mustards and cabbages, onion and garlic), also are good Se accumulators. These S-loving species accumulate Se to fairly high levels (0.1% of dry wt. or 1000 mg Se/kg dry wt.) when supplied with adequate external Se levels, and have been called accumulator species. Se accumulator species likely do not have Se-specific pathways but take up and metabolize Se and S indiscriminately at elevated rates compared to nonaccumulators. Se hyperaccumulators are found in the families Brassicaceae, Fabaceae, and Asteraceae and typically accumulate Se to levels 100-fold higher than surrounding vegetation in the field (Beath et al. 1939a,b). Since hyperaccumulators preferentially take up Se over S and show different patterns for these two elements in terms of seasonal fluctuations and tissue distribution, they likely are able to distinguish between S and Se (Galeas et al. 2007, White et al. 2007). Hyperaccumulators accumulate Se up to 1% of their dry weight (10,000 mg Se/kg dry wt.) from soil typically containing 2–10 mg Se/kg without suffering any toxicity (Neuhierl and Böck 1996, Neuhierl et al. 1999, Persans and Salt 2000, Ellis et al. 2004, LeDuc et al. 2004). Se hyperaccumulators are endemic to seleniferous soils and thus appear to physiologically or ecologically require Se. Perhaps hyperaccumulators need Se as a defense compound against herbivores or pathogens, as discussed in more detail later. It has also been suggested that Se may be essential for hyperaccumulator physiology, since hyperaccumulators grow much better in the presence of Se than without it (more than twofold higher biomass production in some experiments). However, to date there is no proof that hyperaccumulators or any higher plants require Se to complete their life cycle. The positive growth response of hyperaccumulators to Se may also be due to alleviation of phosphorus toxicity, which is much less pronounced when plants are grown at lower phosphorus levels (Broyer et al. 1972). In the following text, we give an overview of Se metabolism in plants in nonhyperaccumulators and hyperaccumulators.

Se Metabolism in Plants

The predominant forms of Se in the environment are inorganic selenate and selenite. Selenate, or Se(VI), is the most oxidized form of Se and the predominant bioavailable form in oxic soils, while selenite, or Se(IV), is more abundant in (more anoxic) wetland conditions. In addition, inorganic elemental Se, Se(0), can become dominant under reducing, anoxic conditions. Both selenate and selenite are bioavailable and readily taken up by plants. Selenate is taken up and mobilized in plants by means of sulfate–proton

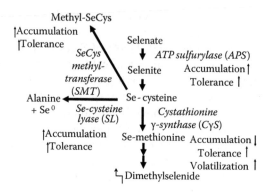

FIGURE 14.1
Overview of genetic engineering approaches that have been used successfully to enhance plant Se tolerance, accumulation, and/or volatilization.

cotransporters (Smith et al. 1995, Leustek 1996, Yoshimoto et al. 2002, 2003, Hawkesford 2003, Maruyama-Nakashita et al. 2004). Selenate assimilation takes place predominantly in the leaf chloroplasts (Pilon-Smits et al. 1999). Most plants supplied with selenate accumulate predominantly selenate, while plants supplied with selenite accumulate organic Se, suggesting that the reduction of selenate to selenite is a rate-limiting step in the Se assimilation pathway (de Souza et al. 1998). The conversion of selenate to selenite is mediated by two enzymes (Figure 14.1). ATP sulfurylase (APS) couples selenate to ATP, forming adenosine phosphoelenate (APSe) (Wilson and Bandurski 1958). APSe is subsequently reduced to selenite by APS reductase (APR). The further reduction of selenite to selenide may happen exclusively in the chloroplast if it is mediated by sulfite reductase, in analogy with sulfite reduction. However, it has also been suggested that nonenzymatic reduction by reduced glutathione (GSH) may be the predominant mechanism for selenite reduction (Anderson 1993, Terry et al. 2000). Selenide can subsequently be coupled to OAS to form SeCys by means of OAS thiol lyase (also called cysteine synthase). OAS is synthesized by the enzyme serine acetyl transferase (SAT) and also functions as a signal molecule that upregulates the activity of sulfate transporters and sulfate assimilation enzymes.

SeCys can be converted to SeMet by means of three enzymes (Figure 14.1). Cystathionine-gamma-synthase (CgS) first couples SeCys to O-phosphohomoserine (OPH) to form Se-cystathionine. Cystathionine-gamma-lyase further converts Se-cystathionine to Se-homocysteine. Se-homocysteine is then converted to SeMet by Met synthase. SeCys and SeMet can also be (mis)incorporated into proteins, replacing Cys and Met; this is thought to be an important reason for the toxicity of Se. SeMet has multiple other possible fates, one of which is to be methylated via methionine methyltransferase (MMT). Methyl-SeMet can be further metabolized to volatile DMSe, which is cleaved off of the intermediate, dimethylselenopropionate (DMSeP), by DMSeP lyase.

SeCys can also be converted to elemental Se (Se(0)), via the action of a SeCys lyase (SL). NifS-like enzymes with SL activity have been found in both chloroplasts and mitochondria (Pilon et al. 2003). Overexpression of the chloroplastic plant SL (called CpNifS) reduced incorporation of Se into proteins and enhanced Se accumulation (Van Hoewyk et al. 2005). Whether this SL activity has any function *in vivo* is questionable: the main function of the NifS-like enzymes in plants is likely to act as Cys desulfurases in S metabolism, providing elemental S for iron–sulfur cluster formation (Van Hoewyk et al. 2007).

Another possible fate of SeCys is to be methylated by SeCys methyltransferase (SMT). SMT enzyme activity is particularly pronounced in hyperaccumulators, and as a result, these species accumulate Se predominantly in the form of methyl-SeCys when supplied with selenate, while most other species accumulate selenate (de Souza et al. 1998, Freeman et al. 2006b). Since Methyl-SeCys does not enter proteins, it can be safely accumulated, explaining in part the Se tolerance of hyperaccumulators. *Brassica oleracea* (broccoli) also has an SMT enzyme, which is only expressed in the presence of Se (Lyi et al. 2005). Thus, some Se accumulators may use the same detoxification mechanism as hyperaccumulators. Methyl-SeCys can be further converted to volatile DMDSe, the predominant volatile form of Se produced by Se hyperaccumulators (Terry et al. 2000, Kubachka et al. 2007). Hyperaccumulators have also been found to couple glutamate to methyl-SeCys, to form gamma-glutamyl-methyl-SeCys, a major storage form of Se in hyperaccumulator seeds (Freeman et al. 2007, Kubachka et al. 2007). The enzyme mediating this reaction is likely gamma-glutamylcysteine synthetase (ECS). In S metabolism, this same enzyme functions in glutathione production (Glu-Cys-Gly). Reduced glutathione (GSH) has many redox functions in cells and also is a negative regulator of sulfate uptake and assimilation.

Manipulation of Plant Se Metabolism Using Genetic Engineering

Various transgenic approaches have been used successfully to enhance Se accumulation, tolerance, and volatilization by plants, particularly through upregulation of key genes involved in S/Se assimilation and volatilization (Figures 14.1 and 14.2). Overexpression in *Brassica juncea* (Indian mustard) of ATP sulfurylase (APS), involved in selenate-to-selenite conversion, resulted in enhanced selenate reduction: the transgenic APS plants accumulated an organic form of Se when supplied with selenate, while wild-type controls accumulated selenate (Pilon-Smits et al. 1999). The APS transgenics accumulated twofold to threefold more Se than the wild type, and 1.5-fold more S. The APS plants also tolerated the accumulated Se better than wild type, perhaps because of the organic form of Se accumulated.

FIGURE 14.2
Se fluxes through plants growing in a field. Plants can locally accumulate and change the speciation of Se in their adjacent soil. They can accumulate Se in their roots and shoots and assimilate inorganic to organic Se in the process. They can also volatilize Se and release it into the atmosphere from their roots and shoots.

Se volatilization rate was not affected in the APS transgenics. Furthermore, overexpression in *B. juncea* of the first enzyme in the conversion of SeCys to SeMet, CgS, resulted in two- to threefold higher volatilization rates compared to untransformed plants (Van Huysen et al. 2003). The CgS transgenics accumulated 40% less Se in their tissues than wild type, probably because of enhanced volatilization. The CgS transgenics were also more Se tolerant than wild-type plants, perhaps due to their lower tissue Se levels.

Another approach to genetically manipulate plant Se metabolism focused on the prevention of the toxicity caused by nonspecific SeCys incorporation into proteins. In one study, a mouse SeCys lyase (SL) was expressed in *Arabidopsis thaliana* and *B. juncea* (Garifullina et al. 2003, Pilon et al. 2003). This enzyme specifically breaks down SeCys into alanine and elemental Se. As expected, the SL transgenics showed reduced Se incorporation into proteins (Pilon et al. 2003). All the transgenic SL plants showed enhanced Se accumulation, up to twofold compared to wild-type plants. Similar results were obtained when an *A. thaliana* homologue of the mouse SL (called CpNifS) was overexpressed: the CpNifS transgenics showed less Se incorporation in proteins and twofold enhanced Se accumulation, as well as enhanced Se tolerance (Van Hoewyk et al. 2005). In another study, SMT from hyperaccumulator *A. bisulcatus* was overexpressed in *A. thaliana* and *B. juncea* (Ellis et al. 2004, LeDuc et al. 2004). The SMT transgenics showed enhanced Se accumulation, in the form of methyl-SeCys, as well as enhanced Se tolerance. The expression of SMT also resulted in increased rates of Se volatilization, with more Se volatilized in the form of DMDSe.

Although the expression of *A. bisulcatus* SMT enhanced Se tolerance, accumulation, and volatilization, the effects were more pronounced when the plants were supplied with selenite as opposed to selenate. This suggests that the conversion of selenate to selenite is a rate-limiting step for the production of SeCys. Therefore, APS and SMT *B. juncea* transgenics were crossed to create double-transgenic plants. The APS × SMT double transgenics accumulated up to nine times higher Se levels than wild type (LeDuc et al. 2006). Most of the Se in the double transgenics was in the form of methyl-SeCys: the APS × SMT plants accumulated up to eightfold more methyl-SeCys than wild type and nearly twice as much as the SMT transgenics. Se tolerance was similar in the single and double transgenics.

These genetic engineering studies demonstrate that the sulfate assimilation and volatilization pathway is responsible for selenate assimilation and volatilization. APS appears to be a rate-limiting enzyme for the assimilation of selenate to organic Se, and CgS is rate-limiting for DMSe volatilization. Enhanced APS expression appears to trigger selenate uptake and Se and S accumulation, perhaps due to upregulation of sulfate transporter expression. The results from the SL and CpNifS transgenics show that specific breakdown of SeCys can reduce nonspecific incorporation of Se into proteins, enhancing Se tolerance. Overexpression of SL or CpNifS led to enhanced Se accumulation, suggesting that introduction of this new sink for Se upregulates Se and S uptake. The results from the SMT transgenics show that SMT is a key enzyme for Se hyperaccumulation, conferring enhanced Se tolerance and accumulation when expressed in nonhyperaccumulators. However, for improved Se assimilation and detoxification, APS needs to be overexpressed together with SMT. APS × SMT double transgenics combine the ability to reduce selenate to selenite and SeCys with the ability to methylate SeCys and thus to detoxify the increased pool of internal Se. While APS × SMT double transgenics show significantly enhanced Se tolerance and accumulation, they do not approach the performance of Se hyperaccumulator species. Further research is needed to identify additional Se tolerance and accumulation genes in these specialized plant species.

Under laboratory conditions, the different transgenics showed enhanced Se tolerance, up to ninefold higher Se accumulation and up to threefold faster Se volatilization. These properties are all useful for phytoremediation. To test the transgenics' potential for phytoremediation, they were analyzed for their capacity to accumulate Se from naturally seleniferous soil and from Se-contaminated sediment. When grown on naturally seleniferous soil in a greenhouse pot experiment, the APS transgenics accumulated Se to threefold higher levels than wild-type *B. juncea*, and the CgS transgenics contained 40% lower Se levels than wild type (Van Huysen et al. 2004), all in agreement with the laboratory results. Plant growth was the same for all plant types in this experiment. Two field experiments were carried out on Se (selenate)-contaminated sediment in the San Joaquin Valley (California) by Bañuelos et al. (2005, 2007). The APS transgenics accumulated Se to fourfold higher levels than wild-type *B. juncea*, and cpSL and SMT transgenics showed twofold higher Se accumulation than wild-type *B. juncea*, all in agreement with earlier laboratory experiments. Biomass production was comparable for the different plant types. Thus, in the field as well as the lab, the various transgenics showed enhanced Se accumulation, volatilization, and/or tolerance, all promising traits for use in phytoremediation or as Se-fortified foods.

New Insight into Se Tolerance and Accumulation Mechanisms from Integrated Genomic, Genetic, and Biochemical Approaches

In order to obtain new insight into key genes that control Se uptake, hyperaccumulation, and volatilization, several studies were done in the past years using either model plant species or hyperaccumulators. First, a comparative study was performed using recombinant inbred lines (RIL) of model species *A. thaliana*. Several quantitative trait loci (QTL) were identified that cosegregated with the higher selenate tolerance in accession Columbia compared with accession Landsberg erecta (Zhang et al. 2006a). Genes in the identified QTL regions include a homologue of SeCys methyl transferase (SMT), ATP sulfurylase, and SAT. The results from the QTL study strongly suggested that tolerance to selenate and selenite is controlled by different loci. In another study, tolerance to and accumulation of Se were found to not be correlated in a study comparing 19 different ecotypes of *Arabidopsis* with variable tolerance to Se (Zhang et al. 2006b). Based on these results, it should be possible to breed plants that both accumulate and tolerate Se well.

In another approach to identify key genes modulating the Se response, a transcriptome study was performed on plants grown on selenate and control media for 10 days (Van Hoewyk et al. 2008). Genes involved in ethylene and jasmonic acid (JA) pathways were upregulated by Se. Furthermore, *Arabidopsis* mutants with a defect in genes involved in ethylene synthesis (acs6), ethylene signaling (ein2-1 and ein3-1), and JA signaling (jar1-1) showed reduced tolerance to selenate, and overexpression of a protein involved in ethylene signaling (ERF1) increased selenate resistance (Van Hoewyk et al. 2008). A similar study by Tamaoki et al. (2008) also implicated the involvement of the two hormones ethylene and JA in selenite resistance (Tamaoki et al. 2008) and additionally suggested that reactive oxygen species (ROS) may have a signaling role. The resistance mechanism appears to involve enhanced sulfate uptake and reduction; this may serve to prevent Se from replacing S in proteins and other S compounds.

The genus *Stanleya*, which is from the same family as *Arabidopsis* (Brassicaceae), contains hyperaccumulators and nonhyperaccumulators. In *Stanleya*, similar Se tolerance mechanisms were found to those described earlier for *Arabidopsis* (Freeman et al. 2010). The plant hormones JA and ethylene, as well as the hormone salicylic acid, appear to play a role in regulating Se stress response in *Stanleya*. Probably as a response to the elevated levels of these hormones, hyperaccumulators have constitutively upregulated expression of S transporters and assimilatory enzymes, and hence higher levels of total S, reduced S compounds (including the antioxidant glutathione), and higher levels of total Se. In addition, the Se hyperaccumulator *Stanleya* species showed interesting Se sequestration patterns that were not observed in nonhyperaccumulators. Around 90% of the accumulated Se was present as methyl-SeCys in vacuoles of leaf epidermal cells (Freeman et al. 2006b, 2010). This may indicate Se-specific transport mechanisms into these specialized cell types in hyperaccumulators. In future research, it will be very interesting to identify such specialized Se transporters, as well as the key genes that bring about the enhanced phytohormone levels that appear to trigger the cascade of reactions that together bring about the hyperaccumulation syndrome.

Ecological Aspects of Se Phytoremediation

Contribution of Microbes to Plant Se Uptake and Volatilization

Most plant species have one or more types of symbiosis with bacteria and fungi, organisms that are thought to play an important role in the biogeochemistry of Se by mineralization, immobilization, or volatilization. The potential effects of associated microbes on the fate of Se in their host plants offer an interesting area of study (Thompson-Eagle et al. 1989, de Souza and Terry 1997, Pankiewicz et al. 2006). Different types of plant–microbial associations may have different effects on plant Se accumulation and volatilization. Endophytic or epiphytic bacteria and fungi may affect Se accumulation within tissues or on plant surfaces, and root-associated fungi and bacteria may volatilize or change the oxidation state of Se, making it more or less available for plant uptake (Garbisu et al. 1996, Dowdle and Oremland 1998). Selenate- and selenite-reducing bacteria have, for instance, been isolated from the rhizosphere of the Se hyperaccumulator *A. bisulcatus* (Di Gregorio et al. 2005, Vallini et al. 2005). The two more oxidized forms, Se(IV) and Se(VI), are relatively bioavailable to plants in the rhizosphere, as opposed to the more reduced elemental form of Se. Thus, Se oxidative bacteria may enhance Se availability to the plant while Se reducers may immobilize Se making it less available to the plant. In addition to bacteria, unicellular, polymorphic, and filamentous fungi have the ability to reduce selenite to elemental Se (Gharieb et al. 1995); volatilization of Se has also been reported in fungi (Fleming and Alexander 1972, Thompson-Eagle et al. 1989).

In several studies, bacteria have shown to contribute to plant Se uptake and volatilization. In broccoli (*B. oleracea*), 95% of Se root volatilization was inhibited when roots were treated with bacterial antibiotics (Zayed and Terry 1994). Also, Indian mustard (*B. juncea*) treated with antibiotics volatilized 30% less Se and accumulated 70% less Se than untreated plants. In addition, Indian mustard plants grown from surface-sterilized seeds that were inoculated with rhizospheric bacteria accumulated fivefold more Se and volatilized fourfold more Se than control plants from seeds that were not inoculated with bacteria. The mechanism for the stimulatory effect by the bacteria appeared to be both stimulation of root growth and stimulation of plant S/Se uptake and assimilation. When inoculated with rhizospheric bacteria, the plants had increased root surface area, and the culture media contained ninefold higher serine levels than control plants. OAS is known to stimulate sulfate uptake and assimilation (de Souza et al. 1999). In another study, Di Gregorio et al. (2006) using *B. juncea* grown in soil spiked with selenite and selenate also showed that rhizobacteria stimulated *B. juncea* Se uptake and volatilization and that the bacteria contributed to the reduction of these oxyanions in the soil.

As for the effects of plant-associated fungi on plant Se uptake and volatilization, much less is known. In one study, ryegrass (*Lolium* spp.) accumulated less Se when treated with the mycorrhizal fungus *Glomus mosseae* compared to controls lacking this fungus (Munier-Lamy et al. 2007). There is also virtually nothing known about the role endophytic microbes play in plant Se uptake, metabolism, and volatilization.

Effects of Plant Se on Ecological Partners

The high Se levels in hyperaccumulators likely play an important role in the ecology of these plants, and even Se accumulated in crop plants may have ecological effects (Figure 14.3). The best-known effects of plant Se hyperaccumulation on other species are the toxic effects of plant Se on livestock herbivores. Se ingestion of hyperaccumulator plants has been reported to be responsible for poisoning and death of cattle, sheep, and horses to the extent of hundreds of millions of U.S. dollars annually in the United States alone (Wilber 1980). Thus, hyperaccumulators may sequester Se as a defense against herbivory. In support of this elemental defense hypothesis (first proposed by Boyd and Martens 1992), laboratory and field studies showed that Se accumulation can protect plants from a range of herbivores and pathogens, from prairie dogs to a variety of arthropods and fungi (Hanson et al. 2003, 2004, Freeman et al. 2006a, 2007, 2009, Quinn et al. 2008, 2011a). Se protected the plants both through deterrence of herbivores and toxicity. Also in support of the elemental defense hypothesis, Se-hyperaccumulating species

FIGURE 14.3
Se hyperaccumulator *S. pinnata* growing in its natural habitat and some of its natural ecological partners. Plant Se accumulation has been shown to protect against herbivory by grasshoppers (top left) and prairie dogs (bottom right) and to reduce infestation with root nematodes (bottom center). Some ecological partners appear to have evolved Se tolerance, such as a diamondback moth (top center) and native bumblebees (top right), as well as certain nematodes and certain neighboring plant species (bottom left).

harbored fewer arthropod species and individuals in their natural habitat than comparable Se nonaccumulators (Galeas et al. 2008).

While methyl-SeCys is not toxic by itself, herbivores that ingest hyperaccumulator plant material readily demethylate it to SeCys (Freeman et al. 2006b), which is toxic because of its nonspecific incorporation into proteins. Thus, Se hyperaccumulation is an effective plant defense mechanism against herbivory. For most plant defenses, over time some herbivores evolve tolerance. This is also true for Se hyperaccumulation. A population of diamondback moths living in a seleniferous area on hyperaccumulator *S. pinnata* was shown to be completely Se tolerant (Freeman et al. 2006b).

The Se-tolerant moth accumulated Se in the ingested form, methyl-SeCys, which is not incorporated into proteins, while a Se-sensitive population of diamondback moths from a nonseleniferous habitat converted the ingested methyl-SeCys to SeCys (Freeman et al. 2006b).

Herbivores may avoid Se-rich plant tissue as another strategy to minimize Se toxicity. Se is not distributed evenly throughout Se-hyperaccumulating plants, and Se levels fluctuate over the growing season. Se concentrations in leaves peak in early spring, and young leaves and reproductive tissues contain much higher Se levels than older leaves (Galeas et al. 2007). Also, flowers of hyperaccumulators have higher Se levels than leaves or roots, and within the flowers of *S. pinnata*, the stamens and pistils have a higher Se concentration than the petals and sepals (Quinn et al. 2011b). As mentioned earlier, hyperaccumulators allocate Se to the periphery of the leaf. In *S. pinnata*, Se is stored primarily in specialized cells in the epidermis, and in *A. bisulcatus*, Se is stored primarily in leaf hairs (Freeman et al. 2006b). Therefore, it appears that hyperaccumulators preferentially allocate Se to areas that first come into contact with attackers, and predominantly in their most valuable tissues, for maximal protection from herbivores and pathogens. Sequestration in the epidermis may also contribute to Se tolerance, since it keeps the Se away from metabolically sensitive processes. Depending on herbivore feeding mode, Se hyperaccumulation may be more or less effective against different herbivores. In view of the particularly high Se levels in the flowers, Se may also have an effect on pollination ecology. This will be an interesting area of further research. Recent studies (Quinn et al. 2011b) indicate that insect pollinators in seleniferous areas can accumulate substantial Se levels and that high-Se plants are visited at similar rates as low-Se plants. It will be interesting to study whether this Se has any toxic effects on the pollinators.

As already discussed earlier, plant-associated microbes may affect plant Se metabolism. Conversely, the elevated Se levels in and around (hyper) accumulator plants also appear to affect local microbial communities. Soil around Se-hyperaccumulating plants has a ~10-fold higher Se concentration than the surrounding bulk soil (El Mehdawi et al. 2011a), and there is evidence that rhizospheric and saprophytic fungi from seleniferous areas have evolved enhanced Se tolerance (Wangeline et al. 2011, Quinn et al. 2011b).

Another class of ecological interaction that may be affected by Se is that between hyperaccumulators and their plant neighbors. As mentioned earlier, Se levels around hyperaccumulators tend to be higher compared to other soil in the area. Recent data (El Mehdawi et al. 2011b) indicate that neighbors of these high-Se plants have significantly higher Se levels when growing next to hyperaccumulators. Depending on whether the neighbors are Se tolerant or sensitive, this increased Se concentration may have a positive or negative effect. If they are tolerant, they may benefit from the Se in a similar way to the hyperaccumulators themselves, enjoying less herbivory. If they are sensitive, on the other hand, they may suffer reduced germination and growth.

Future Prospects

Further studies building on the genomic and biochemical studies described earlier may reveal key genes that trigger the cascade of responses that together provide Se tolerance and accumulation in model plants and hyperaccumulators. These studies may also reveal genes that encode specific transporters of selenocompounds into and within hyperaccumulators. These key genes will be very interesting candidates for overexpression studies, with the potential of transferring the complete Se hyperaccumulator profile into high-biomass crop species.

Se accumulation in plants appears to have important ecological implications. More research is needed on the role microbes play in plant Se uptake and volatilization, and the movement of Se through the food chain via Se hyperaccumulators or Se-fortified crop plants. The role of Se in belowground ecological interactions with microbes and other organisms is also a relatively unexplored area. Plant Se may affect root–microbe interactions and may protect plants from root-feeding herbivores. Moreover, selenocompounds released from hyperaccumulator roots may affect surrounding vegetation. The effects of Se on pollination ecology will also be an interesting field of further study.

Together, better insight into the processes involved in plant metabolism of Se, the limiting factors involved, the contributions of ecological partners, and the effects of Se on ecological partners are all useful in order to minimize any potential harmful effects of Se while benefiting from the positive effects of plant Se on animal and human health. The capacity of plants to accumulate and volatilize Se will be very useful for the phytoremediation of Se-contaminated soils and waters (Bañuelos et al. 2002). When plant Se accumulation is managed well, plants offer an efficient and cost-effective means to remove Se from the environment. Plants are also an effective source of dietary Se, and therefore, Se-enriched plant material from phytoremediation or other sources may be considered fortified food. After being grown on Se-contaminated soil or being irrigated with Se-contaminated water, the Se-laden plant material may be used as a feed supplement for livestock, and/or as a biofuel. The potential of this strategy may be further enhanced by the use of selected transgenic lines. Of course the use of any Se-accumulating wild-type or transgenic plants should be accompanied by careful risk assessment, to avoid escape of transgenes and to minimize any adverse ecological effects of plant-accumulated Se.

Acknowledgment

National Science Foundation grant # IOS-0817748 to EAHPS supported the writing of this chapter.

References

Anderson JW. 1993. Selenium interactions in sulfur metabolism. In: De Kok LJ (ed.) *Sulfur Nutrition and Assimilation in Higher Plants—Regulatory, Agricultural and Environmental Aspects*. SPB Academic publishing, The Hague, the Netherlands, pp. 49–60.

Bañuelos G, LeDuc DL, Pilon-Smits EAH, Tagmount A, Terry N. 2007. Transgenic Indian mustard overexpressing selenocysteine lyase, selenocysteine methyltransferase, or methionine methyltransferase exhibit enhanced potential for selenium phytoremediation under field conditions. *Environmental Science and Technology* 41: 599–605.

Bañuelos GS, Lin ZQ, Wu L, Terry N. 2002. Phytoremediation of selenium-contaminated soils and waters: Fundamentals and future prospects. *Reviews on Environmental Health* 4: 291–306.

Bañuelos G, Terry N, LeDuc DL, Pilon-Smits EAH, Mackey B. 2005. Field trial of transgenic Indian mustard plants shows enhanced phytoremediation of selenium contaminated sediment. *Environmental Science and Technology* 39: 1771–1777.

Beath OA, Gilbert CS, Eppson HF. 1939a. The use of indicator plants in locating seleniferous areas in Western United States. I. General. *American Journal of Botany* 26: 257–269.

Beath OA, Gilbert CS, Eppson HF. 1939b. The use of indicator plants in locating seleniferous areas in Western United States. II. Correlation studies by states. *American Journal of Botany* 26: 296–315.

Boyd RS, Martens SN. 1992. *The raison d'être for metal for metal hyperaccumulation by plants*. In: Baker AJM, Proctor J, Reeves RD (eds.) The vegetation of ultramafic (Serpentine) soils. Intercept, Andover, UK, pp. 279–289.

Broyer TC, Huston RP, Johnson CM. 1972. Selenium and nutrition of Astragalus. 1. Effects of selenite or selenate supply on growth and selenium content. *Plant and Soil* 36: 635–649.

Cartes P, Gianfreda L, Mora ML. 2005. Uptake of selenium and its antioxidant activity in ryegrass when applied as selenate and selenite forms. *Plant and Soil* 276: 359–367.

de Souza MP, Chu D, Zhao M, Zayed AM, Ruzin SE, Schichnes D, Terry N. 1999. Rhizosphere bacteria enhance selenium accumulation and volatilization by Indian mustard. *Plant Physiology* 119: 565–574.

de Souza MP, Pilon-Smits EAH, Lytle CM et al. 1998. Rate-limiting steps in selenium volatilization by *Brassica juncea*. *Plant Physiology* 117: 1487–1494.

de Souza MP, Terry N. 1997. Selenium volatilization by rhizosphere bacteria. *Abstracts of the General Meeting of the American Society for Microbiology* 97: 499.

Di Gregorio S, Lampis S, Malorgio R, Petruzzelli G, Pezzarossa B, Vallini G. 2006. *Brassica juncea* can improve selenite and selenate abatement in selenium contaminated soils through the aid of its rhizospheric bacterial population. *Plant and Soil* 285: 233–244.

Di Gregorio S, Lampis S, Vallini G. 2005. Selenite precipitation by a rhizospheric strain of *Stenotrophomonas* sp. isolated from the root system of *Astragalus bisulcatus*: A biotechnological perspective. *Environment International* 31: 233–241.

Diwadkar-Navsariwala V, Prins GS, Swanson SM et al. 2006. Selenoprotein deficiency accelerates prostate carcinogenesis in a transgenic model. *Proceedings of the National Academy of Sciences USA* 103: 8179–8184.

Djanaguiraman M, Durga Devi D, Shanker AK, Sheeba JA, Bangarusamy U. 2005. Selenium - an antioxidative protectant in soybean during selenscence. *Plant and Soil* 272: 77–86.

Dowdle PR, Oremland RS. 1998. Microbial oxidation of elemental selenium in soils and bacterial cultures. *Environmental Science and Technology* 32: 3749–3755.

Draize JH, Beath OA. 1935. Observation on the pathology of "blind staggers" and "alkali disease." *Journal of the American Veterinary Medical Association* 86: 53–763.

Ellis DR, Sors TG, Brunk DG et al. 2004. Production of Se-methylselenocysteine in transgenic plants expressing selenocysteine methyltransferase. *BMC Plant Biology* 4: 1–11.

El-Mehdawi AF, Quinn CF, Pilon-Smits EAH. 2011a. Effects of selenium hyperaccumulation on plant-plant interactions: Evidence for elemental allelopathy? *New Phytologist* 191: 120–131.

El-Mehdawi AF, Quinn CF, Pilon-Smits EAH. 2011b. Selenium Hyperaccumulators Facilitate Selenium-Tolerant Neighbors via Phytoenrichment and Reduced Herbivory. *Current Biology* 21: 1440–1449.

Fleming RW, Alexander M. 1972. Dimethylselenide and dimethyltelluride formation by a strain of *Penicillium*. *Applied Microbiology* 24: 424–429.

Freeman JL, Lindblom SD, Quinn CF, Fakra S, Marcus MA, Pilon-Smits EAH. 2007. Selenium accumulation protects plants from herbivory by orthoptera due to toxicity and deterrence. *New Phytologist* 175: 490–500.

Freeman JL, Quinn CF, Lindblom SD, Klamper EM, Pilon-Smits EAH. 2009. Selenium protects the hyperaccumulator *Stanleya pinnata* against black-tailed prairie dog herbivory in native seleniferous habitats. *American Journal of Botany* 96: 1075–1085.

Freeman JL, Quinn CF, Marcus MA, Fakra S, Pilon-Smits EAH. 2006a. Selenium tolerant diamondback moth disarms hyperaccumulator plant defense. *Current Biology* 16: 2181–2192.

Freeman JL, Tamaoki M, Stushnoff C et al. 2010. Molecular mechanisms of selenium tolerance and hyperaccumulation in *Stanleya pinnata*. *Plant Physiology* 153: 1630–1652.

Freeman JL, Zhang LH, Marcus MA, Fakra S, McGrath SP, Pilon-Smits EAH. 2006b. Spatial imaging, speciation and quantification of selenium in the hyperaccumulator plants *Astragalus bisulcatus* and *Stanleya pinnata*. *Plant Physiology* 142: 124–134.

Fu LH, Wang XF, Eyal Y et al. 2002. A selenoprotein in the plant kingdom: Mass spectrometry confirms that an opal codon (UGA) encodes selenocysteine in *Chlamydomonas reinhardtii* glutathione peroxidase. *The Journal of Biological Chemistry* 277: 25983–25991.

Galeas ML, Klamper EM, Bennett LE, Freeman JL, Kondratieff BC, Pilon-Smits EAH. 2008. Selenium hyperaccumulation affects plant arthropod load in the field. *New Phytologist* 177: 715–724.

Galeas ML, Zhang LH, Freeman JL, Wegner M, Pilon-Smits EAH. 2007. Seasonal fluctuations of selenium and sulfur accumulation in selenium hyperaccumulators and related non-accumulators. *New Phytologist* 173: 517–525.

Garbisu C, Ishii T, Leighton T, Buchanan BB. 1996. Bacterial reduction of selenite to elemental selenium. *Chemical Geology* 132: 199–204.

Garifullina GF, Owen JD, Lindblom S-D, Tufan H, Pilon M, Pilon-Smits EAH. 2003. Expression of a mouse selenocysteine lyase in *Brassica juncea* chloroplasts affects selenium tolerance and accumulation. *Physiologia Plantarum* 118: 538–544.

Gharieb MM, Wilkinson SC, Gadd GM. 1995. Reduction of selenium oxyanions by unicellular, polymorphic and filamentous fungi: Cellular location of reduced selenium and implications for tolerance. *Journal of Industrial Microbiology* 14: 300–311.

Hansen D, Duda PJ, Zayed A, Terry N. 1998. Selenium removal by constructed wetlands: Role of biological volatilization. *Environmental Science and Technology* 32: 591–597.

Hanson BR, Garifullina GF, Lindblom SD et al. 2003. Selenium accumulation protects *Brassica juncea* from invertebrate herbivory and fungal infection. *New Phytologist* 159: 461–469.

Hanson BR, Lindblom SD, Loeffler ML, Pilon-Smits EAH. 2004. Selenium protects plants from phloem-feeding aphids due to both deterrence and toxicity. *New Phytologist* 162: 655–662.

Harris T. 1991. *Death in the Marsh*. Island Press, Washington, DC.

Hartikainen H. 2005. Biogeochemistry of selenium and its impact on food chain quality and human health. *Journal of Trace Elements in Medicine and Biology* 18: 309–318.

Hawkesford MJ. 2003. Transporter gene families in plants: The sulphate transporter gene family—Redundancy or specialization? *Physiologia Plantarum* 117: 155–163.

Kabata-Pendias A. 1998. Geochemistry of selenium. *Journal of Environmental Pathology, Toxicology and Oncology* 17: 173–177.

Kubachka KM, Meija J, LeDuc DL, Terry N, Caruso JA. 2007. Selenium volatiles as proxy to the metabolic pathways of selenium in genetically modified *Brassica juncea*. *Environmental Science and Technology* 41: 1863–1869.

LeDuc DL, AbdelSamie M, Montes-Bayón M, Wu CP, Reisinger SJ, Terry N. 2006. Overexpressing both ATP sulfurylase and selenocysteine methyltransferase enhances selenium phytoremediation traits in Indian mustard. *Environmental Pollution* 144: 70–76.

LeDuc DL, Tarun AS, Montes-Bayon M et al. 2004. Overexpression of selenocysteine methyltransferase in *Arabidopsis* and Indian mustard increases selenium tolerance and accumulation. *Plant Physiology* 135: 377–383.

Leustek T. 1996. Molecular genetics of sulfate assimilation in plants. *Physiologia Plantarum* 97: 411–419.

Lewis BG, Johnson CM, Delwiche CC. 1966. Release of volatile selenium compounds by plants: Collection procedures and preliminary observations. *Journal of Agricultural and Food Chemistry* 14: 638–640.

Lyi SM, Heller LI, Rutzke M, Welch RM, Kochian LV, Li L. 2005. Molecular and biochemical characterization of the selenocysteine Se-methyltransferase gene and Se-methylselenocysteine synthesis in broccoli. *Plant Physiology* 138: 409–420.

Lyons GH, Genc Y, Soole K, Stangoulis JCR, Liu F, Graham RD. 2009. Selenium increases seed production in *Brassica*. *Plant and Soil* 318: 73–80.

Maruyama-Nakashita A, Nakamura Y, Yamaya T, Takahashi H. 2004. A novel regulatory pathway of sulfate uptake in *Arabidopsis* roots: Implication of CRE1/WOL/AHK4-mediated cytokinin-dependent regulation. *Plant Journal* 38: 779–789.

Munier-Lamy C, Deneux-Mustin S, Mustin C, Merlet D, Berthelin J, Leyval C. 2007. Selenium bioavailability and uptake as affected by four different plants in a loamy clay soil with particular attention to mycorrhizae inoculated ryegrass. *Journal of Environmental Radioactivity* 97: 148–158.

Neuhierl B, Böck A. 1996. On the mechanism of selenium tolerance in selenium accumulating plants. Purification and characterization of a specific selenocysteine methyltransferase from cultured cells of *Astragalus bisulcatus*. *European Journal of Biochemistry* 239: 235–238.

Neuhierl B, Thanbichler M, Lottspeich F, Böck A. 1999. A family of S-methylmethionine dependent thiol/selenol methyltransferases. Role in selenium tolerance and evolutionary relation. *The Journal of Biological Chemistry* 274: 5407–5414.

Novoselov SV, Rao M, Onoshko NV et al. 2002. Selenoproteins and selenocysteine insertion system in the model plant system, *Chlamydomonas reinhardtii*. *EMBO Journal* 21: 3681–3693.

Ohlendorf HM, Hoffman DJ, Salki MK, Aldrich TW. 1986. Embryonic mortality and abnormalities of aquatic birds: Apparent impacts of selenium from irrigation drain water. *Science of the Total Environment* 52: 49–63.

Pankiewicz U, Jamroz J, Schodziński A. 2006. Optimization of selenium accumulation in *Rhodotorula rubra* cells by treatment of culturing medium with pulse electric field. *International Agrophysics* 20: 147–152.

Persans MW, Salt DE. 2000. Possible molecular mechanisms involved in nickel, zinc and selenium hyperaccumulation in plants. *Biotechnology and Genetic Engineering Reviews* 17: 389–413.

Pilon M, Owen JD, Garifullina GF, Kurihara T, Mihara H, Esaki N, Pilon-Smits EAH. 2003. Enhanced selenium tolerance and accumulation in transgenic *Arabidopsis thaliana* expressing a mouse selenocysteine lyase. *Plant Physiology* 131: 1250–1257.

Pilon-Smits EAH, Hwang S, Lytle CM, Zhu Y, Tai JC, Bravo RC, Chen Y, Leustek T, Terry N. 1999. Overexpression of ATP sulfurylase in Indian mustard leads to increased selenate uptake, reduction, and tolerance. *Plant Physiology* 119: 123–132.

Pilon-Smits EAH, Quinn CF, Tapken W, Malagoli M, Schiavon M. 2009. Physiological functions of beneficial elements. *Current Opinion in Plant Biology* 12: 267–274.

Quinn CF, Freeman JF, Galeas ML, Klamper EM, Pilon-Smits EAH. 2008. Selenium protects plants from prairie dog herbivory—Implications for the functional significance and evolution of Se hyperaccumulation. *Oecologia* 155: 267–275.

Quinn CF, Wyant K, Wangeline AL et al. 2011a. Selenium hyperaccumulation increases leaf decomposition rate in a seleniferous habitat. *Plant and Soil* 341: 51–61.

Quinn CF, Prins CN, Gross AM et al. 2011b. Selenium accumulation in flowers and its effects on pollination. *New Phytologist* 192: 727–737.

Rosenfeld I, Beath OA. 1964. *Selenium, Geobotany, Biochemistry, Toxicity, and Nutrition.* Academic Press, New York.

Smith FW, Ealing PM, Hawkesford MJ, Clarkson DT. 1995. Plant members of a family of sulfate transporters reveal functional subtypes. *Proceedings of the National Academy of Sciences USA* 92: 9373–9377.

Sors TG, Ellis DR, Salt DE. 2005. Selenium uptake, translocation, assimilation and metabolic fate in plants. *Photosynthesis Research* 86: 373–389.

Stadtman TC. 1990. Selenium biochemistry. *Annual Review of Biochemistry* 59: 111–127.

Stadtman TC. 1996. Selenocysteine. *Annual Review of Biochemistry* 65: 83–100.

Tamaoki M, Freeman JL, Pilon-Smits EAH. 2008. Cooperative ethylene and jasmonic acid signaling regulates selenite resistance in *Arabidopsis thaliana*. *Plant Physiology* 146: 1219–1230.

Terry N, Zayed AM, de Souza MP, Tarun AS. 2000. Selenium in higher plants. *Annual Review of Plant Biology* 51: 401–432.

Thompson-Eagle ET, Frankenberger WT Jr, Karlson U. 1989. Volatilization of selenium by *Alternaria alternata*. *Applied and Environmental Microbiology* 55: 1406–1413.

Unni E, Koul D, Alfred Yung WK, Sinha R. 2005. Se-methylselenocysteine inhibits phosphatidylinositol 3-kinase activity of mouse mammary epithelial tumor cells in vitro. *Breast Cancer Research* 7: R699–R707.

Vallini G, Di Gregorio S, Lampis S. 2005. Rhizosphere-induced selenium precipitation for possible applications in phytoremediation of se polluted effluents. *Zeitschrift für Naturforschung* 60: 349–356.

Van Hoewyk D, Abdel-Ghany SE, Cohu C et al. 2007. The *Arabidopsis* cysteine desulfurase CpNifS is essential for maturation of iron-sulfur cluster proteins, photosynthesis, and chloroplast development. *Proceedings of the National Academy of Sciences USA* 104: 5686–5691.

Van Hoewyk D, Garifullina GF, Ackley AR et al. 2005. Overexpression of AtCpNifS enhances selenium tolerance and accumulation in *Arabidopsis*. *Plant Physiology* 139: 1518–1528.

Van Hoewyk D, Takahashi H, Hess A, Tamaoki M, Pilon-Smits EAH. 2008. Transcriptome and biochemical analyses give insights into selenium-stress responses and selenium tolerance mechanisms in *Arabidopsis*. *Physiology Plant* 132: 236–253.

Van Huysen T, Abdel-Ghany S, Hale KL, LeDuc D, Terry N, Pilon-Smits EAH. 2003. Overexpression of cystathionine-γ-synthase in Indian mustard enhances selenium volatilization. *Planta* 218: 71–78.

Van Huysen T, Terry N, Pilon-Smits EAH. 2004. Exploring the selenium phytoremediation potential of transgenic *Brassica juncea* overexpressing ATP sulfurylase or cystathionine γ-synthase. *International Journal of Phytoremediation* 6: 111–118.

Wangeline AL, Valdez JR, Lindblom SD, Bowling KL, Reeves FB, Pilon-Smits EAH. 2011. Selenium tolerance in rhizosphere fungi from Se hyperaccumulator and non-hyperaccumulator plants. *American Journal of Botany* 98: 1139–1147.

Whanger PD. 1989. China, a country with both selenium deficiency and toxicity: Some thoughts and impressions. *Journal of Nutrition* 119: 1236–1239.

White PJ, Bowen HC, Marshall B, Broadley MR. 2007. Extraordinarily high leaf selenium to sulfur ratios define 'Se-accumulator' plants. *Annals of Botany* 100: 111–118.

White PJ, Broadley MR. 2009. Biofortification of crops with seven mineral elements often lacking in human diets—Iron, zinc, copper, calcium, magnesium, selenium and iodine. *New Phytologist* 182: 49–84.

Wilber CG. 1980. Toxicology of selenium: A review. *Clinical Toxicology* 17: 171–230.

Wilson LG, Bandurski RS. 1958. Enzymatic reactions involving sulfate, sulfite, selenate and molybdate. *The Journal of Biological Chemistry* 233: 975–981.

Yoshimoto N, Inoue E, Saito K, Yamaya T, Takahashi H. 2003. Phloem-localizing sulfate transporter, Sultr1;3, mediates re-distribution of sulfur from source to sink organs in *Arabidopsis*. *Plant Physiology* 131: 1511–1517.

Yoshimoto N, Takahashi H, Smith FW, Yamaya T, Saito K. 2002. Two distinct high affinity sulfate transporters with different inducibilities mediate uptake of sulfate in *Arabidopsis* roots. *The Plant Journal* 29: 465–473.

Zayed AM, Terry N. 1994. Selenium volatilization in roots and shoots: Effects of shoot removal and sulfate level. *Journal of Plant Physiology* 143: 8–14.

Zhang LH, Ackley AR, Pilon-Smits EAH. 2006b. Variation in selenium tolerance and accumulation among nineteen *Arabidopsis* ecotypes. *Journal of Plant Physiology* 164: 327–336.

Zhang L, Byrne PF, Pilon-Smits EAH. 2006a. Mapping quantitative trait loci associated with selenate tolerance in *Arabidopsis thaliana*. *New Phytologist* 170: 33–42.

15

Microbial Transformation of Phthalate Esters: Diversity of Hydrolytic Esterases

Ji-Dong Gu and Yuping Wang

CONTENTS

Introduction

Phthalic acid esters or phthalate esters (PAEs) are derivatives of phthalic acid (PA). Structures of PAEs are shown in Figure 15.1. When methanol, ethanol, or other alcohols react with the carboxyl groups on the benzene ring of phthalic acids, the corresponding esters are formed with different alkyl chains, depending on the alcohol involved in the reaction, such as dimethyl phthalate (DMP), dibutyl phthalate (DBP), and di(2-ethylhexyl) phthalate (DEHP). Each PAE has three isomers on which the carboxyl groups can be

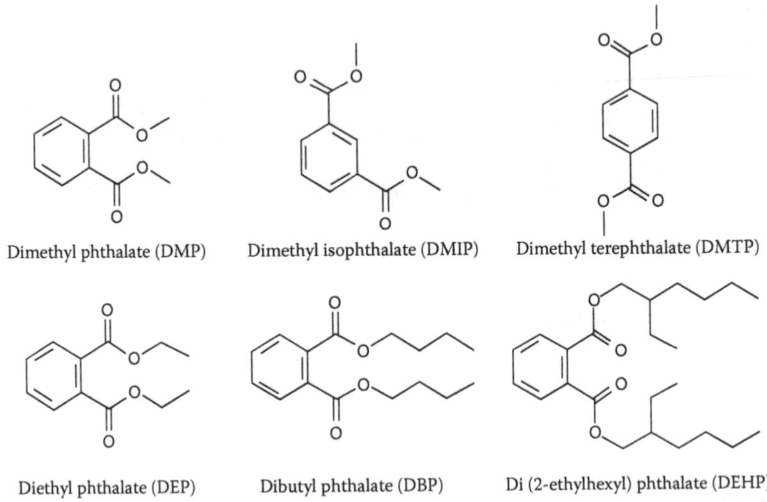

FIGURE 15.1

Chemical structures of several common phthalate esters, including phthalate ester isomers (DMP, DMIP, and DMTP), diethyl phthalate (DEP), dibutyl phthalate (DBP), and di(2-ethylhexyl) phthalate (DEHP).

at the *ortho, meta,* or *para* position. Meanwhile, the esterification forms not only diesters but also monoesters. Various reactions yield a large and diverse group of products of PAEs, with distinctive position and the length of the alkyl side chains.

Different chemical structures can have distinctive physicochemical properties for PAEs. The physical properties of selective PAEs are listed in Table 15.1. What should be noticed is the hydrophilic ability of PAEs, because hydrophilic ability affects the occurrence, distribution, and biotoxicity of a chemical in the environment. The existence of aromatic ring and ester groups on the ring generally results in low solubility of PAEs in water, with a pattern of decrease following the increase of the alkyl chain length. For example, with an increase of alkyl chain: DMP < diethyl phthalate (DEP) < diallyl phthalate (DAP), the solubility of PAEs in water generally follows: DMP (4200 mg/L) > DEP (1100 mg/L) > DAP (182 mg/L) (Staples et al. 1997). However, there are exceptional cases for high-molecular-weight PAEs. As mentioned by Howard et al. (1985), the diisononyl phthalate (DINP) has water solubility of 0.20 mg/L, whereas the solubility of diisooctyl phthalate (DIOP) is 0.09 mg/L. There is a large variation among the water solubility values, especially when measurement is obtained based on different methods. For example, in the case of di-*n*-octyl phthalate (DnOP), it can be 0.02 mg/L by conventional shake-flask/centrifugation method (Hollifield 1979). Staples et al. (1997), however, recommended the value of 0.0005 mg/L for the solubility of DnOP based on the data available. These results suggest that even though the data were published as a

TABLE 15.1

Physical Properties of Selected Phthalate Esters

	Dimethyl Phthalate	Dimethyl Isophthalate	Dimethyl Terephthalate	Di(2-Ethylhexyl) Phthalate	Dibutyl Phthalate
Abbreviation	DMP	DMIP	DMTP	DEHP	DBP
CAS RN	131-11-3	1459-93-4	120-61-6	117-81-7	84-74-2
Position of ester bond	*Ortho*	*Meta*	*Para*	*Ortho*	*Ortho*
Molecular weight	194.1866	194.1866	194.1866	390.5618	278.3474
Density	1.19	1.1477	1.2[a]	0.9732	1.043
Melting point	2°C	66°C–67°C	141°C	−50°C	−35°C
Boiling point	283.7°C	124°C at 12 mm Hg	288°C	386.9°C	340°C
Water solubility	4200 mg/L at 20°C[b]	Insoluble	28.7 mg/L at 20°C[c]	0.003 mg/L[b]	13 mg/L
Physical state	Colorless, oily liquid with a slight ester odor	White flake[d]	Colorless to white crystals	Colorless, oily liquid with almost no odor	Colorless, oily liquid with a very weak, aromatic odor

Source: http://chemfinder.cambridgesoft.com/
[a] http://www.chemicalland21.com/arokorhi/petrochemical/DIMETHYL%20TEREPHTHALATE.htm
[b] Staples et al., *Chemosphere*, 35, 667, 1997.
[c] Watanabe and Hamamura, *Current Opinion in Biotechnology*, 14, 289, 2003.
[d] http://www.chemicalland21.com/arokorhi/specialtychem/perchem/DIMETHYL%20ISOPHTHALATE.htm

result of objective observation, the methods used could greatly influence the final values.

A close relationship exists between water solubility and toxicity to organisms. Biotoxicity of an organic chemical to a particular organism depends on its partition coefficient in the animal or organism body, namely, K_p ($K_p = C_{solvent}/C_{water}$) (Crosby 1998). Normally, K_{ow}, partition coefficient of octanol (an organic solvent) and water, is applied to indicate the water solubility. The higher the value, the lower the solubility. The toxicity of chemicals can be measured through a range of concentrations to extrapolate the concentration at which 50% of the population show mortality (LC_{50}). The relationship between K_{ow} value and LC_{50} was evaluated by the significance of equation $\log_{10}(LC_{50}) = a \log_{10}(K_{ow}) + b$. In a report conducted to test several PAEs' toxicity, the slope a and constant b were calculated via linear regression

(Call et al. 2001a,b). As an example, the LC_{50} values of DMP, DEP, DBP, and butylbenzyl phthalate (BBP) were 28.1, 4.21, 0.63, and 0.46 mg/L, respectively, for the amphipod *Hyalella azteca*.

Industrial Application

PAEs are used as plasticizers in the plastic industries. Specifically, they are incorporated in the polyvinyl chloride (PVC), which has been used for over half a century. PAEs play an important role in increasing flexibility and softness to the products such as home furnishings, daily used containers, food packaging, medical devices, and children's toys (Staples et al. 1997). The most popular PAE, DEHP, has been widely used as the plasticizer in children's toys, toothbrushes, and clothing (National Toxicology Program 1995). In Germany, between 1994 and 1995, DEHP was consumed by more than 250,000 ton, and the consumption of total phthalates reached 400,000 ton (Fromme et al. 2002). DBP is used in the manufacturing of epoxy resins and cellulose esters. DMP, the simplest structure among the phthalate diesters, is utilized as plasticizer in cellulose ester–based plastics. These plasticizers, due to their benzene and alkyl chain structure, combine with the plastic molecules to form the noncovalent bond structure. The long alkyl chain structure helps reduce the rigidity of the polymers and increase the flexibility of their products.

Not only do PAEs soften the long-chain molecules but can also be used as the monomer in the polyester synthesis. For example, both dimethyl isophthalate (DMI) and dimethyl terephthalate (DMTP) are used in polymer manufacturing, for their ability to react with other esters in order to produce polyesters. They are mainly used in the manufacture of electric and electronic products, including video, film, electrical capacitors, and fibers (like polyethylene isophthalate-co-terephthalate) (Monarca et al. 1991, Lee et al. 1999). New applications have also been found recently. Ukielski (2000) has used these monomers to synthesize multiblock terpolymers as well. However, because of their potential carcinogenicity, their application is more restricted by regulations from the 1980s in many countries (Wilkinson and Lamb IV 1999).

Occurrence and Fate in the Environment

When PAEs are largely produced and heavily used in industrial processes, their fate in the environment has been regarded as a problem (Bauer et al. 1998). Intensive production inevitably increases the occurrence rate of this group of chemicals in the ambient environment. For the purpose of monitoring the behavior of PAEs and evaluating their risks to human health,

TABLE 15.2

Occurrence and Concentration of Several Phthalate Esters in the Environment

Phthalate Esters	Occurrence in Environment	Concentration	References
Dimethyl phthalate (DMP)	River	10 mg/L	Fajardo et al. (1997)
Dibutyl phthalate (DBP)	River	45 ppb	Mori (1976)
	Well water	0.19–9.26 µg/L	Hashizume et al. (2002)
	Sea surface microlayer	0.223 µg/L	Cincinelli et al. (2001)
	Subsurface layer	0.065 µg/L	Cincinelli et al. (2001)
Diethyl phthalate (DEP)	Sea sediment	0.049 µg/L	Peterson and Freeman (1982)
	Household waste	18.2 mg/kg	Bauer and Herrmann (1997)
Di(2-ethylhexyl) phthalate (DEHP)	Sea sediment	0.18 µg/L	Peterson and Freeman (1982)
	Household waste	16,820.6 mg/kg	Bauer and Herrmann (1997)
	Landfill leachate	346 µg/L	Paxéus (2000)

efforts have been made to determine the extent of the appearance of PAEs, clarify the sources of PAEs, and understand their fate in the environment. In this case, many reports have been published to advance an understanding of the potential environmental problems posed by the use of PAEs. The environment that these reports studied includes soils, landfills, sediments, receiving waters, and drinking water, which will be discussed in what follows (Table 15.2).

Receiving Waters

The receiving water involves streams, rivers, estuaries, and marine (Gomes and Lester 2003). Since 1976, a huge volume of research reported that phthalate esters were detected in receiving waters (Schouten et al. 1979). Phthalate esters are most likely released from domestic sewage effluents, industrial sewage effluents, or industrial discharges because of inadequate, or the absence of, filtering before discharge (Birkett 2003). Mori (1976) applied high-performance liquid chromatography (HPLC) to detect PAEs in river-water samples. DBP and DEHP were determined to be 45 and 10 µg/L, respectively, in those samples. In Nigeria, Fatoki and Ogunfowokan (1993) reported that DMP, DEP, and DBP were detected in several rivers at concentration from 10 to 1472 mg/L. In Japan, DBP (up to 9.26 µg/L) and DEHP (up to 5.22 µg/L) were also detected in well water, Tempaku River, and other water sources (Hashizume et al. 2002). In another report, DEHP was found in several rivers, including Furu, Jinzu, and Itachi, with the concentrations ranging from 0.5 to 48.4 µg/L (Hata et al. 2004). These studies have invariably recognized DBP and DEHP to be the predominant PAE pollutants in the river waters, a finding supported by the fact that DEHP is the most commonly used phthalate ester (Jeng 1986).

When the PAEs enter rivers or streams, some are resistant to decomposition and may be transported for long distance into the sea. Therefore, it is not surprising to detect PAEs occurring in the marine environment (Jeng 1986, Chen et al. 1999, Cincinelli et al. 2001, Wenger and Isaksen 2002, Mackintosh et al. 2004, Zhao et al. 2004a). Recent studies showed that the distribution pattern of organic chemicals is influenced not only by their physiochemical attributes but also by the receiving water layer structure. A level of 177 µg/L phthalate was detected in the marine environment, especially in the sea surface microlayer that contained a higher concentration than the subsurface water (Cincinelli et al. 2001). According to the report, the reason for the concentration at sea surface microlayer was probably due to the natural hydrophobic property or the absorption of floatable particles. Sea surface microlayer is defined as 1000 µm of the ocean surface (Liss 1997). As the interface of gas and liquid, sea surface microlayer is very important for marine ecosystem, because of its abundant assemblage of microorganisms, such as bacteria (Sieburth 1971). A noticeable consequence of PAEs existence in surface microlayer would result in the exposure of these microorganisms to PAEs.

On the other hand, the water conditions, such as salinity, also affect the distribution of organic chemicals. In some specific areas, like in the junction between freshwater and sea water such as estuaries, which was affected differently by salinity, phthalate esters were found to behave much differently. Turner and Rawling (2000) gave a detailed discussion on the distribution of DEHP and the factors affecting both freshwater and saltwater. The results suggested that the distribution was largely influenced by the particulates and salinity in water. Generally speaking, the particulate organic matters can partly dissolve DEHP in both waters. However, salinity may increase the solvency of particulate organic matters by salting out or structure modification; as a result, the concentration of DEHP in sea water was lowered (Turner and Rawling 2000). The particle concentration also had a relationship to the distribution coefficient (K_D), but the particle concentration in sea water had less effect on the K_D than that in fresh water, which was probably caused by the particle–particle interaction that was weakened in high salinity (Fatoki and Ogunfowokan 1993, Heindel et al. 1995, Paune et al. 1998, Turner and Rawling 2000, Luks-Betlej et al. 2001, Mihovec-Grdic et al. 2002, Cai et al. 2003, Fauser et al. 2003, Gomes and Lester 2003, Gomes et al. 2003, Salafranca et al. 2003, Mackintosh et al. 2004, Zhao et al. 2004a, Zhu et al. 2004).

Sediments and Soils

The PAEs occurring in the sediments or soils probably originated from the corresponding water sources, such as the discharge of rivers, streams, or marines. The investigation on PAEs in sediments can trace back to 1982; Peterson and Freeman published a report on the detection of several PAEs in sea bay, Chesapeake Bay sediments (Peterson and Freeman 1982). The report showed that three phthalate esters, DEP (49 ng/g), DBP (93 ng/g), and DEHP (180 ng/g), respectively, were present in the sediments. The concentration of these PAEs exhibited a particular pattern of distribution over depth but did not follow a clear trend with time, which was probably attributed to the discharge from nearby factories at different times and varied intensities (Peterson and Freeman 1982). However, the investigation also indicated that the movement of organism in sediments such as benthic organisms might have disturbed the concentration gradient distribution, causing some abnormal results. Yuan et al. (2002) published the results of sampled sediments from six rivers in Taiwan. The survey presented that several PAEs were found in their samples, among which DEHP and DEP were detected at very high concentrations. That could be attributed to the large-scale discharge and their high persistence under the normal environmental conditions (Staples et al. 1997).

Except for the contaminated sediments, another main source of PAEs is landfills from which various sorts of wastes are disposed. Plastic products account for a large proportion of the wastes. After getting buried, the plastics release organic additives under the landfill conditions of high pressure and elevated temperature. For identifying the organic chemicals in landfill leachates, Öman and Hynning (1997) investigated a municipal landfill site in Sweden. They discovered that in this landfill three PAEs (DEP, DBP, and BBP) were found in all domestic wastes, where the food packaging materials contained high concentration of DEP. In an article by Bauer and Herrmann (1997), a number of domestic wastes are classified as food wastes, plastic wastes, compound packing waste, and others. In these domestic wastes, the compound material was detected containing high levels of PAEs, such as DMP (3.4 mg/kg), DEP (18.2 mg/kg), DBP (1473 mg/kg), BBP (178.4 mg/kg), and DEHP (16,820.6 mg/kg). Even in food waste, 0.8 mg/kg DMP and 64.3 mg/kg DEHP were also detected. The results suggested that these organic chemicals may appear ubiquitously in the landfill. After the wastes were imbedded, PAEs can be easily leaked into soil and further into the surrounding aquatic environment, such as underground water. An investigation of underground water conducted in Taiwan reported that detectable residue of DBP might come from leaching of disposal materials (Yin and Su 1996). Paxéus (2000) published a report summarizing the results of a survey about three landfills. Several kinds of PAEs were detected in these three landfills, including DEP, DBP, BBP, DEHP, and DnOP. However, the comparison among three landfills on the concentration of PAEs and other organic chemicals showed different distribution patterns based on different origination of domestic wastes. This suggested that the sources of wastes were a main factor influencing the occurrence of organic chemicals. Nascimento Filho et al. (2003) also described in a report that several PAEs were identified in landfill leachate, including dioctyl phthalate (DOP), diisobutyl phthalate (DIBP), and diisopentyl phthalate (DIPP).

In other reports by Schwarzbauer et al. (2002) and Jonsson et al. (2003), several derivatives of PAEs degradation were identified in the seepage and leakage water from landfills (Jonsson et al. 2003). Several commonly used PAEs and their metabolites were detected in a landfill deposit in Germany, including DIBP, DBP, DEHP, and their monophthalate ester as well as phthalic acid, methylisophthalic acid, isophthalic acid, and terephthalic acid (Schwarzbauer et al. 2002). Similarly, Jonsson et al. (2003) also observed that in landfills in Sweden, Denmark, and Italy, different phthalates were found and recognized as DMP, DEP, DBP, DEHP, BBP, their monophthalic acid, and phthalic acid. Among these organic chemicals, phthalic acid (PA) was found at the highest concentration and almost appeared in all of these landfills. Such high levels of PA in the landfill implied that biodegradation by microorganisms had transformed esters to PA but further degradation of PA was not optimized, resulting in high concentration of this intermediate. This is an interesting fact in the degradation of PAEs because most of the previous investigations had not shown the degradative pathway in step and by pure culture of microorganisms. Only by pure culture can the cooperation between bacteria be observed, such as those published in our laboratory (Wang et al. 2003).

In the studies described earlier, almost all landfills have been detected with DEHP, DEP, and BBP at noticeable levels. These three PAEs have very wide applications in various industrial manufactures, including pharmaceuticals, pesticides, textile, leather preservatives, plasticizers, and even food packaging (Öman and Hynning 1993). These daily commodities are essential to our daily life. As a result they are commonly found in the domestic wastes. However, except for the chemical sources, the organisms' activities would also disturb the distribution in the environment. Some phthalate esters showed accumulation in these organisms, before getting fully metabolized. These esters with their metabolic intermediates may exist for a long time and have a potential impact on the human health.

Toxicity of PAEs to Organisms

The persistence of PAEs has caused a serious concern about their toxic effects on human. Most of the adverse effects suggested estrogenic or antiandrogenic activity of PAEs (Sohoni and Sumpter 1998, Zacharewski et al. 1998). The endocrine system in humans and organisms produces various hormones transported to different parts of the body in order to take part in the regulation of development, stimulation of target cell division, and so on (Griffin and Ojeda 1996). Chemicals like PAEs also function as estrogen mimics to stimulate the hormonal responses by competitive bonding to estrogen receptor (ER) (Nakai et al. 1999). Several endogenous estrogens, synthetic estrogens, and endocrine-active compounds (EACs) are listed in Table 15.3.

TABLE 15.3

Estrogens, Synthetic Estrogen, and Environmental Endocrine Active Compounds

Type of Estrogen	Class	Activity	Compound	Structure
Endogenous estrogen	Steroid	Pure agonist	17β-Estradiol (Estrodiol)	
Synthetic "estrogen"	Stilbene	Pure agonist	Diethylstilbestrol (DES)	
Environmental chemicals	Phenol	Agonist	Bisphenol A	

(continued)

TABLE 15.3 (continued)

Estrogens, Synthetic Estrogen, and Environmental Endocrine Active Compounds

Type of Estrogen	Class	Activity	Compound	Structure
	Organochlorines pesticide	Agonist	*o, p′*-DDT	
	PCB/Hydroxy-PCB	Agonist/antagonist	2′,4′,6′-trichloro-4-biphenylol	
	Phthalate	Agonist	Di-*n*-butyl phthalate ester	

Source: Mueller, S.O. and Korach, K.S., Mechanisms of estrogen receptor-mediated agonistic and antagonistic effects, in: *Endocrine Disruptors Part 1* (Metzler M., ed.), Springer-Verlag, Berlin, Germany, 2001.

Basically, the toxicity of estrogen-like PAEs comes from the action called "estrogen receptor-mediated hormone action." The estrogen is bound to the protein carrier to penetrate cell membrane before binding ER. The estrogen dimmer with estrogens binds with its DNA binding to the domain of an estrogen-responsive element and regulates the gene expression with general transcription assembly together (Mueller and Korach 2001). As a result, the endocrine disruptors cause the endocrine system malfunction and even develop cancer (Birkett 2003).

Conjugation of ER and estrogens is closely related to phthalate ester structure, especially to the length of alkyl chain and the chain position on benzene ring (Nakai et al. 1999). The chain length up to eight carbons exhibits a relatively weaker potency to ER than does the length of three to four carbons. Furthermore, for different chain position isomers (namely, *ortho-*, *meta-*, and *para-*), the *ortho-*PAE has higher affinity than *meta-* and *para-*PAE; for example, *meta-* and *para-*dipropyl phthalate have almost twice the IC_{50} value (210 and 147 μM respectively) than *ortho-*dipropyl phthalate (78.8 μM). Evidence supported that the ring hydroxylated PAEs may have higher affinity to ER. The hydroxylated PAEs should also be given some attention because hydroxylation reaction is very common in the environment by organisms (Toda et al. 2004).

After the Second World War, the toxicity of PAEs had been surveyed (Shaffer et al. 1945, Carpenter et al. 1953, Harris et al. 1956). Autian (1973) published a review to recapitulate the toxicity and health threats of PAEs to human beings and reported that the acute toxicity did not show a severe damage to the testing tissues or organisms. The lethal dose (LD_{50}) for commonly utilized compounds like DMP, DEP, and DEHP were 1.58–10.0 mL/kg, 1.0–5.06 mL/kg, and 10.0–50.0 mL/kg, respectively. However, the chronic and subacute toxicity detected was shown to cause irritation to the upper respiratory tract of humans, particularly when its presence was associated with DMP, DEP, and DEHP. The estrogenic activity of several PAEs was observed *in vitro*. Such compounds were ranked for their estrogenic potentialities in the order: BBP > DBP > DIBP > DEP > DINP (Harris et al. 1997).

DEHP was confirmed to be causing possible hepatocellular carcinomas in female rats and male and female mice (National Toxicology Program 1982). Hepatocellular carcinomas and neoplastic nodules induced in male rats with concentration reaching 12,000 mg/L caused rats' liver neoplastic lesion as well as toxic damage in testes and pituitary. The male rats exposed chronically to DEHP developed reproductive damage, including hypospermia in the testis and epididymis (Moore 1996). The reproductive toxicity was also confirmed at a high concentration of 1.7 g/kg/day for 14 days (Wilkinson and Lamb IV 1999). However, according a recent study on the DEHP cancer risk assessment, no genotoxicity was observed at a concentration level that is relevant to the human cancer development (Doull et al. 1999). The author suggested that DEHP should not be classified as a likely carcinogen based on margin of exposure (MOE) approach to the human risk assessment.

Dimethyl phthalate (DMP), one of the PAEs produced on a large scale, was not found to induce carcinogenic activities in both male and female rats (National Toxicology Program 1995). However, a DMP uptake in higher amounts triggered central nervous depression and noticeable kidney damage (Autian 1973). Other bioassays using scud (*H. azteca*), midge (*Chironomus tentans*), and California blackworm (*Lumbriculus variegates*) showed that *H. azteca* was most sensitive to DMP with the mechanism of narcosis response (Call et al. 2001b). At 50 mg/L of DMP, mortality at the rate of 95% was observed in *H. azteca*.

DMTP, a monomer in the polymer industry, was tested for its toxicity as early as 1973 (Krasavage et al. 1973). The initial tests supported that DMTP was basically a nontoxic compound ingested at low levels by organisms. Although a reduced body weight was observed in the assay, a syndrome-like hematology was insignificant. The nongenotoxicity of DMTP was demonstrated by *in vitro* tests: the Ames test, DNA single-strand break assays in CO60 cells and in primary rat hepatocytes, and so on (Monarca et al. 1991). In 1995, the European Commission initiated an investigation on the DMTP recovered from pet bottles (The Scientific Committee for Food 1995). The report claimed that there was no special evidence of harmful effects on humans when using DMTP as the monomer in pet bottle manufacturing.

Degradation of PAEs

PAEs are very persistent in the environment. In the case of DMP, its half-life could be 3.2 years in the aquatic environment (Staples et al. 1997). For the purpose of eliminating such esters, acceleration of the hydrolysis of ester bond is very important for their remediation. There are many ways to perform the degradation test, and they can be classified into two main categories: abiotic and biotic methods.

Abiotic Degradation

Abiotic degradation of PAEs refers to physicochemical processes such as light exposure, artificial irradiation, and chemical reaction, which occur without the participation of any organism. Photolysis is a very important but slower process for decomposition of organic chemicals in the environment. For PAEs, the ester bonds can be broken when they absorb enough ultraviolet radiation (UV) at the range of 290–400 nm. The ester bonds rupture by direct absorption of the energy of UV light, or by their reaction to UV-activated matters, such as oxygen radical and hydroxy radical. For instance, 1 mg/L of butyl benzyl phthalate (BBP) was decomposed over 95% after a direct exposure to sunlight for 28 days (Gledhill et al. 1980). DMP and DEP after UV irradiation were decomposed to several fractions of compounds and radicals through breakage of ester bonds and decarboxylation (Balabanovich and Schnabel 1998).

Oxygen or hydroxy radicals are very active substances with high oxidation ability. Some additives were applied to generate these radicals by UV irradiation, such as hydrogen peroxide. The UV-H_2O_2 system has been applied to produce hydroxy radicals in the degradation of DMP (Balabanovich and Schnabel 1998). The results of experiment indicated that a higher degradation efficiency was obtained when DMP was exposed to UV-H_2O_2 system, with over 98% of DMP removed in 45 min; in contrast, only 60% of removal was achieved under direct irradiation using UV for 60 min.

Other oxidation methods such as Fenton reaction were also under investigation for their potential application (Halmann 1992). The Fenton reaction is based on the system of Fe^{2+}-H_2O_2 and is, accordingly, called Fenton reagent. Under the irradiation condition, by absorbing the photon energy the Fe^{2+} can activate hydrogen peroxide to yield the hydroxy radical, which is highly oxidative in aquatic environment. The hydroxy radical with highly oxidative ability is a main component in the oxidation process involved with the degradation of PAEs. Zhao et al. (2004b) used the Fenton reaction to oxidize DMP and demonstrated that the Fenton system proceeded more efficiently in acidic condition (pH 3.0) than in neutral condition, attaining over 81% removal in 120 min (Figure 15.2).

Biotic Degradation

Apparently, abiotic degradation of PAEs has merits such as faster degradation and higher efficiency. However, the disadvantages are also obvious: high investment costs as well as the cost of long-term maintenance. As a result, in many wastewater treatment plants, the biotic degradation treatments are commonly chosen for organic chemical degradation. While reviewing the biodegradation studies on PAEs, several aspects should be noticed, such as aerobic or anaerobic conditions, the study methods used (inoculum test, water sample degradation test, and soil/sludge sample test), the microorganisms involved, the enzymatic mechanism, and the degradation pathway.

Aerobic Degradation

Many research works that studied PAE degradation in laboratory were based on aerobic condition because of its easy operation and short experimental periods (Chauret et al. 1995, Shah et al. 1998, Wang et al. 2003, Zeng et al. 2004). The experimental instrument and culture medium were exposed in the ambient air containing oxygen. The target chemicals were precisely measured and added to the prepared mineral salt medium (MSM), which contained the essential inorganic elements required for microbial growth. The inoculum was normally enriched from environments, including fresh waters, soils, sediments, active sludge, and so on and incubated using an incubation shaker. The list of microorganisms and their degradation substrates surveyed were listed in Table 15.4.

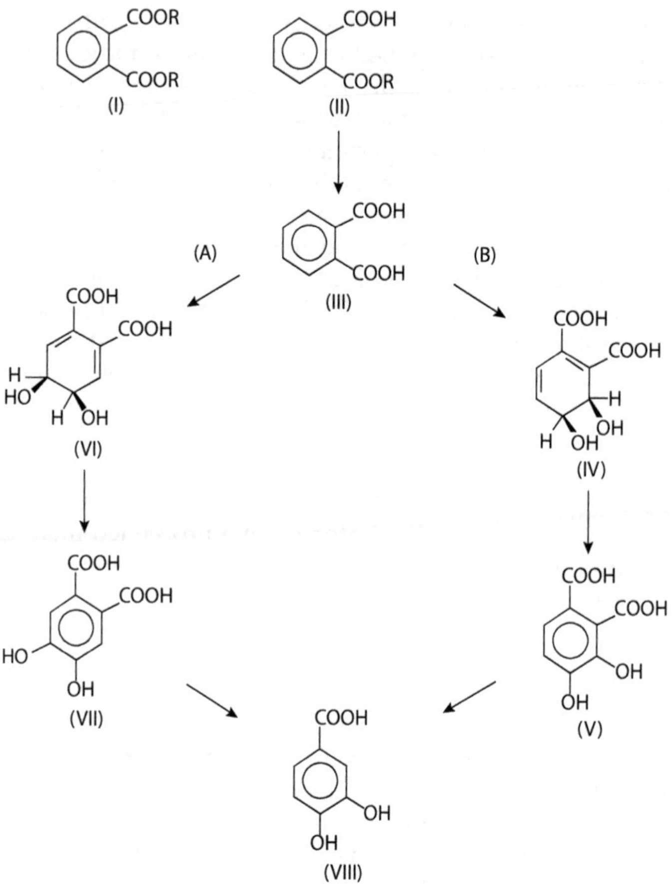

FIGURE 15.2
The scheme of transformation of *ortho*-phthalate ester to protocatechate. (A) *Pseudomonas* spp. transformation pathway (From Keyser, P. et al., *Environ. Health Perspect.*, 18, 159, 1976.); (B) *A. keyseri* 12B transformation pathway. (From Eaton, R.W., *J. Bacteriol.*, 183(12), 3689, 2001.) (I) *ortho*-phthalate ester; (II) Monophthalate ester; (III) Phthalic acid; (IV) *cis*-3,4-dihydroxy-3,4-dihydrophthalate; (V) 3,4-dihydroxyphthalate; (VI) *cis*-4,5-dihydroxy-4,5-dihydrophthalate; (VII) 4,5-dihydroxyphthalate; and (VIII) protocatechuate.

Single-Species Degradation

The enriched bacteria from environment are normally isolated using the selective medium technique. The species with degradation ability can be used to perform biodegradation for the target chemicals. From 1972 to date, there are many reports on the use of single species in the degradation process of PAEs (Keyser et al. 1976, Babu and Vaidyanathan 1982, Eaton 1982, Eaton and Ribbons 1982, Karegoudar and Pujar 1984, Sivamurthy et al. 1991, Williams et al. 1992, Ganji et al. 1995, Jackson et al. 1996, Aleshchenkova et al. 1997, Patel et al. 1998, Yan and Liu 1998,

TABLE 15.4

Microorganisms Capable of Degrading Phthalate Esters Completely or Partially

Substrates	Microorganisms	Degradation	References
DMP	*P. fluorescens, Pseudomonas. Aureofaciens* and *S. paucimobilis*	Complete	Wang et al. (2003)
	Xanthomonas maltophilia and *S. paucimobilis*	Complete	Wang et al. (2003)
	P. fluorescens FS1	Complete	Zeng et al. (2004)
	Micrococcus sp. strain 12B	Complete	Keyser et al. (1976)
	Bacillus sp.	Complete	Niazi et al. (2001)
	Arthrobacter sp.	Partial	Vega and Bastide (2003)
DMTP	*A. niger* (fungus)	Complete	Ganji et al. (1995)
	S. paucimobilis	Complete	Li et al. (2005)
	Sclerotium rolfsii (fungus)	Partial	Sivamurthy et al. (1991)
	C. acidovorans D-4	Complete	Patel et al. (1998)
	Pasteurella multocida	Complete	Li et al. (2005)
DMIP	*Klebsiella oxytoca*	Partial	Li and Gu (2004)
	R. erythropolis 5D	Complete	Aleshchenkova et al. (1997)
	R. ruber 1B	Complete	Aleshchenkova et al. (1997)
DIBP	*P. fluorescens* FS1	Complete	Zeng et al. (2004)
	Micrococcus sp. strain 12B	Complete	Eaton and Ribbons (1982)
DEHP	*Acinetobacter* sp.	Complete	Hashizume et al. (2002)
DBP	*Acinetobacter* sp.	Complete	Hashizume et al. (2002)

Karpagam and Lalithakumari 1999, Lefevre et al. 1999, Sharanagouda and Karegoudar 2000, Eaton 2001, Niazi et al. 2001, Hashizume et al. 2002, Kim et al. 2002, Tserovska and Dimkov 2002).

Kim et al. (2002) reported that one of the fungi species, *Fusarium oxysporum*, exhibited a stronger degradation ability for BBP biodegradation. BBP was hydrolyzed by esterase and formed oxo-bridge structure within the 1,3-isobenzofurandione (IBF). IBF underwent hydrolysis by two pathways, esterase or cutinase, according to the report. However, the process by cutinase indicated a higher efficiency than the esterase. Almost 60% of the BBP was degraded in 7.5 h by the cutinase process. One notable phenomenon was that during the IBF degradation process certain amount of dimethylphthalate was formed.

Rhodococci were isolated from a phthalate-contaminated soil and demonstrated to be capable of degrading DMIP and DAIP (Aleshchenkova et al. 1997). Meanwhile, *Rhodococcus ruber* 1B and *Rhodococcus erythropolis* 100B mineralized DAIP 0.2% in 72 h and 0.5% in 168 h.

In 1982, Eaton reported a study on the DBP biodegradation by *Micrococcus* sp. strain 12B. *Micrococcus* sp. strain 12B isolated from compost obtained at Pennsylvania was found to be capable of causing a complete degradation of DBP. The substrate was transformed to phthalate and then to protocatechuate, which was *meta*-cleavaged in the following process. *Acinetobacter lwoffii* isolated from Tempaku River was investigated by Hashizume et al. (2002). They reported that this strain was able to degrade DBP at an initial concentration of 20 μg/mL.

Micrococcus sp. from soil mineralized DEP quickly, when the DEP was used as the sole carbon and energy source (Karegoudar and Pujar 1984); 0.2% of substrate DEP was mineralized in only 2 days, with the bacterial growth reaching up to OD 1.85 at the same time. DEPs were also found degraded in soil samples containing bacteria of *Aureobacterium saperdae*, *Flavobacterium aquatile*, and *Micrococcus keristinae* (Jackson et al. 1996).

Several bacteria were identified while hydrolyzing diethyl terephthalate (DETP) and included 13 strains of *Rhodococcus rhodochrous* and 1 *Xanthomonas maltophila* (Jackson et al. 1996). However, the DETP was only hydrolyzed to monoester, without further utilization by bacteria.

In the case of DMP, the simplest form of phthalate esters, there are a number of studies describing various microorganisms utilizing this chemical, including bacteria, fungi, and algae. Both *Pseudomonas fluorescens* and *Pseudomonas testosteroni* can use DMP as a sole carbon and energy source, with distinct benzene ring cleavage pathway (Keyser et al. 1976). Eaton and Ribbons (1982) described that *Micrococcus* sp. strain 12B can mineralize DMP with a main pathway and a minor pathway. Another Gram-positive bacterium, *Bacillus* sp., was also found to degrade DMP when DMP is used as a sole carbon and energy source (Niazi et al. 2001). *P. fluorescences* was used to degrade several different PAEs, including DMP (Zeng et al. 2004). Other than bacteria, the fungi and algae presented their capability to hydrolyze DMP. Green alga *Dunaliella tertiolecta* hydrolyzed DMP to the product PA with the maximum tolerance level reaching up to 300 mg/L concentration (Yan and Liu 1998). No further degradation of phthalic acid was mentioned in this report. Another study reported that *Aspergillus niger*, one of the fungal species, can also metabolize DMP (Sharanagouda and Karegoudar 2000). The aromatic intermediates of monomethyl phthalate, phthalate, and protocatechuate were yielded by the strain, followed by *ortho*-cleavage process.

For DMIP, *P. testosteroni* was described as having the ability to mineralize (Keyser et al. 1976). DMIP was transformed to isophthalate by this strain, followed by the intermediate protocatechuate. Before the formation of protocatechuate, one possible intermediate was speculated to be 4-hydroxyisophthalate, which is produced by isophthalic acid grown cells. In another report, *Rhodococcus ruber* 1B and *R. erythropolis* 5D completely degraded DMIP at 0.2% concentration in 168 h (Aleshchenkova et al. 1997).

P. testosteroni was found to be able to degrade all the three PAE isomers. The strain transformed DMTP to protocatechuate, but the process of transformation from terephthalic acid to protocatechuate was not mentioned in the report (Keyser et al. 1976). The bacterium *Comamonas acidovorans* was described in the article (Patel et al. 1998). About 0.5% (w/v) concentration of DMTP was mineralized by this strain in 72 h (Patel et al. 1998). *C. acidovorans* was also described in another report as having the ability to utilize phthalic acid as the sole carbon and energy source (Wang et al. 2003). Some fungi have been described as the species having the ability to hydrolyze and degrade DMTP (Sivamurthy et al. 1991, Ganji et al. 1995). They included *A. niger* and *Sclerotium rolfsii*. *A. niger* metabolized DMTP via monoester, terephthalic acid, protocatechuate, and further benzene ring cleavage. However, *Sclerotium rolfsii* only hydrolyzed DMTP to terephthalic acid without further ability to complete the ring cleavage.

Consortium Degradation

Although many microorganisms are capable of mineralizing PAEs, a majority of them do so only partially. Some can hydrolyze the diester to monoester, others only metabolize acid phthalic and do not have the ability to hydrolyze esters. However, the consortium consisting of various strains together revealed better ability to complete the biodegradation process. Goud et al. (1990) reported about using a mixed culture of bacteria including *Pseudomonas* sp., *Aeromonas* sp., *Arhtrobacter* sp., and *Bacillus* sp., to degrade the waste water effluent from a DMTP plant. The waste water contained chemical oxygen demand (COD) 80,000 mg/L, in which up to 86% COD and 95% biochemical oxygen demand (BOD) were reduced in 48 h. The consortium of *Arthrobacter* sp. and *Sphingomonas paucimobilis* degraded the DMP in concert (Vega and Bastide 2003). *Arthrobacter* sp. was shown to metabolize DMP directly; however, the monomethyl phthalate, a common intermediate, remained unchanged. Another intermediate phthalic acid was able to be catabolized. *S. paucimobilis* was tested for its ability to hydrolyze MMP to PA and completely degrade PA. However, *S. paucimobilis* could not hydrolyze DMP to MMP, which was, however, performed well by *Arthrobacter* sp. Therefore, in order to achieve better results in the elimination of DMP, the application of a consortium of microorganisms was proposed. However, another report showed a conflicting result about *Arthrobacter* sp. used in the hydrolysis of MMP. *Arthrobacter keyseri* 12B was demonstrated capable of transforming a number of phthalate analogues, including monomethyl phthalate, to phthalate (Eaton 2001). Wang et al. (2003) investigated two consortia isolated from activated sludge and found that the common species *S. paucimobilis* played an important role in the consortia degradation. The results were similar to those reached by Vega and Bastide (2003).

Degradation Pathway and Enzymatic Mechanisms

Other than searching for the bacteria with degradation ability, discovery of the pathway and enzymatic mechanism for the biodegradation process is largely important. For one chemical, there can be various pathways existing in different microbial enzymatic systems. Even for the same species of

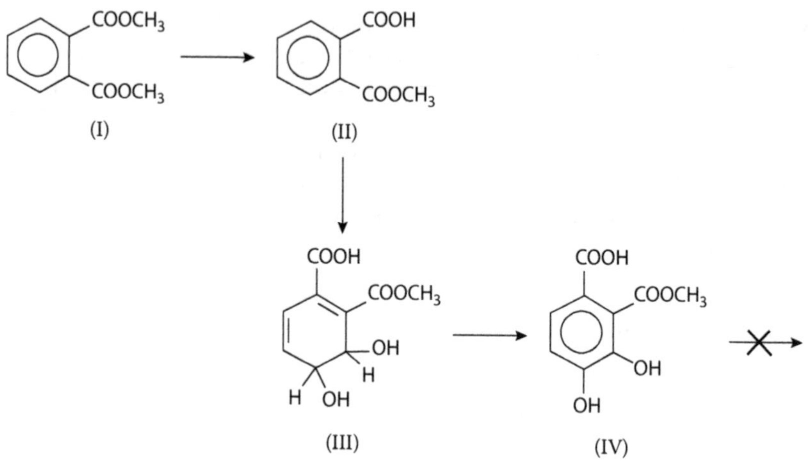

FIGURE 15.3

The pathway of minor DMP biodegradation by *Micrococcus* sp. strain 12B. (I) dimethyl phthalate; (II) monomethyl phthalate; (III) 3,4-dihydro-3,4-dihydroxyphthalate-2-methyl ester; and (IV) 3,4-dihydroxyphthalate-2-methyl ester. (From Eaton, R.W. and Ribbons, D.W., *J. Bacteriol.*, 151(1), 465, 1982.)

bacterium, different studies would exhibit distinctively different mechanisms for the same chemical. As a result, there are still many questions waiting to be answered in terms of microbial behaviors. *Pseudomonads* sp. capable of degrading DMP were isolated from the environment (Keyser et al. 1976), such as *P. fluorescens* PHK and *P. testosteroni*, which provide distinctive pathways to the bacteria. Both strains had esterase to hydrolyze the diester to monomethyl phthalate, and then to *ortho*-phthalic acid (Figure 15.3); *ortho*-Phthalate was transformed to 4,5-dihydroxyphthalate, between which there was dihydrodiol, namely, *cis*-4,5-dihydroxy-4,5-dihydrophthalate yielded by phthalate 4,5-dioxygenase. The 4,5-dihydroxyphthalate was decarboxylated to form protocatechuate, which was processed at different ring cleavage pathways by these two strains. *A. keyseri* 12B adopted a different pathway via *cis*-3,4-dihydroxy-3,4-dihydrophthalate to 3,4-dihydroxyphthalate, then to protocatechuate (Eaton 2001) (Figure 15.3). A minor degradation pathway for DMP was proposed via intermediate 3,4-dihydro-3,4-dihydroxyphthalate-2-methyl ester to form 3,4-dihydroxyphthalate-2-methyl ester (Eaton and Ribbons 1982) (Figure 15.4). According to Keyser et al. (1976), *P. testosteroni* proceeded *via meta*-cleavage to open benzene ring, whereas *P. fluorescens* proceeded via *ortho*-cleavage by protocatechuate 3,4-dioxygenase to form 3-carboxy-*cis, cis*-muconate, which was further transformed generally to β-ketoadipate (Figure 15.5). However, Keyser et al. did not give detailed information about the enzyme producing β-ketoadipate. The detail on the derivatives for 3-carboxy-*cis, cis*-muconate was described in another report (Williams et al. 1992), which mentioned the function of enzyme

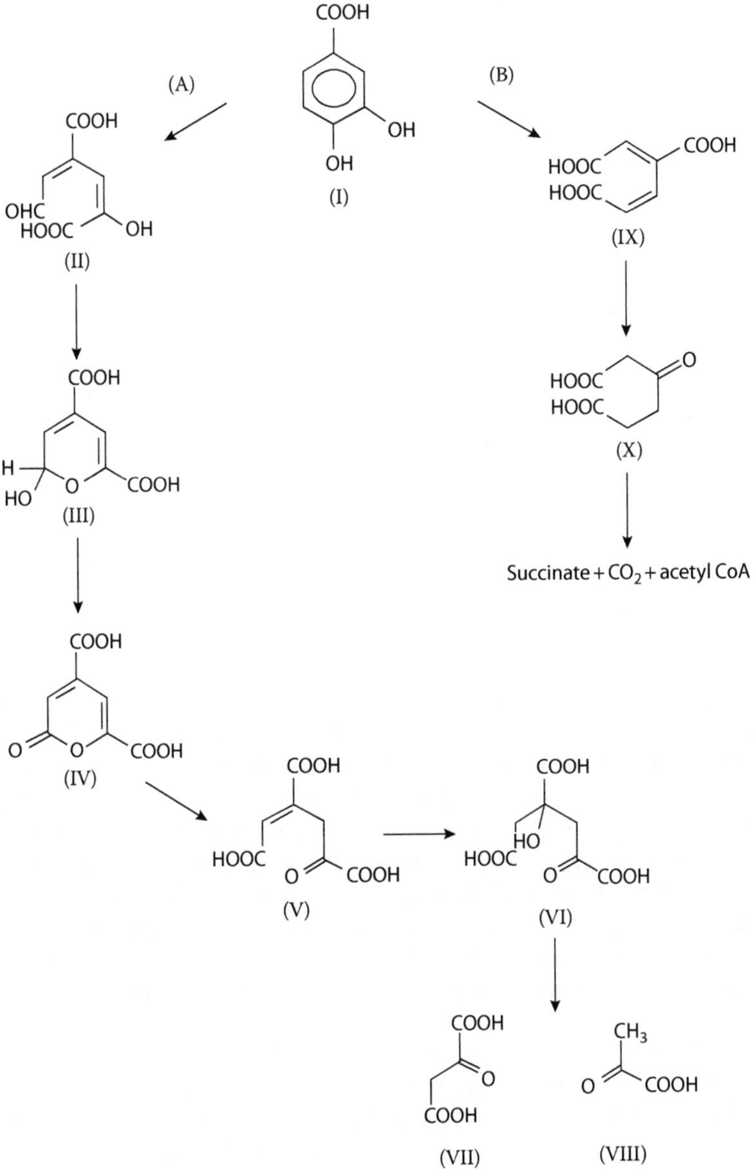

FIGURE 15.4
The degradation pathway of protocatechuate. (A) *A. keyseri* 12B (From Eaton, R.W., *J. Bacteriol.*, 183(12), 3689, 2001.) and *P. testosteroni* (From Keyser, P. et al., *Environ. Health Perspect.*, 18, 159, 1976.) degradation pathway; (B) *P. fluorescens* (From Keyser, P. et al., *Environ. Health Perspect.*, 18, 159, 1976.) degradation pathway. (I) protocatechuate; (II) 2-hydroxy-4-carboxymuconic semialdehyde; (III) 2-hydroxy-4-carboxymuconic semialdehyde-hemiacetal; (IV) 2-pyrone-4,6-dicarboxylate; (V) 4-oxalomesaconate; (VI) 4-oxalocitramalate; (VII) oxaloacetate; (VIII) pyruvate; (IX) 3-carboxy-*cis, cis*-muconate; and (X) β-ketoadipate.

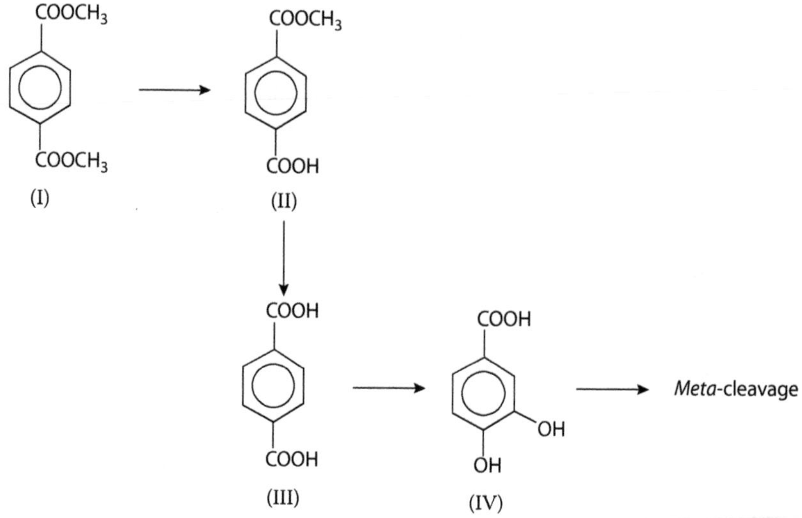

COOCH₃ — (I) dimethyl terephthalate → COOCH₃ / COOH (II) → COOH / COOH (III) → COOH / OH / OH (IV) → *Meta*-cleavage

FIGURE 15.5

The pathway of DMTP biodegradation by *A. niger*. (I) dimethyl terephthalate; (II) monomethyl terephthalate; (III) terephthalic acid; and (IV) protocatechuate. (From Ganji, S.H. et al., *Biodegradation*, 6, 61, 1995.)

3-carboxymuconate cycloisomerase, 4-carboxymuconolactone decarboxylase, and β-ketoadipate enol-lactone hydrolase in *Pseudomonas putida*.

For the nonfluorescent strain *P. testosteroni*, protocatechuate was transformed by *meta*-cleavage by protocatechuate 4,5-dioxygenase to yield intermediate 2-hydroxy-4-carboxymuconic semialdehyde as a product of the ring cleavage. The 2-hydroxy-4-carboxymuconic semialdehyde was oxidized to form three carboxyl groups (Keyser et al. 1976). The process was explained in detail in other articles, for different species of bacteria utilizing protocatechuate (Eaton 2001) (Figure 15.5). In these further oxidations, 4-oxalomesaconate was produced by enzymes and then transformed to pyruvate.

Although most of the evidence gathered indicated that *P. fluorescens* can utilize DMP as the sole carbon and energy source, there is one report that showed contradictory results (Chauret et al. 1995). It was found that this strain indeed transformed di-*n*-butyl phthalate (DBP) to phthalate; however, no degradation of substrate PA was observed, whereas the fragment of butanol was utilized as the sole carbon source. The most interesting part of the report was that the addition of respiration inhibitor, sodium azide, and a cationic agent, polymyxin B, increased the permeability of the outer membrane of cells. As a result, the organic chemical entered into the cell more easily, resulting in a sharp increase of the primary degradation.

Meanwhile *P. fluorescens* was demonstrated to degrade terephthalic acid (TA) (Karpagam and Lalithakumari 1999). According to the report,

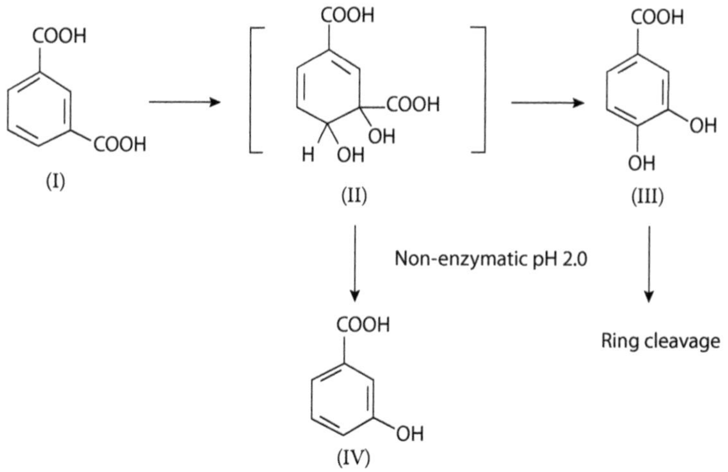

FIGURE 15.6
The pathway of degradation of isophthalic acid. (I) isophthalic acid; (II) 4-hydrogen-3,
4-dihydroxyisophthalate; (III) protocatechuate; and (IV) 3-hydroxybenzoic acid.

TA degradation followed the *meta*-cleavage process similar to other
nonfluorescent bacteria, which meant that under different substrate condition it is possible for *P. fluorescens* to follow the *meta*-cleavage, in con-
tradiction to what was described by Keyser et al. (1976) in which only the
nonfluorescent bacteria could follow the *meta*-cleavage. Some fungal species
were found to metabolize DMTP as a sole carbon and energy source, such
as *A. niger* (Ganji et al. 1995) (Figure 15.6). These species transform DMTP
to monomethyl phthalate and then to the corresponding acid. The acid was
oxidized by dioxygenase to form protocatechuate, which was further cleavaged by 3,4-dioxygenase.

For DMIP degradation, there were many strains that were shown to
secrete esterase to hydrolyze DMIP to monomethyl isophthalate (MMIP)
and further to isophthalate or isophthalic acid (IA). The IA degradation
was a key step to perform in the whole mineralization process. *P. testosteroni* utilized IA as a sole carbon and energy source (Keyser et al. 1976).
In the transformation process, the common intermediate protocatechuate
was produced via an unconfirmed intermediate 4-hydroxyisophthalate,
which was not observed for phthalate or terephthalate degradation. Babu
and Vaidyanathan (1982) isolated a soil bacterium degrading isophthalate and postulated that another degradation pathway exists (Figure 15.7).
Isophthalate was oxidized by isophthalate 3,4-dioxygenase to form the
proposed 4-hydrogen-3,4-dihydroxyisophthalate. Furthermore, the protocatechuate was yielded and cleavaged by *ortho*-fission. One interesting
phenomenon was the production of 3-hydroxybenzoic acid, which was
metabolized via gentisic acid in the bacterium. However, the addition

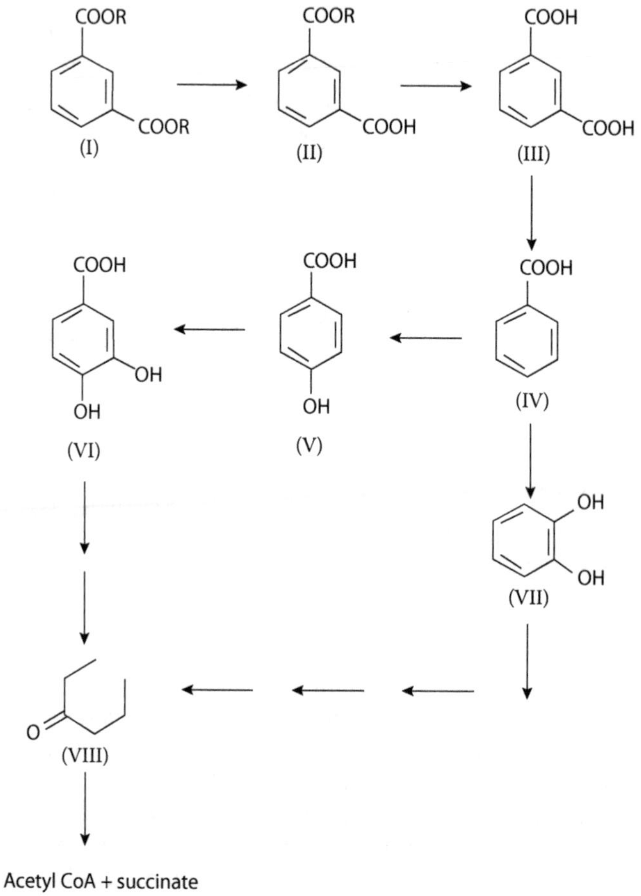

FIGURE 15.7
The pathway of *meta*-phthalate diester biodegradation by *Rhodococcus* spp. (I) *meta*-phthalate diester; (II) *meta*-phthalate monoester; (III) isophthalic acid; (IV) benzoic acid; (V) *para*-hydroxybenzoic acid; (VI) protocatechuate; (VII) pyrocatechol; and (VIII) 3-ketoadipic acid. (From Aleshchenkova, Z.M. et al., *Microbiology*, 66(5), 515, 1997.)

of acid into 4-hydrogen-3,4-dihydroxyisophthalate also resulted in the nonenzymatic degradation to form 3-hydroxybenzoic acid. *Rhodococcus*, including *R. ruber* and *R. erythropolis*, were found to be able to degrade DMIP (Aleshchenkova et al. 1997) (Figure 15.7), but their degradation pathways including the two parts were quite different from what was described earlier. In one of the pathways, the strain was decarboxylated from IA to benzoic acid, instead of being oxidized to protocatechuate directly. The benzoic acid was transformed to 4-hydroxybenzoic acid and then to protocatechuate. For another part, benzoic acid was first transformed to pyrocatechol and found to go through further degradation in the process.

Measurements for Biodegradation

For screening the microbial growth, some studies measured the oxygen uptake in the aerobic degradation process. Aichinger et al. (1992) conducted a research to develop a respirometric protocol for detecting low-soluble organic compounds. They used a modified model to simulate the relationship between substrate concentration, biomass concentration, and oxygen uptakes. The results showed that the protocol can only be applied when the concentration of a chemical was below its solubility limit. The proportion of organic chemicals above the solubility limit cannot be determined with precision. This restricted the protocol's application.

However, if the measurement of oxygen consumption is to be viewed as the sole evidence of microbial growth, it is useful to record oxygen uptakes as demonstrated in other studies (Wang et al. 1999, Juneson et al. 2001, Wang 2004). Wang et al. (1999) used calcium alginate gel beads to immobilize microorganisms to degrade DBP. The oxygen consumption indicated that the immobilized cells absorbed oxygen at a lower rate than the free cells with reduced available oxygen. Juneson et al. (2001) measured oxygen uptakes to confirm whether the sequencing batch reactor (SBR) system has a higher degradation rate than standard batch reactor and found evidence affirming the higher degradation rate possible with using the SBR. Wang (2004) also compared the degradation ability of unacclimated activated sludge versus acclimated activated sludge. The results showed that activated sludge eliminated higher COD when acclimated in 300 mg/L and that it was able to degrade DBP better than the unacclimated activated sludge. The unacclimated activated sludge was found to inhibit the oxygen uptake process when the DBP concentration is at a higher level. However, the opposite was found for to be the case for acclimatedly activated sludge; the uptake rate slightly increased with the rise of substrate concentration. Sharanagouda and Karegoudar (2000) used fungal strain *A. niger* to degrade DMP and proved that the cell grown with DMP as a sole substrate had a higher oxidation ability than those grown in glucose, by measuring the oxygen consumption.

The CO_2 evolution test is frequently used as part of the biodegradation experiment to evaluate the extent of degradation process. Gledill (1975) mentioned CO_2 evolution can assess the ultimate level of biodegradation (Sharanagouda and Karegoudar 2000). Sugatt et al. (1984) determined 14 biodegradable PAEs, on the basis of shake-flask CO_2 evolution. The level of BBP and DTDP found was no more than 80% for primary degradation and 50% for ultimate degradation. The primary degradation was expressed by the loss of PAEs during the degradation process, and the ultimate degradation was expressed by the difference between the percentage of CO_2 actually measured and the theoretical CO_2 value set for the experiment. Comparison of these results indicated the existence of an important relationship between bacterial utilization and chemical degradation. Recently, an improvement on the CO_2 evolution test method was proposed

(Strotmann et al. 2004). The improved test system was found to be more suitable for poorly water-soluble and high-volatile chemicals.

Anaerobic Degradation

Apart from aerobic degradation, some of the studies focused on bioremediation under anaerobic condition. The main anaerobic biodegradation methods involve using activated sludge, acclimated activated sludge, biomass support particles, and methanogenic consortia to degrade several PAEs, including DMP, DMIP, and DMTP. As early as 1985, activated sludge applied to degrade commercial phthalate esters was reported by O'Grady et al. The procedure was designed using two different tests, a semicontinuous activated sludge test and an acclimated 19 day die-away procedure (O'Grady et al. 1985). The results showed that within 5 days most of the PAEs could be eliminated down to less than 20% of the original level, demonstrating that most of the PAEs, including DMP, DEP, and DEHP, were eliminated by the activated sludge. Detailed kinetics of anaerobic degradation by the acclimated activated sludge was presented in another study (Wang et al. 1996b). DMP showed the fastest degradation rate with k_1 (0.033/h) and $t_{1/2}$ (21.0 h), but di-n-octyl phthalate (DnOP) was found having the slowest degradation rate. Apparently, the kinetics constant of biodegradation was found to be correlated to the kinetics constant of hydrolysis rate. Other new techniques like biomass support particles were also introduced in the treatment of organic pollutants. The bacterial biomass was immobilized to be applied in the bioreactors on support particles such as polyurethane reticulated foam. Sharma et al. (1994) reported the existence of high levels of COD on the basis of an experiment conducted using wastewater flux from a DMTP manufacturing plant. The biomass support particles were applied to treat the waste water outflow. The system had a longer sludge retention time (SRT) than the unsupported biomass particles, which implies a higher tolerance to the toxicity of the outflow. Although no concentration of DMTP was stated in the report, biomass support particles could be applied to degrade various PAEs, including DMTP.

Several works regarding the anaerobic degradation of phthalate isomers by consortia were published in 1999 (Kleerebezem et al. 1999a,b,c,d). In one such work, *ortho*-phthalate, isophthalate terephthalate, DMP, DMTP, *para*-toluate, and *para*-xylene were all tested for their biodegradability by the anaerobic consortium (Kleerebezem et al. 1999b). Results showed that three isomers, namely, dimethylphthalate, isophthalate, and terephthalate, are the substrates that had different a lag phase and in the sequence dimethylphthalate < terephthalate < isophthalate occurring over a period of 17–156 days. This sequence might reflect the methanogenic activity of cultures that degrades these three isomers and is closely related to the amounts of these isomers appearing in the environment. For exploring the internal relationship among different species and their contribution in the cooperation, Kleerebezem conducted a broad investigation on the anaerobic biodegradation process, particularly analyzing the role of different species in the system, the energetic influence, and the intermediate benzoate effect. Various species played specific roles in the whole consortia. Certain species only degraded corresponding to the presence of phthalate isomer, regardless of whether the other two isomers are present or not (Kleerebezem et al. 1999c). The syntrophic collaboration between the phthalate-degrading bacteria and the methanogens were confirmed in the report. Because yielding acetate in the transformation process is thermodynamically unfavorable, the acetoclastic methanogens and hydrogenotrophic methanogens, which produce a negative value of Gibbs energy, were found to be indispensable. Phthalate fermentation produced a large amount of acetic acid and hydrogen. This process generated positive Gibbs energy, which meant the normal fermentation reaction could not be carried out successfully. For bacteria, one way to attain the energy is through acetoclastic methanogens and hydrogenotrophic methanogens, through which the whole methanogenic reaction can move from phthalates to acetic acid/hydrogen and then to methane. This supports the point of view that bacteria need methanogens to remove degradation intermediates including acetic acid and hydrogen and further aid the anaerobic biodegradation process, which is a more subtle conclusion drawn from the results. In another report, bromoethanosulfonate was applied as methanogen inhibitor for exploring the benzoate oxidation and reduction equilibrium (Kleerebezem et al. 1999a). The results showed that the benzoate oxidation and reduction were rather comparable. Thus, the Gibbs free energy change for carboxycyclohexane oxidation was close to 0 kJ/mol, which equalized the value of benzoate oxidation minus value of reduction. Meanwhile, the result proved a threshold of about −30 kJ/mol for the oxidation of benzoate. However, it was found that the intermediates produced in the degradation process could inhibit the elimination of parental compounds. The appearance of benzoate and acetate was demonstrated to be capable of inhibiting and inactivating the strain with degradation ability in anaerobic consortium (Kleerebezem et al. 1999d).

Marine Microorganisms

The bacteria mentioned earlier were isolated mainly from the sediments of leaching site, landfill, or rivers. Apart from these sources, marine waters and sediments are other sources possible for isolating microbes, which are able to degrade organic chemicals, like polychlorinated biphenyl (PCB), polycyclic aromatic hydrocarbon (PAH), and phthalic acid esters (PAEs). Marine bacteria were favored for their chemoreception ability in responding to the chemical signals in the marine environment (Mitchell et al. 1972). Thus, bacteria living in the high-salt and low-temperature condition may attenuate and remediate organic pollutants in the ocean naturally. Marine bacteria were isolated and enriched to degrade surfactants, such as sulfonated alkylbenzenes and *para*-sulfophenyl carboxylic acid (Sigoillot and Nguyen 1990). Although these organic surfactants did not fall into the category of persistent polyaromatic hydrocarbons, the results of this study showed that the bacteria from the seawater could be enriched and incubated in the normal medium to utilize alkyl chain hydrocarbons as a sole substrate. Moreover, *Pseudomonas* spp. and *Rhodococcus* spp. strains in the marine environment were demonstrated to be capable of degrading crude oil (Sorkhoh et al. 1990).

Polycyclic aromatic hydrocarbon (PAH) was one class of dominant organic pollutant, reaching to the marine environment mainly through wastewater treatment plant and the combusion of industrial discharge. In the past decades, many bacterial strains were found and were applied in the bioremediation of PAH in the marine environment (Hayes et al. 1999, Poeton et al. 1999, Juhasz and Naidu 2000). A detailed review is available on the microbial communities in the oil-contaminated seawaters (Harayama et al. 2004). Several species, including *Pseudomonas*, *Vibrio*, and *Flavobacterium*, were recognized as the common microbes in the marine environment. Other uncommon species like *Cycloclasticus oligotrophus* were also isolated for their PAH degradation ability (Wang et al. 1996a).

There are many studies based on the microorganisms enriched from marine waters and sediments, some of which were novel species. It is very interesting that marine bacteria having specialized degradation capability can be applied to biodegrade PAEs in the environment (Xu et al. 2005). However, their degradation capability and possible biochemical pathway still need stronger proof by way of further investigation. Future research in this regard might yield critical information that will help close the knowledge gaps between the fields of marine microbiology and bioremediation application.

Conclusions

Phthalate esters as environmental pollutants will likely stay with us for a long time. Therefore, future research in this regard should focus on their biodegradation capacity and effects on the human health and the environment. Newly available information has shown that degradation of organic compounds in the environment is a concerted activity of a group of microorganisms, offering an ecological view of the degradation community. Esterases are much more diversified and specific than what was thought previously.

Acknowledgments

This research is a culmination of the efforts of a handful of graduate students and postdoctoral fellows who had worked at this laboratory. Laboratory assistance was provided by Jessie Lai.

References

Aichinger G, Grady CPL, Tabak HH. 1992. Application of respirometric biodegradability testing protocol to slightly soluble organic-compounds. *Water Environment Research* 64 (7): 890–900.

Aleshchenkova ZM, Smsonova AS, Semochkina NF, Baikova SV, Tolstolutskaya LI, Begel'man MM. 1997. Utilization of isophthalic acid esters by *Rhodococci*. *Microbiology* 66 (5): 515–518.

Autian J. 1973. Toxicity and health threats of phthalate esters: Review of the literature. *Environmental Health Perspectives* 4: 3–26.

Babu BH, Vaidyanathan CS. 1982. Catabolism of isophthalic acid by a soil bacterium. *FEMS Microbiology Letters* 14 (2): 101–106.

Balabanovich AI, Schnabel W. 1998. On the photolysis of phthalic acid dimethyl and diethyl ester: A product analysis study. *Journal of Photochemistry and Photobiology A: Chemistry* 113 (2): 145–153.

Bauer MJ, Herrmann R. 1997. Estimation of the environmental contamination by phthalic acid esters leaching from household wastes. *Science of the Total Environment* 208 (1–2): 49–57.

Bauer MJ, Herrmann R, Martin A, Zellmann H. 1998. Chemodynamics, transport behaviour and treatment of phthalic acid esters in municipal landfill leachates. *Water Science and Technology* 38 (2): 185–192.

Birkett JW. 2003. Scope of the problem. In: *Endocrine Disrupters in Wastewater and Sludge Treatment Processes* (Birkett J.W. and Lester J.N., eds.), pp. 1–34. Lewis Publishers, Boca Raton, FL.

Cai YQ, Jiang GB, Liu JF, Zhou QX. 2003. Multi-walled carbon nanotubes packed cartridge for the solid-phase extraction of several phthalate esters from water samples and their determination by high performance liquid chromatography. *Analytica Chimica Acta* 494 (1–2): 149–156.

Call DJ, Markee TP, Geiger DL et al. 2001a. An assessment of the toxicity of phthalate esters to freshwater benthos. 2. Sediment exposures. *Environmental Toxicology and Chemistry* 20 (8): 1805–1815.

Call DJ, Markee TP, Geiger DL et al. 2001b. An assessment of the toxicity of phthalate esters to freshwater benthos. 1. Aqueous exposures. *Environmental Toxicology and Chemistry* 20 (8): 1798–1804.

Carpenter CP, Weil CS, Smyth HF. 1953. Chronic oral toxicity of di(2-ethylhexyl) phthalate for rats, guinea pigs, and dogs. *American Medical Association Archives of Industrial Hygiene and Occupational Medicine* 8 (3): 219–226.

Chauret C, Mayfield CI, Inniss WE. 1995. Biotransformation of di-*n*-butyl phthalate by a psychrotrophic *Pseudomonas fluorescens* (BGW) isolated from subsurface environment. *Canadian Journal of Microbiology* 41 (1): 54–63.

Chen J, Wiesner MG, Wong HK, Zheng L, Xu L, Zheng S. 1999. Vertical changes of POC flux and indicators of early degradation of organic matter in the South China Sea. *Science in China Series D—Earth Sciences* 42 (2): 120–128.

Cincinelli A, Stortini AM, Perugini M, Checchini L, Lepri L. 2001. Organic pollutants in sea-surface microlayer and aerosol in the coastal environment of Leghorn—(Tyrrhenian Sea). *Marine Chemistry* 76 (1–2): 77–98.

Crosby DG. 1998. *Environmental Toxicology and Chemistry*. Oxford University Press, New York.

Doull J, Cattley R, Elcombe C et al. 1999. A cancer risk assessment of di(2-ethyl-hexyl)phthalate: Application of the new U.S. EPA risk assessment guidelines. *Regulatory Toxicology and Pharmacology* 29 (3): 327–357.

Eaton RW. 1982. Metabolism of dibutylphthalate and phthalate by *Micrococcus* sp. strain 12B. *Journal of Bacteriology* 152 (1): 48–57.

Eaton RW. 2001. Plasmid-encoded phthalate catabolic pathway in *Arthrobacter keyseri* 12B. *Journal of Bacteriology* 183 (12): 3689–3703.

Eaton RW, Ribbons DW. 1982. Metabolism of dimethylphthalate by *Micrococcus* sp. Strain 12B. *Journal of Bacteriology* 151 (1): 465–467.

Fajardo C, Guyot JP, Macarie H, Monroy O. 1997. Inhibition of anaerobic digestion by terephthalic acid and its aromatic by products. *Water Science and Technology* 36 (6–7): 83–90.

Fatoki OS, Ogunfowokan AO. 1993. Determination of phthalate ester plasticizers in the aquatic environment of Southwestern Nigeria. *Environment International* 19 (6): 619–623.

Fauser P, Vikelsoe J, Sorensen PB, Carlsen L. 2003. Phthalates, nonylphenols and LAS in an alternately operated wastewater treatment plant—Fate modelling based on measured concentrations in wastewater and sludge. *Water Research* 37 (6): 1288–1295.

Fromme H, Küchler T, Otto T, Pilz K, Müller J, Wenzel A. 2002. Occurrence of phthalates and bisphenol A and F in the environment. *Water Research* 36: 1429–1438.

Ganji SH, Karigar CS, Pujar BG. 1995. Metabolism of dimethylterephthalate by *Aspergillus niger*. *Biodegradation* 6: 61–66.

Gledhill WE, Kaley RG, Adams WJ et al. 1980. An environmental safety assessment of butyl benzyl phthalate. *Environmental Science and Technology* 14: 301–305.

Gomes RL, Lester JN. 2003. Endocrine disrupters in receiving waters. In: *Endocrine Disrupters in Wastewater and Sludge Treatment Processes* (Birkett J.W. and Lester J.N., eds.), pp. 177–217. Lewis Publishers, Boca Raton, FL.

Gomes RL, Scrimshaw MD, Lester JN. 2003. Determination of endocrine disrupters in sewage treatment and receiving waters. *TrAC Trends in Analytical Chemistry* 22 (10): 697–707.

Goud HD, Parekh LJ, Ramakrishnan CV. 1990. Treatment of DMT (Dimethyl terephthalate) industry waste water using mixed culture of bacteria and evaluation of treatment. *Journal of Environmental Biology* 11 (1): 15–26.

Griffin JE, Ojeda SR. 1996. *Textbook of Endocrine Physiology*. Oxford University Press, New York.

Halmann M. 1992. Photodegradation of di-n-butyl-ortho-phthalate in aqueous solutions. *Journal of Photochemistry and Photobiology A: Chemistry* 66 (2): 215–223.

Harayama S, Kasai Y, Hara A. 2004. Microbial communities in oil-contaminated seawater. *Current Opinion in Biotechnology* 15 (3): 205–214.

Harris CA, Henttu P, Parker MG. 1997. The estrogenic activity of phthalate esters *in vitro*. *Environmental Health Perspectives* 105: 802–811.

Harris RS, Hodge HC, Maynard EA, Blanchet HJ. 1956. Chronic oral toxicity of 2-ethylhexyl phthalate in rats and dogs. *American Medical Association Archives of Industrial Health* 13 (3): 256–264.

Hashizume K, Nanya J, Toda C, Yasui T, Nagano H, Kojima N. 2002. Phthalate esters detected in various water samples and biodegradation of the phthalates by microbes isolated from river water. *Biological and Pharmaceutical Bulletin* 25 (2): 209–214.

Hata N, Yuwatini E, Ando K, Yamada M, Kasahara I, Taguchi S. 2004. Micro-organic ion-associate phase extraction via *in situ* fresh phase formation for the preconcentration and determination of di(2-ethylhexyl)phthalate in river water by HPLC. *Analytical Sciences* 20 (1): 149–152.

Hayes LA, Nevin KP, Lovley DR. 1999. Role of prior exposure on anaerobic degradation of naphthalene and phenanthrene in marine harbor sediments. *Organic Geochemistry* 30 (8, Part 2): 937–945.

Heindel JJ, Chapin RE, George J et al. 1995. Assessment of the reproductive toxicity of a complex mixture of 25 groundwater contaminants in mice and rats. *Fundamental and Applied Toxicology* 25 (1): 9–19.

Hollifield HC. 1979. Rapid nephelometric estimate of water solubility of highly insoluble organic chemicals of environmental interest. *Bulletin of Environmental Contamination and Toxicology* 23: 579–586.

Howard PH, Banerjee S, Robillard KH. 1985. Measurement of water solubilities, octanol/water partition coefficients and vapor pressures of commercial phthalate esters. *Environmental Toxicology and Chemistry* 4 (5): 653–661.

Jackson MA, Labeda DP, Becker LA. 1996. Isolation for bacteria and fungi for the hydrolysis of phthalate and terephthalate esters. *Journal of Industrial Microbiology* 16: 301–304.

Jeng W. 1986. Phthalate esters in marine sediments around taiwan. *Acta Oceanographica Taiwanica* 17: 61–68.

Jonsson S, Ejlertsson J, Ledin A, Mersiowsky I, Svensson BH. 2003. Mono- and diesters from o-phthalic acid in leachates from different European landfills. *Water Research* 37 (3): 609–617.

Juhasz AL, Naidu R. 2000. Bioremediation of high molecular weight polycyclic aromatic hydrocarbons: A review of the microbial degradation of benzo[a]pyrene. *International Biodeterioration and Biodegradation* 45 (1–2): 57–88.

Juneson C, Ward OP, Singh A. 2001. Biodegradation of bis(2-ethylhexyl)phthalate in a soil slurry-sequencing batch reactor. *Process Biochemistry* 37 (3): 305–313.

Karegoudar TB, Pujar BG. 1984. Metabolism of diethylphthalate by a soil bacterium. *Current Microbiology* 11 (6): 321–324.

Karpagam S, Lalithakumari D. 1999. Plasmid-mediated degradation of o- and p-phthalate by *Pseudomonas fluorescens*. *World Journal of Microbiology and Biotechnology* 15 (5): 565–569.

Keyser P, Pujar BG, Eaton RW. 1976. Biodegradation of the phthalates and their esters by bacteria. *Environmental Health Perspectives* 18: 159–166.

Kim YH, Lee J, Ahn JY, Gu MB, Moon SH. 2002. Enhanced degradation of an endocrine-disrupting chemical, butyl benzyl phthalate, by *Fusarium oxysporum* f. sp. *pisi* cutinase. *Applied and Environmental Microbiology* 68 (9): 4684–4688.

Kleerebezem R, Hulshoff Pol LW, Lettinga G. 1999a. Energetics of product formation during anaerobic degradation of phthalate isomers and benzoate. *FEMS Microbiology Ecology* 29 (3): 273–282.

Kleerebezem R, Pol LWH, Lettinga G. 1999b. Anaerobic biodegradability of phthalic acid isomers and related compounds. *Biodegradation* 10 (1): 63–73.

Kleerebezem R, Pol LWH, Lettinga G. 1999c. Anaerobic degradation of phthalate isomers by methanogenic consortia. *Applied and Environmental Microbiology* 65 (3): 1152–1160.

Kleerebezem R, Pol LWH, Lettinga G. 1999d. The role of benzoate in anaerobic degradation of terephthalate. *Applied and Environmental Microbiology* 65 (3): 1161–1167.

Krasavage WJ, Yanno FJ, Terhaar CJ. 1973. Dimethyl terephthalate (DMT): Acute toxicity, subacute feeding and inhalation studies in male rats. *American Industrial Hygiene Association Journal* 34 (10): 455–462.

Lee SW, Ree M, Park CE et al. 1999. Synthesis and non-isothermal crystallization behaviors of poly(ethylene isophthalate-co-terephthalate)s. *Polymer* 40 (25): 7137–7146.

Lefevre C, Mathieu C, Tidjani A et al. 1999. Comparative degradation by micro-organisms of terephthalic acid, 2,6-naphthalene dicarboxylic acid, their esters and polyesters. *Polymer Degradation and Stability* 64 (1): 9–16.

Li JX, Gu JD. 2004. Degradation of dimethyl terephthalate ester and its isomer by mangrove microorganisms. *Chinese Journal of Applied and Environmental Biology* 10 (6): 782–785.

Li J, Gu JD, Yao J. 2005. Degradation of dimethyl terephthalate by *Pasteurella multocida* Sa and *Sphingomonas paucimobilis* Sy isolated from mangrove sediment. *International Biodeterioration and Biodegradation* 56 (3): 158–165.

Liss PS. 1997. *The Sea Surface and Global Change.* Cambridge University Press, Cambridge, U.K.

Luks-Betlej K, Popp P, Janoszka B, Paschke H. 2001. Solid-phase microextraction of phthalates from water. *Journal of Chromatography A* 938 (1–2): 93–101.

Mackintosh CE, Maldonado J, Jing HW, Hoover N, Chong A, Ikonomou MG, Gobas FAPC. 2004. Distribution of phthalate esters in a marine aquatic food web: Comparison to polychlorinated biphenyls. *Environmental Science and Technology* 38 (7): 2011–2020.

Mihovec-Grdic M, Smit Z, Puntaric D, Bosnir J. 2002. Phthalates in underground waters of the Zagreb area. *Croatian Medical Journal* 43: 493–497.

Mitchell R, Fogel S, Chet I. 1972. Bacterial chemoreception: An important ecological phenomenon inhibited by hydrocarbons. *Water Research* 6 (10): 1137–1140.

Monarca S, Pool-Zobel BL, Rizzi R, Klein P, Schmezer P, Piatti E, Pasquini R, De Fusco R, Biscardi D. 1991. In vitro genotoxicity of dimethyl terephthalate. *Mutation Research Letters* 262 (2): 85–92.

Moore MR. 1996. Oncogenicity study in rats with di(2-ethylhexyl) phthalate including ancillary hepatocellular proliferation and biochemical analyses. Corning Hazelton Incorporated (CMV), Vienna, Virginia, 22182–1699.

Mori S. 1976. Identification and determination of phthalate esters in river water by high-performance liquid chromatography. *Journal of Chromatography* 129: 53–60.

Mueller SO, Korach KS. 2001. Mechanisms of estrogen receptor-mediated agonistic and antagonistic effects. In: *Endocrine Disruptors Part 1* (Metzler M., ed.), Springer-Verlag, Berlin, Germany.

Nakai M, Tabira Y, Asai D, Yakabe Y, Shimyozu T, Noguchi M, Takatsuki M, Shimohigashi Y. 1999. Binding characteristics of dialkyl phthalates for the estrogen receptor. *Biochemical and Biophysical Research Communications* 254 (2): 311–314.

Nascimento Filho Id, von Muhlen C, Schossler P, Bastos Caramao E. 2003. Identification of some plasticizers compounds in landfill leachate. *Chemosphere* 50 (5): 657–663.

National Toxicology Program. 1982. Carcinogenesis bioassay of di(2-ethylhexyl) phthalate (CAS No. 117-81-7) in F344 rats and B6C3F$_1$ mice (feed study). National Toxicology Program, Research Triangle Park, NC NTP-80-29, NIH Publication No. 81-1768.

National Toxicology Program. 1995. Toxicology and carcinogenesis studies of diethylphthalate (CAS NO. 84-66-2) in F344/N rats and B6C3F$_1$ mice (dermal studies) with dermal initiation/promotion study of diethylphthalate and dimethylphthalate (CAS NO. 131-11-3) in male swiss (CD-1) mice. National Toxicology Program, Research Triangle Park, NC.

Niazi JH, Prasad DT, Karegoudar TB. 2001. Initial degradation of dimethylphthalate by esterases from *Bacillus* Species. *FEMS Microbiology Letters* 196: 201–205.

O'Grady DP, Howard PH, Werner AF. 1985. Activated sludge biodegradation of 12 commercial phthalate esters. *Applied and Environmental Microbiology* 49 (2): 443–445.

Öman C, Hynning PÅ. 1993. Identification of organic compounds in municipal landfill leachates. *Environmental Pollution* 80 (3): 265–271.

Patel DS, Desai AJ, Desai JD. 1998. Biodegradation of dimethylterephthalate by *Comamonas acidovorans* D-4. *Indian Journal of Experimental Biology* 36: 321–324.

Paune F, Caixach J, Espadaler I, Om J, Rivera J. 1998. Assessment on the removal of organic chemicals from raw and drinking water at a Llobregat river water works plant using GAC. *Water Research* 32 (11): 3313–3324.

Paxéus N. 2000. Organic compounds in municipal landfill leachates. *Water Science and Technology* 42 (7–8): 323–333.

Peterson JC, Freeman DH. 1982. Phthalate ester concentration variations in dated sediment cores from Chesapeake Bay. *Environmental Science and Technology* 16 (8): 464–469.

Poeton TS, Stensel HD, Strand SE. 1999. Biodegradation of polyaromatic hydrocarbons by marine bacteria: Effect of solid phase on degradation kinetics. *Water Research* 33 (3): 868–880.

Salafranca J, Domeno C, Fernandez C, Nerin C. 2003. Experimental design applied to the determination of several contaminants in Duero River by solid-phase microextraction. *Analytica Chimica Acta* 477 (2): 257–267.

Schouten MJ, Peereboom JWC, Brinkman UAT. 1979. Liquid-chromatographic analysis of phthalate-esters in Dutch river water. *International Journal of Environmental Analytical Chemistry* 7 (1): 13–23.

Schwarzbauer J, Heim S, Brinker S, Littke R. 2002. Occurrence and alteration of organic contaminants in seepage and leakage water from a waste deposit landfill. *Water Research* 36 (9): 2275–2287.

Shaffer CB, Carpenter CP, Smyth HF. 1945. Acute and subacute toxicity of di(2-ethylhexyl) phthalate with note upon its metabolism. *Journal of Industrial Hygiene and Toxicology* 27: 130–135.

Shah SS, Desai JD, Ramakrishna C, Bhatt NM. 1998. Aerobic biotreatment of wastewater from dimethyl terephthalate plant using biomass support particles. *Journal of Fermentation and Bioengineering* 86 (2): 215–219.

Sharanagouda P, Karegoudar TB. 2000. Metabolism of dimethylphthalate by *Aspergillus niger*. *Journal of Microbiology and Biotechnology* 10 (4): 518–521.

Sharma S, Ramakrishna C, Desai JD, Bhatt NM. 1994. Anaerobic biodegradation of a petrochemical waste-water using biomass support particles. *Applied Microbiology and Biotechnology* 40: 768–771.

Sieburth J. 1971. Distribution and activity of oceanic bacteria. *Deep-Sea Research* 18: 1111–1121.

Sigoillot JC, Nguyen MH. 1990. Isolation and characterization of surfactant degrading bacteria in a marine environment. *FEMS Microbiology Letters* 73 (1): 59–67.

Sivamurthy K, Swamy BM, Pujar BG. 1991. Transformation of dimethylterephthalate by the fungus *Sclerotium rolfsii*. *FEMS Microbiology Letters* 79: 37–40.

Sohoni P, Sumpter J. 1998. Several environmental oestrogens are also anti-androgens. *Journal of Endocrinology* 158 (3): 327–339.

Sorkhoh NA, Ghannoum MA, Ibrahim AS, Stretton RJ, Radwan SS. 1990. Crude oil and hydrocarbon-degrading strains of *Rhodococcus rhodochrous* isolated from soil and marine environments in Kuwait. *Environmental Pollution* 65 (1): 1–17.

Staples CA, Peterson DR, Parkerton TF, Adams WJ. 1997. The environmental fate of phthalate esters: A literature review. *Chemosphere* 35 (4): 667–749.

Strotmann U, Reuschenbach P, Schwarz H, Pagga U. 2004. Development and evaluation of an online CO_2 evolution test and a multicomponent biodegradation test system. *Applied and Environmental Microbiology* 70 (8): 4621–4628.

Sugatt RH, Ogrady DP, Banerjee S, Howard PH, Gledhill WE. 1984. Shake flask biodegradation of 14 commercial phthalate-esters. *Applied and Environmental Microbiology* 47 (4): 601–604.

The Scientific Committee for Food. 1995. Reports of the Scientific Committee for Food European Commission, Brussels, Luxembourg.

Toda C, Okamoto Y, Ueda K, Hashizume K, Itoh K, Kojima N. 2004. Unequivocal estrogen receptor-binding affinity of phthalate esters featured with ring hydroxylation and proper alkyl chain size. *Archives of Biochemistry and Biophysics* 431 (1): 16–21.

Tserovska L, Dimkov R. 2002. Dimethylterephthalate catabolism by *Pseudomonas* sp. *Journal of Culture Collections* 3 (1): 33–37.

Turner A, Rawling MC. 2000. The behaviour of di-(2-ethylhexyl) phthalate in estuaries. *Marine Chemistry* 68 (3): 203–217.

Ukielski R. 2000. New multiblock terpoly(ester-ether-amide) thermoplastic elastomers with various chemical composition of ester block. *Polymer* 41 (5): 1893–1904.

Vega D, Bastide J. 2003. Dimethylphthalae hydrolysis by specific microbial esterase. *Chemosphere* 51: 663–668.

Wang J. 2004. Effect of di-n-butyl phthalate (DBP) on activated sludge. *Process Biochemistry* 39 (12): 1831–1836.

Wang YY, Fan YZ, Gu JD. 2003. Aerobic degradation of phthalic acid by *Comamonas acidovorans* Fy-1 and dimethyl phthalate ester by two reconstituted consortia from sewage sludge at high concentrations. *World Journal of Microbiology and Biotechnology* 19: 811–815.

Wang J, Horan N, Stentiford E, Yi Q. 1999. The radial distribution and bioactivity of *Pseudomonas* sp immobilized in calcium alginate gel beads. *Process Biochemistry* 35 (5): 465–469.

Wang Y, Lau P, Button D. 1996a. A marine oligobacterium harboring genes known to be part of aromatic hydrocarbon degradation pathways of soil *pseudomonades*. *Applied and Environmental Microbiology* 62 (6): 2169–2173.

Wang J, Liu P, Qian Y. 1996b. Biodegradation of phthalic acid esters by acclimated activated sludge. *Environment International* 22 (6): 737–741.

Watanabe K, Hamamura N. 2003. Molecular and physiological approaches to understanding the ecology of pollutant degradation. *Current Opinion in Biotechnology* 14 (3): 289–295.

Wenger LM, Isaksen GH. 2002. Control of hydrocarbon seepage intensity on level of biodegradation in sea bottom sediments. *Organic Geochemistry* 33: 1277–1292.

Wilkinson CF, Lamb IV JC. 1999. The potential health effects of phthalate esters in children's toys: A review and risk assessment. *Regulatory Toxicology and Pharmacology* 30: 140–155.

Williams SE, Woolridge EM, Ransom SC, Landro JA, Babbitt PC, Kozarich JW. 1992. 3-Carboxy-cis,cis-muconate lactonizing enzyme from *Pseudomonas putida* is homologous to the class II fumarase family: A new reaction in the evolution of a mechanistic motif. *Biochemistry* 31 (40): 9786–9776.

Xu XR, Li HB, Gu JD. 2005. Biodegradation of an endocrine-disrupting chemical di-n-butyl phthalate ester by *Pseudomonas fluorescens* B-1. *International Biodeterioration and Biodegradation* 55 (1): 9–15.

Yan H, Liu YX. 1998. Dimethyl phthalate biodegradation by *Dunaliella tertiolecta*. *Journal of Environmental Sciences* 10 (3): 296–301.

Yin MC, Su KH. 1996. Investigation on risk of phthalate ester in drinking water and marketed foods. *Journal of Food and Drug Analysis* 4 (4): 311–317.

Yuan SY, Liu C, Liao CS, Chang BV. 2002. Occurrence and microbial degradation of phthalate esters in Taiwan river sediments. *Chemosphere* 49 (10): 1295–1299.

Zacharewski TR, Meek MD, Clemons JH, Wu ZF, Fielden MR, Matthews JB. 1998. Examination of the in vitro and in vivo estrogenic activities of eight commercial phthalate esters. *Toxicological Sciences* 46 (2): 282–293.

Zeng F, Cui K, Li X, Fu J, Sheng G. 2004. Biodegradation kinetics of phthalate esters by *Pseudomonas fluorescences* FS1. *Process Biochemistry* 39 (9): 1125–1129.

Zhao XK, Yang GP, Wang YJ. 2004a. Adsorption of dimethyl phthalate on marine sediments. *Water, Air, and Soil Pollution* 157 (1–4): 179–192.

Zhao XK, Yang GP, Wang YJ, Gao XC. 2004b. Photochemical degradation of dimethyl phthalate by Fenton reagent. *Journal of Photochemistry and Photobiology A: Chemistry* 161 (2–3): 215–220.

Zhu K, Zhang L, Hart W, Liu M, Chen H. 2004. Quality issues in harvested rainwater in arid and semi-arid Loess Plateau of northern China. *Journal of Arid Environments* 57 (4): 487–505.

16

Mixed Contamination of Polyaromatic Hydrocarbons and Metals at Manufactured Gas Plant Sites: Toxicity and Implications to Bioremediation

Palanisami Thavamani, Mallavarapu Megharaj,
Kadiyala Venkateswarlu, and Ravi Naidu

CONTENTS

Mixed Contaminants

Any combination of two or more chemical substances, regardless of source or of spatial or temporal proximity that can influence the risk of chemical toxicity in the target population, is defined as a mixture of contaminants

(US EPA 1986). The sites contaminated with mixtures are therefore complex, multicomponent systems with a range of different organic and inorganic chemicals coexisting under various physiochemical conditions (Thavamani et al. 2011). Although some potential environmental hazards involve significant exposure to only a single chemical compound, most instances of environmental pollution involve concurrent or sequential exposures to a mixture of compounds that may induce similar or dissimilar effects over exposure periods ranging from short term to lifetime.

Mixed contamination could be due to the occurrence of organic–organic (e.g., pesticides plus their metabolites), organic–inorganic (e.g., DDT–As), and inorganic–inorganic (e.g., sewage sludge containing several heavy metals) chemicals. Mixtures of compounds that are placed in the same area for disposal or storage have the potential in causing toxicity to biota because of the combined effect of pollutants (US EPA 2000). Multichemical exposures are ubiquitous including air and soil pollution from municipal incinerators, leakage from hazardous waste facilities and uncontrolled waste sites, and drinking water containing chemical substances formed during disinfection. Manufactured gas plant (MGP) sites contain mixed contaminants of organic and inorganic chemicals such as polyaromatic hydrocarbons (PAHs) and metals, and the remediation of these sites is more complex and challenging. This chapter focuses on mixed contamination of MGP sites, with special emphasis on issues such as bioavailability and toxicity associated with bioremediation of those sites.

Sites with Mixed Contaminants

Almost all the contaminated sites have more than one type of pollutants. In fact, 40% of the hazardous waste sites in the National Priority List (NPL) of the U.S. Environmental Protection Agency are contaminated with mixtures of metals and organic pollutants (Sandrin et al. 2000). Metals most frequently found at Superfund sites include arsenic, barium, cadmium, chromium, lead, mercury, nickel, and zinc. Common organic contaminante include petroleum, chlorinated solvents, and pesticides. Following are some of the identified sites contaminated with such mixtures worldwide:

Manufacture gas plant sites	—	PAHs and metals
Petrochemicals units	—	TPHs and metals
Sheep and cattle dip sites	—	DDT and As
Wood preservation sites	—	Creosote (PAHs), Cu, Cr, and As
Electronic waste processing sites	—	PAHs, PCBs, and metals
Military sites	—	Nitro compounds and As

Railway corridors	—	PAHs, TPHs, and metals
Urban runoff	—	TPHs, PAHs, and metals
Roadside soils	—	PAHs and Pb
Contaminated river sediments	—	PAHs, PCBs, and metals
Sewage sludge	—	PAHs and metals

Most sites included in the NPL are complex, requiring remediation for more than one type of contaminant groups, 24% of the sites contain two contaminant groups, and 52% contain three groups (Figure 16.1). These contaminants are not necessarily in the same contaminated medium. Halogenated VOCs (volatile organic compounds) are by far the most common subgroup of organic contaminants, followed by BTEX (benzene, toluene, ethylbenzene, and xylenes), nonhalogenated VOCs, PAHs, nonhalogenated SVOCs (semi-VOCs), phenols, pesticides and PCBs, and metals, including lead, arsenic, chromium, cadmium, zinc, nickel, and other less frequently found metals (US EPA 2004).

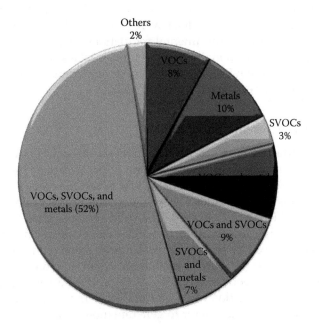

FIGURE 16.1
Frequencies of major contaminant groups at NPL, United States. (Modified from US Environmental Protection Agency, Cleaning up the Nation's wastes sites: Markets and technology trends, Office of Solid Waste and Emergency Response, US EPA, Washington, DC, 2004.)

MGP Sites

History

The MGP is an industrial facility at which gas was produced from coal, oil, and other feedstock. For a period of over 100 years, MGPs were an important part of life in cities and towns throughout the world. By the turn of twentieth century, almost every city in the United States had at least one MGP. According to U.S. EPA report, 50,000 plants were built over 140 years of gas production (US EPA 2004). Most of the MGPs were located adjacent to waterways and rail spurs for easy access to coal (EI Digest 1995). As the technology developed, it became a common source of light, heat, and fuel for a variety of industrial and commercial facilities, and residences. After electricity and natural gas became common afterward, many of the larger MGP properties were converted for new uses by the facilities and other companies that owned them. In addition to the commercial MGPs, many railroad companies, military installations, large institutions (e.g., hotels, hospitals, prisons, and schools), industrial facilities, and large private homes were equipped with gas plants (Heritage Research Center 2002). By the early 1940s, natural gas became more cost-effective compared to MGP gases, and MGP production declined rapidly. At this point, MGPs were closed, sold, or decommissioned. In the absence of strict environmental regulations, MGP process residuals, by-products, and wastes remained at the former MGPs. The manufacturing practices thus left an environmental legacy of hazardous waste contamination in soil and water ecosystems around MGPs.

Typical MGP Process and the Residuals

A typical manufactured gas production involves three processes—coal carbonization, carbureted water gas, and oil gas (New York State Department of Environmental Conservation [NYDEC] 2003). When the coal is heated under anoxic conditions (carbonization), volatile products are driven off as gas that is collected, purified, and delivered. The solid remains after carbonization can be burnt at high temperature to yield coke. Carbureted water gas or "blue gas" is produced by passing steam through a bed of incandescent carbon. The resultant gas (mixture of methane and carbon monoxide) is further reacted with light oil petroleum products to produce a gas of higher calorific value. In the production of oil gas similar to the carbureted water-gas process, oil is added to the reactor to generate more heat. The oil vapors are thermally cracked and fixed into gases (Luthy 1994, US EPA 2004).

All the three gas manufacturing processes described earlier left various hazardous process residuals, and a typical MGP has several remnant structures. Generally, coal is stockpiled on-site for considerable length of time during the periods of less demand. Leachates from these sources contain heavy metals, sulfides, and some hydrocarbons. Surface contamination by coal and coke may also be expected. The retort houses on-site are loaded with coal and heated for several hours to drive off all volatile material, resulting in coke that is removed from the retort. Solid process residuals, left in retort houses, include ash, clinker, and coke. The carbureted water-gas process residuals include clinker and wastewater containing lighter petroleum hydrocarbons and metals (NZ MFE 1997). Nickel, uranium, and vanadium catalysts are often left on-site where oil gas production takes place. Tanks and pipe works left on-site after decommissioning may contain residual material. Old gas mains are sometimes used on-site as convenient receptacles for waste holder oil and water during decommissioning. MGP residuals are found both in surface and subsurface soils; tar often is present in deep subsurface soils due to vertical migration along preferential pathways (Sharon and Dennis Unites 1998).

Contaminants of Concern in MGP Sites

The gas manufacturing processes in the past resulted in a variety of residuals, some of which were converted to by-products, while other hazardous materials were managed as wastes. The detailed processes and generation of wastes are shown in Table 16.1. Contamination of soil and groundwater in MGP sites is dominated by six primary classes of chemicals (Luthy 1994, Middleton 1995, Pradhan and Srivastava 1997), which include the following:

- PAHs
- Volatile aromatic compounds (VACs)
- Metals (arsenic, chromium, copper, lead, nickel, and zinc)
- Phenolic compounds
- Inorganic nitrogen (including cyanide compounds)
- Inorganic sulfur

Each of these wastes is made of hundreds or thousands of different chemicals (Table 16.2). These chemicals are used in manufacturing or are produced as wastes, individually or collectively, in industries such as petroleum refineries, wood treatment and preservation, coke manufacturing, gas plant facilities, and synthetic fuel production (Knight et al. 1999). It was a common practice to dump coal tar and other wastes into on-site pits or ponds, bury, or use as fill to adjust the grade of the gas yard. The gas manufacturing process also involved the use of wood chips and iron filings to remove sulfur and cyanide from the gas. These chips and filings were disposed off in pits or buried. As a result of these practices, wastes from manufactured gas processing can be found in soil, sediment, groundwater, and surface water (Fischer et al. 1999, US EPA 2004).

TABLE 16.1

Summary of the Principal Waste Types at Gasworks Sites

Principal Waste Type	Source	Contaminants
Coal tar Tar oils	Separated from gas and liquors at various stages of the purification processes	PAHs Petroleum hydrocarbons, including BTEX Phenols Complex cyanides
Spent oxides	Used to remove sulfur during gas purification	Free cyanides Metals
Coke, coke breeze, ash, clinker residues	By-products and furnace residues	PAHs Metals
Light oils	Light oils used around all machinery and as scrubbing agent in recovery process	Petroleum hydrocarbons (including BTEX)
Drip oils	Drip oils condensed from gas	
Ammoniacal wastes	Nitrogen removal during gas purification processes	Phenols, nitrates, sulfates, sulfides, PAHs
Asbestos	Used as lagging around many of the "hot" processes and pipes	Asbestos
Pb, mercury, Zn	Pb from batteries, pipelines, paints, etc. Mercury sometimes used in metering switches	Metals

Source: New South Wales Environmental Protection Agency (NSW EPA), Information about assessing gasworks sites, Department of Environment and Conservation, New South Wales, Australia, http://www.environment.nsw.gov.au/clm/gasworksassessment.html, 2005.

Coal or water-gas tar, the primary by-products of gas manufacture from MGP sites, are of particular concern because elevated PAH concentrations and tar form a dense nonaqueous phase liquid (DNAPL) that is capable of migrating downward in the subsurface below the water table. Thus, the subsurface of old gasworks sites consists of a mixed fill contaminated with coal tar and its distillation products, organics, and heavy metals. The contaminants are heterogeneously distributed, but are found throughout the area. Several authors reported the presence of PAHs and metals together in the former MGP sites (Thomas and Lester 1994, Haeseler et al. 1999, Reddy et al. 2006, Thavamani 2010, Thavamani et al. 2011). The PAH concentrations in MGP sites varied from 150 to 40,000 mg/kg, and most of the studies reported a total PAH concentration >10,000 mg/kg (Table 16.3). PAHs are present in high concentrations in coal tars, up to 70%–80% by weight. Tar, pitch, and ash once spilt or deposited in/on the grounds persist for longer periods, mostly in the form of soil-polluted PAHs. The purifier waste of MGP sites often contains significant quantities of metals like cadmium, lead, arsenic, and zinc compounds. Inorganic contamination at the former gasworks sites usually occurs at or near the ground surface. Some metals can leach from

TABLE 16.2

Principal Chemicals of Interest at Gasworks Sites

Inorganic Compounds	Metals and Metalloids	Monocyclic Aromatic Hydrocarbons	Phenolics	Polycyclic Aromatic Hydrocarbons
Ammonia	Aluminum	Benzene	Phenol	Acenaphthene
Cyanide	Antimony	Ethyl benzene	2-Methylphenol	Acenaphthylene
Nitrate	Arsenic	Toluene	4-Methylphenol	Anthracene
Sulfate	Barium	Total xylenes	2,4-Dimethylphenol	Benzo(a)anthracene
Sulfide	Cd			Benzo(a)pyrene
Thiocyanate	Chromium			Benzo(b) fluoranthene
	Copper			Benzo(g,h,i) perylene
	Iron			Benzo(k) fluoranthene
	Pb			Chrysene
	Manganese			Dibenzo(a,h) anthracene
	Mercury			Fluoranthene
	Nickel			Fluorene
	Selenium			Naphthalene
	Silver			Phenanthrene
	Vanadium			Pyrene
	Zn			Indeno (1,2,3-cd) pyrene

Source: New South Wales Environmental Protection Agency (NSW EPA), Information about assessing gasworks sites, Department of Environment and Conservation, New South Wales, Australia, http://www.environment.nsw.gov.au/clm/gasworksassessment.html, 2005.

the waste and contaminate groundwater and subsurface soil. The mobility of inorganic contaminants depends heavily on their solubility, and factors such as pH and presence of other chemical species affect the solubility and binding of contaminants to the soil. Shallow soil contamination by inorganic compounds can be extensive at the former gasworks sites (NSW EPA 2005, US EPA 2004). However, deeper contamination can be variable, depending on the volume and mobility of the source and the presence of preferential migration pathways (Thomas and Lester 1994).

PAHs and Metals as Mixed Contaminants

By far, two of the most abundant environmental pollutants found in large number of contaminated sites are PAHs and heavy metals such as cadmium, lead, zinc, chromium, copper, and arsenic (Fischer et al. 1999). Heavy

TABLE 16.3

Reported Concentrations of the PAHs and Metals Found in the
MGP Site Soil

PAH	Concentration Range (mg/kg)	Metal	Concentration Range (mg/kg)
Acenaphthene	50–222	Lead	88–671
Acenaphthylene	50–623	Cadmium	8–112
Anthracene	55–181	Zinc	64–488
Benz(a)anthracene	66–82[a]	Chromium	33–379
Benzo(a)pyrene	58–738	Copper	18–57
Benzo(b)fluoranthene	31–33[a]	Arsenic	5–9
Benzo(g,h,i)perylene	23–30[a]	Manganese	52–181
Benzo(k)fluoranthene	4.8–33[a]	Nickel	10–230
Chrysene	33–748[a]		
Dibenz(a,h) anthracene	9.1[a]		
Dibenzofuran	7.7[a]		
Fluoranthene	55–1245		
Fluorene	25–726		
Indeno(1,2,3-cd) pyrene	12–21[a]		
Naphthalene	51–660		
Phenanthrene	52–2738		
Pyrene	55–1357		

Source: Thavamani, et al., *Environ. Int.*, 37, 184, 2011.
[a] Reddy et al. (2006).

metals and PAHs are frequently found together as contaminants in soils and groundwater at sites of industrial operations (Figure 16.2) such as MGPs, wood treatment, metal finishing, petroleum refining, and paint and automobile manufacturing plants (Mueller et al. 1989, Luthy 1994, Kong 1998, Fischer et al. 1999, Knight et al. 1999). Former MGPs thus serve as the typical examples for sites with mixed contaminants that contain both heavy metals and PAHs.

The PAHs, consisting of more than 100 organic compounds, contain two or more fused aromatic rings and are released into the environment during incomplete combustion. They have high melting and boiling points, low vapor pressure, high lipophilicity, and very low solubility in water. The PAHs, with their higher molecular weight, are hydrophobic and bind strongly to soil particles and have low solubility. These heavier PAHs are therefore generally found at higher concentrations near the source of contamination, particularly in surface soils. The lighter and more soluble PAHs (e.g., naphthalene) are frequently detected in groundwater, although volatilization and leaching losses reduce their concentrations in surface soils. Although their acute toxic effect is considered to be moderate to low, their affinity to lipids

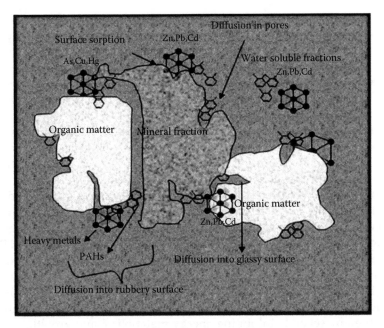

FIGURE 16.2
Conceptual model of the occurrence of PAHs and metals in mixed contaminated soils.

can cause bioaccumulation and even biomagnification through the food chain. Thirty PAHs included in the WHO report (UNEP-ILO-WHO 1998) were identified with genotoxicity to animals, and 17 were suspected to be carcinogenic. Though PAH toxic effects toward animals are widely studied, their direct effects on plants are yet to be completely understood (Efroymson et al. 1997). The sites contaminated with PAHs and metals are considered difficult to clean up using present bioremediation technologies because of the toxicity of heavy metals to microorganisms.

Toxicity of PAHs and Metal Mixtures to Biota

Contaminants in mixtures are known to interact with biological systems in ways that can greatly alter the toxicity of the individual compounds. Some mixtures of contaminants have higher toxicity than predicted (Tsiridis et al. 2006). Thus, even if the characteristics of individual chemicals are known, their behavior in mixtures cannot be easily predicted. This situation subjects organisms in contaminated environments at maximum risk from their integrated effects. The majority of studies investigating metal–PAH joint toxicity toward various organisms suggest that contaminant interactions occur frequently and arise through a variety of mechanisms. The differential retention/efflux of contaminants caused by these mechanisms may alter exposure and hence the overall toxicity to organisms in natural environments. Most of the studies indicated that the occurrence of synergistic toxicity between PAHs and metals may be predominant (Gust 2005). Therefore, metals may facilitate either synergistic or antagonistic toxic effects depending on which reactions they affect during PAH metabolism.

In addition to metals altering the toxicity of hydrocarbons, interactive toxicity may also be elicited when hydrocarbons influence metal toxicity. For example, George and Young (1986) showed that the PAH 3-methylcholanthrene delayed the onset of metallothionein induction in plaice. Also, PAHs have been found to affect the bioaccumulation of metals. Fair and Sick (1983) found that the Black Sea bass, *Centropristis striata*, assimilated higher concentrations of Cd in various tissues from food (spiked oyster tissue) containing Cd and naphthalene compared to food containing Cd alone. Babu et al. (2001) found synergistic inhibitory effects on photosynthetic rate and plant growth in duckweed (*Lemna gibba*) when exposed to an oxygenated PAH (1, 2-dihydroxyanthraquinone) and Cu. The combined contaminants were found to dismantle the function of photosynthetic electron-transport chain, thereby severely inhibiting energy metabolism and growth.

Microbial Mechanisms for Metal Resistance

Microorganisms have coexisted with metals since the beginning of life, and essential metals are used for catalyzing key metabolic reactions and maintaining protein structures. Although metals are thought to inhibit the ability of microorganisms to degrade organic pollutants, several microbial mechanisms of resistance to metal are known to exist (Nies and Silver 1995). A breadth of microbial metabolic function is presumed to enhance resistance to environmental stress and disturbance. For protection against the toxic effects of heavy metals, bacteria can adapt diverse resistance systems that confer upon them a certain range of metal tolerances (Gadd 2010, Valls and Lorenzo 2002). Resistant mechanisms include intracellular and extracellular metal sequestration, metal reduction, metal efflux pumps, and production of metal chelators such as metallothioneins and biosurfactants. Despite the ubiquity and efficacy of microbial mechanisms for metal resistance, a few studies have attempted to exploit them to increase pollutant biodegradation in mixed contaminated systems.

Generally, the decrease in metal concentration in microbial experiments could be the result of several known detoxification mechanisms such as (1) sorption of metals by the bacterial cells (Bollag and Duszota 1984), (2) energy-dependent accumulation of metals by the cells and subsequent interaction

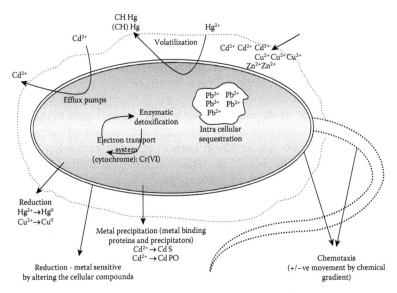

FIGURE 16.3
Microbial mechanisms for metal resistance and sequestration. (Modified from Gadd, G.M., *Microbiology*, 156, 609, 2010.)

with metal-binding proteins (Hettiarachchi et al. 2000), and (3) conversion of soluble metals in the medium to insoluble forms via interaction with metabolites like exopolysaccharides. In response to metal toxicity, many microorganisms have developed unique mechanisms of resistance and detoxification. These mechanisms may be intracellular or extracellular and may be specific to a particular metal or a general metal interaction mechanism (Figure 16.3). Extracellular molecules produced under metal stress by microorganisms may also provide general resistance to metals. For example, siderophores, the iron-complexing low-molecular-weight organic compounds produced by some microorganisms like cyanobacteria in their habitats with low iron content, not only facilitate iron transport into the cell but also interact with other metals similar to iron in chemistry, such as Al, Cr, and Cu. Exploration for microorganisms with pollutant-degradative capabilities is needed for their potential application in remediation of soils contaminated with mixtures.

Microbial Degradation of Organic Pollutants

Direct Microbial Metabolism

Roane and Pepper (1997) showed that the metal-resistant population could protect the metal-sensitive and organic-pollutant-degrading population from metal toxicity. However, this study suggested staggered approach wherein 48 h time must be allowed for metal detoxification to occur before organic degradation could be observed. The viability of the population that degrades organic pollutant decreased when it was added to the system prior to metal detoxification. There are two widely observed processes for microbial detoxification of environmental pollutants: (1) many microorganisms are capable of degrading a variety of organic pollutants, and (2) a number of metal-resistant microorganisms are known to detoxify metals, such as Cd, Hg, Pb, Zn, and Se (Roane et al. 2001). Exploring these two processes further by involving two different organisms (dual bioaugmentation) or one organism (metal-resistant and PAH-degrading) could be the viable bioremediation approach for soils contaminated with both PAHs and metals. The dual-bioaugmentation approach involves coinoculating a metal-detoxifying/resistant organism and an organic-pollutant-degrading organism that cooperatively functions to remediate both metal and organic pollutants in a mixed contaminated system (Roane et al. 2001). This can be achieved only by directing future research toward isolating and characterizing specialized microorganisms with abilities of both metal detoxification and degradation of organic pollutant.

Pepper et al. (2002) reported that in mixed contaminated soils, cell bioaugmentation that allows immediate degradation of the organic pollutant may be the viable technique. The study of Roane et al. (2001) on augmentation reported the potential of Cd-resistant microorganisms to reduce soluble Cd levels and to enhance degradation of 2,4-dichlorophenoxyacetic acid (2,4-D). *Pseudomonas* sp. strain H1 and *Bacillus* sp. strain H9 possessing plasmid-dependent intracellular mechanism for Cd detoxification reduced soluble Cd levels by 36%. Both *Arthrobacter* strain D9 and *Pseudomonas* strain I1 produced an extracellular polymer layer that bound and reduced soluble Cd levels by 22% and 11%, respectively. Although none of the Cd-resistant isolates could degrade 2,4-D, results of dual-bioaugmentation studies conducted with both the pure cultures in laboratory soil microcosms showed that each of the four Cd-resistant isolates supported complete degradation of 500 mg of 2,4-D by the Cd-sensitive 2,4-D-degrading *Ralstonia eutropha* JMP134. Degradation occurred in the presence of Cd up to 24 mg in pure cultures and up to 60 mg in amended soil microcosms.

Degradation by Metal-Resistant Microorganisms

The presence of multiple contaminants may provide a more restricted niche because microbes must tolerate both metals and toxic hydrocarbons. Consequently, metals and hydrocarbons impose selection pressure for specific microbial types. These stresses may cause a reduced phylogenetic diversity due to the elimination of some microbes or the enrichment of particularly successful ecotypes. The surviving bacterial strains in soils contaminated with PAHs and metals for long term, as in the case of former MGP sites and sheep dip sites, could have the capacity of tolerating toxic concentrations of metals and, at the same time, capacity of degrading organic contaminants and utilizing carbon as source. These changes in community structure are important and may lead to the development of successful bioremediation strategies. To our knowledge, there are no reports on the successful isolation and characterization of microorganisms that are metal resistant and capable of degrading organic pollutants.

Inhibition of PAHs Degradation by Metals

Significant associations between metals and PAHs in soil exert an influence on microbes that manifests itself in drastic changes in microbial diversity. It has been reported that the presence of metals impedes the biodegradation of PAHs (Table 16.4), although microorganisms require some of these heavy metals at trace levels for growth (Wong et al. 2005, Atagana 2006). Maslin and Maier (2000) examined the impact of Cd on phenanthrene biodegradation by indigenous soil community in two desert soils over a 9 day period. Results showed an increase in lag period by 5 days for phenanthrene degradation in the presence of 1 and 2 mg solution-phase Cd/L and complete inhibition at 3 mg solution-phase Cd/L. The influence of trace metals on microbial degradation of aromatic and aliphatic hydrocarbons, either in solution (Amor et al. 2001) or in soil (Baldrian et al. 2000), has been investigated. Baldrian et al. (2000) observed degradation of PAH molecules up to five to six aromatic rings by the white rot fungus, *Pleurotus ostreatus*, in the presence of 100 mg/kg Cd in nonsterile soil, but Cd at 500 mg/kg completely inhibited the PAH biodegradation. Although the rate of PAH degradation by the fungus was not affected, presence of Cd was inhibitory to the activity of ligninolytic enzymes, laccase, and Mn-dependent peroxides. The biodegradation of PAHs was reduced by free Cd species between 0.002 and 1 mg/L (Springael et al. 1993, Sandrin et al. 2000, Sandrin and Maier 2003).

Riis et al. (2002) studied the degradation of diesel fuel by a microbial community from a soil polluted with heavy metals such as Cu, Ni, Zn, Pb, Cd, Hg, and Cr. The degradation was greatly inhibited by the concentrations of metals, and toxicity of the metals followed the order: $Hg > Cr(VI) > Cu > Cd > Ni > Pb > Zn$. Fluoranthene degradation by bacteria grown on agar plates was totally inhibited by 10 mg/L of Cd and 50 mg/L of Zn and Cu (Gogolev and Wilke 1997). Since metal complexing ligands are present in agar, the concentration of bioavailable metal in agar plates might be very low or totally zero. Shen et al. (2005) spiked the mixtures of PAHs (phenanthrene, fluoranthene, benzo(a)pyrene) and metals (Cd, Pb, Zn) to study the effect of mixed contaminants on soil urease activity. The results showed that Zn interacted more easily with PAHs than Pb or Cd; the combined effect of

TABLE 16.4

Reported Heavy Metal Concentrations That Cause Inhibition of Biodegradation of PAHs

Heavy Metal	PAHs	Toxic Metal Concentration	Organisms	Study System	pH	Reference
Cd^{2+}	Naphthalene	<25.3–50.6 mg/L	Alcaligenes sp. Pseudomonas sp.	Tris minimal media	7.0	Springael et al. (1993)
	Naphthalene	1 mg/L	Burkholderia sp.	Dilute MSM	6.5	Sandrin et al. (2000)
	Naphthalene	10 mg/L	Pseudomonas putida	MMO medium	7.0	Malakul et al. (1998)
	Phenanthrene	1 mg/L	Indigenous microorganisms	Soil microcosms	7.6	Maslin and Maier (2000)
	Phenanthrene	10 mg/L	Indigenous microorganisms	Soil microcosms	8.18	Shen et al. (2006)
	Pyrene	60 mg/L	Indigenous microorganisms	Soil microcosms	6.2–7.2	Irha et al. (2003)
	Pyrene	100–500 mg/kg	Indigenous microorganisms	Soil microcosms	5.3	Baldrian et al. (2000)
	Fluoranthene	2 mmol	Indigenous microorganisms	Houba–Remarcle media	NA	Riha et al. (1992)
	Benzo(a)anthracene	100–500 mg/kg	Indigenous microorganisms	Soil microcosms	5.3	Baldrian et al. (2000)
	Benzo(a)pyrene	10 mg/kg	Indigenous microorganisms	Soil microcosms	5.3	Baldrian et al. (2000)
Pb^{2+}	Pyrene	6 mg/L	Indigenous microorganisms	Soil microcosms	6.2–7.2	Irha et al. (2003)
	Anthracene	1000 mg/L	Rhodococcus equi IM 6 KB3	Mineral salt media	7.0	Fijałkowska et al. (1998)
	Phenanthrene	1000 mg/L	Rhodococcus equi IM 6 KB3	Mineral salt media	7.0	Fijałkowska et al. (1998)
Zn^{2+}	Naphthalene	<29.5–736 mg/L	Alcaligenes sp. Pseudomonas sp. Moraxella sp.	Tris minimal media	7.0	Springael et al. (1993)
	Phenanthrene	720–1440 mg/kg	Indigenous microorganisms	Soil microcosm	NA	Wong et al. (2005)
	Fluoranthene	20 mmol	Indigenous microorganisms	Houba–Remarcle media	NA	Riha et al. (1992)
Cu^{2+}	Naphthalene	<14.3–71.6 mg/L	Alcaligenes sp. Pseudomonas sp. Moraxella sp.	Tris minimal media	7.0	Springael et al. (1993)

	Contaminant	Concentration	Microorganism	Media	pH	Reference
	Phenanthrene	700 mg/kg	Indigenous microorganisms	Soil suspension	7.3	Sokhn et al. (2001)
	Pyrene	20 mg/L	Indigenous microorganisms	Soil microcosms	6.2–7.2	Irha et al. (2003)
	Fluoranthene	8.5 mmol	Indigenous microorganisms	Houba–Remarcle media	NA	Riha et al. (1992)
Cr^{6+}	Naphthalene	<131 mg/L	*Alcaligenes* sp. *Pseudomonas* sp. *Moraxella* sp.	Tris minimal media	7.0	Springael et al. (1993)
Hg^{2+}	Naphthalene	<45.2–226 mg/L	*Alcaligenes* sp. *Pseudomonas* sp. *Moraxella* sp.	Tris minimal media	7.0	Springael et al. (1993)
	Pyrene	50–100 mg/kg	Indigenous microorganisms	Soil microcosms	5.3	Baldrian et al. (2000)
	Benzo(a)anthracene	50–100 mg/kg	Indigenous microorganisms	Soil microcosms	5.3	Baldrian et al. (2000)
	Benzo(a)pyrene	10 mg/kg	Indigenous microorganisms	Soil microcosms	5.3	Baldrian et al. (2000)
Ni^{2+}	Naphthalene	5.18–10.3 mg/L	*Alcaligenes* sp. *Pseudomonas* sp. *Moraxella* sp.	Tris minimal media	7.0	Springael et al. (1993)
Co^{2+}	Naphthalene	<13.3–1330 mg/L	*Alcaligenes* sp. *Pseudomonas* sp. *Moraxella* sp.	Tris minimal media	7.0	Springael et al. (1993)

PAHs and metals on soil urease activity depended largely on the incubation time. However, concentrations of 0.1 mg/kg phenanthrene and 10 mg/kg Cd applied to soil is insufficient to arrive at a conclusion on the toxicity toward urease activity.

Strongly complexed metals are less toxic than weakly complexed forms, which in turn are less toxic to biota than the free ions (Adriano 2001). The co-occurrence of Cd with Zn or Pb (Cd–Zn or Cd–Pb) is common in certain contaminated environments due to their origin from similar sources. However, the interaction of these metal combinations on PAH remediation is not yet known. Also, the influence of heavy metals on biodegradation of PAHs depends upon the fraction of bioavailable metal concentration rather than the total concentration. Determining the metal speciation is therefore important to understand the real toxicity of metals to biota and thereby bioremediation in PAH–metal cocontaminated soils. Thavamani et al. (2011) used visual MINTEQ (Ver 2.52) computer software (Gustafsson 2009) and determined for the first time the metal speciation of soil solutions (water-extractable) from a former MGP site. The water-extractable Zn was the highest followed by Pb and Cd. More than 80% of total Pb complexed with DOC and the inorganic ligand sulfate, and the calculated speciation showed the lowest free-ion activity for Pb^{2+}. On the other hand, Zn and Cd that are characteristic in having high mobility showed the same pattern of speciation irrespective of soil pH. These results warrant the consideration of free-ion activity of metals while assessing risk of the MGP sites.

Necessity for Remediation of MGP Sites

The MGP sites, which were once welcome additions to the cities and towns, are now sources of complex environmental contamination problems and are slowly being cleaned up. Industrialization led to urbanization, and city growth eventually placed heavy demand for residential, recreational, and commercial areas to be developed in the erstwhile industrial zones. The estimated land use patterns around MGPs are the following: industrial/commercial (50%), residential (30%), and recreational (20%) (US EPA 2004). The abandoned MGPs were therefore revisited, and the waste material contamination in those sites was well recognized. The presence of mixed organic (e.g., PAHs) and inorganic (e.g., metals) contaminants in the sites limits the application of various remediation technologies to cleanup because of the complex nature of the contaminants. Excavation and disposal, containment and *in situ* technologies, and soil washing are the three broad categories of remediation techniques used often in MGP sites. Remediation of MGPs had commenced in the past 30 years, and much focus has been on single contaminants such as PAHs, BTEX, and chlorinated hydrocarbons.

Over the next several decades, federal, state, and local governments and private industry all over the world have to spend billions of dollars annually to clean up sites contaminated with hazardous wastes and petroleum products from a variety of industrial sources. At the current level of site cleanup activity in the United States (about $6–8 billion/annum), it would take at least 35 years to complete most of the cleanup work. In particular, MGP cleanup costs have been documented to range from a few hundred thousand dollars to $86 million for a single site; most tend to be in the $3–10 million range. Should all 30,000–45,000 sites need cleanup, the estimated cost would be $26–128 billion. The increasing cost of these technologies, therefore, forces us to look for the potentially low-cost technologies such as bioremediation.

Approaches for Remediation of PAHs in Metal-Contaminated Systems

Since 1970s, research on biological degradation of PAHs demonstrated that bacteria, fungi, and algae possess catabolic abilities that may be utilized for the remediation of soils contaminated with PAHs. However, co-occurrence of metals and PAHs in soil complicates the bioremediation process. The strategies for remediation of soil contaminated with either hydrocarbons or heavy metals have undergone considerable improvement in recent years; however, in real field situation, cost of treatment offsets the application of commonly used remediation techniques such as land removal, incineration, or land filling. Also, some remediation processes are slow due to the presence of heterogeneous mixture of organic and inorganic contaminants. This is because very few existing techniques (such as chemical oxidation, microbial, or phytoremediation) can deal with both metals and organic compounds. However, the efficiency of these techniques has not been explored completely and optimized for mixed contaminants in soils. This remains to be an unexplored area in remediation research. Bioremediation approaches that involve PAH degradation in the presence of metals must consider microbial mechanisms of metal resistance (Hettiarachchi et al. 2000), microbial degradation (Roane et al. 2001), biosurfactants (Fraser 2000), and abiotic methods like addition of natural materials to reduce metal toxicity (Malakul et al. 1998, Sandrin et al. 2000).

Conclusion

Pollutants generally occur as mixtures at any contaminated site. The former MGP sites are typical examples of such sites contaminated with various inorganic and organic chemicals. The available reports clearly indicate that the presence of heavy metal(loid)s affects the bioremediation of organic contaminants by inhibiting microorganisms that are capable of degrading these pollutants. Successful remediation of such sites contaminated with mixtures requires novel methods based on (1) isolation of microorganisms able to tolerate toxic levels of metal(loid)s and degrade organic pollutants, (2) dual bioaugmentation with two types of microorganisms comprising metal detoxifiers and degraders of organic compounds, (3) combination of natural/modified materials to immobilize metal(loid)s and organic detoxifying microorganisms, and (4) phytoremediation using plants able to accumulate/stabilize metals and detoxify organic pollutants using rhizosphere bacteria. Although few reports exist on the remediation of mixed contaminants such as 2,4-D in the presence of Cd under laboratory conditions, virtually no information is available on remediation of PAHs in the presence of heavy metals in soils. Also, it is important to isolate and characterize specialized microorganisms able to detoxify both toxic metals and organic pollutants, and their survival and efficiency under field conditions should be tested for their use in the field-scale bioremediation. This is now emerging as a promising technology for bioremediation of sites contaminated with mixed pollutants as in MGP sites.

References

Adriano DC. 2001. *Trace Elements in Terrestrial Environments*, 2nd edn. Springer-Verlag, New York.

Amor L, C Kennes, MC Veiga. 2001. Kinetics of inhibition in the biodegradation of monoaromatic hydrocarbons in presence of heavy metals. *Bioresources Technology* 78: 181–185.

Atagana HI. 2006. Biodegradation of PAHs in contaminated soil by biostimulation and bioaugmentation in the presence of copper ions. *World Journal of Microbiology and Biotechnology* 22: 1145–1153.

Babu TS, JB Marder, S Tripuranthakam, DG Dixon, BM Greenberg. 2001. Synergistic effects of a photo oxidized polycyclic aromatic hydrocarbon and copper on photosynthesis and plant growth: Evidence that *in vivo* formation of reactive oxygen species is a mechanism of copper toxicity. *Environmental Toxicology and Chemistry* 20: 1351–1358.

Baldrian P, C In der Wiesche, J Gabriel, F Nerud, F Zadrazil. 2000. Influence of cadmium and mercury on activities of ligninolytic enzymes and degradation of polycyclic aromatic hydrocarbons by *Pleurotus ostreatus* in soil. *Applied and Environmental Microbiology* 66: 2471–2478.

Bollag JM, M Duszota. 1984. Effect of the physiological state of microbial cells on cadmium sorption. *Archives of Environmental Contamination and Toxicology* 13: 265–270.

Efroymson RA, ME Will, GW Suter II. 1997. Toxicological benchmarks for contaminants of potential concern for effects on soil and litter invertebrates and heterotrophic processes: Oak Ridge National Laboratory, Oak Ridge, TN. ES/ER/TM-126/R2. 128 pp.

EI Digest. May 1995. Manufactured gas plants. *EI Digest*, May, 1.

Fair PA, LV Sick. 1983. Accumulations of naphthalene and cadmium after simultaneous ingestion of the Black Sea bass, *Centropristis striata*. *Archives of Environmental Contamination and Toxicology* 12: 551–557.

Fijałkowska S, K Lisowska, J Długoński. 1998. Bacterial elimination of polycyclic hydrocarbons and heavy metals. *Journal of Basic Microbiology* 38: 361–369.

Fischer CLJ, RD Schmitter, LE O'Neil. November 1999. *Manufactured Gas Plants: The Environmental Legacy*. South & Southwest Hazardous Substance Research Center, Georgia Institute of Technology, Atlanta, GA. http://www.hsrc.org/hsrc/html/tosc/sswtosc/mgp.html

Fraser L. 2000. Innovations: Lipid lather removes metals. *Environmental Health Perspectives* 108: A320.

Gadd GM. 2010. Metals, minerals and microbes: Geomicrobiology and bioremediation. *Microbiology* 156: 609–643.

George SG, P Young. 1986. The time course of effects of cadmium and 3-methylcholanthrene on activities of enzymes of xenobiotic metabolism and metallothionein levels in the place, *Pleuronectes platessa*. *Comparative Biochemistry and Physiology C* 83: 37–44.

Gogolev A, BE Wilke. 1997. Combination effects of heavy metals and fluoranthene on soil bacteria. *Biology and Fertility of Soils* 25: 274–278.

Gust AK. 2005. Ecotoxicology of metal-hydrocarbon mixtures in benthic invertebrates. PhD thesis, Louisiana State University, Eunice, LA.

Gustafsson JP. 2009. Visual MINTEQ (Ver 2.52). USEPA publishing, http://www.lwr.kth.se/English/OurSoftware/vminteq/

Haesler F, D Blanchet, V Druelle, P Werner, JP Van de Casteele. 1999. Analytical characterisation of contaminated soils from former manufactured gas plants. *Environmental Science and Technology* 33: 825–830.

Heritage Research Center, Ltd. 2002. A brief history of the manufactured gas industry in the United States. http://www.heritageresearch.com/manufactured_gas_F.htm

Hettiarachchi GM, GM Pierzynski, MD Ransom. 2000. In situ stabilization of soil lead using phosphorus and manganese oxide. *Environmental Science and Technology* 34: 4614–4619.

Irha N, J Slet, V Petersell. 2003. Effect of heavy metals and PAH on soil assessed via dehydrogenase assay. *Environment International* 28: 779–782.

Knight RL, RH Kadlec, HM Ohlendorf. 1999. The use of treatment wetland for petroleum industry effluents. *Environmental Science and Technology* 33: 973–980.

Kong IC. 1998. Metal toxicity on the dechlorination of monochlorophenols in fresh and acclimated anaerobic sediment slurries. *Water Science and Technology* 38: 143–150.

Luthy RG. 1994. Remediating tar-contaminated soils at manufactured gas plant sites. *Environmental Science and Technology* 28: 226–276.

Malakul P, KR Srinivasan, HY Wang. 1998. Metal toxicity reduction in naphthalene biodegradation by use of metal-chelating adsorbents. *Applied and Environmental Microbiology* 64: 4610–4613.

Maslin P, RM Maier. 2000. Rhamnolipid-enhanced mineralization of phenanthrene in organic-metal co-contaminated soils. *Bioremediation Journal* 4: 295–308.

Middleton A. 1995. Historical overview of manufactured gas plant processes used in the United States. *Land Contamination and Reclamation* 3(4): 5-17–5-19.

Mueller JG, PJ Chapman, PH Pritchard. 1989. Action of a fluoranthene-utilizing bacterial community on polycyclic aromatic hydrocarbon components of creosote. *Applied and Environmental Microbiology* 55: 3085–3090.

New South Wales Environmental Protection Agency (NSW EPA). 2005. Information about assessing gasworks sites. Department of Environment and Conservation, New South Wales, Australia. http://www.environment.nsw.gov.au/clm/gasworksassessment.html

New York State Department of Environmental Conservation. 2003. Web site on general information about MGPs. http://www.dec.state.ny.us/website/der/mgp/mgp_faq.html

New Zealand Ministry for the Environment (NZMfE). 1997. Guidelines for assessing and managing contaminated gasworks sites in New Zealand, Ministry for the Environment, and Draft guidelines for assessing and managing petroleum hydrocarbon contaminated sites in New Zealand.

Nies DH, S Silver. 1995. Ion efflux systems involved in bacterial metal resistances. *Journal of Industrial Microbiology* 14: 186–199.

Pepper IL, J Terry, G Deborah, T Newby, TM Roane, KL Josephson. 2002. The role of cell bioaugmentation and gene bioaugmentation in the remediation of co-contaminated soils. *Environmental Health Perspectives* 110: 943–946.

Pradhan S, V Srivastava. 1997. A pilot-scale demonstration of an innovative soil remediation process: Air emissions quality. *Journal of the Air and Waste Management Association* 47: 710–715.

Reddy KR, PR Ala, S Sharma, SN Kumar. 2006. Enhanced electrokinetic remediation of contaminated manufactured gas plant soil. *Engineering Geology* 85: 132–146.

Riha WE, WL Wendorff, S Rank. 1992. Benzo(a)pyrene content of smoked and smoked-flavored cheese products sold in Wisconsin. *Journal of Food Protection* 55: 636–638.

Riis V, W Babel, OH Pucci. 2002. Influence of heavy metals on the microbial degradation of diesel fuel. *Chemosphere* 49: 559–568.

Roane TM, KL Josephson, IL Pepper. 2001. Dual-bioaugmentation strategy to enhance remediation of cocontaminated soil. *Applied and Environmental Microbiology* 67: 3208–3215.

Roane TM, IL Pepper. 1997. Microbial remediation of soils co contaminated with 2,4-dichlorophenoxy acetic acid and cadmium. In: *Proceedings of the 12th Annual Conference on Hazardous Waste Research: Building Partnership for Innovative Technologies*, May 19–22, Kansas City, MO.

Sandrin TR, AM Chech, RM Maier. 2000. A rhamnolipid biosurfactant reduces Cd toxicity during naphthalene biodegradation. *Applied and Environmental Microbiology* 66: 4585–4588.

Sandrin TR, RM Maier. 2003. Impact of metals on the biodegradation of organic pollutants. *Environmental Health Perspectives* 111: 1093–1101.

Sharon OPE, PG Dennis Unites. 1998. A survey of MGP sites remedial technologies. In: *IGT Symposium on Environmental Biotechnology and Oil & Gas Site Remediation Technology*, December, Colorado Springs, CO.

Shen G, L Cao, LY Lu, J Hong. 2006. Combined effect of heavy metals and polycyclic aromatic hydrocarbons on urease activity in soil. *Ecotoxicology and Environmental Safety* 63: 474–480.

Sokhn J, FAAM De Leij, TD Hart, JM Lynch. 2001. Effect of copper on the degradation of phenanthrene by soil micro-organisms. *Letters in Applied Microbiology* 33: 164–168.

Springael D, L Diels, L Hooyberghs, S Kreps, M Mergeay. 1993. Construction and characterization of heavy metal-resistant haloaromatic-degrading *Alcaligenes eutrophus* strains. *Applied and Environmental Microbiology* 59: 334–339.

Thavamani P. 2010. Remediation of mixed contaminated soils with special reference to polyaromatic hydrocarbons and metals. PhD thesis, University of South Australia, Adelaide, South Australia, Australia.

Thavamani P, M Megharaj, GSR Krishnamurti, R McFarland, R Naidu. 2011. Finger printing of mixed contaminants from former manufactured gas plant (MGP) site soils: Implications to bioremediation. *Environment International* 37: 184–189.

Thomas AO, JN Lester. 1994. The reclamation of disused gasworks sites: New solutions to an old problem. *Science of the Total Environment* 152: 239–260.

Tremaroli V, ML Workentine, JJ Harrison, S Fedi, H Ceri, RJ Turner, D Zannoni. 2009. Metabolomics investigation of bacterial 1 response to metal challenge. *Applied and Environmental Microbiology* 75: 719–728.

Tsiridis V, M Petala, P Samaras, S Hadjispyrou, GP Sakellaropoulos, A Kungolos. 2006. Interactive toxic effects of heavy metals and humic acids on *Vibrio fischeri*. *Ecotoxicology and Environmental Safety* 63: 158–167.

UNEP-ILO-WHO. 1998. Selected non-heterocyclic polycyclic aromatic hydrocarbons. Environmental health criteria 202. United Nations Environmental Program-International Labour Organisation, Geneva, Switzerland.

US Environmental Protection Agency. September 24, 1986. Guidelines for health risk from exposure to chemical mixtures. 51(185): 34014–34025, EPA/630/R-98/002.

US Environmental Protection Agency. 2000a. Guidance for conducting health risk assessment for mixtures (EPA/630/R-00/002); Risk Assessment Forum, US EPA, Washington, DC.

US Environmental Protection Agency. 2000b. Superfund sites. http://www.epa.gov/superfund/sites/index.htm (accessed October 15, 2006).

US Environmental Protection Agency. 2004. Cleaning up the Nation's wastes sites: Markets and technology trends, Office of Solid Waste and Emergency Response. US EPA, Washington, DC.

Valls M, V Lorenzo. 2002. Exploiting the genetic and biochemical capacities of bacteria for the remediation of heavy metal pollution. *FEMS Microbiological Reviews* 26: 327–338.

Wong KW, BA Toh, YP Ting, JP Obbard. 2005. Biodegradation of phenanthrene by the indigenous microbial biomass in a zinc amended soil. *Letters in Applied Microbiology* 40: 50–55.

Part IV

Ecological Restoration of Contaminated Sites: Phytoremediation

17

Toxicity and Bioavailability of Heavy Metals and Hydrocarbons in Mangrove Wetlands and Their Remediation

Lin Ke and Nora F.Y. Tam

CONTENTS

Introduction

Mangrove wetlands are coastal wetland ecosystems that dominate the intertidal zone of tropical and subtropical foreshore areas. The ecological, environmental, and socioeconomic importance of mangrove wetlands has been widely recognized. They have an extraordinarily high rate of primary productivity (Alongi 2002), act as both an atmospheric CO_2 sink and as an essential source of oceanic carbon (Cahoon et al. 2003, Chmura et al. 2003), provide nursery grounds and refuge for ecologically and commercially important marine organisms (Primavera 1998, Mumby et al. 2004), protect coastal erosion and maintain shore stability (Dahdouh-Guebas et al. 2005, Danielsen et al. 2005), and filter river-borne sediment and nutrients to minimize their inputs into more sensitive systems, such as seagrass beds and coral reefs (Alongi and McKinnon 2005). Mangrove wetlands are under a serious threat of anthropogenic pollution due to the current population expansion and accompanying increase in the usage of resources. They have long been used as convenient sites for waste disposal and discharge of untreated sewage and livestock wastewater (Clough et al. 1983). As a transit zone between terrestrial and marine environments, mangrove wetlands also inevitably receive contaminants from tidal water, rivers, and storm runoff (Tam and Wong 1993, 1995, 2000, Ke et al. 2002). Mangrove sediments have been reported to serve as reservoirs of contaminants, including nitrogen and phosphorus (e.g., Corredor and Morell 1994, Tam and Wong 1996b, Rivera-Monroy et al. 1999), heavy metals (e.g., Harbison 1986, Silva et al. 1990, Tam and Wong 1993, 1995, 1996a, 1999a), and organic pollutants (e.g., Tam et al. 2001, Maskaoui et al. 2002).

Scientific evidence of mangrove wetlands thriving in a relatively harsh intertidal environment (e.g., periodic changes in temperature, water and salt exposure, and varying degrees of oxygen depletion) over long geological time scales suggests their high tolerance to extreme environmental conditions (Alongi 2008). Mangrove plants have developed physiological, morphological, and anatomical adaptations, such as salt regulation, well-developed aerenchyma, and highly specialized root systems (e.g., knee joints, pneumatophores or aerial roots, cable roots, and buttress/prop roots), to cope with anoxic and saline environments (Lugo 1980, Pi et al. 2009). These adaptive changes in mangrove plants, together with the harsh growth conditions (environmental extremes and pollution), suggest the possibility of using mangrove plants in phytoremediation of contaminated water and sediments.

Heavy metals (e.g., copper, zinc, and lead) and petroleum hydrocarbons (e.g., aliphatic and aromatic hydrocarbons) are ubiquitous pollutants causing serious environmental problems. Due to the close proximity of mangrove wetlands to human activities and along the route of oil transportation, elevated concentrations of heavy metals and hydrocarbons in mangrove sediments have been reported (Tam and Wong 1993, 1995, 1996a, 1999a, Tam et al. 2001, Ke et al. 2005). Accidental oil spills have also been reported in Hong Kong mangrove wetlands from time to time (Ke et al. 2002, Wong et al. 2002, Tam et al. 2005).

We are one of the first groups to employ mangrove wetlands in the phytoremediation of wastewater and contaminated sediments (e.g., Yim and Tam 1999, Ke et al. 2003a,b, Wu et al. 2008, Yang et al. 2008). A prerequisite for the success of phytoremediation of these pollutants is to screen tolerant plants. Our research group has been working on the toxicity of heavy metals and hydrocarbons present in the mangrove plants since the 1990s, and most of the early studies focused on growth response and accumulation in plants (e.g., Yim and Tam 1999, Ke et al. 2003a,b). The toxic effects of these pollutants on mangrove plants, in particular at the morphological (e.g., structural changes in roots), physiological, and biochemical levels, have received little attention until the past 5 years. This chapter, therefore, reviews our research on the structural, physiological, and biochemical changes of mangrove plants, including root anatomy, radial oxygen loss (ROL), iron (Fe) plaque formation, activities of superoxide dismutase (SOD) and peroxidase (POD), superoxide radical (O_2^-) release, malondialdehyde (MDA) content, and phenolic compounds (total polyphenols [TP] and extractable condensed tannins [ECT]), in response to heavy metal and hydrocarbon stress. The goal of this research is to identify the most tolerant plant species for the remediation of the contaminated sediment. The bioavailability of these pollutants and their sorption and desorption, as well as the role of associated bacterial communities in degrading hydrocarbons, are also discussed.

Plant Response to Toxicity

Root Anatomy and Radial Oxygen Loss

To cope with anoxic environments, wetland plants have developed an airy tissue, aerenchyma, which allows exchange of oxygen and other gases between the shoot and the root, to support root aerobic metabolism. Excessive oxygen can diffuse into the rhizosphere, a process defined as ROL (Armstrong et al. 1992). ROL plays an important role for wetland plants to maintain an aerobic environment in the rhizosphere. The ability to oxidize the rhizosphere has long been regarded as a partial explanation for flood tolerability of mangroves and other wetland plants (Youssef and Saenger 1996, McDonald et al. 2002, Jackson and Colmer 2005). The beneficial effects of ROL to wetland plants include (1) oxidizing reduced toxic substances, such as phytotoxins, in the rhizosphere (Armstrong et al. 1992, Pedersen et al. 2004); (2) altering both microbial and chemical processes, such as nitrification and denitrification, to change nutrient availability in rhizosphere sediments (Kirk and Kronzucker 2005); and (3) aerobically degrading and transforming organic pollutants (St-Cyr and Campbell 1996, Visser et al. 2000). However, these beneficial effects of ROL are highly dependent on the locations where ROL occurs. The root tip cells are especially sensitive to anoxia due to their high metabolic activity and thus have a high demand for oxygen. To enhance longitudinal oxygen diffusion toward the root tips, wetland plants have developed a barrier to ROL (Visser et al. 2000, Vasellati et al. 2001, Colmer et al. 2006).

The barrier to ROL can be categorized into three types: "tight" (defined as a low radial permeability to oxygen at the root base and higher levels of oxygen at the root tip), "partial" (with ROL an the root base similar to that in the apex), and "weak" (the rate of ROL at the root apex is much lower than that in the base) barriers (Armstrong 1979, Armstrong and Beckett 1987, Armstrong et al. 2000, Colmer 2003). These different types of barriers result from the differences in root anatomical features (especially exodermal and hypodermal structure) as well as the quantitative variations in suberin composition and their distribution within the exodermal cell walls among different wetland species (Soukup et al. 2007). Pi et al. (2009) conducted a pioneering study on root anatomy and spatial pattern of the ROL of eight native mangrove species in Hong Kong, including *Avicennia marina* (Forsk.) Vierh., *Acanthus ilicifolius* L., *Aegiceras corniculatum* (Linn.) Blanco, *Bruguiera gymnorrhiza* (L.) Poir, *Excoecaria agallocha* L., *Kandelia obovata* Sheue (Liu and Yong; previously known as *Kandelia candel* (L.) Druce), *Heriteria littoralis* Dryand. exW. Ait, and *Lumnitzera racemosa* Willd., and found that all eight species had a similar spatial pattern of ROL, with more oxygen lost from the tip than from the basal and mature zones and had a "tight" barrier. However, the amounts of ROL varied along the root and were species specific (Table 17.1), probably related to the anatomical differences in roots, including cortex aerenchyma air spaces and outer layers (i.e., epidermis and hypodermis) among different species. Roots of *A. marina* and *Acanthus ilicifolius* had the highest proportions of aerenchyma air spaces but had the thinnest outer layers, among the eight species. On the other hand, *H. littoralis* had the least longitudinal oxygen transfer because of its lower proportions of aerenchyma air spaces in the root. These differences may also explain the different tolerance levels of mangrove species to inundation, which followed the declining order of *A. marina* (most foreshore species) > *Acanthus ilicifolius* > *K. obovata* > *Aegiceras corniculatum* > *B. gymnorrhiza* > *E. agallocha* > *L. racemosa* > *H. littoralis* (most landward species).

The ROL spatial pattern in each mangrove species would change as a consequence of the changes in the relative proportion of aerenchyma air

TABLE 17.1

Radial Oxygen Loss along the Lateral Root of Mangrove and Non-Mangrove Wetland Plants

Species	Age (Year)	ROL (μM O$_2$/day/g DW)	Classification
Mangrove plants			
Acanthus ilicifolius	1	43.80 $\pm$ 3.19[a]	Strong
Aegiceras corniculatum	1	34.03 $\pm$ 3.49[ab]	Strong
K. obovata	1	30.74 $\pm$ 4.51[b]	Strong
A. marina	1	19.99 $\pm$ 1.57[c]	Medium
B. gymnorrhiza	1	9.02 $\pm$ 1.31[d]	Weak
L. racemosa	2	8.74 $\pm$ 0.89[d]	Weak
Non-mangrove wetland plants			
Echinodorus amazonicus	1	97.63 $\pm$ 6.39[A]	Strong
Phragmites australis	1	63.32 $\pm$ 4.81[B]	Strong
Schisandra chinensis	1	47.58 $\pm$ 3.57[C]	Strong
Typha latifolia	1	72.17 $\pm$ 8.21[B]	Strong
Zantedeschia aethiopica	1	50.88 $\pm$ 4.89[C]	Strong

Mean and standard deviation of six and four replicates for mangrove and non-mangrove plants, respectively, are shown; different capital or lower case letters indicate significant differences at $P \leq 0.05$ by one-way ANOVA (analysis of variance), followed by Tukey's HSD (honestly significant difference) test.

spaces and outer barriers to the whole root when exposed to pollution, and the changes to the same pollutant were species specific. Pi et al. (2010a) investigated the variability of ROL and root anatomy among three mangrove species under the influence of wastewater discharge for 105 days. ROL at the root tip of *B. gymnorrhiza* increased from 22.44 ng/cm^2/min in the fresh water control to 31.09 ng/cm^2/min when receiving normal synthetic wastewater (NW; with concentrations of dissolved organic carbon [DOC], NH_4^+-N, NO_3^--N, total Kjeldahl N [TKN], and PO_4^{3-}-P, comparable to the primarily settled municipal sewage in Hong Kong) and to 44.22 ng/ cm^2/min when treated with strong wastewater (10 NW; DOC, NH_4^+-N, NO_3^--N, TKN, and PO_4^{3-}-P at concentrations of 60, 25, 0.5, 45, and 5 mg/L, respectively), which however did not change the "tight" barrier pattern of *B. gymnorrhiza* (Figure 17.1A). However, for *E. agallocha*, discharge of wastewater caused a decrease in ROL at the root tip but an increase at the root base, resulting in a shifting in its barrier type from "tight" to "partial" (Figure 17.1B). Such a trend was even more significant for *Acanthus ilicifolius*, leading to its barrier type changing from "tight" to "weak" (Figure 17.1C). The changes in barrier types due to wastewater discharge were related to the root anatomy. Among the three species, *Acanthus ilicifolius* had the highest proportions of the cross-sectional area of aerenchyma air spaces, suggesting that internal oxygen transfer to the root tip was the fastest. However, the area of aerenchyma air spaces in the root tip of *Acanthus ilicifolius* treated with 10 NW was significantly reduced, whereas the area of

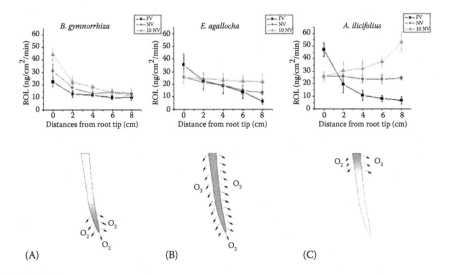

FIGURE 17.1
Changes in ROL pattern along the lateral roots of mangrove species after receiving different strengths of wastewater for 105 days: (A) "Tight" barrier, (B) "partial" barrier, and (C) "weak" barrier. FW, fresh water; NW, normal wastewater; 10 NW, 10 times of NW. (Modified from Pi, N. et al., *Int. J. Phytoremediat.*, 12, 468, 2010a.)

epidermis and hypodermis was increased, leading to a reduced oxygen supply to the root tip. Compared to *B. gymnorrhiza* and *E. agallocha*, the epidermis and hypodermis layers in *Acanthus ilicifolius* were the thinnest, in which the cells without suberized walls had the least protection from the harmful effect of wastewater exposure. These results suggested that *B. gymnorrhiza* was the least affected by wastewater discharge, followed by *E. agallocha*, and *Acanthus ilicifolius* was the most susceptible species and was not suitable for treating strong wastewater.

The ROL spatial patterns of mangrove plants were found to have a direct link to their metal tolerance. Liu et al. (2009) investigated the toxicity of mixed heavy metals of Pb, Zn, and Cu at different concentrations (low, medium, and high: 50–100–50, 100–200–100, and 200–400–200 mg Pb–Zn–Cu/kg dry sand, respectively) to 6 months old mangrove seedlings of *Aegiceras corniculatum*, *A. marina*, and *B. gymnorrhiza*. Heavy metals inhibited plant growth, reduced the amount of ROL, and changed the ROL spatial pattern, which were positively correlated with metal tolerance. *B. gymnorrhiza* had the highest ROL amount and the tightest barrier to ROL, and it was also the most tolerant species to heavy metals. In another study by Cheng et al. (2010), plant uptake and tolerance of 1 year old seedlings of *Aegiceras corniculatum*, *B. gymnorrhiza*, and *Rhizophora stylosa* Griff to Zn toxicity were compared. Zn exposure led to a significant decrease in ROL in roots of all three species, but the degree of reduction varied among different species, with the reduction percentages following the descending order of *R. stylosa* > *B. gymnorrhiza* > *Aegiceras corniculatum* (Figure 17.2). The ROL reduction was suggested to be due to the inhibition of root permeability, including an obvious thickening of outer cortex and a significant increase in lignification of cell walls. *B. gymnorrhiza*, which possessed the "tightest barrier" to ROL, accumulated the least Zn and had the highest Zn tolerance. This study provided new evidence of the structural adaptations of mangrove plants to metal tolerance.

Fe Plaque Formation

Another important consequence of ROL is the induction of Fe precipitation in the rhizosphere, creating iron-rich root coatings or Fe plaque on root surfaces (Otte et al. 1989, Guo et al. 2007, Hu et al. 2007). Fe plaque is a mixture of crystalline and amorphous ferric hydroxides goethite and lepidocrocite (St-Cyr et al. 1993, Hu et al. 2007, Chen et al. 2008a). Its formation is controlled by a number of physiochemical factors, such as the availability of Fe^{2+}, organic matter, soil texture, and redox potential (Mendelssohn et al. 1995). Wastewater-borne pollutants also affect the formation of Fe plaque. Pi et al. (2010a) reported that Fe plaque formation on the root surface of 1 year old seedlings of *B. gymnorrhiza* and *E. agallocha* increased significantly after 105 days of exposure to strong synthetic wastewater (10 NW). However, the degree of increase in different parts of the root differed from species to species, with more significant increases toward the basal and mature zones

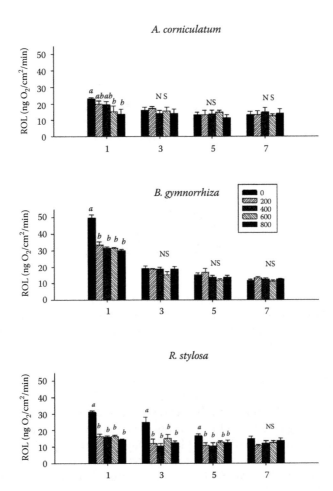

FIGURE 17.2

Spatial patterns of ROL along the lateral roots of three mangrove seedlings receiving different concentrations of Zn (mg/kg) under greenhouse conditions. Different letters in the same group of bars indicate significant differences at $P \le 0.05$ as determined by LSD. NS, not significant. (Data used for plotting are from Cheng, H. et al., *Environ. Pollut.*, 158, 1189, 2010.)

for *B. gymnorrhiza*, but toward the root tip for *E. agallocha*. The relationship between ROL and Fe plaque was not significant, suggesting that although ROL had a significant contribution to Fe plaque formation, too much Fe plaque would serve as a "barrier" to ROL. Actually, Fe plaque not only acts as a "barrier" to ROL but also reduces the plant uptake of heavy metals, including Zn, Cd, As, Ni, Mn, Pb, and Cu, and nutrients (such as P) through immobilization and coprecipitation (Taylor and Crowder 1983, Christensen and Sand-Jensen 1998, Ye et al. 1998, 2003a, Liu et al. 2004, Machado et al. 2005, Liang et al. 2006). The relationships among ROL, Fe plaque formation, and immobilization of heavy metals are shown in Figure 17.3. The Fe plaque

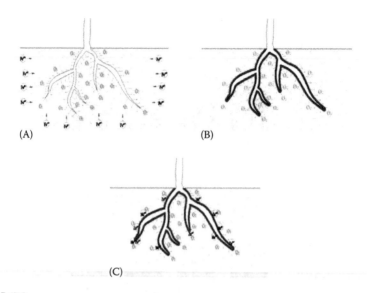

FIGURE 17.3

Formation of Fe plaque and immobilization of heavy metals: (A) oxygen release from the root and oxidation of Fe^{2+}, (B) formation of Fe plaque on the root surface, and (C) immobilization of heavy metals in Fe plaque.

barrier reduced metal mobility and bioavailability, thus preventing trace metals from reaching the more sensitive parts of the plant. Moorthy and Kathiresan (1998) also reported that heavy metal concentrations in *Rhizophora apiculata* seedlings decreased from roots to stems and from stems to leaves. In terms of phytoremediation, Fe plaque may play an important role in the removal and detoxification of wastewater-borne heavy metals, which however is far from fully evaluated because of limited published materials. Pi et al. (2010c) found that after 75 days of exposure, Fe plaque formation on the root of 1 year old seedlings of *B. gymnorrhiza*, *E. agallocha*, and *Acanthus ilicifolius* was enhanced with increasing concentrations of wastewater-borne heavy metals (synthetic wastewater having 5–10 times the strength (5–10 MW) of primarily settled municipal sewage in Hong Kong containing DOC, NH_4^+-N, NO_3^--N, TKN, PO_4^{3-}-P, Fe^{3+}, Ni^{2+}, Cu^{2+}, Zn^{2+}, Mn^{2+}, Pb^{2+}, Cr^{6+}, and Cd^{2+} at 60, 25, 0.5, 45, 5, 30, 1, 2, 5, 5, 1, 0.5, and 0.1 mg/L, respectively), but the extent of enhancement was species specific. The concentrations of trace metals immobilized in Fe plaque were also increased significantly with wastewater strength and were positively correlated with the concentration of Fe plaque formed on the root surface for all three species. These results confirmed that Fe plaque could act as a reservoir to immobilize metals, thus reducing their bioavailability and contributing to the removal and detoxification of wastewater-borne heavy metals. These results, however, also indicated that metal immobilization via Fe plaque formation may have a negative relationship with metal tolerance, as the most metal-tolerant species *B. gymnorrhiza* had the least accumulation of Fe plaque on root surface. All these studies suggested the importance of the interactions between ROL and Fe plaque formation and their effects, which merits further study.

Biochemical Response

Oxidative stress on plants is associated with an excessive generation of reactive oxygen species (ROS), including superoxide (O_2^-), hydrogen peroxide (H_2O_2), and hydroxyl free radicals ($\cdot OH$), which can be induced by various environmental stresses such as extreme temperatures, salt, hypoxia, UV radiation, and toxic pollutants (Bartosz 1997). The highly reactive and nonspecific nature of ROS has been reported to contribute to cell damage, including lipid peroxidation, inactivation of enzymes and other functional proteins, base modifications, and strand breaks of DNA (reviewed by Bartosz 1997). MDA is a final decomposition product of lipid peroxidation. The positive relationship between ROS and MDA accumulation (Smirnoff 1995) suggests that MDA can be used as an indicator of the status of lipid peroxidation. In response to oxidative stress, plants have developed a variety of antioxidative defense systems to scavenge excessive ROS and diminish their harm to plants. The antioxidative defense systems can be classified into two groups, antioxidative enzymatic system and nonenzymatic system. Antioxidative enzymes include SOD, POD, and catalase (CAT) (Scandalios 1993, Vangronsveld and Clijsters 1994). SOD is involved in the first step of ROS elimination, which catalyzes the conversion of O_2^- to H_2O_2, and O_2, and H_2O_2 can be further decomposed by CAT and POD (Parida et al. 2004). Nonenzymatic constituents include glutathione, carotenoids, and ascorbate phenolic compounds (Tukendorf and Rauser 1990). Phenolic compounds, including tannins and derived polyphenols, are among the most widely distributed plant secondary metabolites and can be induced by stress (Dixon and Paiva 1995). Phenolic compounds, mainly due to their redox properties, also play an important role in absorbing and neutralizing free radicals and can function as antioxidants (Lavid et al. 2001). The antioxidant capacity of phenolic compounds is also attributed to their ability to chelate metals (Hernes et al. 2001, Schützendübel and Polle 2002, Kraus et al. 2003).

The activities of antioxidative enzymes and contents of nonenzymatic constituents are associated with plant tolerance to metal stress; however, both positive and negative responses have been observed, depending on the plant species and tissue analyzed, as well as the species and intensity of the metal stress (Schützendübel and Polle 2002). POD activities in the leaves of *A. marina* were positively correlated with the amounts of heavy metals (Cu, Zn, and Pb) accumulated in leaf tissues (MacFarlane 2002). However, Liu et al. (2008) found that both SOD and POD activities in the roots of *Sedum alfredii* were negatively correlated with Pb concentrations in the substrate.

Research on antioxidative defense systems in mangrove plants, in response to environmental stresses, including waterlogging, salt, wastewater, heavy metal, and oil pollution, has been extensively conducted by our group (Ye and Tam 2002, 2007, Ye et al. 2003, 2005, Zhang et al. 2007a, Ke et al. 2010) and others (e.g., MacFarlane and Burchett 2001, Qin et al. 2007, Zhang et al. 2007b). Most of these studies have focused on the biochemical responses in plants subjected to long-term, chronic exposure (in months), which may not accurately reflect the rapid response feature of the antioxidative system. Yan and Tam (2010) investigated the temporal changes (in the scale of days) in the contents of phenolic compounds and activities of SOD and POD in *K. obovata* as well as plant metal accumulation after exposure to Pb and Mn stress and found that both stress time and stress intensity had significant effects, with significant accumulation of Pb and Mn in the root on Day 1 under moderate and high metal stress; however, metal accumulation in the leaf was only found on Day 7. Metal stress had no significant effect on TP and ECT contents in roots and leaves on Day 1; however, a prolonged exposure to metals caused significant decreases in root TP and ECT; however, a substantial decease was observed on Days 7 and 49. In contrast, POD activities in both roots and leaves were significantly decreased from Day 0 to Day 1, but increased thereafter. The enhancement effect was tissue, metal species, and concentration dependent. SOD activities followed a trend similar to that of POD. This study also revealed that antioxidative enzyme activities in *K. obovata* seedlings were more sensitive to Pb or Mn stress than the changes in phenolic compounds.

Research on the physiological and biochemical responses of mangrove plants to toxic, organic pollutants is relatively scarce and is focused mostly on petroleum hydrocarbons. Ke et al. (2010) compared the tolerance of four dominant mangrove species in South China to different doses of spent lubricating oil. Based on the mortality rates, *B. gymnorrhiza* presented the highest tolerance to oil pollution (mortality 16.7% at $15 L/m^2$ oil, $n = 12$ for all species), followed by *Acanthus ilicifolius* (25% at $10 L/m^2$) and *Aegiceras corniculatum* (83.3% at $10 L/m^2$), and *K. obovata* (91.7% at $10 L/m^2$) was the most oil-sensitive species. Biochemical responses, including MDA content, SOD activity, and O_2^- release, in the oil-treated seedlings of all species were enhanced with oil doses. Zhang et al. (2007a) showed that fresh and spent lubricating oil at a single initial dose of $5 L/m^2$ posed an oxidative stress to *B. gymnorrhiza*, leading to significant increases in O_2^- release and MDA content. Spent lubricating oil, even at this low dose, was found to decrease the content of chlorophyll and carotenoid, the activity of nitrate reductase, POD, and SOD, but it was found that it increased the MDA content too; these responses reflected that *A. marina* was more sensitive to spent lubricating oil than *Aegiceras corniculatum* whereas canopy oiling resulted in more direct physical damage and stronger lethal effects than base oiling (Ye and Tam 2007). The most tolerant mangrove species to heavy metals, *B. gymnorrhiza*, was also found to be the most tolerant to spent-oil pollution. However, further research is needed to ascertain whether the effect of oil on root anatomy and the oil tolerance mechanism of this species is similar to the effect caused by heavy metal pollution.

Bioavailability and Remediation

Heavy Metals

Wetland sediments often undergo intermittent flooding and draining, providing an alternating anaerobic and aerobic environment. The high levels of reduction in sulfide, iron, and manganese observed favor the precipitation and immobilization of heavy metals (Ambus and Lowrance 1991, Dunbabin and Bowmer 1992). On the other hand, the ROL of wetland plants would aerate the rhizosphere and increase the mobilization of heavy metals due to sulfide oxidation and dissolution. ROL would also cause the formation of Fe plaque, which provides some extra binding sites for heavy metals on root surface (Pi et al. 2010c). The formation of Fe plaque was found to be positively correlated to the amount of ROL present in the roots of mangrove plants grown in the freshwater control (Pi et al. 2010b) as well as to the immobilization of wastewater-borne heavy metals (Pi et al. 2010c, Figure 17.4).

Fe plaque formation vs. ROL

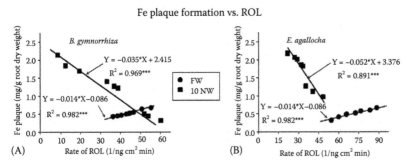

Metal immobilization vs. Fe plaque formation

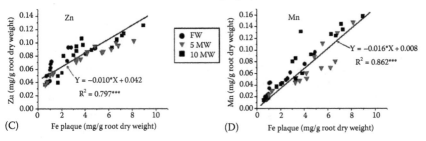

FIGURE 17.4
Relationships between Fe plaque formation and ROL (A and B), and between Fe plaque formation and metal immobilization (C and D). The curves were fitted with the linear regression models ($y = y0 + ax$); R^2, regression coefficients; *** indicates the R^2 values are significant at $P \leq 0.001$; FW, freshwater control; 10 NW, 10 times normal wastewater without heavy metals; 5 MW, 5 times normal wastewater containing heavy metals; 10 MW, 10 times normal wastewater containing heavy metals. Details on the components and concentrations of NW and MW are described in the text. (Data used for plotting are from Pi, N. et al., *Environ. Pollut.*, 158, 381, 2010b; Pi, N. et al., *Mar. Pollut. Bull.*, 63, 402, 2010c.)

Strong wastewater (10 NW) discharge resulted in a shift of the positive relationship between Fe plaque and ROL to a negative one (Figure 17.4), suggesting that excessive Fe plaque served as a barrier to ROL (Pi et al. 2010b). The concentrations of Fe plaque on the root surface could reach 5–10 times the concentrations of the surrounding sediments (Sundby et al. 1998). Fe plaque also acted as a physical barrier to reduce plant uptake of metal toxicity (Otte et al. 1989, Batty et al. 2000). Nevertheless, the role of Fe plaque in the reduction of plant uptake of metal toxicity is still arguable, as the results were not always convincing and conclusive (Ye et al. 1998, Zhang et al. 1998). For phytoremediation of heavy metals using wetland plants, the net effect of the presence of plants, particularly roots and associated features, such as ROL and Fe plaque, on the behavior of heavy metals (e.g., mobilization or immobilization) should be further evaluated.

Other factors, such as sediment texture, organic matter, and salinity, would also affect the bioavailability of heavy metals in mangrove sediments. The clay-like nature of mangrove sediments could provide a physical trap for fine particulates and heavy metals. High contents of organic matter, especially the humic substances in mangrove sediments, also provided strong adsorptive properties in binding heavy metals. Salinity may also affect the binding of pollutants in sediments. Paalman et al. (1994) reported that when river water mixed with seawater, due to increases in chloride concentrations, heavy metals such as Cd get mobilized from the sediment and dissolve later into chloro-complexes. Tam and Wong (1999b) also reported that mangrove sediments receiving wastewater prepared in freshwater had slightly higher concentrations of nutrients and heavy metals and larger enrichment factors than that treated with saline wastewater (salinity 15%).

Hydrocarbons

Mangrove sediments are often anaerobic or anoxic just a few centimeters below the surface (Mitsch and Gosselink 2000), which reduces aerobic degradation of hydrocarbons, especially polycyclic aromatic hydrocarbons (PAHs), and as a result, they accumulate and persist in deep layers of the sediment for up to 20 years (Burns et al. 1994, 2000). ROL of wetland plants has the potential to significantly alter both microbial and chemical processes in the rhizosphere, such as increasing the aerobic respiration (Schussler and Longstreth 1996) and aerobic degradation and the transformation of environmental pollutants such as PAHs (St-Cyr and Campbell 1996, Visser et al. 2000).

Compared to plants, microorganisms may play a more significant role in removing and degrading organic pollutants. Ke et al. (2003a) found that the uptake of pyrene (PYR), a four-ring PAH, by *B. gymnorrhiza* and *K. candel*, only contributed to 0.65%–0.88% of PYR removal and that most of the removal (84.8%–92.5%) was due to microbial degradation. There is increasing published evidence that mangrove sediments are high in the diversity of hydrocarbon-degrading bacteria as well as the genetic diversity of dioxygenase genes and have a high potential of intrinsic bioremediation of hydrocarbons (e.g., Tam et al. 2002, 2003, Guo et al. 2005, 2010, Yu et al. 2005a,b, Zhou et al. 2008, 2009). A total of 11 PAH-degrading strains, belonging to 4 genera, *Mycobacterium*, *Sphingomonas*, *Terrabacter*, and *Rhodococcus*, were isolated from surface mangrove sediments in South China (Zhou et al. 2008). Despite the presence of the degraders in contaminated sediment, natural attenuation of hydrocarbons such as PAHs by indigenous microorganisms was often low, unless the environmental conditions were modified. Bioaugmentation, with the inoculation of isolates with known biodegradation ability, is often used to enhance the efficiency of bioremediation. The significant enhancement effects of *Sphingomonas* sp., a bacterial strain isolated from surface mangrove sediment to degrade a three-ring PAH, phenanthrene (PHE), in contaminated sediment slurry was reported by Chen et al. (2008b; see Figure 17.5).

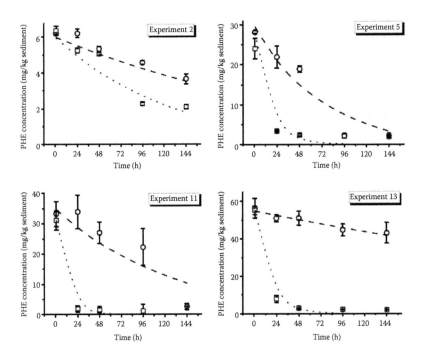

FIGURE 17.5
Effects of bioaugmentation of *Sphingomonas* sp. on phenanthrene (PHE) degradation in sediment slurry. Different initial PHE concentrations are shown. Curves fitted to the first-order-rate models indicate that the rates of degradation in the inoculated system were significantly higher than in the corresponding control system (□: inoculated system [dotted line], ○: control system [dash line]; mean and standard deviation of three replicates are shown). (Adapted from Chen, J. et al., *Mar. Pollut. Bull.*, 57, 695, 2008b.)

Solubility is an important factor limiting biodegradation of toxic organic pollutants, such as oil and PAHs. Ke et al. (2009) showed that the addition of humic acid (0%–1.6%, w/v) significantly enhanced the solubility of all PAHs in both liquid medium and sediment slurry, but only the biodegradation of phenanthrene and pyrene (and not benzo[a]pyrene) in the liquid medium was enhanced. Solubility is related to the sorption and desorption behavior of organic pollutants, which in turn is affected by the physiochemical properties of sediment. Zhang et al. (2007a) found that the lubricating oil in sandy sediment had a more acute toxic effect on the germination and early growth of *B. gymnorrhiza* than muddy sediment, but the lower bioavailability and biodegradation of hydrocarbons in muddy sediment made the oil more persistent, which may lead to more chronic toxicity and pose a long-term risk to the environment. The bioavailability of toxic pollutants such as PAHs was linked to the sorption–desorption behavior, which was closely related to sediment properties. The sorption–desorption behaviors of PAHs in different mangrove sediment slurries were best described by a linear model. The muddy sediment had a higher binding affinity to PAHs than silty sediment. The sandy sediment had the least binding affinity (unpublished data). Sediments with high contents of clay and total organic carbon provided large surface areas for PAH binding, which enhanced the sorption of PAHs into sediments, thus decreasing their bioavailability and biodegradation.

___

Concluding Remarks

Toxic contaminants, such as heavy metals and petroleum hydrocarbons, pose significant adverse effects on the establishment and initial development of mangrove seedlings and are found to alter the root anatomy and physiology. Changes in the areas of aerenchyma tissues and thickness of outer layers in roots due to these contaminants led to the variations in the amount and spatial pattern of ROL, which then affected Fe plaque formation on the root surface and immobilization of heavy metals. The activity of anti-oxidant enzymes and other biochemical responses in the roots and leaves of mangrove plants were also changed when exposed to toxic contaminants. The tolerance level of mangroves to these toxic contaminants was species specific. *B. gymnorrhiza* was the most robust and tolerant species to heavy metals and hydrocarbons, as it had (1) the "tightest" barrier on root surface, (2) the least amount of Fe plaque on root surface, (3) the lowest ROL per root dry weight, and (4) the least modification in root anatomy and biochemical properties. *B. gymnorrhiza* might be the most ideal candidate for phytoremediation of sediment contaminated with heavy metals or hydrocarbons among the mangrove plants found in Hong Kong. The success of bioremediation depended on the bioavailability and the sorption–desorption behavior of the contaminants, which in turn were affected by sediment texture and organic matter. Further research on the bioavailability of toxic chemicals and its association with toxicity and bioremediation is still needed.

Acknowledgment

The authors thank Ms. Na Pi for her inputs on table and figure drawing. The work described in this chapter was supported by a grant from the Innovation and Technology Commission of Hong Kong (Project number: GHP/004/08SZ).

References

Alongi DM. 2002. Present state and future of the world's mangrove forests. *Environmental Conservation* 29: 331–349.

Alongi DM. 2008. Mangrove forests: Resilience, protection from tsunamis, and responses to global climate change. *Estuarine, Coastal and Shelf Science* 76: 1–13.

Alongi DM, McKinnon AD. 2005. The cycling and fate of terrestrially-derived sediments and nutrients in the coastal zone of the Great Barrier Reef shelf. *Marine Pollution Bulletin* 51: 239–252.

Ambus R, Lowrance R. 1991. Comparison of denitrification in two riparian soils. *Soil Science Society of America Journal* 55: 994–997.

Armstrong W. 1979. Aeration in higher plants. *Advance in Botany Research* 7: 225–332.

Armstrong J, Armstrong W, Beckett PM. 1992. *Phragmites australis*: Venturi- and humidity-induced convections enhance rhizome aeration and rhizosphere oxidation. *New Phytologist* 120: 197–207.

Armstrong W, Beckett PM. 1987. Internal aeration and the development of stellar anoxia in submerged roots: A multishelled mathematical model combining axial diffusion of oxygen in the cortex with radial losses to the stele, the wall layers and the rhizosphere. *New Phytologist* 105: 221–245.

Armstrong W, Cousins D, Armstrong J, Turner DW, Beckett PM. 2000. Oxygen distribution in wetland plant roots and permeability barriers to gas-exchange with the rhizosphere: A microelectrode and modelling study with *Phragmites australis*. *Annals of Botany* 86: 687–703.

Bartosz G. 1997. Oxidative stress in plants. *Acta Physiologiae Plantarum* 19: 47–64.

Batty LC, Baker AJM, Wheeler BD, Curtis CD. 2000. The effect of pH and plaque on the uptake of Cu and Mn in *Phragmites australis* (Cav.) Trin ex. Steudel. *Annals of Botany* 86: 647–653.

Burns KA, Codi S, Duke NC. 2000. Gladstone, Australia field studies: Weathering and degradation of hydrocarbons in oiled mangrove and salt marsh sediments with and without the application of an experimental bioremediation protocol. *Marine Pollution Bulletin* 41: 392–402.

Burns KA, Garrity SD, Jorissen D, MacPherson J, Stoelting M, Tierney J, Yelle-Simmons L, 1994. The Galeta oil spill. II. Unexpected persistence of oil trapped in mangrove sediments. *Estuarine, Coastal and Shelf Science* 38: 349–364.

Cahoon DR, Hensel PR, Rybczyk J, McKee KL, Proffitt EE, Perez BC. 2003. Mass tree mortality leads to mangrove peat collapse at Bay Islands, Honduras after Hurricane Mitch. *Journal of Ecology* 91: 1093–1105.

Chen XP, Kong WD, He JZ, Liu WJ, Smith SE, Smith FA, Zhu YG. 2008a. Do water regimes affect iron-plaque formation and microbial communities in the rhizosphere of paddy rice? *Journal of Plant Nutrition and Soil Science* 171: 193–199.

Chen J, Wong MH, Wong YS, Tam NFY. 2008b. Multi-factors on biodegradation kinetics of polycyclic aromatic hydrocarbons (PAHs) by *Sphingomonas* sp. a bacterial strain isolated from mangrove sediment. *Marine Pollution Bulletin* 57: 695–702.

Cheng H, Liu Y, Tam NFY, Wang X, Li SY, Chen GZ, Ye ZH. 2010. The role of radial oxygen loss and root anatomy on zinc uptake and tolerance in mangrove seedlings. *Environmental Pollution* 158: 1189–1196.

Chmura GL, Anisfeld SC, Cahoon DR, Lynch JC. 2003. Global carbon sequestration in tidal, saline wetland soils. *Global Biogeochemical Cycles* 17: 22-1–22-14.

Christensen KK, Sand-Jensen K. 1998. Precipitated iron and manganese plaques restrict root uptake of phosphorus in *Lobelia dortmanna*. *Canadian Journal of Botany* 76: 2158–2163.

Clough BF, Boto KG, Attiwill PM. 1983. Mangroves and sewage: A re-evaluation. In: Teas, H. (Ed.), *Tasks for Vegetation Science*, vol. 8. Dr. W. Junk Publishers, The Hague, the Netherlands, pp. 151–161.

Colmer TD. 2003. Long-distance transport of gases in plants: A perspective on internal aeration and radial loss from roots. *Plant, Cell and Environment* 26: 17–36.

Colmer TD, Cox MCH, Voesenk LACJ. 2006. Root aeration in rice (*Oryza sativa*): Evaluation of oxygen, carbon dioxide, and ethylene as possible regulators of root acclimatizations. *New Phytologist* 170: 767–778.

Corredor JE, Morell JM. 1994. Nitrate depuration of secondary sewage effluents in mangrove sediments. *Estuaries* 17: 295–300.

Dahdouh-Guebas F, Jayatissa LP, Di Nitto D, Bosire JO, Lo Seen D, Koedam N. 2005. How effective were mangroves as a defence against the recent tsunami? *Current Biology* 15: R443–R447.

Danielsen F, Sorensen MK, Olwig MF, Selvam V, Parish, F, Burgess N.D, Hiraishi T. et al. 2005. The Asian tsunami: A protective role for coastal vegetation. *Science* 310: 643.

Dixon RA, Paiva NL. 1995. Stress-induced phenylpropanoid metabolism. *The Plant Cell* 7: 1085–1097.

Dunbabin S, Bowmer KH. 1992. Potential use of constructed wetlands for treatment of industrial wastewaters containing metals. *The Science of the Total Environment* 111: 151–168.

Guo C, Ke L, Dang Z, Tam NFY. 2010. Temporal changes in *Sphingomonas* and *Mycobacterium* populations in mangrove sediments contaminated with different concentrations of polycyclic aromatic hydrocarbons (PAHs). *Marine Pollution Bulletin* 62: 133–139.

Guo CL, Zhou HW, Wong YS, Tam NFY. 2005. Isolation of PAH-degrading bacteria from mangrove sediments and their biodegradation potential. *Marine Pollution Bulletin* 51: 1054–1061.

Guo W, Zhu YG, Liu WJ, Liang YC, Geng CN, Wang SG. 2007. Is the effect of silicon on rice uptake of arsenate (As^V) related to internal silicon concentrations, iron plaque and phosphate nutrition? *Environmental Pollution* 148: 251–257.

Harbison P. 1986. Mangrove muds—A sink and a source for trace metals. *Marine Pollution Bulletin* 17: 246–250.

Hernes PJ, Benner R, Cowie GL, Goni MA, Bergamaschi BA, Hedges JI. 2001. Tannin diagenesis in mangrove leaves from a tropical estuary: A novel molecular approach. *Geochimica et Cosmochimica Acta* 65: 3109–3122.

Hu ZY, Zhu YG, Li M, Zhang LG, Cao ZH, Smith FA. 2007. Sulfur (S)-induced enhancement of iron plaque formation in the rhizosphere reduces arsenic accumulation in rice (*Oryza sativa* L.) seedlings. *Environmental Pollution* 147: 387–393.

Jackson MB, Colmer TD. 2005. Response and adaptation by plants to flooding stress. *Annals of Botany* 96: 501–505.

Ke L, Bao W, Chen L, Wong YS, Tam NFY. 2009. Effects of humic acid on solubility and biodegradation of polycyclic aromatic hydrocarbons in liquid media and mangrove sediment slurries. *Chemosphere* 76: 1102–1108.

Ke L, Wang WQ, Wong TWY, Wong YS, Tam NFY. 2003a. Potential of mangrove microcosms to remove pyrene from contaminated sediments. *Chemosphere* 51: 25–34.

Ke L, Wong TW, Wong YS, Tam NF. 2002. Fate of polycyclic aromatic hydrocarbon (PAH) contamination in a mangrove swamp in Hong Kong following an oil spill. *Marine Pollution Bulletin* 45: 339–347.

Ke L, Wong TWY, Wong AHY, Wong YS, Tam NFY. 2003b. Negative effects of humic acid addition on phytoremediation of pyrene-contaminated mangrove wetland. *Chemosphere* 52: 1581–1591.

Ke L, Yu KSH, Wong YS, Tam NFY. 2005. Spatial and vertical distribution of polycyclic aromatic hydrocarbons (PAHs) in mangrove sediments. *Science of the Total Environment* 340: 177–187.

Ke L, Zhang C, Wong YS, Tam NFY. 2010. Dose and accumulative effects of spent lubricating oil on mangrove seedlings. *Ecotoxicology and Environmental Safety* 74: 55–66.

Kirk GJD, Kronzucker HJ. 2005. The potential for nitrification and nitrate uptake in the rhizosphere of wetland plants: A modelling study. *Annals of Botany* 96: 639–646.

Kraus TEC, Dahlgren RA, Zasoski RJ. 2003. Tannins in nutrient dynamics of forest ecosystems—A review. *Plant and Soil* 256: 41–66.

Lavid N, Schwartz A, Yarden O, Tel-Or E. 2001. The involvement of polyphenols and peroxidase activities in heavy-metal accumulation by epidermal glands of the waterlily (Nymphaeaceae). *Planta* 212: 323–331.

Liang Y, Zhu YG, Xia Y, Li Z, Ma Y. 2006. Iron plaque enhances phosphorus uptake by rice (*Oryza sativa*) growing under varying phosphorus and iron concentrations. *Annals of Applied Biology* 149: 305–312.

Liu D, Li TQ, Jin XF, Yang XE, Islam E, Mahmood Q. 2008. Lead induced changes in the growth and antioxidant metabolism of the lead accumulating and non-accumulating ecotypes of *Sedum alfredii*. *Journal of Integrative Plant Biology* 50: 129–140.

Liu Y, Tam NFY, Yang JX, Pi N, Wong MH, Ye ZH. 2009. Mixed heavy metals tolerance and radial oxygen loss in mangrove seedlings. *Marine Pollution Bulletin* 58: 1843–1849.

Liu WJ, Zhu YG, Smith FA, Smith SE. 2004. Do phosphorus nutrition and iron plaque alter arsenate (As) uptake by rice seedlings in hydroponic culture? *New Phytologist* 162: 481–488.

Lugo AE. 1980. Mangrove ecosystems—Successional or steady-state. *Biotropica* 12: 65–72.

MacFarlane GR. 2002. Leaf biochemical parameters in *Avicennia marina* (Forsk.) Vierh as potential biomarkers of heavy metal stress in estuarine ecosystems. *Marine Pollution Bulletin* 44: 244–256.

MacFarlane GR, Burchett MD. 2001. Photosynthetic pigments and peroxidase activity as indicators of heavy metal stress in the Grey Mangrove *Avicennia marina* (Forsk.) Veirh. *Marine Pollution Bulletin* 42: 233–240.

Machado W, Gueiros BB, Lisboa-Filho SD, Lacerda LD. 2005. Trace metals in mangrove seedlings: Role of iron plaque formation. *Wetlands Ecology and Management* 13: 199–206.

Maskaoui K, Zhou JL, Hong HS, Zhang ZL. 2002. Contamination of polycyclic aromatic hydrocarbons in the Jiulong River Estuary and Western Xiamen Sea, China. *Environmental Pollution* 118: 109–122.

McDonald MP, Galwey NW, Colmer TD. 2002. Similarity and diversity in adventitious root anatomy as related to root aeration among a range of wetland and dryland grass species. *Plant, Cell and Environment* 25: 441–451.

Mendelssohn IA, Kleiss BA, Wakeley JS. 1995. Factors controlling the formation of oxidized root channels—A review. *Wetlands* 15: 37–46.

Mitsch WJ, Gosselink JG. 2000. *Wetlands*. John Wiley & Sons, Inc., New York, pp. 920.

Moorthy P, Kathiresan K. 1998. UV-B induced alternations in composition of thylakoid membrane and aminoacids in leaves of *Rhizophora apiculata* Blume. *Photosynthetica* 35: 321–328.

Mumby PJ, Edwards AJ, Arias-González JE, Lindeman KC, Blackwell PG, Gall A, Gorczynska ML et al. 2004. Mangroves enhance the biomass of coral reef fish communities in the Caribbean. *Nature* 427: 533–536.

Otte ML, Rozema J, Koster L, Haarsma MS, Broekman RA. 1989. Iron plaque on roots of *Aster tripolium* L.: Interaction with zinc uptake. *New Phytologist* 111: 309–317.

Paalman MAA, Van der Weijden CH, Loch JPG. 1994. Sorption of cadmium on suspended matter under estuarine conditions: Competition and complexation with major sea-water ions. *Water, Air and Soil Pollution* 73: 49–60.

Parida AK, Das AB, Mohanty P. 2004. Defense potentials to NaCl in a mangrove, *Bruguiera parviflora*: Differential changes of isoforms of some antioxidative enzymes. *Journal of Plant Physiology* 161: 531–542.

Pedersen O, Binzer T, Borum J. 2004. Sulphide intrusion in eelgrass (*Zostera marina* L.). *Plant, Cell and Environment* 27: 595–602.

Pi N, Tam NFY, Wong MH. 2010a. Effect of wastewater discharge on root anatomy and radial oxygen loss (ROL) patterns of three mangrove species in Southern China. *International Journal of Phytoremediation* 12: 468–486.

Pi N, Tam NFY, Wong MH. 2010b. Effects of wastewater discharge on formation of Fe plaque on root surface and radial oxygen loss of mangrove roots. *Environmental Pollution* 158: 381–387.

Pi N, Tam NFY, Wong MH. 2010c. Formation of iron plaque on mangrove roots receiving wastewater and its role in immobilization of wastewater-borne pollutants. *Marine Pollution Bulletin* 63: 402–411.

Pi N, Tam NFY, Wu Y, Wong MH. 2009. Root anatomy and spatial pattern of radial oxygen loss of eight true mangrove species. *Aquatic Botany* 90: 222–230.

Primavera JH. 1998. Mangroves as nurseries: Shrimp populations in mangrove and non-mangrove habitats. *Estuarine, Coastal and Shelf Science* 46: 457–464.

Qin G, Yan C, Liu H. 2007. Influence of heavy metals on the carbohydrate and phenolics in mangrove, *Aegiceras corniculatum* L., seedlings. *Bulletin of Environmental Contamination and Toxicology* 78: 440–444.

Rivera-Monroy VH, Torres LA, Bahamon N, Newmark F, Twilley RR. 1999. The potential use of mangrove forests as nitrogen sinks of shrimp aquaculture pond effluents: The role of denitrification. *Journal of the World Aquaculture Society* 30: 12–25.

Scandalios JG. 1993. Oxygen stress and superoxide dismutase. *Plant Physiology* 101: 7–12.

Schussler EE, Longstreth D. 1996. Aerenchyma develops by cell lysis in roots and cell separation in leaf petioles in *Sagittaria lancifolia* (Alismataceae). *American Journal of Botany* 83: 1266–1273.

Schützendübel A, Polle A. 2002. Plant responses to abiotic stress: Heavy metal induced oxidative stress and protection by mycorrhization. *Journal of Experimental Botany* 53: 1351–1365.

Silva CAR, Lacerda LD, Rezende CE. 1990. Heavy metal reservoirs in a red mangrove forest. *Biotropica* 22: 339–345.

Smirnoff N. 1995. Antioxidant systems and plant response to the environment. In: Smirnoff, N. (Ed.), *Environment and Plant Metabolism: Flexibility and Acclimation*. Bios Scientific Publishers, Oxford, U.K., pp. 217–243.

Soukup A, Armstrong W, Schreiber L, Franke R, Votrubova O. 2007. Apoplastic barriers to radial oxygen loss and solute penetration: A chemical and functional comparison of the exodermis of two wetland species, *Phragmites australis* and *Glyceria maxima*. *New Phytologist* 173: 264–278.

St-Cyr L, Campbell PGC. 1996. Metals (Fe, Mn, Zn) in the root plaque of submerged aquatic plants collected in situ: Relations with metal concentrations in the adjacent sediments and in the root tissue. *Biogeochemistry* 33: 45–76.

St-Cyr L, Fortin D, Campbell PGC. 1993. Microscopic observation of the iron plaque of a submerged aquatic plant (*vallisneria americana michx*). *Aquatic Botany* 46: 155–167.

Stigliani WM. 1995. Global perspectives and risk assessment. In: Salomons, W. and Stigliani, W.M. (Eds.), *Biogeodynamics of Pollutants in Soil and Sediments*. Springer Verlag, Berlin, Germany, pp. 331–334.

Sundby B, Vale C, Cacador I, Catarino F, Madureira MJ, Caetano M. 1998. Metal-rich concretions on the roots of salt marsh plants: Mechanism and rate of formation. *Limnology and Oceanography* 43: 245–252.

Tam NFY, Guo CL, Yau C, Ke L, Wong YS. 2003. Biodegradation of polycyclic aromatic hydrocarbons (PAHs) by microbial consortia enriched from mangrove sediments. *Water Science and Technology* 48: 177–183.

Tam NFY, Guo CL, Yau WY, Wong YS. 2002. Preliminary study on biodegradation of phenanthrene by bacteria isolated from mangrove sediments in Hong Kong. *Marine Pollution Bulletin* 45: 316–324.

Tam NFY, Ke L, Wang XH, Wong YS. 2001. Contamination of polycyclic aromatic hydrocarbons in surface sediments of mangrove swamps. *Environmental Pollution* 114: 255–263.

Tam NFY, Wong YS. 1993. Retention of nutrients and heavy metals in mangrove sediment receiving wastewater of different strengths. *Environmental Technology* 14: 719–729.

Tam NFY, Wong YS. 1995. Mangrove soils as sinks for wastewater-borne pollutants. *Hydrobiologia* 295: 231–242.

Tam NFY, Wong YS. 1996a. Retention and distribution of heavy metals in mangrove soils receiving wastewater. *Environmental Pollution* 94: 283–291.

Tam NFY, Wong YS. 1996b. Retention of wastewater-borne nitrogen and phosphorus in mangrove soils. *Environmental Technology* 17: 851–859.

Tam NFY, Wong YS. 1999a. Mangrove swamps as pollutant sink and wastewater treatment facility. In: Chua, T.E. and Bermas, N. (Eds.), *Challenges and Opportunities in Managing Pollution in the East Asian Seas*. Partnerships in Environmental Management for the Seas of East Asia (PEMSEA), Quezon City, Philippines, pp. 471–483.

Tam NFY, Wong YS. 1999b. Mangrove soils in removing pollutants from municipal wastewater of different salinities. *Journal of Environmental Quality* 28: 556–564.

Tam NFY, Wong YS. 2000. Spatial variation of heavy metals in surface sediments of Hong Kong mangrove swamps. *Environmental Pollution* 110: 195–205.

Tam NFY, Wong TWY, Wong YS. 2005. A case study on fuel oil contamination in a mangrove swamp in Hong Kong. *Marine Pollution Bulletin* 51: 1092–1100.

Taylor GJ, Crowder AA. 1983. Use of the DCB technique for extraction of hydrous iron oxides from roots of wetland plants. *American Journal of Botany* 70: 1254–1257.

Tukendorf A, Rauser WE. 1990. Changes in glutathione and phytochelatins in roots of maize seedlings exposed to cadmium. *Plant Science* 70: 155–166.

Vangronsveld J, Clijsters H. 1994. Toxic effects of metals. In: Farago, M.E. (Ed.), *Plants and The Chemical Elements-Biochemistry, Uptake, Tolerance and Toxicity*. VCH, Weinheim, Germany, pp. 149–177.

Vasellati V, Oesterheld M, Medan D, Loreti J. 2001. Effect of flooding and drought on the anatomy of *Paspalum dilatatum*. *Annals of Botany* 88: 355–360.

Visser EJW, Colmer TD, Blom CWPM, Voesenek LACJ. 2000. Changes in growth, porosity, and radial oxygen loss from adventitious roots of selected mono- and dicotyledonous wetland species with contrasting types of aerenchyma. *Plant, Cell and Environment* 23: 1237–1245.

Wong TWY, Ke L, Wong YS, Tam NFY. 2002. Study of the sediment contamination levels in a mangrove swamp polluted by marine oil spill. In: *Proceedings of the 25th Arctic and Marine Oil Spill Technical Seminar*, Environment Canada, Ottawa, Ontario, Canada, pp. 73–90.

Wu Y, Tam NFY, Wong MH. 2008. Effects of salinity on treatment of municipal wastewater by constructed mangrove wetland microcosms. *Marine Pollution Bulletin* 57: 727–734.

Yan ZZ, Tam NFY. 2010. Temporal changes of polyphenols and enzyme activities in seedlings of *Kandelia obovata* under lead and manganese stresses. *Marine Pollution Bulletin* 63: 438–444.

Yang Q, Tam NFY, Wong YS, Luan TG, Su WS, Lan CY, Shin PKS, Cheung SG. 2008. Potential use of mangroves as constructed wetland for municipal sewage treatment in Futian, Shenzhen, China. *Marine Pollution Bulletin* 57: 735–743.

Ye ZH, Baker AJM, Wong MH, Willis AJ. 1998. Zinc, lead and cadmium accumulation and tolerance in *Typha latifolia* as affected by iron plaque on the root surface. *Aquatic Botany* 61: 55–67.

Ye ZH, Cheung KC, Wong MH. 2003a. Cadmium and nickel adsorption and uptake in cattail as affected by iron and manganese plaque on the root surface. *Communications in Soil Science and Plant Analysis* 34: 2763–2778.

Ye Y, Tam NFY. 2002. Growth and physiological responses of *Kandelia candel* and *Bruguiera gymnorrhiza* to livestock wastewater. *Hydrobiologia* 479: 75–81.

Ye Y, Tam NFY. 2007. Effects of used lubricating oil on two mangroves *Aegiceras corniculatum* and *Avicennia marina*. *Journal of Environmental Sciences* 19: 1355–1360.

Ye Y, Tam NFY, Lu CY, Wong YS. 2005. Effects of salinity on germination, seedling growth and physiology of three salt-secreting mangrove species. *Aquatic Botany* 83: 193–205.

Ye Y, Tam NFY, Wong YS, Lu CY. 2003b. Growth and physiological responses of two mangrove species (*Bruguiera gymnorrhiza* and *Kandelia candel*) to waterlogging. *Environmental and Experimental Botany* 49: 209–221.

Yim MW, Tam NFY. 1999. Effects of wastewater-borne heavy metals on mangrove plants and soil microbial activities. *Marine Pollution Bulletin* 39: 179–186.

Youssef T, Saenger P. 1996. Anatomical adaptive strategies to flooding and rhizosphere oxidation in mangrove seedlings. *Australian Journal of Botany* 44: 297–313.

Yu SH, Ke L, Wong YS, Tam NFY. 2005a. Degradation of polycyclic aromatic hydrocarbons (PAHs) by a bacterial consortium enriched from mangrove sediments. *Environment International* 31: 149–154.

Yu KSH, Wong AHY, Yau KWY, Wong YS, Tam NFY. 2005b. Natural attenuation, biostimulation and bioaugmentation on biodegradation of polycyclic aromatic hydrocarbons (PAHs) in mangrove sediments. *Marine Pollution Bulletin* 51: 1071–1077.

Zhang CG, Leung KK, Wong YS, Tam NFY. 2007a. Germination, growth and physiological responses of mangrove plant (*Bruguiera gymnorrhiza*) to lubricating oil pollution. *Environmental and Experimental Botany* 60: 127–136.

Zhang FQ, Wang YS, Lou ZP, Dong JD. 2007b. Effect of heavy metal stress on antioxidative enzymes and lipid peroxidation in leaves and roots of two mangrove plant seedlings (*Kandelia candel* and *Bruguiera gymnorrhiza*). *Chemosphere* 67: 44–50.

Zhang XK, Zhang FS, Mao DR. 1998. Effect of iron plaque outside roots on nutrient uptake by rice (*Oryza sativa* L.). Zinc uptake by Fe-deficient rice. *Plant and Soil* 202: 33–39.

Zhou HW, Luan TG, Zou F, Tam NFY. 2008. Different bacterial groups for biodegradation of three- and four-ring PAHs isolated from a Hong Kong mangrove sediment. *Journal of Hazardous Materials* 152: 1179–1185.

Zhou HW, Wong AHY, Yu RMK, Park YD, Wong YS, Tam NFY. 2009. Polycyclic aromatic hydrocarbon-induced structural shift of bacterial communities in mangrove sediment. *Microbial Ecology* 58: 153–160.

18

Origin of Paddy Fields and Functions for Environmental Remediation in the Urbanized Areas of Yangtze River Delta

Zhihong Cao

CONTENTS

Introduction

Rice is one of the staple foods for more than 3 billion of the world's population, particularly among the Asian communities (Darr 2010, De data 1981, Li 1992, Loftas 1995). Therefore, it is important to improve the fertility of soils, the supply of water, and the management of diseases in order to maintain successful rice cultivation, thus ensuring food security in the world (Immerzeei et al. 2010, Sanchez 2010, UNDP 2008).

The origin of rice paddy fields and paddy soils is a topic of debate among scientific scholars around the world (An 1998, Bale 2001, Crawford and Lee 2003, Ikehasi 2002, Kawaguchi and Kyuma 1976, Tsude 2001, You 1995, Zong et al. 2007). The middle and lower Yangtze River Basin is a floodplain with many rivers, lakes, ponds, and irrigation channels, which are connected together known as the water network. The spread of wild rice growing on the wetlands and coast marshes of this region deem them as favorable conditions for rice planting, where the richest and oldest archaeological fossil rice grain sites are also distributed within this vicinity (about 73.7% of total sites discovered in China) (An 1998, You 1995). The Yangtze River Delta was previously the most important region for rice production and generation of higher rice yields; it used to be one of the granary barns in China. However, the rapid economic development and growth in urbanization over the past 30 years have taken their toll on rice planting areas, which are now dramatically reduced (Table 18.1), resulting in severe environmental and ecological problems (Cao et al. 2004, Chinese Agricultural annals of the year 1980, 1990, 2000, 2008).

Results from Table 18.1 show that about 41.2% of rice paddy fields were lost during the past 30 years, where most of the losses occurred in places on the Yangtze River Delta, including Shanghai, southern Jiangsu, and northern Zhejiang provinces. This loss is now the driving force behind the economic development in China, which poses the question of what effects will be incurred in this region in the event of rapid and extreme reductions of large areas of rice paddy fields? Many scientists have stated that rice paddy soil is a strategy for sustainable land use with ecological functions that help restore the biological diversity of the floodplain (Greenland 1998, Masaaki 2009, Washitani 2006). But is it a lesson that can be learned in time by the current situation in the Yangtze River Delta?

This chapter reviews major results from research on paddy soil in China and abroad conducted over the past 20 years, intending to show that the Yangtze River Delta may be the origin of paddy soils in China, where paddy fields and rice cultivation are essential to promoting and preserving a harmonious ecosystem in this region. In the past 30 years, fast economic development has caused a dramatic reduction in the proportion of rice planting areas and thus rice output. This is not only harmful for food security but also for ecological safety; therefore, it is important, from an environmentally

TABLE 18.1

Status of Rice Production in Jiangsu, Zhejiang, and Shanghai Region during the Past 30 Years

Region	1980			1990			2000			2008		
Unit[a]	Area	Yield	T. Yields	Area	Yield	T. Yields	Area	Yield	T. Yields	Area	Yield	T. Yields
Jiangsu	2703.1	5.71	1301.5	2414.9	7.35	1780.2	2398.5	8.08	1937.3	2228.1	7.91	1716.1
Zhejiang	2475.1	5.26	1301.0	2380.1	5.60	1331.2	1940.4	5.84	1132.5	954.3	6.68	636.9
Shanghai	324.6	5.60	181.5	258.9	7.02	181.3	200.8	7.69	154.3	109.1	7.88	86.0
Total[b]	5502.8	5.51	2784.0	5058.5	6.66	3292.6	4539.7	7.20	3224.1	3291.5	7.21	2439.0

Source: Data from Chinese Agricultural annals of the year 1980, 1990, 2000, and 2008—Chinese Agriculture Press, Beijing, China.

[a] Unit of area, 10^3 hm²; Yield, t/hm²; Total yields, 10^4 t.

[b] Total area and total yield are the subtotal of three provinces; yield per unit is average of three provinces, t/hm².

protective and ecologically balanced perspective, to keep adequate areas of rice paddy fields in the Yangtze River Delta in order to achieve sustainable development in both urban and rural areas.

Origin of Irrigated Rice Paddy Fields and Chinese Paddy Soils

Archeologists generally accept that irrigated wetland rice cultivation originated in China, where numerous excavated sites showed sufficient evidence supporting this finding. At the Caoxieshan site, for example, 22 pieces of Neolithic (Majiabang culture) paddy fields were jointly excavated by Chinese and Japanese archeologists in 1996 (Fujiwara 1996), and 24 pieces of Neolithic paddy fields were excavated from the Chuodun site (located only 15 km away from the Caoxieshan site)(Ding 2004, Gu 2003, Tang 2003). Although Korean archeologists unearthed paddy fields (ca. 3500–2000 BC) in the Peninsula and thought they dated back to a period earlier than the Caoxieshan site (Bale 2001, Crawford and Lee 2003), the Japanese scientists reconsidered the time period for rice cultivation of the Yayoi period and believed that it may be as early as the Korean Peninsula (Tsude 2001). A team of soil scientists and archeologists revisited a total of 11 sites located in the middle and lower reaches of the Yangtze River valley, where there were reports of fossil rice grains being previously discovered; 7 of these sites were sampled or reexcavated. During the sixth excavation of the Chuodun site (N31°24′ E120°50′) on November 5, 2004, a total of 44 pieces of rice paddy fields, consisting of varying sizes and shapes, were unearthed. Each piece of paddy field was clearly surrounded with ridges, irrigation ditches, channels, and/or water pools (Figure 18.1), and water flow outlet/inlet remains. More than 400 fossil rice grains and some irrigation tools (Pottery Jar, Basin, etc.) were collected from these paddy fields (Cao et al. 2006, 2007, Ding 2004, Lu et al. 2006). In addition, 24 pieces of paddy fields were unearthed during the last

FIGURE 18.1
Forty-four buried ancient paddy fields at the Chuodun site.

5 excavations of the same site, and, together with the 22 pieces of rice paddy fields unearthed at the Caoxieshan site, there were a total of 90 paddy field pieces assigned to the Neolithic period (Majiabang culture) (Ding 2004). These rice paddy fields were the ones that were excavated the earliest and are located in the settlements of Caoxieshan and Chuodun of Neolithic age. These are just 15 km away from the location where rice cultivation was believed to have started in the northeast of present Suzhou, an industrial city near Shanghai. Three paddy soil profiles overlapping one another vertically have been diagnosed in P-01 (left of Figure 18.2), which include a modern paddy soil profile of 0–43 cm, a buried ancient paddy soil profile of 43–100 cm dating back to 3320 BP, and another buried ancient paddy soil profile of 100–200 cm dating back to 6280 BP; however, in P-02 (as shown on the right of Figure 18.2) only one buried ancient paddy soil profile of 43–100 cm was detected (Cao et al. 2006, 2007, Lu et al. 2006). During the Neolithic Age, only very primitive tools and techniques existed; therefore, no large areas of paddy fields (the largest one was 40 m²) were built by the ancient inhabitants. Consequently, only small pieces of paddy field were constructed, with simple artificial ridges following the original shape of the lowlands along the lakes, where a large amount of charred rice grains was sieved from these buried ancient paddy fields. Opals or phytoliths of rice plants were detected at very high densities in which the unit of measurement used is grains per

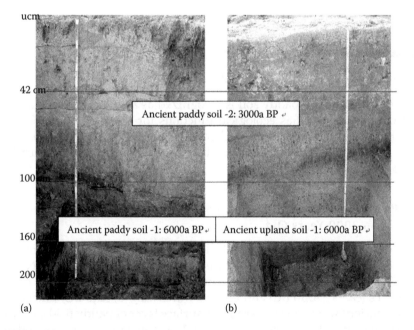

(a)   (b)

FIGURE 18.2
Buried profiles of two ancient paddy soils at P-01 (a) and no ancient paddy soil-1 but ancient upland soil in P-02 (b) in the Chuodum site.

gram of soil (grain/g soil). Archeologists have found that when the phytolith is greater than 5000 grain/g soil then this is a strong indication of rice
having been grown there in the past. The results obtained in our studies far
exceeded this measure, with a maximum reading reaching more than 10^5
grain/g soil at a depth between 100 and 200 cm in the P-01 profile, which
indicates the long history of rice planting within the area; however, the same
was not found at the same depth for P-02 profile. According to ^{14}C dating,
the profiles of the buried paddy soil portrayed for soil organic matter (SOM)
and fossil rice grain had a calibrated age of 3000 BP (43–100 cm) and 6280
BP (100–200 cm), respectively, which placed them in the categories of ancient
paddy soil and prehistoric paddy soil, respectively. These findings correlate
with the results of archeological interpretations reported from potteries frequently discovered in this site, for example, irrigation pots, jars, and so on,
from the Majiabang culture. It can be deduced that the prehistoric paddy soil
profile (100–200 cm in P-01) had already developed horizons, as A, Ap, B, and
C contained various SOM, certain clay distributions and soil pH, which corresponded to other properties found in reports regarding buried paddy soils
(Cao 2008, Cao et al. 2006, 2007, 2010, Dong et al. 2006, Heike et al. 2006, Li
et al. 2007, Lu et al. 2006, 2009). Our findings provide strong support showing
that China may be one of the birthplaces for irrigated rice cultivation in the
world and the buried prehistoric paddy soil found at the Chuodun site is the
origin of paddy soils to date in China (Cao 2008, Cao et al. 2006, 2007).

Evidence for Rice Paddy Soils' Sustainability

The utilization of rice paddy soil is recognized as a strategy for sustainable
land use, which also has ecological functions for the restoration of floodplain biodiversity (Cao et al. 2004, Greenland 1998, Masaaki 2009, Washitani
2006); however, more proof is needed to further investigate the implications
of adopting this approach.

Rice Paddy Fields Located at the Loujiajiao Site Are More Than 3000 Years Old

Approximately 1.2 hm^2 of "Box Rice Fields," surrounded by mulberry gardens
lying in the highlands (Figure 18.3), were found at the Luojiajiao site (N36°65'
E120°55'), Zhejiang Province. Large quantities of more than a tonne of animal bones (ca. ^{14}C, 4710 BC) and pottery wares (ca. ^{14}C, 5210 BC) (Majiabang
culture period) were collected from the surface layer of paddy fields. Further
to this, many burnt straw ash pits (Figure 18.4), fossil rice grains, as well as a
grind stone for polishing tools were found in the cultivation layer (0–25 cm)

FIGURE 18.3
The "box field" found at the Luojiajiao site, Zhejiang province.

FIGURE 18.4
Pits and residual bones found in the sublayer soil of the "box field" (15–30 cm).

of this site. This implies that the current cultivated layers of rice paddy fields were originally utilized by primitive communities approximately 6700–7200 years ago, which were subsequently buried by floods due to prehistoric climate changes. According to the records from the local annals, the land was then used for cultivating mulberry trees and raising silkworms from around 4700 BP. The soil was dug and stacked up to form small hills in order to induce mulberry tree growth, resulting in higher yields of mulberry leaves and, thus, finally a greater harvest of cocoons. After many years of continual digging, the original ancient buried layers of soil were gradually exposed as the surface layer of this polder land and have been continuously used for irrigated rice cultivation even today, a practice in existence since at least 3000 years ago. There were plenty of these types of box paddy fields in the Jiaxing region up until 20 years ago; however, most of them have since been leveled out during the urbanization period. The Luojiajiao site is now the only place protected by law (as a national cultural relic site); thus, it remains a region of vibrantly productive paddy fields, with rice grain yields of 9 t/hm² today. Therefore, it may well be the world's oldest paddy fields of continuous rice cultivation at present, which also provides the most convincing evidence in supporting the efficiency of paddy soils as a means for sustaining the land (Dong et al. 2006, Lu et al. 2009).

Productivity Increased as Rice Cultivation Was Prolonged within the Paddy Soil Chronosequence

At the South bank of Hangzhou Bay in the Yuyao and Cixi counties (N30°–30.50°, E121°–122°) (Figure 18.5), a total of 11 sea dikes were continuously constructed between 1074 and 1980, and the marshland that was created in this process was reclaimed for rice cultivation. For this reason, a unique paddy soil chronosequence with different rice cultivation years from 50 to 2000 years old (recorded by local chorography) were identified. This chronosequence was developed from the same parent materials, under the same ecological conditions with similar cropping systems (Cao 2007, Cao et al. 2007, 2010, Cheng et al. 2009, Kögel-Knabner et al. 2010), where the natural fertility of surface soils increased as cultivation time was prolonged (Figure 18.6).

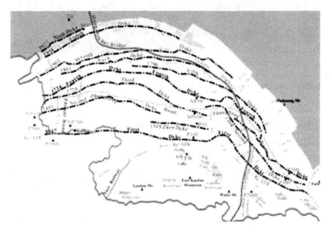

FIGURE 18.5
Coast sediments and dikes on South bank of Hangzhou Bay.

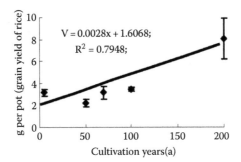

FIGURE 18.6
Natural fertility of the surface layer of the paddy soil chronosequence.

Ecological Functions of Rice Paddy Fields

Sequestrate More Carbon

China is now the world's largest contributor of CO_2 emissions, having superseded the United States as "Polluter Number 1" due to the rapid development in recent years; hence, CO_2 emissions have to be reduced and carbon sequestration increased for the planet's sustainability. Irrigated rice cultivation controls the oxidative/reductive conditions in paddy soils, where a bulk of the paddy soil body is under anaerobic conditions. However, some parts of soil were under oxidative conditions when either near the surface of a very thin (2–3 mm) layer or within the rhizosphere close to the roots or during the dry period created as a result of exposure to oxygen (Cao and Lin 2006, Cao and Zhou 2008, De Data 1981, Kögel-Knabner et al. 2010, Li 1992, Li et al. 2007, Xu et al. 1998). SOM is less mineralized and more stable under anaerobic conditions; therefore, paddy soils contain much more SOM than upland soils (Cao and Lin 2006, Cao and Zhou 2008, De Data 1981, Kögel-Knabner et al. 2010, Li 1992, Xu et al. 1998), which was the situation revealed by the last two National Soil Surveys conducted 40 and 25 years ago. Organic carbon stock has been found to be higher in paddy soils than in upland soils across China, with the average of SOM being 16.70 ± 7.53 g/kg and 10.89 ± 8.36 g/kg, respectively. Since the 1980s, soil organic carbon (SOC) of paddy soils has increased significantly (151 ± 24 Tg C) and still has a large potential (estimated at 112.1 ± 23.1 Tg C in the paddy soils of subtropical region in China) to increase (Liu et al. 2006). When the SOC content in the surface layer of paddy soils reaches the ecological balance or saturation point of the conditions associated with a locality (usually about 21–23 g/kg in the Yangtze River Delta), dissolved organic carbon [DOC] will increase and move down to the deeper layers of the profile, leading to a steady increase in SOC storage. The results of

TABLE 18.2

Comparison of Soil Carbon Density in a Long-Term Field
Experiment of 25 Years Conducted in Paddy Soils
Derived from Red Soil (Yintang, Jiangxi China,
1981–2006)

Type	Treatments	0–25 cm	%	0–50 cm	%
Upland soil	CK	3009		4059	
Paddy soil	CK	4983	65.6	8345	105.6
Upland soil	NPK	2928		4278	
Paddy soil	NPK	5312	81.4	8424	96.9

SOC distribution in the profiles of the paddy soil chronosequence (50–1000 years) showed that as the number of years of cultivation increased so did the reserved organic carbon in the deeper layers (Ci et al. 2007).

Table 18.2 shows the results from a long-term field experiment stretching over 25 years, which indicates that paddy soils derived from red earth appear to have higher organic carbon stocks and carbon sequestration than upland red soils. This also seemed to be true even for buried prehistoric paddy soils when compared with prehistoric nonpaddy soils (Cao et al. 2006, 2007, Lu et al. 2006). Paddy soils can sequestrate more C than upland soils, resulting in a much larger C sink, with greater benefits in coping with the present threats of global warming and climate change. Management of increased SOC content, protection of rice paddy fields, and utilization of these fields' ecological functions may contribute as vital steps needed in achieving CO_2 emission reductions and help with China's environmental diplomacy.

Mutual Contention between Methane and NO_2 Emissions

The growing of rice has traditionally been associated with adverse environmental impacts because of the large quantities of methane gas it generates. It was estimated by IPCC-EF, from 1999 to 2000, that the annual average methane production from Chinese paddy fields was 8.19 Tg CH_4 per year. This level of greenhouse gas generation is a large component of the global warming threat; however, this estimation was based on data from winter-flooded paddy fields, which only occupied 10%–12% of the total Chinese paddy area. Our studies have demonstrated that methane emissions can be significantly reduced by two practices, which also boost crop yield: (1) draining the paddies allowing for the soil to aerate, and (2) implementing drainage for the upland crops during the whole winter season, which in turn interrupts methane production (Burton 2003, Cai et al. 2000, 2003, Xu et al. 2003). Chinese farmers have traditional management practices for paddy soil such as "Kao Tian," which involves the paddy fields being drained once or twice,

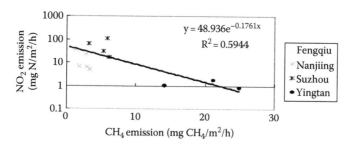

FIGURE 18.7

Relationship between NO_2 and CH_4 emission from paddy. (Adapted from Cai, Z.C. et al., *Plant Soil*, 196(1), 7, 1997.)

each lasting for 2–3 days, after the productive tillering stage. As a result, many fine pores will appear in the surface soil through which oxygen can directly get into the sublayer of soils. The most commonly adopted practice for paddy field systems in China is the annual cropping rotation of summer rice and winter wheat/oil rape seed, which leads to dried-up paddy fields for the entire winter. These practices reduce methane emissions during the summer significantly, ranging from 33% to 68%. According to these results, the annual methane production from Chinese paddy soil (1990–2000) was recalculated with an average of 5.93 Tg CH_4 (Cai et al. 2000). Meanwhile, NO_2 emissions from paddy soil were much less than from upland soils, with an estimated value of about 88 Gg N/year, being approximately 22% of the total NO_2 emissions from China. Results have revealed that the NO_2 emission rate may not be necessarily related to the N fertilizer rate, but to the water regime instead, where the emission is seen occurring mostly during the seasons involving cultivation of crops other than rice (61%). The NO_2 emission factor (amount of NO_2 emission from unit Nitrogen applied) of paddy soil was only one third of what was found in upland soils (Xing et al. 2002, Xing and Zhu 2002); thus, it can be inferred that the emission of methane and NO_2 in paddy soils show a mutually contentious relationship (Figure 18.7) (Cai et al. 1997).

Coping with Disasters of Flooding and Water Lodging

During the end of May 2010, there was a disastrous flood that occurred in a large area of Southern China with the local media reporting several hundred fatalities, extensive damage to roads and buildings, and huge economic losses. It happened not long after a severe drought disaster that occurred during the spring season in Yunnan and Guizhou provinces. The central meteorological observatory issued warnings of flooding, from June to August (the monsoon season), in the middle and lower reaches of the Yangtze River Valley. Every year during the monsoon season, the central and local governments are on high alert in order to prevent and control flooding, and water lodging in Huihe, Yangtze, and Pearl Rivers. The Yangtze River Delta in China is a floodplain with an average altitude of 2–2.5 m above sea level, where the Taihu Lake is in the center with a water network system connecting the whole region. Since the prehistoric times, fertile soils on flat land and annual warm temperatures have provided good conditions for rice cultivation. There is approximately $250 \times 104\,\text{hm}^2$ of irrigated paddy fields that account for more than 70% of the total arable land of the Delta (Cao and Lin 2006, Xu et al. 1998, Zhao et al. 1991). The growing stage of rice is when the flooding period of the Yangtze River and the time of the monsoon season coincide. Paddy fields can usually keep a layer of surface water at a depth of 7–10 cm for growing rice, but it can retain water up to a depth of 15 cm within the fields if needed, during the rainstorm season (Figure 18.8). Therefore, approximately $1500\,\text{m}^3/\text{hm}^2$ more water can be stored in paddy fields than in upland fields, in which the former can impound a total of approximately $37.5 \times 109\,\text{m}^3$ of water from the Yangtze River Delta $(250 \times 104\,\text{hm}^2)$, which is equal to about 85% of the total water that can be impounded by the Taihu Lake $(44.3 \times 109\,\text{m}^3)$. In addition to the standing water layer in paddy fields, there is also a water saturated profile within 0–60 cm depth, which shows paddy soils hold about 50% more water content than that found in upland soils. In total, there is a difference of about $75 \times 109\,\text{m}^3$ between the water storage profiles of paddy soils and upland soils in this delta region. Altogether, paddy soils of the Yangtze River Delta can impound $112.5 \times 109\,\text{m}^3$ of water, which is 2.5 times higher than the water reserve in Taihu Lake (Cao 2007, Cao and Lin 2006, Cao and Zhou 2008, Xu et al. 1998). Paddy fields and soils play a very important role in the prevention and control of flooding and water-lodging disasters by storing excess water. If half of the rice paddy fields were to be lost (see Table 18.1), then the surplus water of $56.25 \times 109\,\text{m}^3$ in the region would have essentially lost its containment, which may lead to catastrophic flooding and water-lodging consequences.

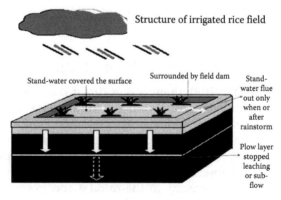

Structure of irrigated rice field

Stand-water covered the surface   Surrounded by field dam   Stand-water flue out only when or after rainstorm

Plow layer stopped leaching or sub-flow

FIGURE 18.8
Schematic diagram of paddy fields and runoff features.

Water and Soil Conservation and Prevention of Nonpoint Pollution

From 1999 to 2004, research was conducted on the migration of substances in the soil-water systems of Taihu Lake region, which demonstrated that irrigated rice fields play an important role in the conversation of soil and water in the Yangtze River Delta (Boo et al. 2003, Cao et al. 2002, 2005, Cao and Lin 2006, Cao and Zhang 2004, Cao and Zhou 2008, Xu et al. 1998, Zhang et al. 2003a,b). This is due to the fact that it is surrounded by 15–20 cm high ridges, covered with 7–10 cm of surface standing water as well as with a plow pan under a 15–20 cm layer of top soil; therefore, it is a relatively closed system for runoff (Figure 18.8). According to the measurements from field experiments at four different sites during 1999 to 2004 in paddy fields of this region, only one or two events of run off flow were recorded with less volume and carried much fewer solid particles as compared with runoff flow in upland fields (Cao et al. 2005, Cao and Lin 2006, Cao and Zhang 2004, Zhang et al. 2003a,b). On the other hand, NH_4^+–N is the major form of nitrogen (N) that is not easily lost in the runoff flow of paddy soils. Due to blockage by the plow pan, there was minimal leaching of a small amount of nitrate (NO_3^-–N) found in paddy soils, resulting in minor N losses via paddy soil runoff (Figure 18.9 and Table 18.3) (Cao and Lin 2006, Xing and Zhu 2002, Zhu et al. 2000). Hence, the N losses through runoff or leaching from paddy soils to the water body are not the main contributors responsible for the eutrophication of surface water body (Table 18.3).

Phosphorus (P) content in water is the key factor in controlling shallow water eutrophication (Cao 2003, Cao and Lin 2006, Foy and Withese 1995, Lagreid et al. 1999, Schaffner et al. 2009, Sharpley et al. 1994), where once

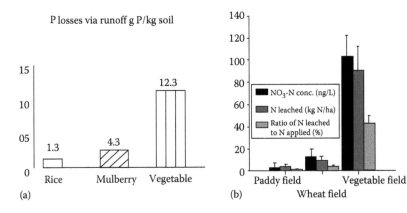

FIGURE 18.9
Phosphorus (P) (a) and nitrogen (N) (b) losses via runoff and leaching from various land use in the Taihu Lake Region (1999–2004).

TABLE 18.3

The Origin of Various Contributors to N and P Contents in the
Water of Taihu Lake Region

Origins	Human Excrement	Domestic Waste	Animal Manure	Fishery Waste (Including Sludge)	
P (%)	26.7	23.8	57.6	11.9	5.7
N (%)	30.7	/	27.8	10.2	7.5

the ratio of N to P content (N/P) in water is over 7 then it will dramatically
cause algae blooming. When the P available in soil reaches a certain concentration, its movement from agricultural soil to water rapidly increases
either via runoff and/or leaching; therefore, a warning value was adopted
(known as the breakpoint), in which higher availability of P in soil is often
used to evaluate the threats of the surrounding quality of the shallow body
of water (Cao et al. 2005, Cao and Lin 2006, Edwards and Withers 1998). In the
4 years of continual field experiments, which were executed in four various
types of paddy soils in the Taihu Lake region, the breakpoint of 25–30 mg/kg
Olsen-P (Figure 18.10) was obtained for the available P in the soil of this
region. However, when P is above this value, it will be dramatically removed
from the soil through an increase in runoff and/or leaching. At present, the
available P in the soil of the Taihu Lake area was observed to be in the range

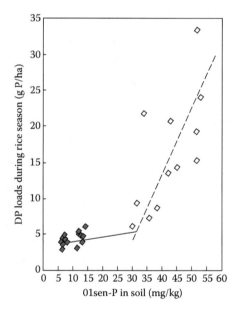

FIGURE 18.10
The breakpoint of soil P runoff from the paddy fields in the Taihu Lake Region.

of 12–15 mg/kg, suggesting that under the existing fertilization regime P loss from paddy soils will not be a serious risk to water bodies for at least the next 5–10 years. But there is still room for phosphorus contents to be increased in paddy soils (Cao et al. 2005, Cao and Lin 2006, Lin et al. 2004).

Treatment of Domestic Wastewater

Paddy fields can be adopted as artificial wetlands (Cao and Lin 2006, Xu et al. 1998) and be used for domestic sewage treatment. Results from field experiments (Figures 18.11 and 18.12) (Li et al. 2009a,b) show that both total N and P in the floodwater of paddy fields tend to reach zero before harvest, when irrigated with both gray and black water. It has been revealed that rice plants and paddy soils are very efficient in removing most of the N and P from domestic wastewater. Another similar study (Zhang et al. 2008) indicated that during the rice growing season, the removal of NH_4^+–N, NO_3^-–N, and total P from wetland paddy fields were 10.7 to 12.3, 6.8 to 9.2, and –1.2 to 2.0 kg/hm², respectively. The output of NH_4^+–N, NO_3^-–N, and total P from drainage was significantly lower than the input during the irrigation of sewage water, which was attributed to the uptake of rice as well as the absorption, transformation, fixation, and filtration of paddy soils. This suggested that paddy fields can be used as a resourceful

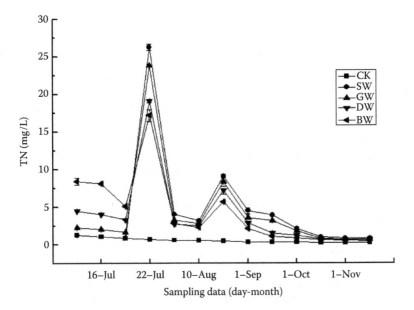

FIGURE 18.11
Total N (TN) content in floodwater during the rice growing season, with various treatments of wastewater and fertilizers.

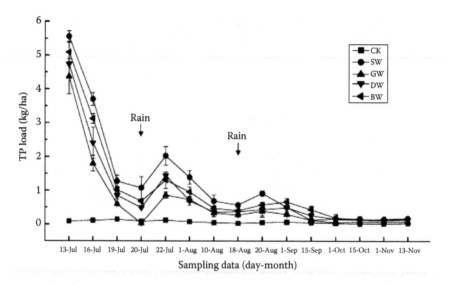

FIGURE 18.12
Variations of total phosphorous (TP) load from floodwater in a rice paddy field with different treatments. The vertical bars are standard deviations of the means, n = 3.

means for domestic wastewater treatment, reducing in turn the level of environmental damage.

Remediation of Organic and Inorganic Pollutants in Soil

Rice fields, similar with natural wetlands, are capable of purifying, consuming, and stabilizing a variety of organic and inorganic pollutants. Besides N and P as described earlier, results have suggested that paddy soils have a specific feature in removing POPs. Due to the reductive conditions and biochemical processes of anaerobic microorganisms in paddy soils, the degradation of chlorinated POPs gets accelerated and the residual from this process is generally found to be lower in paddy soils than in upland soils (An et al. 2006, Yao et al. 2006, 2007, 2008, Wang et al. 2007). Based on these findings, it can be concluded that contaminated soils can be easily remediated and reused through submersion in water or changing from upland to paddy rice cultivation provided there are sufficient water resources in the region. It has been suggested that this approach may be incorporated with the paddy field compartment of the National Institute for Agro-Environmental Sciences' Multimedia Environmental Fate Model (NIAES-MMM-Global), which was developed in Japan. NIAES-MMM-Global is used to estimate the long-term fate of current-use pesticides (CUPs) and POPs emitted into the global environment, specifically from Japan and the rest of Asia (Wei et al. 2008).

In general, available Selenium (Se) is relatively low in paddy soils, for example, as found in the Taihu Lake region, where an average of 1025 soil samples showed only 0.010–0.030 mg/kg belonging to Se-deficient soil type (Cao and Lin 2006). Under anaerobic conditions the availability of Se exhibits decreasing trends due to the reduction of Se^{6+} to Se^{4+} and Se^0 to Se^{2-} (Cao et al. 2001, John et al. 1991, Liu et al. 2004, Mikelsen et al. 1989); hence, the remediation of Se in polluted soils was proposed (Kharakal et al. 2008).

Other Benefits for the Environmental Protection of Urbanized Regions

As a man-made wetland system, paddy fields can provide potential advantages in addressing other environmental degradation issues, which have resulted from the urbanization of the surrounding region. These benefits include the reduction of the urban heat island effect, supplementation of the groundwater table in minimizing land subsidence, maintaining the biodiversity of the suburban area, and cleansing the air to prevent the dangers that may be encountered if dust or sand storms were to arise in urbanized areas.

Groundwater Supplementation in Preventing Land Subsidence

In 2007, CCTV news reported that land subsidence was very severe in the center of the Yangtze River Delta, including Shanghai City, Hangzhou–Jiaxing–Huzhou plain, and Suzhou–Wuxi–Changzhou belt (CCTV 2007). The most severely reported subsidence was located along the Huangpu River of Shanghai, from 1921 up to the ground surface subsidence for 2004 of about 2.63 m. For the past 80 years, the average land subsidence was 1.69 m for the whole of Shanghai City, whereas Huangpu River is considered to be a "hanging river" during the flooding period of the monsoon season. In Lujiazhui, the annual land subsidence for the concentrated area of skyscrapers is reported to be in the range of 12–15 mm. In the Suzhou–Wuxi–Changzhou belt, which is a fast-growing economic development zone of Jiangsu province, the land subsidence level during the past 20 years was over 20 cm in an area totaling 6000 km². A groundwater funnel of around 45 m in depth was also discovered surrounding Jiaxing city, which also happened to be the land subsidence center of Hangzhou–Jiaxing–Huzhou plain in Zhejiang Province. The factors that may have induced such rapid and severe land subsidence are excessive groundwater extraction and the construction of many high-rise buildings and skyscrapers. However, half of the rice fields lost in the region over the past 30 years may also be regarded as another causative factor, ending with a swift deterioration in land subsidence and groundwater levels. Consequently, around 56.25×109 m³ of water may have been wasted and could not be used as groundwater anymore. It is estimated by the Chinese Bureau of Geological Survey that the total financial losses sustained by land subsidence in the Yangtze River Delta, over the past 20 years, is about 3150×10^9 RMB. Shanghai is the most severely affected area, with direct financial losses of 145×10^9 Yuan and 2754×10^9 Yuan of indirect losses (CCTV 2007); therefore, it is vital that certain urbanized areas of paddy fields in the Yangtze River Delta are protected.

Minimizing the Heat Island Effect for Urban and Rural Areas

Dramatic increases in the number of high-rise buildings and skyscrapers (more than 2000 built during the past 10 years, and more than 100 among them being taller than 100 m) (Bureau 2005), wide roads, private cars, air pollution as well as reduced green lawn areas and trees per capita, and rapid expansion of modern living (rapid rise in electronic equipment used in daily life) by the urban population are all contributing causative factors to the heat island effect. It is now becoming an increasingly apparent trend for the Yangtze River Delta. During the summer, the temperatures of downtown Shanghai, Nanjing, and Hangzhou and 14 other medium-sized cities were 3°C–5°C higher than their suburban areas. According to the broadcast on May 24, 2010, by CCTV News, the temperature of downtown Chengdu was 8°C higher than its suburban area where no economic and urbanization developments have taken place (CCTV 2010). The rapid and significant losses of rice cultivation areas in both the Yangtze River Delta and the Chengdu Plain may be primarily responsible for the heat island effect. Thus, the protection of certain areas of flooded rice fields (also referred to as basic farm land) around urban areas is necessary, which is not only for food security but also for minimizing the heat island effect in both urban and/or suburban areas.

Biodiversity System and Clean Air

Biological diversity, landscape, culture, and food in higher urbanized areas are essential to the improvement of citizens' daily life, where clean air is also an important factor to consider when determining the quality of life. During the urbanization process, the lawn areas and tree numbers per capita are reduced; therefore, the integration of paddy fields in urban areas can potentially have many benefits. These include beautiful aesthetic landscapes and a wide range of biodiversity in vegetation, animals (birds and beasts), insects (worms and moths), aquatic creatures (frogs, fishes, loaches, arthropods, etc.) as well as artificially raised cultured ducks, all of which may transform the region into a place of interest for tourists.

The biological diversity of paddy fields can also have significant impacts on insects and disease control from a plant pathology and entomology point of view (News of Hainan University 2010, Yu et al. 2004)

Dust pollution is another problem for urban residents when air quality is concerned, as there are many ongoing construction projects in the outlying urban areas. Therefore, the ignition of dust needs to be avoided at all costs in order to prevent possible threats to the lives of urban residents. Paddy fields with standing water coverage and moist atmospheric conditions produce no dust but fresh clean air and so may help with counteracting this problem.

Summary

Buried irrigated paddy fields and paddy soils were discovered at a Chuodun site nearby to Suzhou that dated back to the Neolithic age of 6280 BP (ca.^{14}C). The paddy fields located at the Luojiajiao site nearby Jiaxing called "Box type field" have shown enough evidence that rice cultivation was taking place in this region dates back to more than 3000 years. A rice cultivation chronosequence of 50–2000 years has been identified at the South bank of Hangzou Bay with the fertility of paddy soils increasing with rise cultivation years, these results strongly indicate that the Yangtze River Delta may be the origin of irrigated paddy fields and paddy soils in China. These findings also suggest that rice cultivation and paddy soils are the most suitable and sustainable method for this region. Paddy fields are important not only for food security but can also be used to solve many environmental problems that have appeared in the region due to rapid urbanization that has been taking place for the past 30 years. Paddy fields are of natural selection and historical interest, and together with rice cultivation, they work as a constituent of harmonious development benefiting both human health and the environment in the urbanized areas of the Yangtze River Delta. This could also be the case for regions with similar conditions and properties, such as the Pearl and Yellow River Delta regions.

Acknowledgments

This work was supported by Chinese Natural Science Foundation with the grant of NSFC No. 40335047 and Sino-Germany Scientific Center (Beijing) grant GZ518. Great acknowledgment to my colleagues: Professors Yang Linzhang, Lin Xiangui, Hu Zhengyi, Dong Yuanhua from the Institute of Soil Science Chinese Academy of Sciences; Professor Ding Jinlong from Museum of Sunzhou; Archeology Research Institute of Suzhou, Jiangsu Province China; Professor Fu Jianrong from Zhejiang Provincial Academy of Agricultural Sciences; and Dr. Zhen Yunfei from the Institute of Relic protection and Archaeology, Zhejiang Province, China. Many thanks also go to Dr. Yin Rui, Dr. Li Jiuhai, Dr. Lu Jia, Dr. Shen Weishou, Dr. Hu Junli, Dr. Chen Yuqin, and postdoctor Li Chunhai. They have all contributed greatly to this work during their work and study at the Institute of Soil Science, Chinese Academy of Sciences.

References

An Z. 1998. Origin of Chinese rice cultivation and its spread east. In *International Symposium Celebrating the Opening of the Biguti Exhibition Hall of Water Conservation Folklore and Cultural Relics*, Jingti City, Korea, July 10, 1998.

An Q, Dong YH, Wang H. et al. 2006. Organochlorine pesticide residues in cultivated soils in the south of Jiangsu, China. *Acta Pedologica Sinica* 41(3):414–419 (in Chinese).

Bale Martin T. 2001. Archeology of early agriculture in Korean—An update on recent development. *Bulletin of the Indo-Pacific Prehistory Association* 21(5):77–84.

Boo ZH, Tang WL, Yang LZ et al. 2003. Advance in quantitative measurements of water and soil losses by remote sensing and application in Taihu Lake region. *Acta Pedologica Sinica* 40(1):1–9 (in Chinese).

Bureau for municipal plan and management of Shanghai Government. 2005. Economic statistics data.

Burton A. 2003. Draining rice paddies cuts methane emissions (Dispatches). *Frontier in Ecology and Environment* 1(2):64.

Cai ZC, Tsuruta H, Gao M et al. 2003. Options for mitigation methane emission from permanently flooded rice field. *Global Change Biology* 9:37–45.

Cai ZC, Tsuruta H, Minami K. 2000. Methane emissions from rice paddy in China: Measurements and influencing factors. *Journal of Geophysical Research* 105 D-13:17231–17242.

Cai ZC, Xing GX, Yan XY et al. 1997. Methane and nitrous oxide emission from rice paddy fields as affected by nitrogen fertilizers and water management. *Plant and Soil* 196(1):7–14.

Cao ZH. 2003. Fertilization and water environmental quality—Effect of fertilization on environment. *Soil Science* 35:353–363 (in Chinese).

Cao ZH. 2007. Rice paddy ecosystem is a sustainable use ecosystem. In *International Symposium on Sustainable Development for East Asia*, JSPA, Beijing, China, March 8–10, 2007.

Cao ZH. 2008. Advance of study on prehistoric irrigated paddy fields and paddy soils. *Acta Pedologica Sinica* 45(5):784–791 (in Chinese).

Cao ZH, Fu JR, Huang JF et al. 2010. Origin and chronosequence of paddy soils in China. In *19th World Congress of Soil Sciences: Soil Solutions for a Changing World* (published on CDROM), Brisbane, Australia, August 1–6, 2010.

Cao ZH, Heike H, Hu ZY et al. 2006. Examination of ancient paddy oils from the neolithic age in China's Yangtze River delta. *Naturwissenchaften* 145(1):35–40.

Cao ZH, Huang JF, Zhang CS et al. 2004. Soil quality evolution after land use change from paddy soil to vegetable land. *Environmental Geochemistry and Health* 26:97–103.

Cao ZH, Lin XG. 2006. *Substance Migration in Soil-Water Systems and Water Environmental Quality in Taihu Lake Region* (in Chinese), Science Press, Beijing, China, pp. 75–97.

Cao ZH, Lin XG, Yang LZ et al. 2005. Ecological functions of "Paddy Field Ring" to urban and rural environment. I. Characteristics of soil P losses from paddy fields to water bodies with runoff. *Acta pedologica Sinica* 42(45):799–804.

Cao ZH, Wang XC, Yao DH et al. 2001. Selenium geochemistry of paddy soils in Yangtze River Delta. *Environment International* 26:335–339.

Cao ZH, Yang LZ, Lin XG et al. 2007. Characteristics of buried paddy fields, prehistoric paddy soils, rice phytolith and fossil grains of neolithic age in Chuodun site. *Acta Pedologica Sinica* 44:838–847 (in Chinese).

Cao H, Yang H, Zhao QG. 2002. Soil erosion and nutrients losses at sloping area of various land use in hill area of Taihu Lake region. *Journal of Limnology* 14(3):243–246.

Cao ZH, Zhang HC. 2004. Phosphorus losses to water from lowland rice fields under rice–wheat double cropping system in the Tai Lake region. *Environmental Geochemistry and Health* 26:229–236.

Cao ZH, Zhou JM. 2008. *Soil Quality of China* (in Chinese). Science Press, Beijing, China, pp. 257–267, 508–511.

CCTV News. February 02, 2007, 22:00, Economic News network.

CCTV News. May 24, 2010, 13:34, and Wang YT, transmit in China broadcast net at 18:46.

Cheng YQ, Yang LZ, Cao ZH et al. 2009. Chronosequential changes of selected properties in paddy soils as compared with original soils. *Geoderma* 151:31–41.

Chinese Agricultural annals of the year 1980, 1990, 2000 and 2008—Chinese Agriculture Press, Beijing, China.

Ci E, Yang LZ, Ma L et al. 2007. Distribution of organic carbon and characteristics of stable carbon isotopes in long term cultivated paddy soils. *Water and Soil Conservation* 05:72–75 (in Chinese).

Crawford GW, Lee Gyoung-AH. 2003. Agricultural origin in the Korean Peninsula. *Antiquity* 77(295):87–95.

Darr S. 2010. Rice—A vital staple food. In Microsoft Internet Explorer www.helium. com/zoone/2397-rice-a vital staple-food, Helium Inc., Andover, MA, June 25, 2010.

De data SK. 1981. *Principles and Practice of Rice Production.* John Willey & Sons, New York, pp. 88–97, 297–331.

Ding JL. 2004. Origin of neolithic paddy fields and rice farming in lower reaches of Yangtze river. *Southeast Culture* 2:19–23 (in Chinese).

Dong YH, Cao ZH, Li JH et al. 2006. Molecular ratio of PAH as a tool to reveal ancient farming practice from pale—Paddy soils in the Yangtze river delta of China. In *Proceedings of 18th World Congress of Soil Sciences*, Philadelphia, PA, July 15, 2006.

Edwards AC, Withers PJA. 1998. Soil phosphorus management and water quality: A UK perspective. *Soil Use and Management* 14:124–130.

Foy RH, Withese PJA. 1995. The contribution of agricultural phosphorus to eutrophication. *The Fertilizer Society Proceedings* No. 365:1–32.

Fujiwara H. (ed.) 1996. *Search for the Origin of Rice Cultivation—Caoxieshan Site in China*, Miyazaki Society for Scientific Studies on Cultural Properties, November 1996, Miyazaki, Japan (in Japanese and Chinese).

Greenland DJ. 1998. *The Sustainability of Rice Farming.* CAB International publication in Association with the International Rice Research Institute, London, U.K., pp. 23–28.

Gu JX. 2003. Irrigated rice paddy fields of Majiangbang culture at Chudun archaeo-logical site. *Southeast Culture* (Suppl. 1):42–45.

Heike K, Cao ZH, Hu ZY, Kögel-Knabner I. 2006. Soil organic matter composition in ancient and prehistoric paddy soils from the lower Yangtze. In *Proceedings of International Urban Soil Conference*. Cairo, Egypt, November 2006.

Huang YP. 2001. *Water Environment and Pollution Control of Taihu Lake*. Science Press, Beijing, China, pp. 29–33.

Ikehasi H. 2002. Paddy field rice cultivation originated from rest cultivation: Reconsideration on origin problem. 1. *Agriculture and Horticulture* 37(10):1056–1064.

Immerzeei WW, Van Beek LPH, Bierkens MFP. 2010. Climate change will affect the Asia water towers. *Science* 328:1382–1385.

John LF John LF, Roger F, Deverel SJ. 1991. Selenium mobility and distribution in irrigated and non irrigated Alluvial soils. *Soil Science Society of America Journal* 55:1313–1320.

Kawaguchi K, Kyuma K. 1976. Paddy soils in tropical Asia. *Southeast ASIAN Studies* 14(3):334–336.

Kharaka YK, Kakouros EG, Thordsen JJ, Naftz DL. 2008. Selenium contamination in the Colorado river basin, western USA: Remediation by leaching and nanofil-tration membranes. *Chinese Journal of Geochemistry* 25(Suppl. 1):100–101.

Kögel-Knabner I, Wulf A, Cao ZH et al. 2010. Biogeochemistry of paddy soils *Geoderma* 157(1–2):1–14.

Lagreid M, BOckman OC, Kaarstad EO. 1999. *Agriculture, Fertilizer and Environment*. CABI Publishing in association with Norsk Hydro ASA, Wallingford, U.K., pp. 132–135.

Li QK. 1992. *Paddy Soils of China*. Science Press, Beijing, China, pp. 145–176, 216–217, 351–357 (in Chinese).

Li S, Li H, Liang XQ et al. 2009a. Phosphorus removal of rural wastewater by the paddy-rice-wetlands system in Tai Lake basin. *Journal of Hazardous Materials* 171:301–308.

Li S, Li H, Liang XQ et al. 2009b. Rural wastewater irrigation and nitrogen removal by the paddy wetland system in the Taihu Lake region of China. *Journal of Soils and Sediments* 9:433–442.

Li CH, Zhang GY, Yang LZ et al. 2007. Pollen and phytolith analyses of ancient paddy fields at Chuodun site of the Yangtze river delta. *Pedosphere* 17(2):209–218.

Lin XG, Yin R, Zhang HY et al. 2004. Cao changes of soil microbiological properties caused by land use change from rice–wheat rotation to vegetable cultivation. *Environmental Geochemistry and Health* 26:119–128.

Liu QH, Shi XZ, Weindorf DC et al. 2006. Soil organic carbon storage of paddy soils in China using the 1:1000,000 soil database and their implications for C sequestra-tion. *Global Biogeochemistry Cycles* 20:1324–1330.

Liu Q, Wang DJ, Jiang XJ et al. 2004. Effects of the interaction between selenium and phosphorus on the growth and selenium accumulation in rice (*Oryza Sativa*). *Environmental Geochemistry and Health* 26:325–330.

Loftas T. 1995. *Dimensions of Need—An Atlas of Food and Agriculture*. FAO, UN, Rome, Italy.

Lu J, Hu ZY, Cao ZH et al. 2006. Characteristics of soil fertility of buried ancient paddy at Chuo-dun site in Yangtze river delta. *Agricultural Sciences in China* 5(6):441–450.

Lu J, Hu ZY, Xu ZH et al. 2009. Effects of rice cropping intensity on soil nitrogen mineralization rate and potential in buried ancient paddy soils from the neolithic
  age in China's Yangtze river delta. *Journal of Soils and Sediments* 9:526–536.
Masaaki H. 2009. Role of rice paddy fields in our finite planet earth. *Clean Technologies
  and Environmental Policy* 11(2):137–141.
Mikelsen RL, Mikkelesen DS, Abshahi A. 1989. Effects of soil flooding on selenium
  transportations and accumulation by rice. *Soil Science Society of America Journal*
  53(1):122–127.
News of Hainan University. 2010. Protection of biodiversity and ecological serve in
  paddy fields of Hainan province—An joint project between Hainan University
  and IRRI, supported by Hong Kong Jiadaorii (嘉道理)i Foundation. Reported
  on March 23, 2010.
Sanchez PA. 2010. Tripling crop yields in tropical Africa. *Nature Geoscience* 3:299–300.
Schaffner M, Bader HP, Scheidegger R. 2009. Modeling the contribution of point
  sources and non-point source to Thachin river water pollution. *Science of the
  Total Environment* 407:4902–4915.
Sharpley AN, Chapra SC, Wedepohi R et al. 1994. Managing agricultural phospho-
  rus for protection of surface waters: Issues and options. *Journal of Environmental
  Quality* 23:437–451.
Tang LH. 2003. Original rice culture remains at Chuodun archaeological site. *Southeast
  Culture* 1:46–49.
Tsude H. 2001. YaYoi farmers reconsidered new perspectives on agricultural develop-
  ment in East Asia. *Bulletin of the Indo-Pacific Prehistory Association* 21(5):53–59.
UNDP. 2008. Secure access to land for food security. In *OGC Workshop on Land
  Governance and Emerging Development Agendas: Legal Empowerment, Climate
  Change and Food Security*. Oslo, Norway, November 24–25, 2008.
Wang F, Jiang X, Bian YR et al. 2007. Schroll, organochlorine pesticides (OCPs) in
  soils under different land usage in the Taihu Lake region, P. R. China. *Journal of
  Environmental Sciences* 19(6):756–760.
Washitani I. 2006. Restoration of biologically diverse floodplain wetlands including
  paddy fields. http://www.airies.Or.JP/publication/ger/pdf/11-2-2006 pdf,
  powered by Geogle.Docs
Wei YF, Nishimori M, Akiyama T et al. 2008. Development of global scale multimedia
  contaminant fate model: Incorporating paddy field compartment. *Science of the
  Total Environment* 406(1–2):219–226.
Xing GX, Shen SL, Shen GY et al. 2002. Nitrous oxide emission from paddy soils in
  three rice-based cropping. *Nutrient Cycling in Agroecosystem* 64:135–143.
Xing GX, Zhu ZL. 2002. Regional nitrogen budgets for China and its major water-
  sheds. *Biogeochemistry* 57/58:405–427.
Xu H, Cai ZC, Tsuruta H. 2003. Soil moisture between rice-growing seasons affects
  methane emission, production, and oxidation. *Soil Science Society of America
  Journal* 67:1147–1157.
Xu Q, Yang LZ, Dong YH. 1998. *Ecological System of Chinese Paddy Fields*. Beijing
  Agriculture Press, Beijing, China, pp. 50–69, 72–80, 83–93, 96–102.
Yao FX, Jiang X, Yu GF et al. 2006. Evaluation of stimulated dechlorination of
  p,p'-DDT in acidic paddy soil. *Chemosphere* 64:628–633.
Yao F, Yu GF, Bian YR et al. 2007. Bioavailability of aged and fresh DDD and DDE in
  soils to rice. *Chemosphere* 68:78–84.

Yao FX, Yu GF, Wang F et al. 2008. Aging activity of DDE in dissimilar rice soils in a greenhouse experiment. *Chemosphere* 71:1188–1195.

You XL (ed). 1995. *History of Rice Culture of China* (in Chinese). China Agriculture Press, Beijing, China, pp. 2–17, 139–143.

Yu SM, Jin QY, Ou YN et al. 2004. Biological control of disease and insect pests in paddy fields by multiple culture of Duck and Rice. *Biological Control in Chinese Agriculture* 2:246–249.

Zhang HC, Cao ZH, Shen QR et al. 2003a. Effect of phosphorus fertilizer application on P loads from paddy soils to water system in Tai Lake region in China. *Chemosphere* 50(6):365–369.

Zhang HC, Cao ZH, Wang GP et al. 2003b. Winter runoff losses of phosphorus from paddy soils in the Taihu Lake region of South China. *Chemosphere* 52(9):1461–1466.

Zhang ZY, Feng ML, Yang LZ. 2008. Comparison of simulated wetlands efficiency of clear up nitrogen and phosphorus from domestic wastewater. *Acta Pedologica Sinica* 45(3):466–475 (in Chinese).

Zhao QG, Gong ZT, Xu Q et al. 1991. *Soil Resources of China* (in Chinese). Nanjing University Press, Nanjing, China, pp. 145–172.

Zhu JG, Han Y, Liu G et al. 2000. Nitrogen in percolation water in paddy fields with rice/wheat rotation. *Nutrient Cycling in Agroecosystems* 57(1):75–82.

Zong YZ, Chen JB, Innes et al. 2007. Fire and flood management of coastal swamp enable first rice paddy cultivation in East China. *Nature* 449:459–462.

19

Remediation and Reuse of Stormwater by Ecotechnology

R. Brian E. Shutes, D. Michael Revitt, J. Bryan Ellis, and Lian Lundy

CONTENTS

Introduction

The United Nations has predicted that the global population will increase from 6.8 billion to 9.1 billion by 2050 (UN, 2009). Changes are urgently required in our current approach to water resource management and the development and implementation of measures to ensure the sustained supply of safe clean water to people throughout the world. Urban water supplies are under increasing pressure to meet the demands of growing populations and changing industries. In many parts of the world, this is occurring within the context of an increased frequency of extreme events, such as flooding or extended droughts. The International Water Association (IWA, 2010) notes that cities will be required to, at least, double the overall efficiency in the use of water and the reduction of pollution. The IWA "Cities of the Future," program states that "vanguard programmes are required for stimulating advances in the leading edge of urban water efficiency, resource recovery and ecological sustainability."One of its five goals is to optimise the design and operation of existing systems in the built environment in the short term (IWA, 2010).

In the European Union (EU) sixth Framework project "Sustainable Water Management Improves Tomorrow's Cities Health" (SWITCH, 2010), Stormwater Management is one of the six research themes. Its objectives include the development of concepts of sustainable stormwater resource use which cities can utilize for their own stormwater management strategies, and to identify catchment-scale stormwater management strategies for integration into urban land management planning. The research theme has investigated and produced reports on the use of alternative hybrid and retrofit technological approaches, including constructed wetlands for stormwater control which contribute aesthetically to the urban environment and provide acceptable levels of prevention/protection against flooding, water pollution, and water shortage when exposed to extreme events and conditions (Scholes and Shutes, 2007; Shutes, 2009).

To ensure a secure water supply into the future, cities must become "water sensitive" by reducing water use per capita, minimizing wastewater, encouraging water recycling, and mitigating anthropogenic impacts on aquatic ecosystems. Water Sensitive Urban Design (WSUD) involves a proactive process which recognizes the opportunities for urban design, landscape architecture, and stormwater management infrastructure to be intrinsically linked (Wong, 2006). Urban stormwater and treated effluents will be reused for landscape irrigation and groundwater recharge in the cities of the future. All three components, that is, water supply, stormwater, and wastewater will be considered and managed in a closed loop (Novotny and Brown, 2007). One of the features of future cities will be localized drainage networks comprising more surface rather than underground systems. The increase in stormwater storage volumes and infiltration volumes provided by best management practices (BMPs), including constructed wetlands, will reduce the volumes of stormwater discharged to the sewer system and to wastewater treatment plants (WWTPs), thus lowering the energy costs of operating these plants. The introduction of stormwater reuse contributes to Integrated Urban Water Management (IUWM) within the urban water cycle by conserving the quality of drinking water supplies (by using the lowest quality of water for lowest quality needs) and generating water supplies for urban agriculture and other uses.

A summary of the benefits to an IUWM program of introducing or retrofitting a drainage system with constructed wetlands and other BMP types is shown in Table 19.1. Cost-benefits will accrue from a reduction in water pollution and flooding. The enhancement of urban biodiversity and the landscape from the use of wetlands, ponds, and basins will also have a cost-benefit in terms of increasing property values. Direct and indirect improvements in the quality of life and health of the urban population will result from an integrated program of stormwater, wastewater, water supply and demand management, and environmental education.

Flood Prevention and Pollution Control

The European Water Framework Directive (WFD) was the main driver behind the review of drainage systems and the implementation of BMPs

TABLE 19.1

IUWM Benefits from the Development of BMP Drainage Systems

Technical	Environmental	Community	Costs	Planning
↓Pollution and flooding	↑Maintaining receiving water volumes	↑Environmental education, information and training	↓Drainage system costs	↑Landscape and flood management planning
↓Run-off volumes to CSOs, WWTPs	↑Water quality		↓O&M costs	
↓Impermeable surface area	↑Wildlife habitats	↑Stakeholder consultation	↓WWTP run-off treatment costs	↑Control of impermeable surfaces
↑Stormwater storage volumes	↑Biodiversity and landscape	↑Community participation	↓Retrofit costs	↑Surface water drainage

↑ represents an increase; ↓ represents a decrease.

in the following examples of European cities. The Greater Dublin Strategic Drainage study recommended the use of BMPs to be greatly increased in the city and its surrounding region. In addition to the benefits of stormwater control and the prevention of flooding, BMPs will bring benefits to developers as well as the public by enhancing the value of existing and new properties (Doyle et al., 2003). On the river Tolka, which is prone to flooding in the city and region of Dublin, retrofit wetlands have been introduced since 2001 on problem surface water sewers prior to discharge into the river (D'Arcy and Chouli, 2007). In Tolka Valley Park, the retrofitting of a pond receiving stormwater and misconnected domestic wastewater as a constructed wetland has reduced nutrient discharges and algal scum and enhanced biodiversity with the reintroduction of the common frog, one of the three amphibian species in Ireland (Dublin City Council, 2010).

Augustenborg is an inner city suburb of Malmo, southern Sweden, which previously experienced flooding during heavy storms from Combined Sewer Overflows (CSOs). An open stormwater system was introduced in 2001 and drainage to the combined sewer system was disconnected. Community participation was an important factor in the implementation of stormwater disconnection. Stormwater now passes through a complex system of green roofs, swales, channels, detention ponds, and small wetlands (Villarreal et al., 2004). The ponds and small wetlands are designed to attenuate 10 year event rainfall, and the green roofs are effective at reducing total run-off. The city of Nijmegen aimed to disconnect 20% of paved surfaces in urban areas within a decade commencing 1997. Grass swales, infiltration trenches, and wetland systems have been developed, and visual water arts projects implemented to raise public awareness of stormwater. Emscher is a region (named after the river) of Germany which is drained by surface water channels that act as combined sewers. The system is currently being replaced over a period of 15 years with sanitary sewers and retrofit source-based stormwater management systems such as constructed wetlands, where feasible. The total amount of stormwater carried by the system is aimed to be reduced by 15%. A GIS-based planning tool has been used to highlight and prioritize the feasible level of disconnection in each subarea. In the United Kingdom, constructed wetlands for the treatment of stormwater comprise around 5% of the probable total number but they are increasing in response to the WFD and the promotion of Sustainable Urban Drainage Systems (SUDS) (Shutes et al., 2005). A survey of 182 drainage sites in Glasgow and Edinburgh, Scotland, indicated appropriate individual BMP techniques or short BMP treatment trains for retrofitting existing drainage systems and for future developments (Scholz, 2007). Glasgow has higher volumes of rainfall run-off and considerably more regeneration sites with potential for developing constructed wetlands than Edinburgh, which has a lack of affordable space and will therefore rely on retrofitting other BMP types. A BMP Decision Support Matrix was prepared comprising dominant criteria including the area available for the BMP and the quality of the run-off and it specifies the technical conditions for the implementation of the corresponding BMP. The supplementary criteria included catchment size and land value and were weighted according to their relative importance for each BMP technique.

A system for the selection of BMP types has been developed in the European Union (EU) Day Water project. The Adaptive Decision Support System (ADSS) for Stormwater Pollution Control involves a multi-criteria decision-making system, the multi-criteria comparator (MCC), which includes technical, socio-economic, and environmental criteria for evaluating and identifying the most preferred BMP option, and a shortlist or ranking of options (Figure 19.1; Ellis et al., 2008). The Day Water ADSS website address is www.daywater.in2p3.fr/EN (user name "guest" and password "guest").

The different drainage options can be scored by stakeholders against each of the identified indicators on a scale of 0–5. For example, if it is considered that a particular option (e.g., constructed wetland) does not contribute anything toward the "pollution control" indicator; it is awarded a score of "0." However, if constructed wetlands are considered to offer the best opportunity for pollution control, a score of "5" is awarded. When developing scores, it is often easiest to decide on the "best" and "worst" options with respect to a particular indicator, and then agree on how well the "intermediate" options contribute to meeting the indicators relative to these identified "best" and "worst" options. The desired percentage weightings for the indicators need to be entered to reflect the importance that the stakeholder places on each of the six criteria. A weighting of 0% indicates that an indicator or criterion is not to be considered within the MCC. The system handles a large amount of information in a consistent way and facilitates discussions between the various stakeholders involved with the BMP decision-making process.

Criteria	Indicators	Swales	Filter strip	Filter drain	Soakaways	Infiltration trench	Infiltration basin	Settlement tank	Lagoon	Retention ponds	Detention basins	Extended detention basin	Constructed wetland	Porous asphalt	Porous paving	Green roofs	Weighting Indicators	Weighting Criteria
Technical	Flood control	2	2	2	2	3	4	4	6	5	6	5	4	1	3	1	30	30
	Pollution control	3	2	2	3	3	5	1	1	2	2	3	4	1	4	2		
	Adaptability to urban growth	3	2	1	2	3	4	2	2	5	5	4	5	1	3	3		
Environmental	Impact on receiving water volume	4	3	4	5	5	5	2	1	2	3	2	2	1	4	4	10	17
	Impact on receiving water quality	4	3	2	2	3	4	1	2	5	4	4	5	1	6	3	0	
	Ecological impact	3	2	1	1	2	3	1	3	4	3	4	6	1	2	1	7	
Operation & Maintenance	Maintenance & servicing requirements	3	4	5	4	4	4	4	3	2	3	2	1	5	3	4	5	10
	System reliability and durability	4	2	2	3	4	4	1	2	5	4	3	3	1	3	3	5	
Social and urban community benefits	Public H & S risks	3	5	5	5	5	3	2	2	1	2	3	1	4	4	5	5	35
	Sustainable development	3	4	2	2	2	2	2	3	4	4	5	6	1	2	3	15	
	Public/community information & awarness	2	2	1	0	1	3	1	3	4	4	4	6	0	1	3		
	Amenity & aesthetics	3	3	2	1	2	3	0	1	5	4	4	6	2	3	3	15	

Instructions — If you would like to return to the instructions please click here

FIGURE 19.1
Part of the MCC page on the DayWater website.

In the United States, several states provide stormwater and BMP management manuals online. The Iowa Stormwater Management Manual (2008) notes that stormwater wetlands require a minimum drainage area of 10 ha in order to maintain the vegetation and a wetland surface area of 3%–5% of the drainage area in order to reduce peak flows and attenuate floods. The City of Portland Environmental Services provides advice on its Sustainable Stormwater Management program. This project aims to maximize the retention, treatment, and infiltration of street run-off, while providing improved safety and a visual amenity for the neighborhood. A Green Street program has introduced pocket wetlands to the sides of streets in order to reduce impermeable surface area, and to enable stormwater to infiltrate and recharge groundwater and surface water. A constructed wetland system named Glencoe Rain Garden, located adjacent to a street, provides 80% reduction in peak flows, 88% flow volume retention annually over 3 years, and protects downstream properties from backups (Kurtz, 2007). There is also a program of disconnection of roof-draining downspouts from the mains sewers, allowing irrigation of gardens and drainage into the soil. In addition to the water quality and ecological benefits to the local Willamette River from reduced CSO discharges, residents experienced an enhancement of their street landscape and improved irrigation of their gardens (Portland Environmental Services, 2009).

The rapid urbanization of Kaohsiung city in south Taiwan, its location in a plain below mountains and sub-tropical climate with high rainfall, has increased its susceptibility to flooding. The city and county governments have constructed flood detention ponds, artificial lakes, and wetlands to reduce the discharge to stormwater drainage systems. The wetland park systems not only regulate the water flow rate and reduce the buildup of sediment sludge in wet weather but also provide recreation, amenities and environmental education. Some of the restored and created wetlands in the lower terrains can also be used as detention ponds to regulate floods in wet weather. The river passing through Shezihlinpi Wetland Park was heavily polluted by discharges of domestic wastewater and contaminated stormwater and by garbage dumping. It was restructured as a wetland park in which the contaminated water is treated by a floating constructed wetland within the river planted with Water Hyacinth (*Eichhornia crassipes*). The wetland park has an area of 3.75 ha, an average water depth of 1.5 m, and a storage volume of approximately 56,250 m^3. The park land next to the river provides a focal point for the community to meet and relax and participate in its management. Singapore introduced a program for water sensitive urban design in 2006 which includes constructed wetlands for the treatment of stormwater and other wastewaters. The Sungai Buloh surface flow stormwater treatment constructed wetland was completed in 2009 with an area of 1000 m^2 and 6000 plants of six species and a nutrient removal performance of 64.4% for total nitrogen and 24.4% for total phosphorus (Sim, 2010).

The first Malaysian urban drainage manual advised on conventional drainage systems to provide rapid transport of stormwater run-off from catchments to receiving waters (DID, 1975). However, these systems have led to an increase in the occurrence of flash floods and levels of water pollution downstream of the catchments. The increase in the urban population of Malaysia from 26.8% in 1970 to 50.7% in 1991 and a projected value of 65% by 2020 has exacerbated the problem of flash flooding. The current Urban Stormwater Management Manual for Malaysia (DID, 2001) promotes the application of BMPs to control the quantity and quality of stormwater and eliminate the impact of future development. A stormwater management plan in the city of Ipoh, Malaysia, where catchment planning, urban infrastructure and services, and stakeholder participation have been addressed, illustrates the importance of Integrated Urban Water Management (Parkinson and Mark, 2005).

One solution to the problem of introducing BMPs in urban areas is to create new cities; for example, the city of Putrajaya, Malaysia, with 70% of the land area allocated as public open green space, illustrates this solution. The Putrajaya wetland system, constructed in 1997 and consisting of 24 constructed wetlands (200 ha) and a downstream lake (400 ha), is located in Putrajaya, the new Federal Government Administrative Center in Malaysia (Figure 19.2). The wetland system and the lake provide an ecological corridor and enhancement of the visual landscape and biodiversity. It treats stormwater in the catchment area, provides flood control, and demonstrates the concept of sustainable development in an urban area (Sim et al., 2008).

A total of 12.3 million individual plants of 62 wetland species were planted in the wetland zone. Of these species, 27 are emergent marsh species including *Phragmites karka*, *Typha angustifolia*, *Lepironia articulata*, *Eleocharis dulcis*, and *Scirpus grossus*. This planted marsh area covers about 40% of the

FIGURE 19.2
Constructed wetland cell, Putrajaya, Malaysia.

wetland area. Putrajaya Wetland with its high diversity of plants and fish has attracted various types of wildlife including more than 100 species of birds and insects. The introduction of insectivorous fish has controlled mosquitoes, a potential hazard of constructed wetlands in hot climates.

The wetlands have generally shown good water quality improvement, but silt run-off from surrounding development land has resulted in low pollutant removal in some months especially during the wet seasons. Siltation has resulted in the wetland bed becoming shallower and terrestrial weeds have invaded the wetland cells. To ensure the sustainability of the wetland system, Putrajaya Corporation has implemented the Polluters Pay Principle, where developers who cause siltation in one wetland cell will be responsible for the desilting work and wetland plant replacement.

BMPs can be retrofitted under a number of conditions including; at the time of building refurbishment; during drainage improvement for large areas such as trading estates or where there are unsatisfactory CSOs; and through incentives to property owners to disconnect roof or driveway run-off from the public drainage system (Gordon-Walker et al., 2007). When selecting ponds or wetlands for retrofitting or construction on new developments, the requirements for flow attenuation, water storage, or pollution treatment should be considered. Systems with a high degree of permanent pool volume have dissolved oxygen and redox potential conditions that may lead to remobilization of contaminants in the sediment. These factors are overcome in constructed wetlands, with their characteristic cycles of filling and draining (Wong et al., 1999). Headley and Tanner (2006) suggest that floating treatment wetlands may be suitable for incorporating into stormwater wetland systems and especially those suffering vegetation decline due to inappropriate water depths and excessive inundation. A Water Environment Research Foundation (2009) study found that the level of maintenance specified had a pronounced effect on the whole-life cost for most BMP facilities. For instance, the level of maintenance for retention ponds and wetlands had a much greater influence on whole-life cost than construction cost. Also, the model developed by the Water Environment Research Foundation (WERF) study predicted that small sites with a high level of maintenance would have a greater whole-life cost compared with facilities that were 10 times as large, but maintained at low or medium levels. The costs of developing or retrofitting, operating, and maintaining a drainage system are normally lower for BMPs than for a traditional system, but the high land-take required by wetlands is a major cost factor in their selection.

Stormwater Reuse

Stormwater can be used for aquifer and groundwater recharge and reused for non-potable use in homes, for example, garden watering, toilet flushing, hot water supply, and car washing, and for commercial/industrial use in cooling towers, cleaning processes, and electricity generation. The benefits of stormwater reuse include freeing-up capacity within sewerage systems, thus facilitating further development; decreased volumes of treated effluent flows to receiving waters; reducing water bills for the community; and in less economically developed countries, reducing the incidence of water-related diseases as reused stormwater is usually of better quality than water from traditional sources.

Rousseau et al. (2008) note that large-scale applications of water reuse of effluent from constructed wetlands are widespread in Australia and the United States but less common in Europe. Australia uses 320 L of water per capita per day for domestic purposes, and its cities will have a predicted 15% average decrease in rainfall by 2030. A significant volume of rainwater in Australia becomes stormwater run-off, and in Sydney, it is estimated that 420 GL of stormwater is discharged into the sea every year. The Metropolitan Adelaide stormwater reuse project (Australia Government Department of the Environment, Water, Heritage and the Arts, 2009) uses constructed wetlands to pretreat stormwater before pumping it through bores to underground water supplies beneath the city. It will save 1000 ML of water per year by using stormwater to replace water drawn from underground supplies for the irrigation of parks and gardens. In Northern Adelaide, the City of Playford's parks and reserves will be irrigated by 570 million liters of water per year from 4 ha of constructed wetlands (Australian Academy of Science, 2008).

In general, stormwater BMPs can potentially offer improved safety over conventional drainage systems, given the reduced presence of gully pots and drain covers, while the total separation of surface drainage systems means fewer routes for vermin to enter piped systems. In addition, the reduction in volume and peak flow rates will reduce the hazards of downstream flash floods to the public and sewer operators. If designed appropriately, BMPs should present fewer risks in terms of health and safety, although there are undoubtedly concerns associated with the provision and operation of drainage facilities in public open spaces which have the potential for health effects due to the presence of insect vectors, vermin and algal blooms in open water or the margins of ponds and wetlands.

The Australian National Guidelines for Water Recycling: Managing Health and Environmental Risks, Phase 2 (2009) address urban stormwater reuse and managed aquifer recharge for end uses including drinking water supply, non-drinking purposes and ecosystem protection. An aquifer storage transfer and recovery (ASTR) project is aimed to determine if stormwater from an urban mixed residential catchment could be harvested through an engineered wetland and recovered at a potable quality (Figure 19.3). However, the project partners are not currently advocating the use of stormwater as drinking water. In the demonstration project in Salisbury, a suburb of Adelaide, stormwater was treated by passing it through a constructed wetland and injecting it via a well into a brackish limestone aquifer 160 m belowground for storage. After 12 months, the stored water was recovered and treated to drinking water quality (CSIRO, 2010).

Stormwater harvesting captures 70% of surface run-off in Salisbury, 300 ML/year or 2% of demand, of which 40% is passed to water supply to municipal irrigation and third pipe systems and 60% injected to ASTR. Salisbury has been a demonstration site for ASTR of stormwater in the European Union (EU) water reclamation technologies for safe artificial groundwater recharge project, with the objective of developing hazard mitigation technologies for water reclamation and providing safe and cost-effective routes for artificial groundwater recharge (RECLAIM, 2006). However, because stormwater supply is intermittent in comparison to the continuous supply of reclaimed

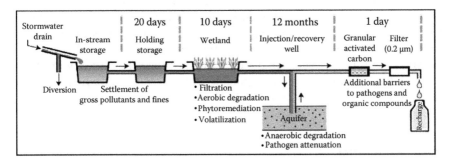

FIGURE 19.3
Diagram of system processing stormwater to drinking water. (From Commonwealth Scientific and Industrial Research Organisation (CSIRO), Storage and treating stormwater for reuse using aquifers, http://www.clw.csiro.au/research/urban/reuse/projects/ASTRbrochure.pdf, 2010. Copyright CSIRO Land and Water.

wastewater, the cost of its storage infrastructure and overall costs are higher than for reclaimed wastewater (McCann, 2010).

The design of Sydney Olympic Park (SOP) was based on the concept of an integrated urban water cycle and includes land rehabilitation, flood alleviation, aquatic habitat restoration, stormwater storage and reuse, pollution control, and recreational provision. Thirty different constructed wetlands were created onsite, a water storage basin was developed in a disused brick pit, and the main stadium has 3 ML/day storage for rainwater. The source of reclaimed water is a "sewer mine" supplemented by surface run-off from the brick pit. Water use at SOP is 48% recycled, 46% direct stormwater reuse, and 6% potable. Of the recycled water, 60% is used for irrigation and washdown (Scholes and Shutes, 2007). The World Games 2009 stadium in Kaohsiung City, Taiwan, is a recent example of the application of sustainable design principles to a sports stadium. The total area, including the stadium and surrounding facilities, is 19 ha of which 7 ha were used to construct green belt, bike track, sports-park, constructed wetlands and ponds. The stadium was designed to collect rainwater from the roof and store it in tanks, and to recycle and reuse the gray water, which is used for irrigating grassland, washing solar cells on the roof, and providing a water source for eco-ponds. The wetlands and ponds in the surrounding parks were designed with gradual slopes and planted with trees, shrubs, and aquatic plant species along the lake shore to imitate a natural river with upstream, middle stream, and downstream sections.

The Chongqing gray water demonstration project will be located in the New Huxi Campus of Chongqing University (CQU) (SWITCH, 2010). There are two large lakes in the teaching area, and one in the residential area. In the first phase, about 250,000 m^2 of lecture, laboratory and student building have been constructed. An integrated scheme combining gray water reuse, landscaping and aesthetics is proposed (Figure 19.4). The gray water from 21 high-rise buildings will be collected and treated onsite by a constructed wetland, and the rainwater will be captured and treated onsite with shallow grass trenches, swales, and constructed wetlands. The reclaimed water (both gray water and rainwater) will be used as the source of landscape irrigation and will complement the lake water. The gray water demonstration project for the new campus of CQU will reduce the consumption of potable water by 150 ML/year. If this project can be replicated at 10 sites in each city in China, it will reduce the consumption of potable water by 990 billion liters per year or the future water supply shortage in China by 2.5%.

Conclusion

BMPs for urban drainage will increase stormwater storage and infiltration volumes and reduce the volumes of stormwater discharged to the sewer

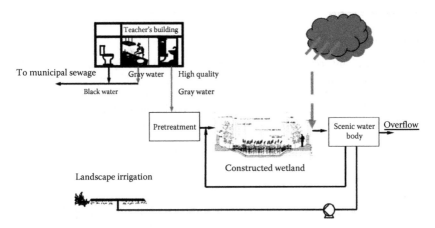

FIGURE 19.4
Flow diagram of sustainable water system in new campus, CQU.

system, receiving waters, and to wastewater treatment plants. Cost-benefits will also accrue from a reduction in water pollution and flooding. The costs of developing or retrofitting and operating and maintaining a drainage system are normally lower for BMPs than for a traditional drainage system, and although there is a high land cost required by wetlands, the whole-life cost of their maintenance is lower than for small BMPs requiring a high level of maintenance.

The reuse of water stored in wetlands contributes to Integrated Urban Water Management (IUWM) within the urban water cycle by conserving the quality of drinking water supplies and thus reducing water treatment volumes, costs and energy consumption. Stormwater supply is intermittent in comparison to the continuous supply of reclaimed wastewater, and the cost of its storage infrastructure and overall costs are higher than for reclaimed wastewater. An aquifer storage transfer and recovery (ASTR) program in Australia will provide drinking water from urban stormwater, and several cities globally are developing the reuse of water stored in wetlands for the irrigation of public parks.

Acknowledgments

We wish to acknowledge the support of the EU FP6 SWITCH (Sustainable Water Management Improves Tomorrow's Cities' Health) project and thank the Commonwealth Scientific and Industrial Research Organisation (CSIRO) for permission to use Figure 19.3.

References

Australian Academy of Science. 2008. Stormwater—Helping to tackle Australia's water crisis. http://www.science.org.au/nova/105/105print.htm

Australian Government Department of the Environment, Water, Heritage and the Arts. 2009. Water for the future. Metropolitan Adelaide stormwater reuse project. http://www.environment.gov.au/water/policy-programs/water-smart/projects/sa)1.html

Australian National Guidelines for Water Recycling: Managing Health and Environmental Risks, Phase 2. 2009. http://www.ephc.gov.au/ephc/water_recycling.html

Commonwealth Scientific and Industrial Research Organisation (CSIRO). 2010. Storage and treating stormwater for reuse using aquifers. http://www.clw.csiro.au/research/urban/reuse/projects/ASTRbrochure.pdf

D'Arcy B, Chouli E. 2007. Europe's drainage gains—Practical experiences of innovation in integrated urban infrastructure. *Water* 21: 29–31.

Department of Irrigation and Drainage (DID), Malaysia. 1975. *Planning and Design Procedure No. 1: Urban Drainage Standards and Procedure for Malaysia.* DID, Malaysia.

Department of Irrigation and Drainage (DID), Malaysia. 2001. *Urban Stormwater Management Manual for Malaysia.* DID, Malaysia.

Doyle P, Hennelly B, McEntee D. 2003. SUDS in the Greater Dublin Area. *National Hydrology Seminar*, Dublin, Ireland, pp. 77–82.

Dublin City Council. 2010. Tolka Valley Park. http://www.dublincity.ie/RecreationandCulture/DublinCityParks/VisitaPark/Pages/TolkaValleyPark.aspx

Ellis JB, Revitt DM, Scholes L. 2008. The daywater multi-criteria comparator. In: Thevenot, D. (Ed.) *Daywater: An Adaptive Decision Support System for Urban Stormwater Management*. IWA Publishing, London, U.K.

Gordon-Walker S, Harle T, Naismith I. 2007. Cost-benefit of SUDS retrofit in urban areas. Science Report-SC060024. Environment Agency for England and Wales, Bristol, U.K.

Headley TR, Tanner CC. 2006. *Application of Floating Wetlands for Enhanced Stormwater Treatment: A Review.* National Institute of Water & Atmospheric Research Ltd, Hamilton, New Zealand.

International Water Association. 2010. Cities of the Future. http://www.iwahq.org/Home/Themes/Cities_of_the_Future/Introduction/

Iowa Stormwater Management Manual. 2008. 2H-1 General Information for Stormwater Wetlands. www.iowasudas.org/stormwater

Kurtz T. 2007. Peak flow and flow volume reductions for urban retrofit projects. Presentation to *2nd National Low Impact Development (LID) Conference*, 12–14 March, Portland, Oregon.

McCann B. 2010. Stormwater reclamation boosts Australia's reserves. *Water* 21: 25–27.

Novotny V, Brown P. 2007. *Cities of the Future: Towards Integrated Water and Landscape Management.* International Water Association (IWA) Publishing, London, U.K.

Parkinson J, Mark O. 2005. *Urban Stormwater Management in Developing Countries.* International Water Association (IWA), London, U.K.

Portland Environmental Services. 2009. Portland Green Street Program. http://www. portlandonline.com/bes/index.cfm?c=4407&

RECLAIM. 2006. Water reclamation technologies for safe artificial groundwater recharge project. EU 6th Framework RECLAIM project. http://www.reclaim-water.org

Rousseau DPL, Lesage E, Story A, Vanrolleghem PA, De Pauw N. 2008. Constructed wetlands for water reclamation. *Desalination* 218: 181–189.

Scholes L, Shutes B. 2007. *Catalogue of Options for the Reuse of Stormwater. Deliverable Task 2.2.1 Part A.* EU 6th Framework SWITCH project, Sustainable Water Management in the City of the Future. http://www.switchurbanwater.eu

Scholz M. 2007. Development of a practical best management practice decision support model for engineers and planners in Nordic countries. *Nordic Hydrology* 38(2): 107–123.

Shutes B. 2009. *A Design Manual Incorporating Best Practice Guidelines for Stormwater Management Options and Treatment Under Extreme Conditions. Part B: The Potential of BMPs to Integrate with Existing Infrastructure (i.e. Retro-Fit/Hybrid Systems) and to Contribute to Other Sectors of the Urban Water Cycle. Deliverable Task 2.1.2 Part B. August 2008.* EU 6th Framework SWITCH Project, Sustainable Water Management in the City of the Future. http://www.switchurbanwater.eu

Shutes RBE, Ellis JB, Revitt DM, Scholes LNL. 2005. Constructed wetlands in UK urban surface drainage systems. *Water Science and Technology* 51(9): 31–37.

Sim CH. 2010. Implementation of wetland systems in Singapore. IWA Specialist Group on Use of Macrophytes in Water Pollution Control Newsletter No. 36, pp. 22–27.

Sim CH, Yusoff MK, Shutes RBE, Ho SC, Mansor M. 2008. Nutrient removal in a pilot and full scale constructed wetland, Putrajaya city, Malaysia. *Journal of Environmental Management* 88: 307–317.

Sustainable Water Management Improves Tomorrow's Cities Health (SWITCH). 2010. EU 6th Framework SWITCH project. http://www.switchurbanwater.eu

UN. 2009. World population prospects: 2008 revision. http://esa.un.org/unpd/ wpp2008/peps_documents.htm

Villarreal EL, Semadeni-Davie A, Bengtsson L. 2004. Inner city stormwater control using a combination of best management practices. *Ecological Engineering* 22: 279–298.

Water Environment Research Foundation (WERF). 2009. Performance and Whole-Life Costs of BMPs and SUDS (01CTS21TA). Water Environment Research Foundation. www.werf.org

Wong THF. 2006. An overview of water sensitive urban design practices in Australia. *Water Practice and Technology* 1(1), doi: 10.2166/WPT.2006018.

Wong THF, Breen PF, Somes NLG. 1999. Ponds vs wetlands—Performance considerations in stormwater quality management. *Proceedings of the 1st South Pacific Conference on Comprehensive Stormwater and Aquatic Ecosystem Management,* Auckland, New Zealand, February 22–26, 1999, vol. 2, pp. 223–231.

20

Phytostabilization for Sustainable End-Use of Arsenic-Rich Mine Tailings in the Victorian Goldfields, Australia

Augustine Doronila, Ron T. Watkins, and Alan J.M. Baker

CONTENTS

Introduction

Anthropogenically generated arsenic (As) contamination is an ever-increasing problem worldwide at many mine tailings storage facilities (TSFs). Phytostabilization provides an attractive technology for long-term remediation of areas for many mine TSFs which contain large amounts of arsenic.

Important characteristics of vegetation for phytostabilization include As tolerance and low As accumulation in order to restrict food chain transfer. In parallel with geochemical studies on the oxidation and formation of secondary minerals of sulfide tailings, the life cycle of the vegetation needs to be monitored long term to determine if this strategy is sustainable and self-perpetuating; this is rarely done. In the case of Stawell Gold Mine, Victorian Goldfields of southeastern Australia, early planning for mine closure strategies has provided a unique opportunity to adequately test regulatory capping prescriptions for tailings surfaces prior to final implementation.

This chapter provides an overview of ecological and geochemical research into rehabilitation of sulfidic tailings undertaken on purpose-built facilities at Stawell Gold Mine, from May 2001 onward. The field trials were preceded by glasshouse studies that indicated the rehabilitation strategies most likely to prove successful. Results of these studies into the growth of grasses and eucalypts, the influence of *mycorrhizae* on growth, and the uptake of arsenic and metals into plant tissues are not reported in detail here, although their relation to the field trials and implications for rehabilitation are discussed. The experimentation was designed to test the efficacy of thin, permeable covers in future rehabilitation of a 100 ha TSF. The trialed covers, which were designed to provide a more environmentally sustainable alternative to the proscribed impermeable cover, involved amendment of fresh tailings with locally derived materials to generate substrates suitable for plant growth. Within the time limitations of the field studies, it is concluded that cultivation of selected native Australian trees and grasses is practical on sulfidic tailings at Stawell Gold Mine after simple amendment of the surface, and this may provide a sustainable vegetative cover able to restrict surface erosion and the release of acidic drainage, whilst imparting value to the rehabilitated land for the local community.

Background

Environmentally sound storage of sulfide-bearing tailings is a critical component of sustainable gold mining in the Victorian Goldfields, where the generation of acid mine drainage (AMD) containing arsenic and heavy metals in solution may seriously affect groundwater and the surface environment in a region of competing land use. Capping of sulfidic tailings with an impermeable cover is frequently proscribed for mine closure to avert such environmental impacts. This may be difficult to engineer for large volume tailings and is prohibitively expensive where natural impermeable materials (e.g., clays) are locally unavailable. Furthermore, an impermeable cover makes self-sustaining revegetation difficult to instate leaving the structure prone to physical degradation (e.g., stream gullying) and restricting the possibilities for end land use. Whereas tailings in the Victorian Goldfields are commonly rich in sulfides, they typically have significant acid neutralizing capacity

(ANC). The resulting self-neutralizing characteristic provides opportunity for environmentally sound surface storage without encapsulation.

Tailings

Stawell Gold Mine, 230 km northwest of Melbourne (Figure 20.1), is depositing tailings into a TSF (Tailings Dam No. 2) of ~100 ha which, at the time of mine closure, will contain a minimum of 10 million tons of tailings with sulfide content of around 2–6 wt%. The predominant sulfide is pyrite, although there are lesser amounts of pyrrhotite, arsenopyrite, and sphalerite. Characteristically of the Victorian Goldfields, the tailings contain significant quantities of arsenic, present in solid solution in pyrite as well as in the arsenopyrite. Acid/base accounting of tailings deposited during the period of this study shows them to be "non-acid producing" or occasionally very weakly acid-producing. Net Acid Producing Potential (NAPP) of 48 samples from the main experimental facility Tailings Experimental Research Facility (TERF) had an average maximum potential acidity of 106.23 kg H_2SO_4/ton and an acid neutralizing capacity (ANC) of 147.82 kg H_2SO_4/ton, i.e., NAPP = -41.58 kg H_2SO_4/ton. The inherent ability of the tailings to neutralize acid solutions produced by sulfide oxidation is readily observed within excavations made into two older TSFs, which show a sharply fronted oxidized

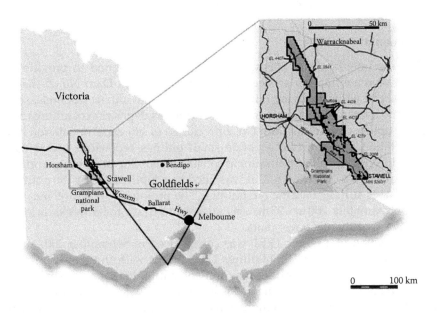

FIGURE 20.1
Location of Stawell Gold Mine in the Victorian Goldfields.

zone extending downward from the surface. Minor carbonate contributes to the ANC, which is however dominantly accredited to the large (>40%) content of chlorite and chloritoid able to exchange cations, including H^+.

Tailings Management

The active TSF (Tailings Dam No. 2) has been developed within a shallow dammed valley and is fed by marginal spigots. Processed lime gives the tailings slurry a pH ~10.5 at the time of deposition and buffers any immediate acidification. Evidence of sulfide oxidation is limited to minor iron-staining at the edges of the TSF and rarely in thin encrustations on dried portions of the surface. The entire surface is utilized by sequential operation of the spigots and will continue to be so throughout mine life, which is expected to be a minimum of 9 years. The initial mine closure plan called for the capping of the TSF with a minimum of 10 cm of impermeable material (clay) and a further 30 cm cover of compacted inert material, and then soil and other fill material of sufficient depth to enable root development. While milled oxide ore and waste rock could provide a large part of the latter, clay is not locally available. With a general preference for revegetation with native "box and ironbark" species a relatively thick cover, of minimum 1 m, would be required, although little research has been undertaken into such rehabilitation and it is unlikely to be successful on a potentially intermittently waterlogged cover resulting from the impermeable clay.

Experimental Facilities

Field experimentation at Stawell Gold Mine was designed to provide an alternative rehabilitation strategy for the TSF (Tailings Dam No. 2) that is cost-effective while ensuring minimal environmental impact from release of acidic and contaminated drainage. The continuous deposition of tailings and regular uplifts of Tailings Dam No. 2 precluded *in situ* experimentation into rehabilitation. Instead, two separate small tailings repositories were constructed. A first, small impoundment constructed on the northern margin of Tailings Dam No. 2 in March 2001 was filled to a depth of 2 m by the pouring of 5700 m³ of sulfidic basaltic ore tailings. The prototype structure, measuring approximately 200 m × 25 m, was allowed to drain for 6 weeks prior to creating experimental cover plots.

Construction of a larger TERF was commenced in September 2001 upon the surface of an older TSF (Tailings Dam No. 1) with the aim of providing a physical and geochemical environment similar to that envisaged at the time of rehabilitation of the current TSF (Figure 20.2). *In situ* surface tailings were bulldozed to form ~3 m high walls of two experimental cells each with dimensions of 75 × 155 m. Fresh sulfidic tailings were poured in the same manner as in the current TSF from spigots positioned around the periphery of each cell. Filling of TERF took 4 months and was completed in May 2002.

FIGURE 20.2
Aerial photograph of TERF constructed upon disused Tailings Dam No. 1 at Stawell Gold Mine and showing vegetation on plots.

Perforated PVC pipes were laid at the base of TERF before filling to allow collection of drainage water and to speed the time of drying of the tailings fill. Experimental plots were created from September 2002 when the surface of the tailings had sufficiently consolidated to enable the use of earth-moving machinery.

Methods

Rehabilitation Strategies

The focus of experimentation was the development of rehabilitation protocols that would

- Provide a self-sustaining vegetation cover that would stabilize the TSF surface by preventing wind and stream erosion
- Inhibit rapid sulfide oxidation and release to the environment of AMD containing dissolved arsenic or heavy metals
- Provide for one, or more, end-uses for the rehabilitated TSF that were sustainable and able to provide a potential economic benefit that could offset future costs of land management
- Yield a rehabilitated TSF that is more utilitarian, easier to construct, and more cost-effective than the originally proscribed impermeable cover

Projected End-Use

The potential for two main end-uses of the rehabilitated TSF were investigated. The first involved the growing of eucalypts for the production of firewood (in short supply in Victoria), and the extraction of cineole (the largest producer of refined eucalyptus oil is presently China, while Australia produces only 5%–10% of world production). The second strategy to be trialed was the growth of native grass species, which can provide more drought-resistant and low-maintenance pasture than introduced grasses, and could serve either as grazing or as a source of native grass seed that is increasingly difficult to obtain for use in environmental rehabilitation and is a high priced commodity (>Aus$ 200/kg).

Cover Materials

Previous attempts to cultivate *Eucalyptus* species and *Echinochloa esculenta* (Japanese millet) directly upon the oxidized surface of Tailings Dam No. 1, with no amendment other than addition of slow release fertilizer, had proved unsuccessful. Planting on unamended tailings was included in the present study to provide control for the main protocols that involved application of one, or more, locally available materials to produce a cover of <50 cm thickness to support vegetation.

The materials employed in the trial covers were as follows:

- *Oxide waste*: A waste rock stockpile resulting from former open-cut mining of the oxidized zone with an estimated volume of 2.5 million m^3. This represents the largest local source of future capping material and is sufficient to create a 25 cm thick surface cover over the final area of the TSF. The oxidized and heavily leached waste rock was milled to produce a fine material that was shown in glasshouse trials to be chemically benign.

- *Stabilized biosolids*: Municipal sewage sludge from the Northern Grampians Water Corporation wastewater treatment plant at Stawell. The biosolids had been stockpiled in a paddock for a minimum of 2 years after removal from anaerobic digestion and maturation ponds. After transport to the experimental sites, the biosolids were spread in small stockpiles to allow aeration prior to application on trial plots.

- *Topsoil*: A limited amount of topsoil stored on the mine site. Unfortunately, former mining practice did not recognize the value of topsoil for future rehabilitation, and the amounts stockpiled are insufficient to cover the TSF (Tailings Dam No 2) with a layer 10 cm thick, as employed in treatments in this study.

Plant Species

Four species of *Eucalyptus* were used as seedlings supplied as tube stock by commercial suppliers: *Eucalyptus cladocalyx*, *Eucalyptus melliodora* (yellow-box), *Eucalyptus polybractea*, and *Eucalyptus viridis*. Native grasses *Microlaena stipoides* (weeping grass), *Austrodanthonia caespitosa* (wallaby grass), *Themeda triandra* (kangaroo grass), and *Bothriochloa macra* (red grass) were purchased from the Ballarat native seed bank and commercial suppliers as caryopses or clean seed. In addition, a native seed collector was contracted to harvest seed from a mixed *Austrodanthonia* grassland.

Experimentation

The surface of Proto-TERF was pegged to a randomized block design, with three blocks replicating 10 different covers. Each of the 30 experimental plots had dimensions of 5×20 m with a 0.5 m buffer zone between each plot. The plots accommodated the following covers/treatments:

1. Tailings
2. Tailings + NPK fertilizer
3. 5 cm biosolids fertilizer
4. 10 cm biosolids
5. 30 cm oxide waste
6. 30 cm oxide waste + NPK fertilizer
7. 30 cm oxide waste + 10 cm topsoil
8. 30 cm oxide waste + 10 cm topsoil + NPK
9. 30 cm oxide waste + 5 cm biosolids
10. 30 cm oxide waste + 10 cm biosolids

Native grasses, as caryopses, were seeded at rates equivalent to 4 kg/ha for *Austrodanthonia* sp., 20 kg/ha for *M. stipoides* and 10 kg/ha for *T. triandra*.

Field trials on TERF commenced in September 2002. Three replicated blocks (75×155 m) were created, with each containing five treatment plots (16×37.5 m) which were prepared and cultivated to a depth of 15 cm with a rotary hoe. The treatments were as follows:

1. Ploughed tailings
2. 10 cm topsoil
3. 10 cm biosolids
4. 30 cm oxide waste + 10 cm topsoil
5. 30 cm oxide waste + 10 cm biosolids

In the final two treatments, milled oxide waste that had been allowed to dry for 2 months was spread on the tailings surface, the topsoil or biosolids were spread on top, and the surface was ploughed. Native grass seed was bulked with topsoil and then handsewn at the same rates as previously employed on proto-TERF (10 kg/ha for *B. macra*). Plots allocated to tree seedlings were further prepared by creating a furrow to a depth of 30 cm with a ripping plough. Ten eucalypt seedlings were planted at 1.5 m spacing along the rip line and cereal rye was seeded in parallel bands alongside the rip to provide a cover nurse crop to reduce wind blasting.

The survival, establishment, and growth of trees and grasses were monitored in parallel with physical and chemical changes occurring in the sulfidic tailings determined from sampled cores. Field plots were excavated to measure the extent of root proliferation by tree species.

Results

Establishment of Perennial Pastures

Native Grasses

Native grass establishment was generally poor on all treatments in the Proto-TERF trial. Of the three species, only *M. stipoides* was initially able to establish. No seedlings of the other two species were observed in December 2001, 5 months after seeding. After 10 months, the exotic grass cover was very dense preventing reliable estimation of emergence of the native species. Germinability of seed of *A. caespitosa* and *T. triandra* was <10% and was the main reason for lack of establishment of these species. Competition from the exotic species may have also limited growth of the native grasses. Only the coverage of *M. stipoides* was monitored and the heterogeneity of variance of measurements was considerable. Data for consecutive May assessments of *M. stipoides* are summarized in Figure 20.3b. The treatments with oxide waste plus biosolids resulted in the greatest cover; however, there were very large standard errors. Estimated percentage cover in the oxide plus low biosolids treatment was 36.7% ± 20.5% in 2002 and 32.3% ± 31.3% in 2003, while the equivalent coverages for the oxide plus high biosolids were 45.0% ± 18.9% and 50.0% ± 28.9%. Adding biosolids directly to the tailings produced less native pasture coverage with further reduction apparent after the second year. Estimated cover on tailings treated with low biosolids was 10.0% ± 2.9% in 2002 and undetected in 2003, while the equivalent figures for tailings treated with high biosolids were 12.0% ± 7.0% and 8.3% ± 8.3%. Covers formed of oxide alone, or oxide plus topsoil, produced minimal foliage cover (<2%) in both years.

Leaf material for native grasses was harvested from most trialed covers with the exception of unamended tailings or oxide treatments. Biomass yields ranked: oxide with high biosolids $(0.64 \pm 0.25 \, kg/m^2)$ > oxide waste with low biosolids $(0.39 \pm 0.19 \, kg/m^2)$ > high biosolids $(0.13 \pm 0.06 \, kg/m^2)$ > low biosolids $(0.06 \pm 0.03 \, kg/m^2)$ > fertilized oxide waste with topsoil $(0.04 \pm 0.02 \, kg/m^2)$ > oxide waste with topsoil $(0.03 \pm 0.01 \, kg/m^2)$ = fertilized oxide waste $(0.03 \pm 0.01 \, kg/m^2)$. The concentration of arsenic in *M. stipoides*

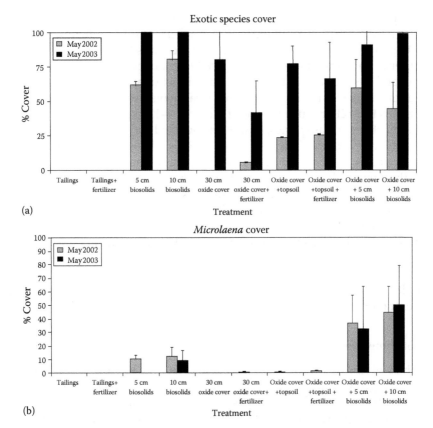

FIGURE 20.3
Percentage plant cover recorded on tailings treatments in May 2002 and May 2003: (a) exotic species and (b) *M. stipoides*.

herbage was not significantly different from differently amended plots, and ranged from 2.0 to 4.4 mg/kg.

Exotic Species

The biosolids contained a viable source of exotic pasture species as a result of having been stockpiled alongside improved pastures. Seed from this source has been volunteered, established, and continually reseeded over many years of stockpiling at the waste water treatment plant. Data for consecutive May assessments of coverage by exotic pasture are given in Figure 20.3a. Exotic species became established within the first year on all trialed covers except bare tailings and unfertilized oxide waste, with maximum coverage >75% recorded for the tailings treated with high biosolids. Exotic pasture cover increased further on all treatments at the end of the second growing season, except on the unamended tailings.

The species which produced the greatest amount of herbage was the pasture grass, *Phalaris aquatica*. Yield in three of the four biosolids amendments was >0.60 kg/m² with the exception of the high biosolids treatment (0.32 kg/m²). Another exotic grass species, *Holcus lanatus*, contributed approximately 3% of the total dry herbage in the biosolid amended treatments with yield ranging from 0.13 to 0.38 kg/m². Oxide only treatments produced 0.06 kg/m² and oxide plus topsoil produced the least yield, 0.04–1.10 kg/m².

Two dicotyledonous species, *Rumex acetosella* and *Spergularia rubra*, were also major volunteer species in the amended tailings. These are not valued animal fodder species. *R. acetosella* produced the least biomass of the four species, with the greatest biomass harvested from the oxide treatment (0.25 kg/m²). All other treatments produced less than 0.05 kg/m². *S. rubra* was present in all the treatments with yield greatest in the biosolids amended treatments.

Most samples of leaf tissue dry matter from exotic species contained <5 mg/kg As, the detection limit of our analytical protocol. Only samples of *S. rubra* had As concentrations >10 mg/kg, in three of five samples. This is a highly prostrate species with stipulate leaves tapering to a short hair-like awn. It is likely that surface contamination accounted for the elevated As concentration.

Establishment of Eucalypts

Survival of the fast-growing *E. cladocalyx* at the beginning of summer (7 months growth) was significantly greater in oxide covers amended with either topsoil or high biosolids (100%) compared with tailings with fertilizer or oxide with fertilizer (33%). Survival rates of this species on all other treatments were not significantly different. The trend in survival was oxide cover with topsoil or biosolids > oxide only cover > biosolids ploughed into tailings > unamended tailings. Mortality increased in all treatments after 12 months. As variation within treatments was significant, no parametric comparison between treatment means was possible. However, the observed trends showed that most of the seedlings planted into tailings (73%–93%) did not survive. Seedlings planted in oxide wastes or biosolids ploughed into tailings had less than 50% survival. The oxide covers with either topsoil or biosolids had >50% survival after the first year of establishment. The majority of seedlings which had survived after 1 year then persisted for the duration of the trials.

The slow-growing *E. melliodora* showed similar trends in survival to the larger, fast-growing *E. cladocalyx*. Of the two mallee species, *E. polybractea* had the greatest number of survivors. At the beginning of the first summer, topsoil and oxide treatments produced significantly greater survival (93%) than all other treatments, except the oxide with high biosolids cover (73%). Other treatments had <50% survival. Unlike the previous two species, further mortalities occurred in all the treatments at the onset of the first autumn. By the second summer, only topsoil treatments had greater than 50% survival. Biosolids treatments were not so effective in enhancing survival with a rate of only 27% in the oxide with low biosolids and 47% in the oxide with high biosolids. *E. viridis* was the least successful of the eucalypts. Survival of planted seedlings in December 2001 was greatest in the topsoil and oxide cover (70%) and then in the biosolids-amended oxide (60%). Other treatments had <50% survival. At the end of the first season, survival had declined in the best treatments such that topsoil and oxide cover was the only one with >50% survival. Mortalities in the biosolids-enriched waste rock cover continued into the second year, whereas all seedlings which had survived in the other treatments continued to persist to the end of the trial.

Leaf samples of eucalypts were taken for analysis from seedlings from several of the experimental covers. Equivalent leaf samples were collected from eucalypt seedlings from a nearby farming property for comparison. Arsenic concentration in dried leaf tissues of *E. cladocalyx* was not significantly different between the control and the experimental covers with concentration ranging from 0.9 to 2.7 mg/kg. The other three eucalypt species had leaf As concentrations significantly greater in the plots growing on the amended tailings than in the farm-grown control plants. Leaves of *E. melliodora* from farm controls had As concentrations of 0.08 mg/kg, while samples from treatment plots ranged from 1.3 to 2.5 mg/kg. Leaves from the oil mallee, *E. polybractea*, had As concentrations of 0.09 mg/kg from the control site and 0.7–1.8 mg/kg in the treatment plots. The least successful of the eucalypts, *E. viridis*, was only sampled in three plots: topsoil and oxide; low biosolids and high biosolids-amended oxide and in the farm area. The mean concentration of As in farm samples was 0.09 mg/kg and 1.1–1.8 mg/kg in the treatment plots. Translocation of As, Cu and Zn into leaf tissue occurred in the biosolids treatments, but no phytotoxic effects could be observed, with low As take up of <5 mg/kg. Uptake of As into plants in the field trials was an order of magnitude less than recorded in previous pot trials, indicating the natural attenuation of As toxicity (Doronila et al. 2004). Plant growth on the thin covers has encouraged oxidation and weathering of the upper profile and has also created conditions that reduce mobility of As, namely, the formation of abundant iron oxyhydroxides that strongly adsorb the metalloid.

Geochemical Development of Substrate

A fundamental requirement for the success of the thin, permeable covers is that they allow for continuous downward oxidation of the sulfidic tailings while not enhancing the mobility and bioavailability of toxic metals and arsenic. The field trials on proto-TERF and TERF showed no evidence of AMD after 3 years. In parallel leaching column tests in the laboratory, employing wetting/drying cycles, the leachate was consistently neutral to alkaline over 3 years. The only exception was in covers with biosolids that produced leachate with pH of approximately 3 on start-up, although this too

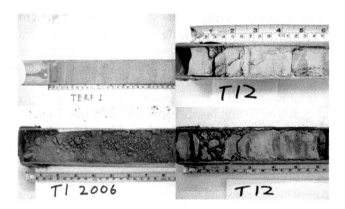

FIGURE 20.4
Downward extension of oxidation in the profile from TERF exhibited in cores taken at time of first planting in 2003, and again in 2006.

quickly evolved to the higher pH, suggesting initial effects of organic acids and/or microbial activity. On TERF, the geochemically open systems to 0.5 m allowed ingress of oxygen and water but did not result in elevated soil acidity, nor was there an increase in the concentrations of labile As and toxic metals. Progressive oxidation of the tailings occurred beneath the amendments (Figure 20.4) either uniformly, or preferentially along layers of coarser tailings that resulted from the original settling.

Chemical changes in the tailings cover were monitored over two summers. Soil sampling of the experimental plots was regularly performed and pH and electrical conductivity measured. After two summers, the pH of tailings ranged from 6.4 to 6.7, and oxide covers pH 5.8 were still slightly acidic (Figure 20.4). The surface of exposed tailings was observed to have changed in color from dark gray to light red in different sections of the experimental plots. Biosolid amended tailings or oxide cover were acidic (pH 4–6). Unamended oxide cover had electrical conductivity levels which could only be tolerated by very salinity-tolerant plant species. Biosolid incorporation into oxide material reduced salinity levels by increasing water infiltration and reducing capillary movement of saline tailings water. Sequential extraction procedures has shown that as tailings particles weather and decompose the predominant form in which it is associated is in the amorphous colloidal phase and minimal contribution to bioavailable forms (Figure 20.5).

Recent excavation of root profiles through the test plots have shown that shrinkage and desiccation cracks have formed in the covered tailings. These form preferential pathways for water to percolate through as well as allowing proliferation of plant roots to the deeper levels of the tailings. The formation of oxidized cutans or skins was also pronounced where there was extensive cracking of the tailings. Geochemical analysis of these boundary layers is currently identifying components of the secondary minerals formed

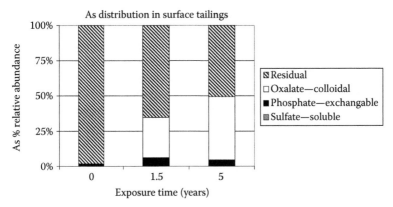

FIGURE 20.5
Relative abundance of arsenic in the different phase of the four step sequential extraction procedure.

through the oxidation process. The pH of deeper unoxidized tailings were consistently >pH 7 and were oxidation had occurred it was pH 6.5–7. This indicates that there is significant acid neutralizing capacity in the tailings. Proliferation of cracks enhanced washing of the secondary minerals, e.g., gypsum, down the profile away from the root zone.

The average density of different size classes of roots at different depths underneath *E. cladocalyx* seedlings is illustrated in Figure 20.6. This species is deep-rooted and the fastest growing of the eucalypts in the trial and would show greater stress responses from limiting factors after 2.5 years. Proliferation of roots in the deeper reaches of the profile appears to be strongly related to the presence of cracks in the tailings. Preferential pathways are created for water to percolate through. Formation of fine and very fine roots was greatest in the upper layers. These are predominantly roots of pasture species; however, eucalypt roots were evident in the deeper cracks down to almost 1 m depth.

Discussion

The compounding effects of drought, salinity, and weed competition limited the establishment of eucalypts on sulfide mine tailings during the experimentation at Stawell Gold Mine. Mortality of all eucalypts in biosolids treatments may have been due to the initial salinity as well as competition from vigorous exotic pasture growth. A major limitation in establishment of eucalypts in reafforestation programs has been attributed to inadequate weed control (Venning 1988, Dalton 1993, Falkiner et al. 2004). Several factors are important in ensuring high survival rates of tree seedlings in field experiments. Establishment is enhanced by deep ripping which allows roots to proliferate deep into the soil and adequately covering roots to remove air pockets which can cause desiccation (Falkiner et al. 2004).

Survival of *E. viridis* was most detrimentally affected by growth in the biosolids amendments. This species has the smallest leaves and tends to have the most prostrate form of the four trial species. Vigorous pasture growth

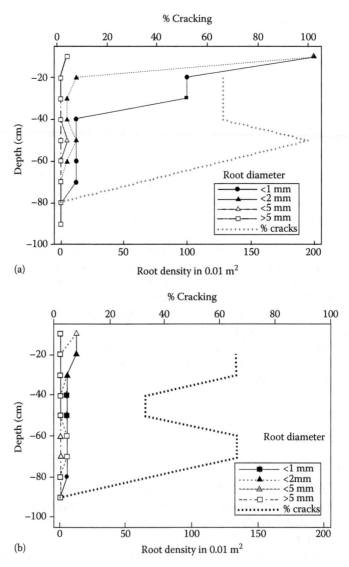

(a)

(b)

FIGURE 20.6
Depth distribution of *E. cladocalyx* roots in (a) 10 cm biosolids; (b) 5 cm biosolids;

may have smothered this species as well as providing competition for moisture and nutrients. The impact of pasture competition is further underlined by the fact that lower pasture establishment in topsoil amendments resulted in the lowest mortality rates of candidate species. Biosolids used in the field experiments had been stockpiled around farm paddocks and contained an

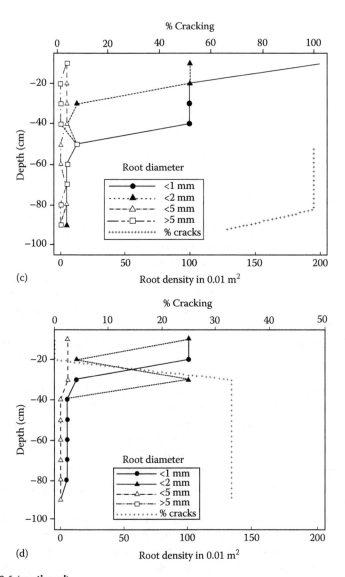

FIGURE 20.6 (continued)
(c) 10 cm biosolids + oxide; and (d) 5 cm biosolids + oxide. Data are abundance of different size classes of roots in different abundance classes 1–10, 10–25, 25–200, >200 for very fine <1 mm and, 1–2 mm, fine roots and for medium and, 2–5 mm, and coarse roots >5 mm. Density scale is logarithmic.

abundance of pasture seed. To maximize its value as a medium to support growth of eucalypts and preferred native grasses, it is essential to minimize the establishment of the highly competitive exotic pasture species. As the majority of the species present in this weed seed bank are exotic grasses, it is possible to apply selective grass herbicides to eliminate germinating seedlings (pre-emergent herbicides) or foliar herbicides to eliminate established grasses. Such use of selective herbicides, especially pre-emergent herbicides, to control monocot weeds, has been effective in regeneration of weed-infested Australian bushland (Buchanan 1989). Adequate weed control and leaching of salts from the biosolids by rainfall prior to planting could provide a cover more suitable to eucalypt survival and establishment.

Applying a thin layer of crushed oxide waste rock over tailings was effective in reducing initial exposure of plant roots to excessive salinity and arsenic in the tailings. The physical structure, however, of this substrate was not favorable for good root development because of the poor hydraulic conductivity and the increased resistance due to its tendency to indurate during the dry months. Biosolids amendments created a substrate which was beneficial for the establishment of both pasture and eucalypts because amended oxide covers and tailings had lower bulk density as well as increased hydraulic conductivity and moisture-holding capacity. These findings are consistent with the ever-increasing number of reports of the improvements in soil physical properties from incorporation of biosolids into mine wastes.

Pasture species and eucalypts did not accumulate significant amounts of arsenic in their leaves. While uptake of As was greater than from non-mining soils with negligible As, the concentration of the metalloid in leaves of plants from the experimental covers was similar to values recorded from other mine sites which had stabilized arsenic tailings (Costello et al. 2004). It is possible however for eucalypts (Ashley and Lottermoser 1999) and pasture species (DeKoe 1994) to take up more As from highly acidic soils. The low uptake of As by the suite of species in the experiments reflects the low bioavailability of the toxic metalloid in the sulfidic tailings covers, where it appears that the geochemistry of the system attenuates the release of more mobile forms of arsenic. In the most recent assessment of the research plots it was showed that after 5 year's growth all four tree species accumulated low As concentrations (King et al. 2008), the highest being recorded in mature leaves, ranging from 0.3 to 5.1 μg/g As. *E. polybractea* had significantly higher foliar As than the other three species but there was also great variation within the species. Between 5 and 10 times lower concentrations were recorded in stem samples and no As was detected in young leaf tips. There was also significant variation in the growth of trees upon the site. *E. cladocalyx* grew taller than the other species although greater variation was detected within the species than between. Variation in tree heights was not correlated with As concentrations in either stems or leaves. Arsenic bioavailability was determined to depths of 2.2 m and found to be low when compared with total As in the tailings.

Pasture vegetation can regress once their roots penetrate into toxic substrates beneath benign inert covering materials. This scenario may occur in direct application of biosolids on tailings which created a thin organic layer of either 5 or 10 cm depth. The mixture was sufficient to attenuate the negative effects of salinity. Exotic pasture immediately established, successfully seeding and regenerating for two growing seasons for both application rates of biosolids. Furthermore, roots had proliferated down to a depth of 30 cm into the tailings without toxicity effects apparent in the leaf tissues. Most of the pasture species present were annuals which completed their life cycle during the wetter months. It is likely that salinity is not a major limiting factor to growth during this period. Stabilization of the tailings in these treatments by exotic pasture species was demonstrated. Further work is required to replace exotic species with native species. Extensive root growth was also observed in the biosolids-amended oxide cover. Proliferation of fine roots through the oxide would improve its physical properties.

Formation of a well-defined oxidation layer did not result in release of acidic water because of the high acid neutralizing capacity of the tailings material. Tree species have been able to persist and grow successfully over three growing seasons. Plant growth has also encouraged weathering of the upper profile and also created conditions which reduce release of arsenic. Biosolids amendments have enhanced the biological activity of soil microflora as well as the improved soil structure which has encouraged weathering. The proliferation of eucalypt roots was shown to be associated with the presence of shrinkage cracks within the tailings. Some roots were observed to penetrate to the base of the tailings layer in ProtoTERF (>1.5 m). It is possible that the largest examples of *E. cladocalyx* and *E. polybractea*, in the low-rate of biosolids-amended tailings had a greater reservoir of available moisture from extensive fracturing underneath the trees. Assessment of crack frequency was performed only on a two-dimensional scale but nonetheless it was possible to show the importance of these voids in root distribution. It is evident that these fractures are the preferential avenues for root growth and extension. Some roots were observed to penetrate to the base of the tailings layer (>1.5 m). Through the release of oxygen and other oxidants plant roots can cause oxidation of iron sulfides in soils to precipitate Fe oxides or iron plaques on plant roots (Armstrong 1967, Batty et al. 2000, Armstrong and Drew 2002). Moreover, it appears that the passage of oxygen into these narrow voids can create thin oxide skins which may form a seal to the release of saline waters from the tailings. As a result, water which accumulates in the cracks may not be as saline as that initially present in the upper layers of the ameliorative covers. Further research is necessary to understand this important root property particularly of the tree species used in this study. This mechanism may be important in the natural attenuation of As toxicity in this mine substrate.

Quantification of the plant-available water is necessary to predict the optimum density of vegetation which the tailings covers can support. Further work should investigate the effect of deep ploughing on water harvesting which is necessary to allow satisfactory growth of tree species. The presence of deep cracks is also important for the structural stability of trees whereby deep roots are required for anchorage to minimize the risk of windthrow (Bengough 2003), particularly in the windy conditions frequently present on the exposed surface of the TSF. It has been shown that deep ripping of reconstructed soil profiles after bauxite mining was a major factor in the survival of the main tree species (Rokich et al. 2001). This work showed that significant differences in the root architecture occurred in deep-ripped substrates. Root proliferation at depth would also improve downward transfer of water when the surface layers of soils with low permeability are wetted after rainfall (Burgess et al. 2001). Another effect of root proliferation at depth would be further oxidation in the tailings. Through the release of oxygen and other oxidants plant roots can cause oxidation of iron sulfides in soils to precipitate Fe oxides or iron plaques on plant roots (Armstrong 1967, Batty et al. 2000, Armstrong and Drew 2002). Further research is necessary to understand this important root property particularly of the tree species used in this study. This mechanism may be important in the natural attenuation of As in this mine substrate. This is particularly relevant as it has been shown recently that formation of the amorphous colloids on the surface of rice roots enhanced arsenite and decreased arsenate uptake (Chen et al. 2005). They showed that the presence of iron plaque diminished the effect of phosphate on arsenate uptake, possibly through a combined effect of arsenate desorption from iron plaque.

The value of using drought-tolerant eucalypts, which do not take up As, as part of the phytoremediation prescription has been clearly demonstrated. The study has shown that growing high value species such as the ones selected for the trials can definitely enhance the value of the TSF as a land resource for agro-forestry. This is in contrast to a pasture only cover which would be far less productive and would have to be continually maintained to minimize the establishment of tree species. The fact that the trees have successfully penetrated deep into the tailings indicate that there is little risk of windthrow. As a result, there is minimal risk of elevated arsenic levels appearing on the surface after phytostabilization. The downward progression of oxidation within the tailings profile beneath the vegetated thin covers indicates that this method of rehabilitation is sustainable when applied to sulfidic tailings with excess acid neutralizing capacity, such as occur widely in the Victorian Goldfields.

Acknowledgments

The authors are indebted to Graham Brock without whose vision and appreciation of the value, rather than the cost, of environmental research at Stawell

Gold Mine the research would not have been possible. Thanks go to staff of Stawell Gold Mine for their continued support of this research through the University Research Alliance. The research was jointly funded by the ARC (Grant no. C00107720), M.P.I. Pty Ltd. and Stawell Gold Mine (now Northgate Minerals).

References

Armstrong W. 1967. The oxidizing activity of roots in waterlogged soils. *Physiologia Plantarum* 20: 920–926.

Armstrong W, Drew M. 2002. Root growth and metabolism under oxygen deficiency. In *Plant Roots: The Hidden Half*, 3rd edn. (revised and expanded), Y. Waisal, A. Eshel, and U. Kafkafi (eds.), Marcel Dekker, Inc., New York, pp. 717–728.

Ashley PM, Lottermoser BG. 1999. Arsenic contamination at the Mole River mine, northern New South Wales. *Australian Journal of Earth Sciences* 46: 861–874.

Batty LC, Baker AJM, Wheeler BD, Curtis CD. 2000. The effect of pH and plaque on the uptake of Cu and Mn in *Phragmites australis* (Cav.) Trin ex. Steudel. *Annals of Botany* 86: 647–653.

Bengough AG. 2003. Root growth and function in relation to soil structure, composition and strength. In *Root Ecology*, H. de Kroon and E.J.W. Visser (eds.), Springer, Berlin, Germany, pp. 151–168.

Buchanan RA. 1989. *Bush Regeneration: Recovering Australian Landscapes*. TAFE New South Wales, Sydney, New South Wales, Australia.

Burgess SO, Adams MA, Turner NC, White DA, Ong CK. 2001. Tree roots: Conduits for deep recharge of soil water. *Oecologia* 126: 158–165.

Chen Z, Zhu YG, Liu WJ. 2005. Direct evidence showing the effect of root surface iron plaque on arsenite and arsenate uptake into rice (*Oryza sativa*) roots. *New Phytologist* 165: 91–97.

Costello J, Lacy H, Rengel Z, Jasper D, Quaghebeur M. 2004. Uptake of arsenic by native plants on gold tailings in Western Australian rangelands. In *Proceedings of the Goldfields Environmental Management Group 2004 Workshop on Environmental Management in Arid and Semi-arid Areas*, May 26–29, 2004, Kalgoorlie, Western Australia, Australia, pp. 79–90.

Dalton G. 1993. *Direct Seeding of Trees and Shrubs: A Manual for Australian Conditions*. Primary Industries, South Australia, Adelaide, Australia, 123pp.

DeKoe T. 1994. *Agrostis castellana* and *Agrostis delicatula* on heavy metal and arsenic enriched sites in NE Portugal. *Science of the Total Environment* 145: 103–109.

Doronila AI, Baker AJM, Watkins R, Osborne JM, Chibnall LM. 2004. Revegetation strategies for sustainable end-use of arsenic rich tailings in the Victoria goldfields. In *Proceedings of the Goldfields Environmental Management Group 2004 Workshop on Environmental Management in Arid and Semi-arid Areas*, May 26–29, 2004, Kalgoorlie, Western Australia, Australia, pp. 73–78.

Falkiner R, Theiveyanathan T, Marcar N, Myers B, Stewart L. 2004. Heartlands low to medium rainfall farm forestry: A report of the heartlands Initiative. Publication HL10–04. CSIRO Forestry & Forest Products. 62pp.

King DJ, Doronila AI, Feenstra C, Baker AJM, Woodrow IE. 2008. Phytostabilisation of arsenical gold mine tailings using four *Eucalyptus* species: Growth, arsenic uptake and mobility after five years. *Science of the Total Environment* 406: 35–42.

Rokich DP, Meney KA, Dixon KW, Sivasithamparam K. 2001. The impact of soil disturbance on root development in woodland communities in Western Australia. *Australian Journal of Botany* 49: 169–183.

Venning J. 1988. *Growing Trees for Farms, Parks and Roadsides: A Revegetation Manual for Australia*. Lothian Publishing, Melbourne, Victoria, Australia, 126pp.

21

Field Application of the Stabilization Process for Arsenic-Contaminated Soil

Myoung-Soo Ko, Ki-Rak Kim, and Kyoung-Woong Kim

CONTENTS

Introduction

Soil is a main component of the environment and the base of human activities such as agriculture, construction, and so forth. A soil can be described as an open, multicomponent chemical system that contains solid, liquid, and gaseous phases influenced by living organisms (Thornton 1983). Soils can be contaminated with heavy metals derived from various artificial sources (Adriano 1986). Especially, inorganic contaminants, such as heavy metals and metalloid, have attracted interest from many researchers due to their nondegradable characteristics, life cycle in ecosystems, and high toxicity, even at low levels.

Arsenic has gained great notoriety because of the toxic properties of a number of its compounds (Alloway 1995). Arsenic contamination in soil is generated by geothermal systems, including hot springs and geysers. Mining and metallurgical operations can also concentrate the arsenic into tailings and soil. Remediation of arsenic-contaminated soil can be achieved by many technologies, including biological treatment, phytoremediation, solidification/stabilization, fixation, soil flushing, and electrokinetic remediation (Mulligan et al. 2001, US EPA 2002). Stabilization is a process where additives are mixed with waste to convert the hazardous constituents into a form that minimizes their rate of migration and level of toxicity. Stabilization is regarded as one of the most effective and efficient methods for the immobilization of contaminants in waste. This chapter introduces the characteristics of arsenic in soil techniques for its stabilization and some of their field applications.

Arsenic in Mining Area Soils

Arsenic in Environment

Arsenic is a naturally occurring element that ranks 20th in abundance in the earth's crust. The average concentration of arsenic in the earth crust is reported to be 1.5–2 mg/kg. As the level of arsenic differs with respect to the origin of soils, the background level in agricultural soils has been reported to be 6.3 mg/kg (Adriano 2001). Sources of arsenic are divided into two categories (Adriano 1986): first, from natural phenomena, such as weathering and volcanic activity, which are the main source, and, second, from anthropogenic sources, such as mine tailings, landfill sites, waste water discharges, and agricultural activities, which account for 60% of arsenic contamination (Cullen and Reimer 1989, Bhumbla and Keefer 1994).

Arsenic is often found in contaminated soils as a result of natural phenomena. The normal concentration range of arsenic in soils is 1–40 mg/kg, with a mean of 5 mg/kg, but in most soils it is in the lower half of this range (Bowen 1979, Kabata-Pendias and Pendias 1984, Bañuelos and Ajwa 1999). There are five levels of arsenic from different types of parent rock, especially high levels in sedimentary rocks (shales, mudstones, and slates). The National Research Council of Canada reported that soils originating from sulfide ore deposits contain arsenic at high concentrations, even up to several hundreds of mg/kg (NRCC 1978). The main source of arsenic in the environment is the parent material from which the soil is derived. The most common arsenic mineral in the environment is arsenopyrite (FeAsS), which is frequently associated with Au, Cu, Sn, Ag, Zn, and Pb ore deposits (Smedley and

Kinniburgh 2002). Table 21.1 shows the normal arsenic minerals in nature. Rock-forming minerals, such as sulfide minerals, oxidation minerals, silicate minerals, carbonate minerals, and sulfide minerals, have different arsenic concentrations. Especially, sulfide minerals contain large concentrations of arsenic (Table 21.2). Arsenic contamination in Japan, Chile, and the United States has occurred due to geothermal sources, and arsenic concentration in the ground water in Bangladesh, India, China, and the United States have increased due to arsenic-bearing minerals (Smedly and Kinniburgh 2002, Wang and Mulligan 2006, Yang et al. 2007). The background level of As in Korea has not been reported, but the soil contaminations have been investigated in detail, with the most frequently contaminated sites being abandoned mine areas, especially old gold mines (Jung 2001). There are approximately 1000 abandoned metal mines in South Korea, where the tailings left behind

TABLE 21.1

Major Arsenic Minerals in Nature

Mineral	Composition	Occurrence
Native arsenic	As	Hydrothermal veins
Niccolite	NiAs	Vein deposits and norites
Realgar	AsS	Vein deposits, often associated orpiment, clays and limestones, also deposits from hot springs
Orpiment	As_2S_3	Hydrothermal veins, hot springs, volcanic sublimation product
Cobalite	CoAsS	High-temperature deposits, metamorphic rocks
Arsenopyrite	FeAsS	The most abundant Arsenic mineral, dominantly mineral veins
Tennantite	$(Cu,Fe)_{12}As_4S_{13}$	Hydrothermal veins
Enargite	Cu_3AsS_4	Hydrothermal veins
Arsenolite	As_2O_3	Secondary mineral formed by oxidation of realgarm arsenopyrite and other arsenic minerals
Claudetite	As_2O_3	Secondary mineral formed by oxidation of realgarm arsenopyrite and other arsenic minerals
Scorodite	$FeAsO_4 \cdot 2H_2O$	Secondary mineral
Annabergite	$(Ni,Co)_3(AsO_4)_2 \cdot 8H_2O$	Secondary mineral
Hoernesite	$Mg_3(AsO_4)_2 \cdot 8H_2O$	Secondary mineral, smelter wastes
Haematolite	$(Mn,Mg)_4As(AsO_4)(OH)_8$	
Conichalcite	$CaCu(AsO_4)(OH)$	Secondary mineral
Pharmacosiderite	$Fe_3(AsO_4)_2(OH)_3 \cdot 5H_2O$	Oxidation products of arsenopyrite and other arsenic minerals

Source: Smedley, P.L. and Kinniburgh, D.G., Source and behavior of arsenic in natural water, United Nations Synthesis Report on Arsenic in Drinking Water, p. 61, 2001.

TABLE 21.2

Typical Arsenic Concentrations in Rock Forming Minerals

Mineral	Arsenic Concentration Range (mg/kg)
Sulfide minerals	
Pyrite	100–77,000
Pyrrhotite	5–100
Marcasite	20–126,000
Galena	5 to –10,000
Sphalerite	5–17,000
Chalcopyrite	10–5,000
Oxide minerals	
Hematite	Up to 160
Fe oxide(undifferentiated)	Up to 2,000
Fe(III) oxyhydroxide	Up to 76,000
Magnetite	27 to 41
Ilmentite	<1
Silicate minerals	
Quartz	0.4–1.3
Feldspar	<0.1–2.1
Biotite	1.4
Amphibole	1.1–2.3
Olivine	0.08–0.17
Pyroxen	0.05–0.8
Carbonate minerals	
Calcite	1–8
Dolomite	<3
Siderite	<3
Sulfate minerals	
Gypsum/anhydrite	<1–6
Barite	<1–12
Jarosite	34–1,000
Other minerals	
Apatite	<1–1,000
Halite	<3–30
Fluorite	<2

Source: Smedley, P.L. and Kinniburgh, D.G., Source and behavior of arsenic in natural water, United Nations Synthesis Report on Arsenic in Drinking Water, p. 61, 2001.

contain high levels of arsenic and heavy metals and, thereby, cause serious risks to humans and the ecosystem (Jung and Thornton 1996, Kim et al. 1998, Choi et al. 2009). The following anthropogenic As sources have to be considered when dealing with As fluxes: high-temperature combustion (oil- and coal-burning power plants, waste incineration, cement works); wastes from intense husbandry (disinfectants); compost and dung (surplus As from animal feeding); household waste disposal; glassware production (decoloring agent); electronics industries (admixture in semiconductor production, arsenide as laser material to convert electrical energy into coherent light); ore production and processing (melting and roasting in nonferrous smelters, melting in iron works); metal treatment (admixture in bronze production, lead and copper alloys); galvanizing; ammunition factories (hardening and improvement of flight characteristics of projectiles); chemistry (dyes and colors, wood preservatives, pesticides, pyrotechnics, drying agent for cotton, oil and dissolvent recycling); and pharmaceutical works (medication) (Reimann and Caritat 1998, Matschullat 2000).

Geochemical Characteristics of Arsenic

Arsenic belongs to group 15 (N, P, As, Sb, Bi) of the periodic table elements, and it has an outer electronic configuration of $4s^2 4p^3$ (Alloway 1995). Geochemically, arsenic is present in close association with the transition metals, namely, Au, W, Sb, Bi, and Mo, and is commonly present as an impurity in varieties of metallic ferrous. Arsenic is a toxic element and is classified by the U.S. Environmental Protection Agency as a human carcinogen. Arsenic occurs in the environment through weathering and volcanism (Juillot et al. 1999) and can form various oxidation states, such as −3, 0, +3, and +5. Arsine (AsH_3) is a poisonous and flammable gas, but rarely occurs in nature (Cheng et al. 2009). Arsenic is toxic to both plants and animals, and inorganic arsenicals are proven carcinogens in humans (Ng 2005). The toxic effects of arsenic on human health range from skin lesions to cancer of the brain, liver, kidney, and stomach (Smith et al. 1992). Arsenite (As^{3+}) and arsenate (As^{5+}) are the most common species in natural environments. Arsenite is reported to be more mobile than arsenate and is 25–60 times more toxic (Conner 1990, Pantsar-Kallio and Manninen 1997, Corwin et al. 1999, Moon et al. 2008). Monomethyl arsenic (MMA) and dimethyl arsenic (DMA), which are organic forms of arsenic, have low toxicity compared with their inorganic forms.

Arsenic is mainly present as arsenite and arsenate under the natural conditions, but its oxidation state can change with pH, Eh, and microbial activity. Figure 21.1 shows the Eh–pH diagram for arsenic. Arsenite species predominate under reducing conditions and is more soluble than arsenate because it is not ionized and adsorbs less strongly than arsenate species (Nordstrom and Archer 2003). Thermodynamic calculations and experimental results indicate that at high redox levels (pe + pH > 10), arsenate is the predominant arsenic species whereas under moderately reducing and reducing conditions (pe + pH < 8), arsenite is the most abundant form of arsenic (Masscheleyn et al. 1991, Sadiq 1997, Cheng et al. 2009). The equilibria for arsenous acid (As(III)) and arsenic acid (As(V)) in aqueous solution are given as follows (Alloway 1995).

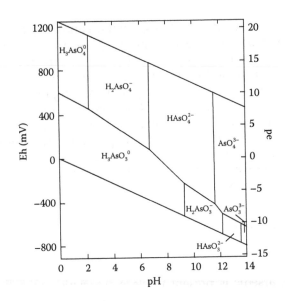

FIGURE 21.1
Eh–pH diagram for arsenic species at 25°C and 1 bar. (From Smedley, P.L. and Kinniburgh, D.G., *Appl. Geochem.*, 17(5), 517, 2002.)

Arsenic acid

$$H_3AsO_4 + H_2O \leftrightarrow H_2AsO_4^- + H_3O^+ \quad pK_a\ 2.20$$

$$H_2AsO_4^- + H_2O \leftrightarrow HAsO_4^{2-} + H_3O^+ \quad pK_a\ 6.97$$

$$HAsO_4^{2-} + H_2O \leftrightarrow AsO_4^{3-} + H_3O^+ \quad pK_a\ 11.53$$

Arsenous acid

$$H_3AsO_4 + H_2O \leftrightarrow H_2AsO_3^- + H_3O^+ \quad pK_a\ 9.22$$

$$H_2AsO_3^- + H_2O \leftrightarrow HAsO_3^{2-} + H_3O^+ \quad pK_a\ 12.13$$

$$HAsO_3^{2-} + H_2O \leftrightarrow AsO_3^{3-} + H_3O^+ \quad pK_a\ 13.4$$

When the Eh value drops below $+300\,mV$ at pH 4 and $-100\,mV$ at pH 8, H_3AsO_3 becomes the thermodynamically more stable species (NRCC 1978). The rate of change in oxidation state with change in Eh/pH conditions does not always appear to be very fast in aqueous systems. Therefore, the proportion of the various arsenic species present in soil pore waters may not correspond to the expected distribution. The occurrence, distribution, and mobility of arsenic are dependent on the interplay with geochemical factors such as redox potential (Eh), pH or acidity, redox pairs, competing anions, and reaction kinetics.

Stabilization of Arsenic

General Principle and Mechanisms of Stabilization

In general terms, stabilization is a process where additives are mixed with waste to convert the hazardous constituents into a form that minimize their rate of migration and level of toxicity. Stabilization is conducted by the addition of reagents that improve the handling and physicochemical properties of the waste, decrease the surface area, which transfers pollutants, and limit the solubility and toxicity of the contaminants. Although limiting the solubility and toxicity is not highly anticipated in practical application, stabilization is regarded as one of the most effective and efficient methods for the immobilization of contaminants in waste.

The fundamental mechanisms of stabilization are mainly one or more of the following phenomena: macroencapsulation/microencapsulation, adsorption, precipitation, and detoxification. With macroencapsulation, contaminants in the pores of waste are physically held in a large structural matrix whereas microencapsulation is the phenomenon where the pollutants are entrapped in the crystalline structure of the matrix. Adsorption is a mechanism where contaminants are electronically bonded to additives by van der Waals forces or hydrogen bonding. In the precipitation process, contaminants from the waste form precipitates with additives, such as hydroxide, carbonate, silicate, and sulfide. These precipitates then become a part of the material structure contained in the stabilized mass. Detoxification is a mechanism that changes the constituents of a toxic chemical into less toxic or nontoxic forms. For example, the change of oxidation state from As(III) to As(V) during stabilization lowers the solubility and toxicity of arsenic. The use of these mechanisms with different stabilizing agents will be discussed in the next chapter.

Stabilizing Agent

Cement

Cement is used as the principal reagent in stabilization, with Portland cement, which is a mixture of limestone and clay formed at high temperature, being the most commonly used for stabilization. Portland cement is composed of particles with a size ranging between 1 and 50 μm, mainly comprising tri-calcium silicate (45%–60%), di-calcium silicate (15%–30%), tri-calcium aluminate (6%–12%), and tetra-calcium aluminoferrite (~8%). For cement stabilization, wastes are mixed with cement and hydrated with water to form a crystalline structure of calcium aluminosilicate. The hydration reaction can be expressed by the following equations:

Tri-calcium silicate : $2(3CaO \cdot SiO_2) + 6H_2O \rightarrow 3CaO \cdot SiO_2 \cdot 3H_2O + 3Ca(OH)_2$

Di-calcium silicate : $2(2CaO \cdot SiO_2) + 4H_2O \rightarrow 3CaO \cdot SiO_2 \cdot 3H_2O + Ca(OH)_2$

Tri-calcium aluminate : $3CaO \cdot Al_2O_3 + 6H_2O \rightarrow 3CaO \cdot Al_2O_3 \cdot 6H_2O + heat$

Cement is suitable for inorganic waste stabilization, especially heavy metals. Increasing the pH of waste by adding cement changes the form of the heavy metals into insoluble hydroxides or carbonate salts. In addition, physical microencapsulation is also one of the main mechanisms for fixing the contaminants present in waste.

Pozzolans

A pozzolan is a material that produces cement by reacting with lime and water at ambient temperature. Most general pozzolanic materials are obtained from fly ash, blast furnace slack, and cement kiln dust. Fly ash is the most common pozzolan, which mainly contains SiO_2, Al_2O_3, Fe_2O_3, CaO, and unburned carbon. Like cement, pozzolans can be applied to the stabilization of inorganic contaminants by creating a high-pH environment. With fly ash, lime or quicklime can be used to enhance the stabilization efficiency.

Lime

Lime, or calcium hydroxide $Ca(OH)_2$, is mainly obtained from the hydration reactions of materials containing calcium silicate, calcium alumina, or calcium aluminosilicates. Lime can be added to increase the pH of waste, with other additive, such as fly ash, providing the main stabilization reactions, which is best suited to inorganic contaminants.

Leaching Test

A reduction in the rate of migration from contaminants into the environment is the most important reason for using stabilization techniques. There are numerous leaching test methods available for testing the leachability of stabilized materials. Leaching tests may be conducted to provide the basis for regulation, for the generation of data to model contamination migration, or to understand the basic mechanism of stabilization. Due to these different purposes, different test methods have been developed. There are several factors that can affect the concentration of leachate:

- Specific surface area of waste
- Volumetric ratio of leachant to waste
- Type and properties of leachant
- Contact time
- Agitation
- Temperature

The extraction procedure toxicity test (EP TOX), toxicity characteristic leaching test (TCLP), synthetic precipitation leaching procedure (SPLP), equilibrium leach test, sequential leach test, and multiple extraction procedure are widely used to test leachability.

Arsenic Stabilization

Cement/Lime/Fly Ash

In soil treated by a stabilizing agent containing Ca ions, such as Portland cement, lime, or fly ash, As stabilization is mainly controlled by the formation of insoluble Ca–As precipitates. It is suggested that the precipitation of $Ca_3(AsO_4)_2$ As (V) and $CaHAsO_3$ controls the stabilization of As in contaminated soils (Dutre and Vandecasteele 1995, 1998, Dutre et al. 1999).

As(V) precipitation can be expressed by the following equations:

$$3Ca^{2+} + 2H_3AsO_4 + 4H_2O \rightarrow Ca_3(AsO_4)_2 \cdot 4H_2O_{(s)} + 6H^+$$

$$5Ca^{2+} + 3H_3AsO_4 + H_2O \rightarrow Ca_5(AsO_4)_3(OH)_{(s)} + 10H^+$$

$$Ca^{2+} + H_3AsO_4 + H_2O \rightarrow CaHAsO_4 \cdot H_2O_{(s)} + 2H^+$$

As(III) precipitation can be expressed by the following equation:

$$Ca^{2+} + H_3AsO_3 + H_2O \rightarrow CaHAsO_3 \cdot H_2O_{(s)} + 2H^+$$

The oxidation state of arsenic is an important factor in controlling the leachability of arsenic from stabilized waste because As(III) is more mobile and toxic than As(V). Due to its higher mobility, As(III) is easily leached from stabilization-treated waste. For example, 95% of the total leached arsenic fraction after stabilization of arsenic baring fly ash waste was As(III)

(Dutre and Vandecasteele 1995). In addition, Dutre et al. (1999) concluded that oxidation of As(III) waste before stabilization into As(V), using H_2O_2, decreases the leaching of arsenic from the stabilized product because the calculated solubility of the $Ca_3(AsO_4)_2$ precipitate is lower than that of the $CaHAsO_3$ precipitate. Arsenic can also be stabilized by other minor mechanisms, such as physical encapsulation and chemical inclusion. Physical encapsulation can be achieved by creating a solidified monolith whereas chemical inclusion is the incorporation of arsenic in a binder hydration product by isomorphous substitution (Dermatas et al. 2004). Yoon et al. (2010) conducted a study on the stabilization of arsenic in contaminated soil using Portland cement and cement kiln dust. The Portland cement and cement kiln dust were added into the As-contaminated soil with dosed ranges of 10–30 wt% and 20–50 wt%, respectively. Stabilized soils were cured for 1, 7, and 28 days, followed by extraction with 1 N HCl to determine if the treated soils were suitable for disposal. The results of the stabilization using Portland cement and cement kiln dust are presented in Figure 21.2. For arsenic, as presented in Figure 21.2a, the concentrations in contaminated soil extracted using 1 N HCl gradually decreased on increasing the dose of Portland cement compared with the control arsenic concentration of approximately 130 mg/L. When 30% Portland cement was added, the arsenic removal efficiency and final pH were significantly increased. The lowest arsenic concentration with a 20 wt% dose was 42.8 mg/L after 28 days of curing and less than 2.44 mg/L after 1 day of curing. When cement kiln dust was added to the soil, a gradual decrease in the concentration of arsenic was also observed (Figure 21.2b). In addition, when 50% cement kiln dust was added to the sample, a sharp increase of arsenic removal was observed, with a pH similar to that with Portland cement. Increasing the pH with regard to the arsenic removal efficacy indicates that pH has a prevailing influence on arsenic mobility. The mechanism of stabilization was investigated using the step-scanned x-ray diffraction (XRPD) patterns. The XRPD patterns from the As(III) and As(V) treated by Portland cement and cement kiln dust are shown in Figure 21.3. From the results of the XRPD, it was concluded that the precipitation of As(III) as calcium arsenite (Ca–As–O) and As(V) as $NaCaAsO_4 \cdot 7.5H_2O$ significantly controlled the immobilizations of As(III) and As(V).

Iron (hydr)oxides

Several ferric (hydr)oxides are widely used as adsorbents for the removal of arsenic from aqueous solutions or contaminated soils. For poorly crystalline iron oxyhydroxides and iron oxide minerals in aqueous solution, the kinetics and equilibrium of arsenic sorption, as well as the factors affecting the sorption behavior, have been studied (Ford 2002, Zeng 2003, Richmond et al. 2004, Zhang et al. 2004). In addition, iron salts, with or without pH buffering, iron oxyhydroxides, and FeOOH have been added

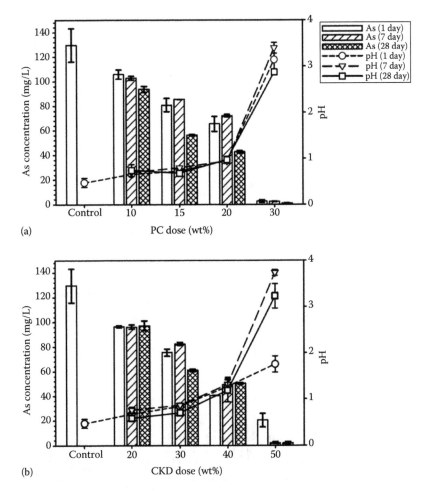

FIGURE 21.2
As concentration and pH after KST method for (a) PC and (b) CKD treatments after 1, 7, and 28 days of curing. (From Yoon, I.H. et al., *J. Environ. Manag.*, 91(11), 2322, 2010.)

to arsenic contaminated soil to prevent arsenic immobilization via chemical fixation (Moore et al. 2000, Garcia-Sanchez et al. 2002, Warren and Alloway 2003).

The main mechanism for As stabilization using iron (hydr)oxide is adsorption. For example, Jain et al. (1999) suggested As(III) and As(V) adsorption onto ferrihydrite from their study of surface charge reduction and net OH$^-$ release (Jain et al. 1999). Table 21.3 indicates the adsorptions of As(III) and As(V) at pH 4.6 and 9.2. The predominant species of As(V) at these pHs are $H_2AsO_4^-$ and $HAsO_4^{2-}$ and that of As(III) at same pHs are $H_3AsO_3^0$ and a mixture of $H_3AsO_3^0$ and $H_2AsO_3^-$, respectively.

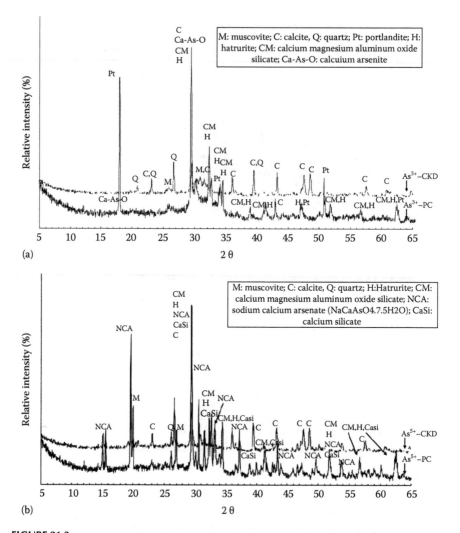

FIGURE 21.3
XRPD patterns of 10 wt% (a) As(III) and (b) As(V) stabilized by Portland cement and cement kiln dust aged for 24 h. (From Yoon, I.H. et al., *J. Environ. Manag.*, 91(11), 2322, 2010.)

The functional groups shown in Table 21.3 are only A-type $Fe-OH_2$ and $Fe-OH$ bound to a single structural Fe via ligand exchange reactions. Both $Fe-OH_2$ and $Fe-OH$ exist in appreciable concentrations at pH 4.6, and $Fe-OH$ dominates at pH 9.2. The surface OH_2 and OH^- groups are bonded in a bidentate edge-sharing complex $(Fe-O_2-As)$ or corner-sharing complex $((Fe-O)_2-As)$.

Iron can also form insoluble iron–arsenic compounds, such as $FeAsO_4$. The amorphous ferric arsenate $(FeAsO_4)$ and scorodite are an insoluble As(V)

TABLE 21.3

Possible Reactions of Adsorption of As(III) and As(V) by Ferrihydrite

As (III) at pH 4.6	$Fe - OH_2]^{+1/2} + H_3AsO_3^0 \rightarrow Fe - OAs(OH)_2]^{-1/2}$
	$Fe - OH_2]^{+1/2} + H_3AsO_3^0 \rightarrow Fe - O(H)As(OH)_2]^{+1/2}$
	$Fe - OH]^{-1/2} + H_3AsO_3^0 \rightarrow Fe - OAs(OH)_2]^{-1/2}$
	$Fe - OH]^{-1/2} + H_3AsO_3^0 \rightarrow Fe - O(H)As(OH)_2]^{+1/2}$
	$Fe\,|\,(OH_2)_2]^{+1} + H_3AsO_3^0 \rightarrow Fe\,|\,O_2(H)_2AsOH]^{+1}$
	$Fe\,|\,(OH_2)_2]^{+1} + H_3AsO_3^0 \rightarrow Fe\,|\,O_2(H)AsOH]^0$
	$Fe\,|\,(OH_2)_2]^{+1} + H_3AsO_3^0 \rightarrow Fe\,|\,O_2AsOH]^{-1}$
	$Fe\,|\,(OH_2)(OH)]^0 + H_3AsO_3^0 \rightarrow Fe\,|\,O_2(H)_2AsOH]^{+1}$
	$Fe\,|\,(OH_2)(OH)]^0 + H_3AsO_3^0 \rightarrow Fe\,|\,O_2(H)AsOH]^0$
	$Fe\,|\,(OH_2)(OH)]^0 + H_3AsO_3^0 \rightarrow Fe\,|\,O_2AsOH]^{-1}$
	$Fe\,|\,(OH)_2]^{-1} + H_3AsO_3^0 \rightarrow Fe\,|\,O_2(H)_2AsOH]^{+1}$
	$Fe\,|\,(OH)_2]^{-1} + H_3AsO_3^0 \rightarrow Fe\,|\,O_2(H)AsOH]^0$
	$Fe\,|\,(OH)_2]^{-1} + H_3AsO_3^0 \rightarrow Fe\,|\,O_2AsOH]^{-1}$
As (III) at pH 9.2	$Fe - OH]^{-1/2} + H_2AsO_3^- + H_3AsO_3^0 \rightarrow Fe - OAs(OH)_2]^{-1/2}$
	$Fe - OH]^{-1/2} + H_2AsO_3^- + H_3AsO_3^0 \rightarrow Fe - OAs(O)(OH)]^{-3/2}$
	$Fe\,|\,(OH)_2]^{-1} + H_2AsO_3^- + H_3AsO_3^0 \rightarrow Fe\,|\,O_2(H)AsOH]^0$
	$Fe\,|\,(OH)_2]^{-1} + H_2AsO_3^- + H_3AsO_3^0 \rightarrow Fe\,|\,O_2AsOH]^{-1}$
As (V) at pH 4.6	$Fe - OH_2]^{+1/2} + H_2AsO_4^- \rightarrow Fe - OAs(O)(OH)_2]^{-1/2}$
	$Fe - OH_2]^{+1/2} + H_2AsO_4^- \rightarrow Fe - OAs(O)_2(OH)]^{-3/2}$
	$Fe - OH]^{-1/2} + H_2AsO_4^- \rightarrow Fe - OAs(O)(OH)_2]^{-1/2}$
	$Fe - OH]^{-1/2} + H_2AsO_4^- \rightarrow Fe - OAs(O)_2(OH)]^{-3/2}$
	$Fe\,|\,(OH_2)_2]^{+1} + H_2AsO_4^- \rightarrow Fe\,|\,O_2As(OH)_2]^0$
	$Fe\,|\,(OH_2)_2]^{+1} + H_2AsO_4^- \rightarrow Fe\,|\,O_2As(O)(OH)]^{-1}$
	$Fe\,|\,(OH_2)(OH)]^0 + H_2AsO_4^- \rightarrow Fe\,|\,O_2As(OH)_2]^0$
	$Fe\,|\,(OH_2)(OH)]^0 + H_2AsO_4^- \rightarrow Fe\,|\,O_2As(O)(OH)]^{-1}$

(continued)

TABLE 21.3 (continued)

Possible Reactions of Adsorption of As(III) and As(V) by Ferrihydrite

$$Fe\,|\,(OH)_2\,]^{-1} + H_2AsO_4^- \rightarrow Fe\,|\,O_2As(OH_2)]^0$$

$$Fe\,|\,(OH)_2\,]^{-1} + H_2AsO_4^- \rightarrow Fe\,|\,O_2As(O)(OH)]^{-1}$$

As (V) at pH 9.2 $\quad Fe - OH]^{-1/2} + HAsO_4^{2-} \rightarrow Fe - OAs(O)_2(OH)]^{-3/2}$

$$Fe - OH]^{-1/2} + HAsO_4^{2-} \rightarrow Fe - OAs(O)_3]^{-5/2}$$

$$Fe\,|\,(OH)_2\,]^{-1} + HAsO_4^{2-} \rightarrow Fe\,|\,O_2As(O)(OH)]^{-1}$$

$$Fe\,|\,(OH)_2\,]^{-1} + HAsO_4^{2-} \rightarrow Fe\,|\,O_2As(O_2)]^{-2}$$

Source: Jain, A. et al., *Environ. Sci. Technol.*, 33(8), 1179, 1999.

phase in mine tailings and play a significant role in limiting the release of As from tailings to the environment (Langmuir et al. 2006). In the work of Langmuir et al., the solubility product of scorodite was established with regard to the As and Fe concentrations (Figure 21.4). The decreasing solubility product of scorodite with increasing pH and Fe/As ratio can increase the fraction of As released into water.

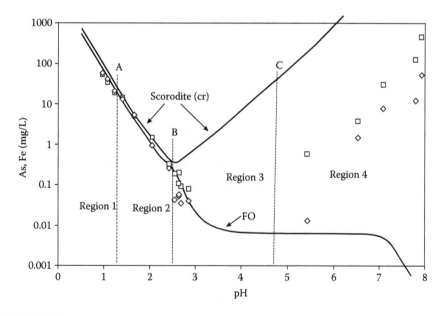

FIGURE 21.4
The solubility of scorodite in terms of As (upper solid line) and Fe (lower solid line). (From Langmuir, D. et al., *Geochim. Cosmochim. Acta*, 70(12), 2942, 2006.)

Field Application of the Stabilization

Stabilization of Arsenic in Mine Tailings

Stabilization techniques are considered the best treatment method for mine tailings. Because mine tailings contain extremely high amounts of arsenic geometrically, other treatments cannot reduce the leachability of arsenic as well as stabilization. Many studies have been conducted using the stabilizing agents introduced earlier, such as Portland cement, fly ash, lime, or iron oxyhydroxides. As an example of field application, arsenic stabilization using magnetite nanoparticles was applied to Samgwang mine tailings in Korea.

Samgwang mine has $226,467 \, m^3$ of mine tailings that are highly contaminated with arsenic, cadmium, and lead. The total arsenic concentration of the mine tailing is approximately $20,000 \, mg/kg$, which is extremely high compared with common soil ($1–40 \, mg/kg$). However, the arsenic extracted by TCLP solution was around $1 \, mg/kg$, indicating the amount of weakly bonded arsenic was relatively lower than that allowed by the regulations ($5 \, mg/kg$).

In order to enhance the surface area and mobility of mine tailings, synthesized magnetite nanoparticles were injected using the Jumbo Special Pattern (JSP) method, as shown in Figure 21.5 (Sheen 2001).

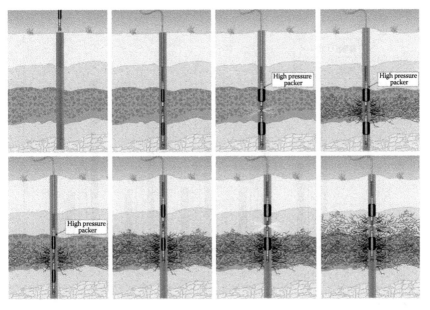

FIGURE 21.5
Jumbo special pattern method. (From Sheen, M.S., *Ground Improv.*, 5(4), 155, 2001.)

The resulting stabilization using the synthesized magnetite (iron/mine tailings ratio of 0.335% and 0.670%) for 28 days is shown in Figure 21.6. Figure 21.6a indicates that the arsenic concentrations in the mine tailings leached by pH 5.8 DI water were significantly reduced. After 1 day of stabilization, more than 90% of the As was stabilized for both 0.335% and 0.670% iron/mine tailings ratios and was stable after 28 days. The difference in the arsenic treatment efficiencies with different iron/mine tailings

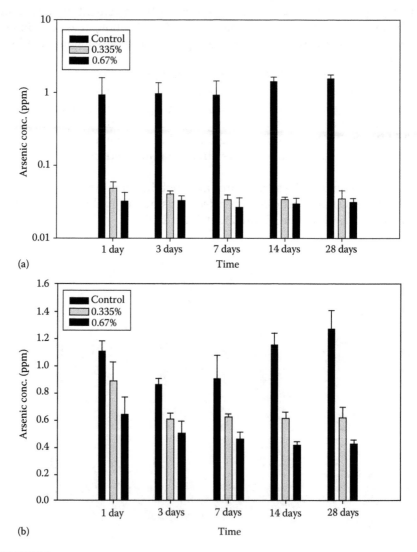

FIGURE 21.6
As concentrations in mine tailings for 28 days of stabilization leached by (a) pH 5.8 DI water and (b) TCLP solution.

ratios was not significant. The arsenic concentration leached by the TCLP solution (Figure 21.6b) indicated relatively low arsenic stabilization efficiency compared with pH 5.8 DI water. The sample treated using magnetite with a 0.335% iron/mine tailings ratio had 45% arsenic stabilization efficiency after 3 days and stable until 28 days of treatment. In the use of magnetite with a 0.670% iron/mine tailings ratio, the efficiency continuously increased and reached 60% after 7 until 28 days. The maximum stabilization efficiency was 45% and 62% at the iron/mine tailings ratio of 0.335% and 0.670%, respectively. As a result of the field application test, synthesized magnetite was shown to have high As treatment efficiency and, therefore, was considered a good stabilizing agent, especially from the results for extraction by pH 5.8 DI water.

Stabilization of Arsenic in Soil

Stabilization is one of the best remediation techniques for the reuse of contaminated soil. However, the application of arsenic stabilization in soil is harder than that for base metals due to the unique geochemical behavior of arsenic. The choice of additives is an important factor for arsenic stabilization in soil.

In order to measure the soil characteristics, soil samples were collected from the Chungyang area, where abandoned Au–Ag mines are located. The pH of the soil samples was 5.07, and the soil texture was silt loam (sand 33%, silt 63%, clay 4%). The CEC value and LOI were 19.8 meq/100 g and 25.6%, respectively. From the results of the SPLP and TCLP tests, the arsenic concentrations in the studied soil were 0.23 and 1.14 mg/kg, respectively. These results show a lower value for the TCLP than that proposed by the U.S. EPA (5 mg/kg). The total arsenic concentration via aqua regia digestion of the soil was 145 mg/kg, which was significantly higher than the normal level proposed by Bowen (1979). From the sequential extraction results, approximately 50% of the arsenic measured was in the easily mobile fraction via steps I and II. This result suggested that the major forms of arsenic in the soil samples would be easily extracted and transferred from soil to the water-plant system. Rice grain samples were collected and the arsenic concentration measured to evaluate the correlation between the soil and rice grain arsenic levels. The arsenic concentrations in the rice grains ranged from 0.07 to 0.59 mg/kg, with an average of 0.19 mg/kg in studied area. The rice grains cultivated in highly contaminated soil contained higher concentrations of arsenic. This result indicates that arsenic can move from soil to rice grains, and soil contamination can lead to plant contamination.

A field study on the stabilization of arsenic in contaminated soil was performed using two types of additives: limestone and mine sludge collected from an acid mine drainage (AMD) treatment system. The mine sludge was the by-product of AMD treatment, for which the oxidized iron compounds were identified by XRD analysis (Figure 21.7). Figure 21.8 shows the pH variation in the pore water collected from the field experimental plot after

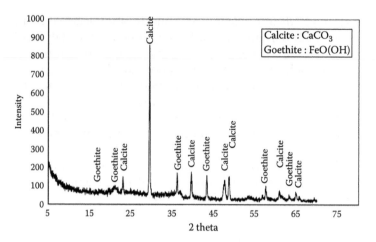

FIGURE 21.7
XRD patterns of mine sludge.

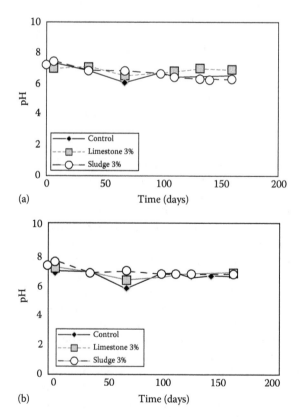

FIGURE 21.8
pH variation in pore water from field experimental plot (a) 0.3 m depth and (b) 0.8 m depth.

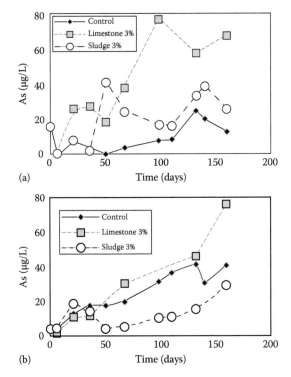

FIGURE 21.9
Arsenic concentration in pore water from field experimental plot (a) 0.3 m depth and (b) 0.8 m depth.

stabilization. The pH was 7.0 ± 0.5 in all experimental plot pore waters at depths of 0.3 and 0.8 m indicating that the pH of the pore water was not increased by the lime stone and mine sludge additives. However, different arsenic concentrations were observed with the two additives (Figure 21.9). Limestone showed low stabilization effects at both depths. The arsenic concentration in the pore water on the addition of mine sludge was slightly higher than the control at a depth of 0.3 m but the lowest arsenic value was observed at a depth of 0.8 m from mine sludge plot.

Summary

Stabilization is a remediation technique that reduces the mobility and toxicity of hazardous contaminants in soil using additives. Stabilization is regarded as one of the most effective and efficient methods for the immobilization of contaminant in waste.

Stabilization can be achieved by adsorption and co-precipitation between additives and contaminants. Hence, it is important to know the characteristics of both the additives and contaminants. Also, the choice of additives is an important factor for stabilization. There are many additives that can be used for stabilization, such as cement, pozzolan, lime, fly ash, and iron (hydr)oxide. In the case of arsenic stabilization, several ferric (hydr)oxides are widely used as adsorbents for the removal of arsenic from aqueous solutions or contaminated soils. For poorly crystalline iron oxyhydroxides and iron oxide minerals in aqueous solution, the kinetics and equilibrium of arsenic sorption, as well as the factors affecting the sorption behavior, have been studied.

References

Adriano DC. 1986. *Trace Elements in the Terrestrial Environment.* Springer-Verlag, New-York.

Adriano DC. 2001. *Trace Elements in Terrestrial Environments; Biogeochemistry, Bioavailability, and Risks of Metals.* Springer-Verlag, New York.

Alloway BJ. 1995. *Heavy Metals in Soil* (2nd edn.) Blackie Academic & Professional, London, U.K.

Bañuelos, GS and Ajwa, HA. 1999. Trace elements in soils and plants: An overview. *Journal of Environmental Science and Health, Part A: Toxic/Hazardous Substances and Environmental Engineering* 34: 951–974.

Bhumbla DK, Keefer RF. 1994. *Arsenic Mobilization and Bioavailability in Soil.* John Wiley & Sons, New York.

Bowen HJM. 1979. *Elemental Chemistry of the Elements.* Academic Press, London, U.K.

Cheng H, Hu Y, Luo J, Xu B, Zhao J. 2009. Geochemical processes controlling fate and transport of arsenic in acid mine drainage (AMD) and natural systems. *Journal of Hazardous Materials* 165:13–26.

Choi WH, Lee SR, Park JY. 2009. Cement based solidification/stabilization of arsenic-contaminated mine tailings. *Waste Management* 29:1766–1771.

Conner JR. 1990. *Chemical Fixation and Solidification of Hazard Wastes.* Van Nostrand Reinhold, New York.

Corwin DL, David A, Goldberg S. 1999. Mobility of arsenic in soil from the rocky mountain arsenal area. *Journal of Contaminant Hydrology* 39:35–38.

Cullen BE, Reimer KJ. 1989. Arsenic speciation in the environment. *Chemical Reviews* 89:713–764.

Dermatas D, Moon DH, Menounou N, Meng X, Hires R. 2004. An evaluation of arsenic release from monolithic solids using a modified semi-dynamic leaching test. *Journal of Hazardous Materials* 116:25–38.

Dutre V, Vandecasteele C. 1995. Solidification/stabilisation of hazardous arsenic containing waste from a copper refining process. *Journal of Hazardous Materials* 40(1):55–68.

Dutre V, Vandecasteele C. 1998. Immobilization mechanism of arsenic in waste solidified using cement and lime. *Environmental Science and Technology* 32(18):2782–2787.

Dutre V, Vandecasteele C, Opdenakker S. 1999. Oxidation of arsenic bearing fly ash as pretreatment before solidification. *Journal of Hazardous Materials* 68(3):205–215.

Ford RG. 2002. Rates of hydrous ferric oxide crystallization and the influence on coprecipitated arsenate. *Environmental Science and Technology* 36(11):2459–2463.

Garcia-Sanchez A, Alvarez-Ayuso E, Rodriguez-Martin F. 2002. Sorption of As(V) by some oxyhydroxides and clay minerals. Application to its immobilization in two polluted mining soils. *Clay Minerals* 37(1):187–194.

Jain A, Raven KP, Loeppert RH. 1999. Arsenite and arsenate adsorption on ferrihydrite: Surface charge reduction and net OH-release stoichiometry. *Environmental Science and Technology* 33(8):1179–1184.

Juillot F, Ildefonse P, Calas G, deKersabiec AM, Benedetti M. 1999. Remobilization of arsenic from buried wastes at an industrial site: Mineralogical and geochemical control. *Applied Geochemistry* 14:1031–1048.

Jung MC. 2001. Heavy metal contamination of soils and waters in and around the Imcheon Au–Ag mine, Korea. *Applied Geochemistry* 16:1369–1375.

Jung MC, Thornton I. 1996. Heavy metal contamination of soils and plants in the vicinity of a lead-zinc mine, Korea. *Applied Geochemistry* 11:53–59.

Kabata-Pendias A, Pendias H. 1984. *Trace Elements in Soils and Plants* (3rd edn.). CRC, Boca Raton, FL.

Kim KW, Lee HK, Yoo BC. 1998. The environmental impact of gold mines in the Yugu–Kwangcheon Au–Ag metallogenic province, Republic of Korea. *Environmental Technology* 19:291–298.

Langmuir D, Mahoney J, Rowson J. 2006. Solubility products of amorphous ferric arsenate and crystalline scorodite ($FeAsO_4H_2O$) and their application to arsenic behavior in buried mine tailings. *Geochimica et Cosmochimica Acta* 70(12):2942–2956.

Masscheleyn PH, Delaune RD, Patrick Jr WH. 1991. Arsenic and selenium chemistry as affected by sediment redox potential and pH. *Journal of Environmental Quality* 20:522–537.

Matschullat J. 2000. Arsenic in the geosphere—A review. *Science of the Total Environment* 249:297–312.

Moon DH, Wazne M, Yoon IH, Grubb D. 2008. Assessment of cement kiln dust(CKD) for stabilization/solidification(S/S) of arsenic contaminated soils. *Journal of Hazardous Materials* 159:512–518.

Moore TJ, Rightmire CM, Vempati RK. 2000. Ferrous iron treatment of soils contaminated with arsenic-containing wood-preserving solution. *Soil and Sediment Contamination* 9(4):375–405.

Mulligan CN, Yong RN, Gibbs BF. 2001. Remediation technologies for metal-contaminated soils and groundwater: An evaluation. *Engineering Geology* 60:193–207.

National Research Council of Canada. 1978. Effects of arsenic in the Canadian Environment, NRCC No. 15391, Ottawa, Ontario, Canada.

Ng JC. 2005. Environmental contamination of arsenic and its toxicological impact on humans. *Environmental Chemistry* 2:46–60.

Nordstrom D, Archer D. 2003. Arsenic thermodynamic data and environmental geochemistry. In *Arsenic in Ground Water Geochemistry and Occurrence*, A.H. Welch and K.G. Stollenwerk (eds.), Springer, New York, pp. 1–25.

Pantsar-Kallio M, Manninen PKG. 1997. Speciation of mobile arsenic in soil samples as a function of pH. *Science of the Total Environment* 204:193–200.

Reimann C, deCaritat P. 1998. *Chemical Elements in the Environment*. Springer, Berlin, Germany.

Richmond WR, Loan M, Morton J, Parkinson GM. 2004. Arsenic removal from aqueous solution via ferrihydrite crystallization control. *Environmental Science and Technology* 38(8):2368–2372.

Sadiq M. 1997. Arsenic chemistry in soils: An overview of thermodynamic predictions and field observations. *Water, Air, & Soil Pollution* 93:117–136.

Sheen MS. 2001. Grouting pressure and damaged adjacent buildings. Part 1: Behaviour analysis. *Ground Improvement* 5(4):155–162.

Smedley PL, Kinniburgh DG. 2001. Source and behavior of arsenic in natural water, United Nations Synthesis Report on Arsenic in Drinking Water, p. 61.

Smedley PL, Kinniburgh DG. 2002. A review of the source, behaviour and distribution of arsenic in natural waters. *Applied Geochemistry* 17(5):517–568.

Smith AH, Hopenhayn C, Bates MN, Goeden HM, Picciotto IH, Duggan HM. 1992. Cancer risks from arsenic in drinking water. *Environmental Health Perspectives* 97:259–267.

Thornton I. 1983. *Applied Environmental Geochemistry*. Academic Press, London, U.K.

US EPA. 2002. Arsenic Treatment Technologies for Soil, Waste, and Water EPA-542-R-02-004. http://www.epa.gov/tio/download/remed/542r02004/arsenic_report.pdf

Wang S, Mulligan CN. 2006. Effect of natural organic matter on arsenic release from soil and sediments into groundwater. *Environmental Geochemistry and Health* 28:197–214.

Warren GP, Alloway BJ. 2003. Reduction of arsenic uptake by lettuce with ferrous sulfate applied to contaminated soil. *Journal of Environmental Quality* 32(3):767–772.

Yang L, Donahoe RJ, Redwine JC. 2007. In situ chemical fixation of arsenic-contaminated soils: An experimental study. *Science of the Total Environment* 387:28–41.

Yoon IH, Moon DH, Kim KW, Lee KY, Lee JH, Kim MG. 2010. Mechanism for the stabilization/solidification of arsenic-contaminated soils with Portland cement and cement kiln dust. *Journal of Environmental Management* 91(11):2322–2328.

Zeng L. 2003. A method for preparing silica-containing iron(III) oxide adsorbents for arsenic removal. *Water Research* 37(18):4351–4358.

Zhang W, Singh P, Paling E, Delides S. 2004. Arsenic removal from contaminated water by natural iron ores. *Minerals Engineering* 17(4):517–524.

22

Bioavailability of Metals and Arsenic on Brownfield Land: Remediation or Natural Attenuation for Risk Minimization?

Nicholas M. Dickinson, William Hartley, Amanda Black, and Cameron S. Crook

CONTENTS

Introduction

Soil concentrations of metals and arsenic (As) elevated above background levels have raised considerable concern in restoration projects, particularly in the context of potential risks to human health. Certainly, there is no shortage of available regulatory documentation and guidance. Routes of transfer of pollutants from soil to receptor are the most critical part of risk assessment, but the extent to which bioavailability of metals and As is considered is variable. When taken into account, predictors and assays of lability, mobility, and bioavailability are well known to be of restricted efficacy in predicting uptake and impact on biota, largely due to differing soil types and environmental conditions. This means that broad generalizations and accurate interpretation of the risks posed by elevated and potentially toxic soil trace elements in soils are difficult, requiring consideration on the basis of each particular trace element, site, and environmental conditions. It also raises concern because expensive and inappropriate risk-based decisions may be taken to remove potential risk, for example, involving extensive but unnecessary site engineering. Indeed, justification and securement of funding for restoration or remedial actions may be entirely driven by mitigation of risk which itself is informed by a misguided interpretation of science.

Remedial actions have moved from dig-and-dump or concealment, to physicochemical amelioration, phytomanagement, and monitored natural attenuation (CL:AIRE 2008). The costs of these more modern remedial actions and technologies often represent only a small fraction of site engineering costs and can be justified on the basis of (1) the restoration of land and property values or (2) the establishment of safe and accessible green space (reserves or gardens) for urban dwellers. Restored land and green space creates a safe and healthy environment, providing improved biodiversity and provision of ecosystem services. We question whether these frequently defined but seldom quantified aspirational outcomes are realistic. Based on a selection of case studies, this paper considers whether elevated trace elements do present a significant risk in the urban environment and what benefits are achieved following concerted site restoration. We focus, in particular, on one recently completed restoration project in which a contaminated canal sediment and adjacent industrial land was converted to an urban nature reserve.

Bioavailability

Using a meta-analysis of 12 agricultural soils contaminated by 4 metals for different periods of time up to 13 years, Black (2010) showed that considerable differences existed between 6 different estimates of bioavailability (Table 22.1). All of the data referred to prediction of uptake into wheat and ryegrass. While Ca $(NO_3)_2$ provided reliable estimates of Zn and Ni uptake, a range of between only 46% and 84% could be explained by this extractant

TABLE 22.1

Coefficients of Variation (r^2) Ranges for Six Measures of Metal Bioavailability[a] versus Uptake into Wheat Seedling or Pasture Ryegrass, and the Best Predictors of Uptake in Each Case

	Cd	Cu	Ni	Zn	n
Wheat seedlings	0.13–0.63	0.27–0.34	0.53–0.71	0.46–0.66	335–502
Best predictor	EDTA	Pseudo-total	Ca(NO₃)₂	Ca(NO₃)₂	
Pasture ryegrass	0.37–0.70	0.21–0.45	0.51–0.84	0.46–0.68	236–325
Best predictor	Rhizon	Pseudo-total	Ca(NO₃)₂	Ca(NO₃)₂	
All data combined	0.20–0.49	0.27–0.34	0.56–0.72	0.49–0.64	538–754
Best predictor	DGT	Pseudo-total	Ca(NO₃)₂	Ca(NO₃)₂	

Data are from studies on 12 soil types in New Zealand, between 6 months and 13 years after spiking and application of biosolids (2.3%–9.8%C, pH 5.1–6.1).

[a] Pseudo-total (HNO₃-extractable), EDTA, Ca(NO₃)₂, Rhizon soil moisture samplers, DGT, Free ion activity (WHAM).

for either plant species. The best explanation of Cd uptake was obtained using EDTA (63%) or rhizon samplers (70%), but when data were combined for both plant species, only 49% of the variability was explained by the best method (DGT). No single method could predict uptake of Cu with these efficiencies. Of course, the scientific explanation is that pH, organic matter, and other physicochemical variables also partially determine the uptake of metals. Furthermore, biotic variables play a role, as illustrated by the difference in how well these techniques reflect uptake in the two species of grasses (Figure 22.1). Other less closely related species would be likely to respond to the four metals in yet different ways.

In terms of restoration of contaminated urban soils, phytotoxic concentrations of Cu, Zn, and Ni are of less concern where human health is the main priority. However, enhanced uptake of the more zootoxic metal Cd through food chains could compromise animal or human health (Alloway 1990, Rodrigue et al. 2007); Cd is highly mobile in environmental media and biological tissues (Dickinson and Pulford 2005). Other less mobile elements, particularly Pb and As, frequently present concern in soils polluted by industrial and urban fallout; both of these elements are potentially hazardous to human health (Hamilton 2000, Datta and Sarkar 2005, Hooker and Nathanail 2006). An urban town and woodland subjected to aerial fallout from 100 years of Cu, Cd, Zn, and Pb processing at Prescot, United Kingdom, was found to present very limited risk to biota or human health (Dickinson et al. 1996). The limited risk was largely due to very restricted pathways that

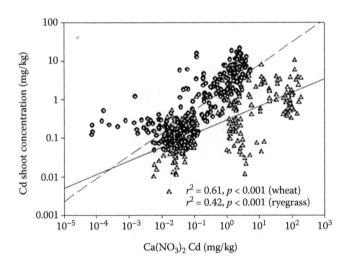

FIGURE 22.1
Relationship between uptake of Cd by wheat seedlings (O) and pasture ryegrass (Δ) and Ca(NO$_3$)$_2$ extractability of Cd from soil. (From Black, A., Bioavailability of cadmium, copper, nickel, and zinc in soils treated with biosolids and metal salts, PhD thesis, Lincoln University, Lincoln, New Zealand, 2010.)

would allow human ingestion of metals. Another site, a former chemical waste dump adjacent to a residential area in St. Helens, United Kingdom, had been converted to a recreational park land, and was found to contain elevated arsenic (As) in surface soil at concentrations 200 times higher than soil-guideline values. Despite this, the existing vegetation cover was apparently healthy and was likely to minimize re-entrainment of dust-blown particulates that presented the only potential risk to human health. In this case, the results of extensive detailed study suggested it was advisable to avoid disturbance and exposure of soil surface, and compost and phosphate application that may result in the solubilization or mobilization of As (Figures 22.2 and 22.3). Otherwise, in terms of management of the site, the effectiveness of monitored natural attenuation and the feasibility of reliance on phytostabilization for long-term management were demonstrated to be quite adequate.

A number of other studies have similarly demonstrated the importance of understanding the significance of bioavailability in the context of sustainable rehabilitation of metal-contaminated brownfield land. This is initially reflected in the effectiveness of natural vegetation processes in withstanding and ameliorating potential toxicity. Rawlinson et al. (2004) found that metal contamination did not limit vegetation establishment in field trials at a

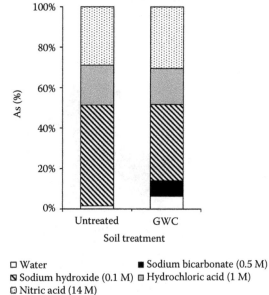

FIGURE 22.2
Arsenic fractionation in an industrially contaminated soil amended with green waste compost (30% v/v), using the method of Shiowatana et al. (2001). (From Hartley, W. et al., *Environ. Pollut.*, 157, 847, 2009.)

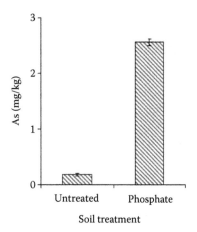

FIGURE 22.3
Change in As mobility as affected by phosphate addition (4000 mg/kg) to an industrially As contaminated soil. (From Hartley, W. et al., *Environ. Pollut.*, 157, 847, 2009.)

range of urban brownfield sites, despite previous identification of pollutants and concerns about contamination. Attenuation of metal and As toxicity has been demonstrated under various types of planted and natural vegetation (French et al. 2006, King et al. 2006, Clemente et al. 2008). An industrially contaminated site in North-West England (Lepp and Dickinson 2003) was one of the most contaminated sites reported (3,238 mg Cu kg^{-1}, 56,250 mg Pb kg^{-1} and 15,530 mg Zn kg^{-1}) but a woodland established on the site over a 50 year period, with no evidence of phytotoxicity in the vegetation; only Zn was significantly bioavailable. The site had stabilized, with minimal transfer of toxic metals (Pb) to the biotic community. In this case, there was very little public access to the woodland, reducing the possibility of direct transfer from soil to humans.

Case Study

An example of a brownfield restoration project to evaluate in the context of this chapter is a recently completed U.K. £1.25 M restoration project of brownfield land and a contaminated canal in Warrington, North-West England. The site is being redeveloped into an urban ecology park, creating a green corridor and a demonstration site for remediation techniques, with public recreational access (Figure 22.4). The justification of the project was the management of contamination risk, based on earlier published scientific research (King et al. 2006) combined with the benefits of public access and amenity

FIGURE 22.4
Case study. The regeneration of a contaminated canal and brownfield land site to create an urban ecology park.

FIGURE 22.5
The brownfield restoration site, outlined with derelict canal shown beneath (www.google. com).

use of the land within an urban residential area. Approximately 6.4 ha of brownfield land consists of former industrial land, a landfill site, and a section of derelict canal (Figure 22.5).

The derelict canal contained sediment consisting of a wet, black, odorous and oily mud, approximately 1.5–1.7 m deep. This contained a wide range of elevated contaminants (Table 22.2) that were largely immobile in the anoxic conditions and stable pH (about 5.7) of the sediment. However, the canal and embankments did support a rich diversity of vegetation, invertebrates, and nesting birds (Figure 22.6). Once exposed to the atmosphere, the chemistry of the sediment changes and contaminants become extremely mobile; bioavailable metal(loid)s then account for as much as 40% of the total concentrations. The driver for the project was largely based on the fact that the canal has become derelict and the sediment is drying out, thus potentially

TABLE 22.2

Mean Pseudo-Total (HNO₃-Extractable) Concentrations of Trace Elements in Soil at One Location on the Warrington Brownfield Site and in the Canal Sediment (mg/kg)

	Brownfield Site	Canal Sediment
As	37	145
Cr	40	1315
Cd	4.3	<1
Ni	38	102
Zn	172	4858
Pb	179	1887
Cu	83	887

FIGURE 22.6
The canal before and after engineering work.

providing a pollution linkage to receptors (groundwater contamination and entrainment of airborne dust particulates). Adjacent brownfield land contained mean trace element concentrations that were only marginally elevated (Table 22.2), although spatial patterns were heterogeneous with localized hotspots of trace elements (Figure 22.7) as usually found on brownfield land (French et al. 2006).

The restoration work involved extensive engineering of the canal, which was narrowed through dredging, stabilization, and relocation of the sediment to the edges of the canal basin, retained by geotextile-lined gabions, and covered by recycled soil forming materials, including green waste compost (Figure 22.4). A water flow was created in the new open water section of the narrowed canal and an artificial wetland was established to receive downstream outflow. Habitat improvement of the adjacent brownfield land was restricted to raised walkways and some import of green waste compost. The project was completed in 2010 but it may be too early

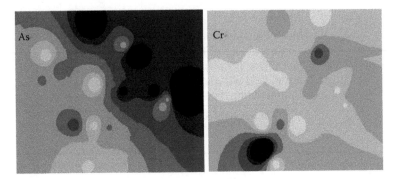

FIGURE 22.7
Spatial dispersion of As and Cr in surface soil at the brownfield site at the location same location referred to in Table 22.2. Different shading categories are arbitrary and operationally defined using ArcView GIS 3.2; spatial interpolation of inverse distance weighted (IDW) was applied with 12 neighboring samples for estimation of each grid point. Maximum As values (darkest shading) in the range of 75–85 mg/kg, Maximum Cr values (darkest shading) in the range of 60–70 mg/kg (not drawn to scale).

FIGURE 22.8
Contaminated sediment from the canal being dredged and deposited behind gabions, before sealing with geotextile and coverage with recycled wastes.

to comment on the success and measurable outputs of this substantial and costly remediation effort (Figure 22.8). It would, however, appear to be appropriate to question whether the remediation effort was justifiable and will the aspirational outcomes become a reality. Firstly, in terms of reducing risk and the consequences of not taking remedial action. As the sediment continued to dry with an increasing period of dereliction, would mobility of potentially toxic trace elements have become of increasing concern to human health or nature conservation? Secondly, is there likely to be a marked improvement of the nature conservation value of the brownfield land site or the canal in the longer term? It is quite possible that this case study represents an unrealistic template for best restoration practice on brownfield land.

Habitat Creation on Brownfield Land

Natural regeneration is a valid and viable alternative to conventional habitat creation practices; the potential exists for a high diversity of habitats and plant species to become established virtually anywhere on almost any substrate, including former agricultural land and industrial land (Garbutt and Wolters 2008). Crook (2008) argued that a great deal of time, money, and effort is wasted in habitat creation in development schemes and environmental projects, with many projects having limited success or failing totally due to poor planning and design, overambition, or neglect. Instead, many sites revert to the vegetation that would have developed naturally had nothing been done in the first place. He recorded 48 species of less-common vascular plants and 34 NVC (UK National Vegetation Classification) plant communities recorded on derelict or unmanaged land in the United Kingdom during a five-year period. These plants were mostly associated with long-established semi-natural habitat, yet had arisen by mostly natural means without any land preparation, management, or control of invasive species. Crook (2008) argues that Habitat Priming or Precursive Habitat Creation is likely to be the most effective restoration option for habitat creation: this involves sparsely seeding the site with a low diversity of suitable (i.e., non-aggressive/invasive) "nurse" species together with wide-spaced tree and shrub planting, and then simply allowing natural regeneration.

Conclusions

Risk management relies on avoiding a linkage or pathway to receptors, largely referring to direct or indirect transfer to humans. This chapter has questioned whether greening of polluted urban land raises or lowers the risk of contamination (i.e., causing harm). Advances in scientific research have provided an understanding of the impacts of elevated inorganic pollutants in soils. We now largely understand the physicochemical conditions that determine bioavailability and mobility of metals and As, even though this is frequently difficult to express in terms of single measures of bioavailability. This has meant that risk assessment is still entwined with total concentrations of metals and arsenic, and worst-case scenarios of exposure. This can be wasteful in terms of site redevelopment, greening, and human usage of brownfield land. The cited case studies point to a natural resilience of the ecological component that tends to stabilize or reduce potential toxicity pathways. Solutions involving engineering, major land disturbance, intensive landscaping, and vegetation establishment are expensive. Furthermore, environmental engineering solutions frequently delay site redevelopment and may be wasteful, unnecessary, and counter-productive. Longer-term success is frequently compromised by a lack of after-management of the restored or remediated site. More sensitive and less costly management of natural regeneration, natural succession, and normal ecological processes may provide more effective risk minimization and a better solution for brownfield land. In some situations, minimal intervention may be the most ecologically beneficial form of management of contaminated land, presenting both negligible risk to human health and benefits to nature conservation.

References

Alloway B.J. 1990. (Ed.) *Heavy Metals in Soils*. Blackie, London, U.K., pp. 100–124.

Black A. 2010. Bioavailability of cadmium, copper, nickel and zinc in soils treated with biosolids and metal salts. PhD thesis, Lincoln University, Lincoln, New Zealand.

CL:AIRE. 2008. Guidance on comparing soil contamination data with a critical concentration, in CIEH, T.C.I.o.E.H. (Ed.), London, U.K., p. 66.

Clemente R., Dickinson N.M., Lepp N.W. 2008. Mobility of metals and metalloids in a multi-element contaminated soil 20 years after cessation of the pollution source activity. *Environmental Pollution*, 155: 254–261.

Crook C. 2008. Habitat creation: What's the point? *Institute of Ecology and Environmental Management Autumn Conference on Mitigation: Smoke and Mirrors or Biodiversity*, Glasgow, U.K., November 18–20 [www.ieem.net/docs/06%20Cameron%20 Crook.pdf].

Datta R., Sarkar D. 2005. Consideration of soil properties in assessment of human health risk from exposure to arsenic-enriched soils. *Integrated Environmental Assessment and Management* 1: 55–59.

Dickinson N.M., Pulford I.D. 2005. Cadmium phytoextraction using short-rotation coppice *Salix*: The evidence trail. *Environment International* 31: 609–613.

Dickinson N.M., Watmough S.A., Turner A.P. 1996. Ecological impact of 100 years of metal processing at Prescot, North-West England. *Environmental Reviews* 4: 8–24.

French C.J., Dickinson N.M., Putwain P.D. 2006 Woody biomass phytoremediation of contaminated brownfield land. *Environmental Pollution* 141: 387–395.

Garbutt A., Wolters M. 2008. The natural regeneration of salt marsh on formerly reclaimed land. *Applied Vegetation Science* 11: 335–344.

Hamilton E.I. 2000. Environmental variables in a holistic evaluation of land contaminated by historic mine wastes: A study of multi-element mine wastes in West Devon, England using arsenic as an element of potential concern to human health. *Science of the Total Environment* 249: 171–221.

Hartley W., Dickinson N.M., Clemente R., French C., Piearce T.G., Sparke S., Lepp N.W. 2009. Arsenic stability and mobilization in soil at an amenity grassland overlying chemical waste (St. Helens, UK). *Environmental Pollution* 157: 847–856.

Hooker P.J., Nathanail C.P. 2006. Risk-based characterization of lead in urban soils. *Chemical Geology* 226: 340–351.

King R.F., Royle A., Putwain P.D., Dickinson N.M. 2006. Changing contaminant mobility in a dredged canal sediment during a three-year phytoremediation trial. *Environmental Pollution* 143: 318–326.

Lepp N.W., Dickinson N.M. 2003. Natural bioremediation of metal polluted soils—A case history from the UK. In: *Risk Assessment and Sustainable Land Management using Plants in Trace Element Contaminated Soils* (Eds. Mench, M. and Mocquut, B.), INRA, Bordeaux, France.

Rawlinson H., Dickinson N., Nolan P., Putwain P. 2004. Woodland establishment on closed old-style landfill sites in N.W. England. *Forest Ecology and Management* 202: 265–280.

Rodrigue J., Champou L., Leclair D., Duchesne J.F. 2007. Cadmium concentrations in tissues of willow ptarmigan (*Lagopus lagopus*) and rock ptarmigan (*Lagopus muta*) in Nunavik, Northern Quebec. *Environmental Pollution* 147: 642–647.

Shiowatana J., McLaren R.G., Chanmekha, N., Samphao, A. 2001. Fractionation of arsenic in soil by a continuous-flow sequential extraction method. *Journal of Environmental Quality* 30: 1940–1949.

Index

Milton Keynes UK
Ingram Content Group UK Ltd.
UKHW021918071024
449327UK00022B/1679